T0245234

CAMBRIDGE LIBRARY COLLECTION

Books of enduring scholarly value

Physical Sciences

From ancient times, humans have tried to understand the workings of the world around them. The roots of modern physical science go back to the very earliest mechanical devices such as levers and rollers, the mixing of paints and dyes, and the importance of the heavenly bodies in early religious observance and navigation. The physical sciences as we know them today began to emerge as independent academic subjects during the early modern period, in the work of Newton and other 'natural philosophers', and numerous sub-disciplines developed during the centuries that followed. This part of the Cambridge Library Collection is devoted to landmark publications in this area which will be of interest to historians of science concerned with individual scientists, particular discoveries, and advances in scientific method, or with the establishment and development of scientific institutions around the world.

Life of James Ferguson, F.R.S.

James Ferguson (1710–76) was a Scottish self-taught astronomer, instrument maker and artist. Of humble background, he became a highly successful lecturer on experimental philosophy and science. He became a Fellow of the Royal Society in 1763, received a royal pension, and is particularly remembered as an inventor and improver of astronomical and other scientific apparatus. These include a new type of orrery, clocks, and his astronomical rotula. His lectures and books were noted for their clear explanations for a general audience, and *Astronomy Explained upon Sir Isaac Newton's Principles and Made Easy for Those Who Have Not Studied Mathematics* (1756) was a bestseller. This autobiographical memoir, expanded by Ebenezer Henderson in 1867, also contains a full description of Ferguson's principal inventions, with many illustrations.

Cambridge University Press has long been a pioneer in the reissuing of out-of-print titles from its own backlist, producing digital reprints of books that are still sought after by scholars and students but could not be reprinted economically using traditional technology. The Cambridge Library Collection extends this activity to a wider range of books which are still of importance to researchers and professionals, either for the source material they contain, or as landmarks in the history of their academic discipline.

Drawing from the world-renowned collections in the Cambridge University Library, and guided by the advice of experts in each subject area, Cambridge University Press is using state-of-the-art scanning machines in its own Printing House to capture the content of each book selected for inclusion. The files are processed to give a consistently clear, crisp image, and the books finished to the high quality standard for which the Press is recognised around the world. The latest print-on-demand technology ensures that the books will remain available indefinitely, and that orders for single or multiple copies can quickly be supplied.

The Cambridge Library Collection will bring back to life books of enduring scholarly value (including out-of-copyright works originally issued by other publishers) across a wide range of disciplines in the humanities and social sciences and in science and technology.

Life of James Ferguson, F.R.S.

In a Brief Autobiographical Account, and Further Extended Memoir

EDITED BY EBENEZER HENDERSON

CAMBRIDGE UNIVERSITY PRESS

Cambridge, New York, Melbourne, Madrid, Cape Town, Singapore,
São Paolo, Delhi, Dubai, Tokyo

Published in the United States of America by Cambridge University Press, New York

www.cambridge.org
Information on this title: www.cambridge.org/9781108021265

© in this compilation Cambridge University Press 2010

This edition first published 1867
This digitally printed version 2010

ISBN 978-1-108-02126-5 Paperback

This book reproduces the text of the original edition. The content and language reflect
the beliefs, practices and terminology of their time, and have not been updated.

Cambridge University Press wishes to make clear that the book, unless originally published
by Cambridge, is not being republished by, in association or collaboration with, or
with the endorsement or approval of, the original publisher or its successors in title.

Townsends Hall

James Ferguson

A. Fullarton & C° London & Edinburgh.

THIS work is illustrated with One Hundred and Fifteen Woodcuts, representing the various inventions of Ferguson, and other subjects connected with his life and adventures.

LIFE

OF

JAMES FERGUSON, F.R.S.,

IN A BRIEF

AUTOBIOGRAPHICAL ACCOUNT,

AND FURTHER

EXTENDED MEMOIR.

WITH NUMEROUS NOTES AND ILLUSTRATIVE ENGRAVINGS.

BY E. HENDERSON, LL.D.

A. FULLARTON & Co.,
EDINBURGH, LONDON, AND GLASGOW.

1867.

TO

SIR DAVID BREWSTER,

K. H., D. C. L., L L. D., F. R. S., V.-P. R. S. E., &c.

VICE-CHANCELLOR AND PRINCIPAL OF

THE UNIVERSITY OF EDINBURGH,

THIS MEMOIR IS RESPECTFULLY

DEDICATED

IN CONSIDERATION OF HIS FORMER SERVICES AS

EDITOR OF FERGUSON'S LECTURES, ASTRONOMY, ESSAYS,

AND OTHER WORKS;

AS WELL AS IN ADMIRATION OF

HIS OWN DISTINGUISHED LABOURS

IN THE FIELD OF SCIENCE.

PREFACE.

In the summer of 1827 the Editor of this Memoir purchased, at a book-stall in Edinburgh, a copy of Ferguson's "Select Mechanical Exercises;" and at this distance of time is still retained a vivid recollection of the pleasure experienced in the perusal of the "Short Account of the Author" prefixed to the volume, and also of the disappointment felt at its coming so abruptly to a close. Ferguson, in his preliminary remarks, promises to give the reader "a faithful and circumstantial detail of his whole proceedings from his first obscure beginning to the present time,"—viz. from 1710—the year of his birth—till 1773, the year in which the "Account" was written. It occupies 43 pages, 36 of which refer to the details of his life in Scotland, while only 7 pages are devoted to his life in England; the latter extending over a space of 30 years, of which next to nothing is there related.

To supply, if possible, more ample information regarding Ferguson, the writer began so early as the year 1829 to collect, from all available sources, as opportunity offered, whatever particulars could be found and were not generally known. In an examination lately made, with a view to publication, of correspondence, memoranda and notes, these were found to form a heterogeneous collection, comprising much interesting matter little known, as well as certain particulars apparently of trivial importance, but which, as these also indicated Ferguson's pursuits at the time, have been retained.

With respect to arrangement, it was seen that if the autobiographical, or "Short Account of the Author," was to be republished, a departure from the usual biographical method should be adopted: that the proper arrangement would then be to reprint the "Short Account," and throw all the materials at command into annalistic sections, which has this advantage, —it exhibits, in unbroken succession, year after year, the events of his life, as far as these have been ascertained. In accordance with this plan, the reader will find Ferguson's autobiographical Memoir, illustrated with engravings and explanatory notes, forms the first part of the work. This part begins with 1710, and although written in 1773, may be said to terminate in 1748, as so little of himself is related in it subsequent to this year.

The second part, or "Extended Memoir," begins with 1743, the year in which he went to London: here it was necessary to go back a little way into his autobiography, to notice and supply omitted incidents from that year till 1773, when his Memoir was concluded, in order that the autobiography and the Extended Memoir should be duly connected. This portion, commencing with 1743 and ending with 1776—the year of Ferguson's death—is chiefly compiled from his own works;—the engravings and descriptions in which of his apparatus and curious mechanical inventions are reproduced, explained, and illustrated—from manuscript letters and philosophical papers and drawings extant; from his lately discovered Common Place Book; from reviews and notices of his publications and inventions;—found in the Records of the Royal Society, and in the magazines and newspapers of the last century—and lastly, from materials collected by the Editor during a period of thirty-six years.

The Editor's correspondents have been numerous; the names of many of them are given in the notes, and he avails himself of this opportunity to offer to such of each and to all as are still alive his grateful thanks. He now closes his labours, and will feel much gratified if the work receives the approbation of the public.

ASTRAL VILLA,
MUCKHART, *August*, 1867.
(Perthshire.)

CONTENTS.

PART II.—EXTENDED MEMOIR, 1743 TO 1776.

APPENDIX.

☞ NOTE.—Those having letters, writings, drawings, machines, or other relics of Ferguson, would much oblige the Editor by communicating particulars, addressed to ASTRAL VILLA, MUCHART, PERTHSHIRE.

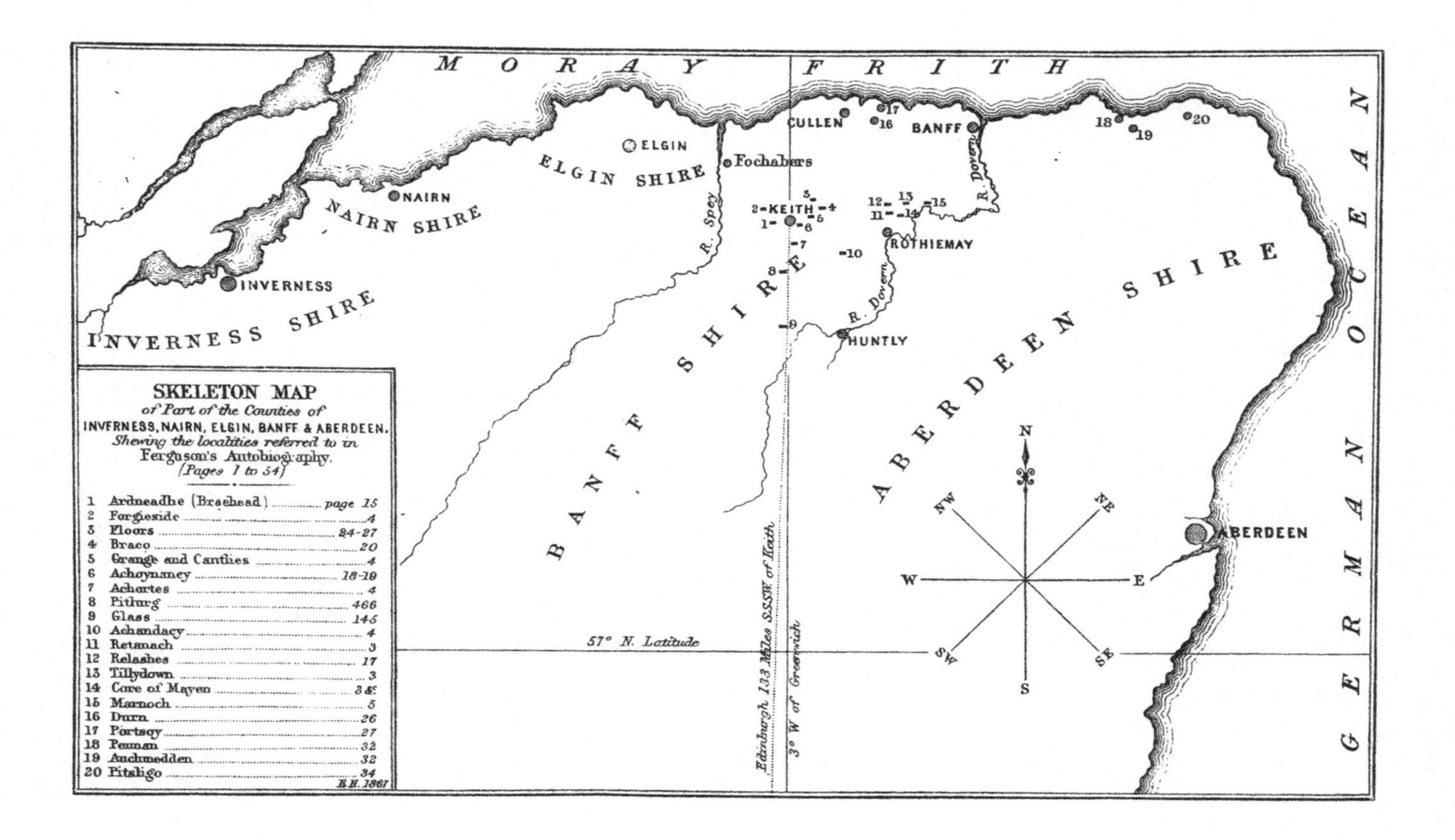
MORAY FRITH
GERMAN OCEAN
INVERNESS
INVERNESS SHIRE
NAIRN
NAIRN SHIRE
ELGIN
ELGIN SHIRE
Fochabers
R. Spey
CULLEN
BANFF
R. Dovern
KEITH
ROTHIEMAY
HUNTLY
BANFF SHIRE
ABERDEEN SHIRE
ABERDEEN
N
S
E
W
NE
NW
SE
SW
57° N. Latitude
Edinburgh 133 Miles S.S.W. of Keith
3° W of Greenwich
SKELETON MAP
of Part of the Counties of
INVERNESS, NAIRN, ELGIN, BANFF & ABERDEEN.
Shewing the localities referred to in
Ferguson's Autobiography.
(Pages 1 to 54)
1 Ardneadhe (Braehead) page 15
2 Forgieside 4
3 Floors 24-27
4 Braco 20
5 Grange and Canthes 4
6 Achoynaney 18-19
7 Achortes 4
8 Pitlurg 466
9 Glass 145
10 Achandacy 4
11 Retanach 3
12 Relashes 17
13 Tillydown 3
14 Core of Mayen 3 &c
15 Marnoch 5
16 Durn 26
17 Portsoy 27
18 Pennan 32
19 Auchmedden 32
20 Pitsligo 34
B.H. 1861

A SHORT ACCOUNT

OF THE

LIFE OF THE AUTHOR.

1710.—"As this is probably the last book I shall ever publish,[1] I beg leave to prefix to it a short account of myself, and of the manner I first began, and have since prosecuted my studies.—For, as my setting out in life from a very low station, and in a remote part of the island, has occasioned some false and indeed very improbable particulars to be related of me,[2] I therefore think it the better way, instead of contradicting them one by one, to give a faithful and circumstantial detail of my whole proceedings, from my first obscure beginning to the present time;[3] wherein if I should insert some particulars of little

[1] The book here alluded to is his "Select Mechanical Exercises," pp. 272; octavo, London, 1773. He, however, lived to publish "The Art of Drawing in Perspective Made Easy," &c., London, 1775:—"The Description and Use of the Astronomical Rotula," London, 1775:—"Three Letters to the Rev. John Kennedy:" London, 1775:—and a sheet, "Table of the Equation of Time," London, 1776. (Copies of all these publications are in our possession.—EDIT.)

[2] Many self-taught, and self-made men, in their latter days have been much annoyed by the *false* reports and improbable particulars, which had been put into circulation regarding them. To such *annoyances* and *false* reports, however, biography is greatly indebted for many interesting memoirs, which, otherwise, never would have been published. In the present instance, it is indebted for "*A short account of the life of the author:*"—of which "*short account*" it has been justly observed, "*that there are few more interesting narratives in any language*" (vide Lib. Entg. Knowl., vol. i. p. 196).

[3] It is to be regretted that Mr. Ferguson did not fulfil the promise he here makes. The "*Short Account*" of his life which he prefixed to his "Select Mechanical Exercises" embraces 43 octavo pages nearly,—36 of which are devoted to the details of his early life, in Scotland, from 1710 to 1743; (a period of 33 years)—whilst only a small portion of the remaining 7 pages have been made to suffice for recording details of his life in London,—from 1743 to 1773,—a period of 30 years! Had his life in England been written out as minutely as was his life in Scotland, an additional 36 pages at least would have been required for it, instead of a portion of 7 pages only. What had made Ferguson alter his mind, and thus to have hurried over the details of his life in London so very briefly, cannot now with certainty be known. We may, however, throw out a conjecture which may probably account for the *hiatus:*—Our memoranda make it evident that he finished writing his

moment, I hope the good-natured reader will kindly excuse me.

I was born in the year 1710, a few miles from Keith, a little village in Banffshire, in the North of Scotland,[4] and can with

Ferguson's Birth-Place—Core of Mayen.

"Select Mechanical Exercises," (to which he prefixed the short account of his life) about the middle of August 1773; and that during the time he was correcting its proof-sheets, he was slowly progressing with his memoir,—that his wife, who had long been unwell, became alarmingly ill whilst he was writing it, and died on September 3d, 1773. This melancholy occurrence, taken in connection with his publishers being in pressing want of the memoir in order to have the work published, would naturally induce him to hurry over the latter half of his life by recording a few incidents only. Probably he had got the length of the 36th page of his memoir when his wife died. His "Select Mechanical Exercises, with a Short Account of the Life of the Author" was published about Sept. 16th, 1773.

[4] Ferguson has here mistaken the locality of his birth-place. He was not born "*a few miles from Keith,*" comparatively speaking, nor anywhere in Keith parish, but at a place called the "*Quoir,*" or Core of Mayen, nearly 11 miles east of Keith, in the parish of Rothiemay, and about 2½ miles NE of Rothiemay village. This we shall place beyond doubt by evidence from two sources. We find in Sir John Sinclair's Statistical Account of Scotland, vol. xix., p. 392, Edin. 1797, an article on "THE PARISH OF ROTHIEMAY," from the pen of the then minister of the parish, the Rev. Dr. Simmie, who notes that "*James Ferguson, well known among men of Science for his publications on Astronomy, was a native of this parish*" (Rothiemay); and in a foot-note he adds, "*To certify this, it may be proper to subjoin, that my information of Mr. Ferguson's being a native of this parish, I received from his brother John Ferguson, who was an elder of this parish, and who died very lately. He told me that his father's name was John Ferguson, his mother's Elspet Lobban; that they dwelt at the Core of Mayen; that both he and his brother James were born there; that he himself was born in* 1708, *and that his brother James was two years younger. Accordingly, the Session Record, which is now before me, confirms this information in every particular;—James Ferguson was born April 25th,* 1710." Dr. S. concludes by saying, "*I do not recollect every particular related in the Memoirs prefixed to his works, not hav-*

pleasure say, that my parents, though poor, were religious and honest; lived in good repute with all who knew them, and died with good characters.[5]

ing seen them for several years; but John, his brother, who had read these memoirs, said, that though generally agreeable to fact, they were not equally correct in every particular."

Here we have evidence direct from two undoubted authorities,—from John, *the elder brother*, and from Rothiemay Session Record,—testifying that James Ferguson was born at the Core of Mayen, parish of Rothiemay, Banffshire, Scotland, on April 25th, 1710, and not at a subsequent residence of his father's, "*a few miles from Keith;*" also, we here have his brother John assuring us that his brother James's Memoir, "*though generally agreeable to fact,*" was "*not equally correct in every particular;*" no doubt this remark refers not only to the mistake he makes regarding the locality of his birth, but also to several other inaccuracies in the original Memoir, which will be found corrected in future notes. Thinking that it would be well to have a copy of Ferguson's birth as it stands in the Session Records of Rothiemay, we addressed a note to the present minister of the parish, the Rev. Mr. Moir, soliciting him to favour us with an extract. Mr. Moir not only complied with our request, but also kindly furnished us with a great many very interesting particulars relating to Ferguson, which will be found recorded in future notes. Rev. Mr. Moir, in his reply observes, that "*from the time of the Reformation to the middle of the 18th century, the Records of the parish of Rothiemay form, I am sorry to say, a mass of loose leaves, the penmanship in which has become, in many instances, almost illegible, through the carelessness of those who had the charge of them. After a long and somewhat tedious search I have been fortunate in laying my hands on the leaf containing the baptism of James Ferguson the Astronomer, which I give below in the orthography of the period:*"—viz.

"Apryle 25. 1710."

John Ferguisone in Quoir had a son be his wife Elspet Lobban, baptized, and called James:—before witnesses, James Horn in Rattannach:—Jam: Lobban yr. William Wilson yr.:—John Horn in Tilldown:—Elizabeth Johnstone yr.:—James Lorimer yr."

Rattannach is about $\frac{3}{4}$ths of a mile NW of the Core of Mayen:—Tilldown, or Tillydown, about $\frac{1}{4}$ of a mile to the NE of it.

We may here remark that Ferguson wrote his *Memoir* in 1773, when in the 64th year of his age, the 31st year of his residence in London, and at a period of nearly 40 years after he had ceased being a resident in the home of his parents. The greater part of this long period was chequered, and a struggle with poverty, which, taken in connection with each other, inclines us to conclude that such would have a tendency to obliterate, or render faint, in the evening of his days, the impressions of the "*mind-picture*" of his early years, thus making him mistake the place of his nativity. Be this as it may, it is certain that Ferguson could never have the slightest tangible recollection of the cot at the Core of Mayen, as his parents removed from it in the year 1712 to some place *now unknown*, within "*a few miles of Keith,*" at which period James would be an infant, and could not possibly remember anything about it; hence, we think there can be no doubt that it is to this *after residence* of his parents, "*a few miles from Keith,*" he refers here as the place of his birth. It would undoubtedly be the home associated with the earliest dawn of his memory,—perhaps this was the very cot he alludes to in connection with his 7th or 8th year, the roof of which becoming decayed he saw his father repair by raising it to its original position by means of a lever and upright spar, the operation exciting in his young mind "wonder, astonishment, and terror." The after musings and experiments relative to said incident, were his first starting points as an experimental philosopher, and the bases on which he reared his future fame; if so, this *after residence*, "a few miles from Keith," connected with, to him, such a remarkable occurrence, would be certain to remain firmly fixed on his mind during all the ups and downs of his after-life. If this is the cottage he had in view as the place of his birth,

As my father had nothing to support a large family but his daily labour, and the profits arising from a few acres of land which he rented, it was not to be expected that he could bestow

then, it becomes a question where was it situated? This cannot now be known in consequence of the indefinite way he refers to it; it was "*a few miles from Keith,*" he says, but does not mention in what direction from that village. Tradition and conjecture are now therefore the only guides, and they at present point to the following places, situated in nearly every direction round Keith,—viz.

1st, To Forgieside, 3 miles NW of Keith.
2d, To Achorties, or Denhead, 1 mile S of ditto.
3d, To Drum, or Greenslades, 1¼ mile E of ditto.
4th, To Achandacy (vicinity), 4 miles SE of ditto.
5th, To Cantly, about 2½ miles ENE of ditto,

and even to *Old Keith* itself! But as Ferguson himself assures us that it was "*a few miles*" distant from Keith, therefore, the site must be sought in *some direction*, between 3 and 5 miles of Keith.

Had we been certain that Ferguson, when alluding to his father's residence in his Memoir, had always in view *one* and the *same cot*, we could have fixed the site with tolerable exactness, as we would then have entered his own Memoir, and made the site correspond to the following conditions he gives us:—At page 2d of his Memoir he mentions that the cot was "*a few miles from Keith.*" 2d, At page 15, that it was "12 *miles distant*" from one of the residences of the Earl of Fife. 3d, At page 16, that it stood "*close by a public road;*" and 4thly, At page 19, that it was "*about seven miles from Durn*" (seventeen miles?). But as there is no certainty that Ferguson refers in his Memoir to *one* and the *same* residence of his parents, we cannot draw correct conclusions from these quotations; and we regret that *the site* must still be left in charge of tradition and conjecture. It is, however, evident, from our memoranda on the point, that this cottage to which Ferguson's father removed when he left the *Core of Mayen* for Keith parish, stood somewhere in one of the Cantlies, between 3 and 4 miles to the ENE of Keith.

KEITH.—The Keith to which Ferguson here alludes is at the present day called "*Old Keith,*" to distinguish it from two modern Keiths in the immediate vicinity which have sprung up since the middle of last century:—viz. 1st, New Keith, now Keith proper, is situated on a rising ground a few hundred yards to the south of Old Keith; and 2d, Fife Keith, adjacent to it on the west. As none of these modern Keiths is connected wth Ferguson's history, we pass them by, and confine our remarks to a few observations on Old Keith, as it was in his youth.

OLD KEITH, or what may here appropriately be designated "*Ferguson's Keith,*" is a village of unknown antiquity,—authentic records, relative to its church, date as early as 1203. The village is situated on the southern bank of the rivulet Isla and in the centre of an amphitheatre of hills. In early times, the magnitude of the village corresponded to the importance of its judicial authority. It "enjoyed the power or privilege of Pit and Gallows." Its court of Regality sat in the church. Being built in an inconvenient situation, it was gradually abandoned after 1750, in favour of New Keith on the south, and is now dwindled down into an insignificant hamlet. It is situated on the main road leading from Huntly, Rothiemay, and Grange, to Fochabers, Elgin, and Speyside. In Ferguson's young days it appears to have consisted of about 60 clay-built and rubble cottages, with 300 inhabitants, who were employed in agricultural pursuits, in weaving, spinning, flax-dressing, &c. The old church of Keith stood on the north side of the church-yard, and with it are associated "some memorable deeds of other days," which the reader will find well detailed in the "*Legends of Strathisla,*" a very entertaining little work by our friend Mr. Robert Sim, Keith. About 7 yards from the west gable of the old church, and in the church-yard, stood the "*old Grammar School of Keith,*" in which Ferguson had "*about* 3 *months*' instruction. The old church and school were removed in 1819. About 40 yards SW of the old grammar school stood the Manse of the Rev. Mr. Gilchrist, which Ferguson mentions in his Memoir. The Manse has also been removed; so at the present day

much on the education of his children, yet they were not neglected; for at his leisure hours he taught them to read and write:———And it was while he was teaching my elder brother to read the Scotch Catechism that I acquired my reading.

there is no building standing connected with the subject of our memoir. Keith is situated in lat. 57° 34′ 48″ N, and long. 2° 55′ 10″ W, at a distance of 133 miles NNE of Edinburgh, by Huntly; 49 miles NW of Aberdeen; 20 SW of Banff; 12 SSW of Cullen; 17 ESE of Elgin; 59 ENE of Inverness; 8 SE of Fochabers; and 10 NW of Huntly. RAIL distances are different.

CORE OF MAYEN:—The Rev. Mr. Moir of Rothiemay, in a note, informs us that "*The Core of Mayen is in the central portion of the estate of Mayen. It is a hollow surrounded by hills of considerable elevation on the north, west, and south sides, and open to the east. It is so completely enclosed by hills that the crops in a late season are in danger of not ripening. It is about a quarter of a mile broad, and nearly three quarters of a mile in length. A* BURN *or rivulet flows down the middle of the hollow, and forms the boundary between the parishes of Rothiemay and Marnoch, so that one half of the hollow, that on the north-east, called Tilliedoun, is in Marnoch, the other half, that on the south-west, is called the Core, and is in the parish of Rothiemay. In this portion stood the cottage in which the astronomer was born. About twelve years ago* (1848) *the ruins of the cottage were removed, and its site is now within a corn field. The well of the Ferguson family, however, is still to be seen, and a solitary tree indicates the site of the cottage.* The Core *lies to the north-east of Rothiemay parish church, and is distant from it fully two miles, and is about eleven miles due east of Keith.*" We may here note, that the site of the cottage in which Ferguson was born is within a few hundred yards of the western boundary of Aberdeenshire; hence, Banffshire was exceedingly near losing the honour of registering the astronomer as one of its illustrious sons.

We annex a view of the cottage at the Core of Mayen in which he was born, taken from an old pen and ink sketch of date about 1780. Referring to one of our notes, we observe that "*Ferguson's birth-place cot*" *stood unroofed in* 1807, *and that previous to the entire removal of its walls, they were nearly on a level with the ground—immediate cause or urgency for the removal of the foundations being a want of stones to build a neighbouring dike.* A stone pillar ought to be erected on the site, to inform future generations, that on this spot stood a cot, wherein the rustic astronomer was born.

[5] DEATH OF FERGUSON'S PARENTS.—We have made a great many inquiries respecting the date of death, the age at death, and place of interment of Ferguson's parents, but without, we are sorry to say, any satisfactory result. However, we find him in his Memoir, at page 27, referring to the period he resided in Edinburgh, viz. 1734-1736; and in alluding to his success in taking likenesses there, he mentions, that he was enabled thereby "not only to put a good deal of money in his pocket, but also to spare what was sufficient to help to *supply my father and mother in their old age.*" Here we ascertain that both parents were alive and in their old age in 1735; and it is evident, from the opening lines of his Memoir, that they were both dead in 1773, when he wrote it; hence, all we learn from him respecting this question is, that they died somewhere between the years 1735 and 1773. On carefully going over our notes on this point, we think that this interval can be very much narrowed. Our notes make it appear evident that the parents were born about the year 1670, and died about 1741, in their 71st year; consequently, when Ferguson was in Edinburgh, in 1735, his parents would then be in their 65th year;—and when he left Scotland for London, in May 1743, they would be both dead. His wife's father, George Wilson, died in 1742,—thus, within two years, 1741-1742, three tender cords were broken, and probably such sad events may have had considerable weight with him when he determined to leave Scotland for London. We may add, that the Rev. Mr. Annand, Minister of Keith, in his reply to our inquiries respecting the dates of deaths of Ferguson's parents, says, "*It is currently reported here that Mr. Ferguson's parents, in their latter days, left Keith for*

1716.—Ashamed to ask my father to instruct me, I used, when he and my brother were abroad, to take the Catechism and study the lesson which he was teaching my brother; and when any difficulty occurred, I went to a neighbouring old woman, who gave me such help as enabled me to read tolerably well before my father had thought of teaching me.[6]

Rothiemay, but that the father afterwards returned to Keith and died there. The inference from this appears to be, that his mother had died at Rothiemay and had been buried there, and that his father had been buried in Keith churchyard." Dr. Cruickshanks, of Marischal College, Aberdeen, in his note to us on this matter, observes, "*It is very probable that Rothiemay was the burial-place of the family, and that his body was brought thither from the parish, for corpses are often carried much farther to be deposited beside those of relations. In those days there were no regular registering of deaths or funerals.*" Grave-stones appear to have been the usual and only record of deaths at the period referred to. There are no memorial stones of Ferguson's parents at Keith or Rothiemay; but, as before mentioned, as far as we can judge from the notes before us, the parents died between the years 1740 and end of 1741.

[6] It may appear singular to many that Ferguson should have felt *ashamed* to ask his father to instruct him. We have always thought that the word *ashamed* is too strong a word for expressing a tender feeling of bashfulness. The old Scotch way of expressing such a feeling would have been for him to have said "*I didna like til ask my faither ta learn me.*" *He* would see his father in the evenings, after a hard day's toil, sit down on his chair and instruct his elder brother. He would see the trouble his father had as an instructor, and he therefore "did not like," or did not wish to throw any additional labour on his father's hands. Being at this time in his 6th year, he would be left at home with his mother, and often would he peep into the old Catechism and endeavour to master its lessons. Difficulties would arise before him, but when they did, *he* informs us that he "*went to a neighbouring old woman,*" who helped him over them. But why did he go to the old woman for help? Why did he not apply for assistance to his own mother who was constantly by his side? Such questions cannot now be answered with certainty. We may, however, venture to remark, that either his mother was unable to render him any assistance in the way of self-tuition; or, that he "went to the neighbouring old woman" by stealth for aid, in order to give his parents a surprise at being able to read without having been taught by them.

He here alludes to the "Scotch Catechism" (sometimes called the "*Single Book*"). It was a duodecimo tract of a few pages only. One page had the A B C repeated in different sorts of letters, and a few letters thrown into monosyllables, and then followed the questions and answers approved of by the Church of Scotland, and which had to be committed to memory. It was a kind of *Primer* in schools in the early part of last century, and was sold at a penny.

He also here speaks of his father and brother when "*they were abroad,*" i. e. out in the fields; but as this "*elder brother*" could not at this time be more than 8 years old, we cannot suppose that such a mere child would then be set to any out-door work, but that he merely went along with his father to bear him company and wile away the time.

His father appears to have been a superior man for his station in life. To be able to write in his young days (about 1680), was an uncommon accomplishment, and perhaps not one in a thousand of his class could either write, or read what was written.

The "*elder brother*" was named JOHN. He was born at the Core of Mayen, March 12th, 1708, and died in 1796, aged 88. He was consequently 2 years older than James. The Rev. Mr. Moir, of Rothiemay, has kindly furnished us with an extract of the birth of his "*elder brother,*" from Rothiemay Baptismal Record.

> "John Ferguisone in Quoir had a son be his wife Elspet Lobban, baptized and called JOHN, before witnesses John Lobban in Rettanach;—John

1717.—Some time after, he was agreeably surprised to find me reading by myself; he thereupon gave me further instruction, and also taught me to write; which, with about three months I afterwards had at the grammar school at Keith, was all the education I ever received.[7]

Old Grammar School, Keith.——(Built 1695—Removed 1819).

My taste for mechanics arose from an odd accident. When about 7 or 8 years of age, a part of the roof of the house being decayed, my father, desirous of mending it, applied a prop and lever to an upright spar to raise it to its former situation; and, to my great astonishment, I saw him, without considering the reason, lift up the ponderous roof, as if it had been a small

George, in Glennie House:—John Horn in Tillidown:—Isabel Davidsone, in Rettanach;—James Lobban yr. and Jean Ferguisone."

This brother, John, resided at Relashes at the time of his death in 1796, which place lies about 1½ mile NW of the Core of Mayen, and was at the time one of the elders of Rothiemay parish. He thus lived 20 years after the death of James.

[7] Grammar School of Keith.—The old Grammar school of Keith, to which Ferguson here alludes, and in which he had "*about* 3 *months*' " instruction (about 1718), was built in the year 1695, became ruinous about the year 1780, and was then disused as a school. Shortly after 1782 it was repaired and used as the session-house of the church. In 1819 it was entirely removed, so that there is now not a stone standing to mark the site on which it stood. We may note, however, that it stood in the NW corner of the old churchyard, at a distance of 20 feet from the west gable of the church, and about the same distance NE of the back wall of the present manse. The old grammar school was a small building, being only about 30 feet long, 16 feet broad, and 12 feet high. The view we give of the old school is taken from a drawing sent to us from Keith.

weight.[8] I attributed this at first to a degree of strength that excited my terror as well as wonder; but thinking further of the matter, I recollected that he had applied his strength to that end

Ferguson's Father raising the Roof of his Cottage.

of the lever which was furthest from the prop; and finding, on inquiry, that this was the means whereby the seeming wonder was effected, I began making levers (which I then called bars);

[8] The lever used on this occasion by Ferguson's father was evidently one of the "*first kind*," that is, having the *prop* or fulcrum placed between the weight and the power, but very much nearer to the *weight*. The velocity of each point of such a lever being directly as the distance of the point from the prop, the power gained will be in proportion to the different lengths of the arm of the lever on each side of the prop or fulcrum. We shall illustrate this *roof-raising* case to our young readers by giving them a short calculation relative to the probable weight of the cottage roof, and the length of lever which would be required to raise it:—1st, The total weight of the roof, composed of timber, turf, heather, and straw, could not exceed a ton, (or 2240 lbs.) But as only "a part" had to be lifted, the estimated weight of the *lift* would not probably exceed half a ton, or say 1200 lbs. Some of our old notes regarding this roof make it evident that a large portion of the back part of it, near the top, had begun to *sink*, thrusting the front part of the roof forward so as to overhang the front wall. In such a case the ends of the several rafters resting on the back wall would have to be secured so as to prevent them slipping out of their places during the operation of lifting the front part. Let us therefore assume the weight to be 1200 lbs., the lever 16 feet in length, and *the prop* 1 foot distant from the weight. In this case, we have the end of the long arm of the lever 15 feet from the prop, and the short arm, a length of one foot only; hence, without taking into account the weight of the lever itself, we have the power as 15 to 1; that is, by using such a lever, a man can lift 15 times more than by his unaided strength. Suppose Ferguson's father exerted a power of 160 lbs. at the end of the long arm of such a lever, then the power he would gain would be

and by applying weights to them different ways, I found the power gained by my bar was just in proportion to the lengths of the different parts of the bar on either side of the prop. I then thought it was a great pity that, by means of this bar, a

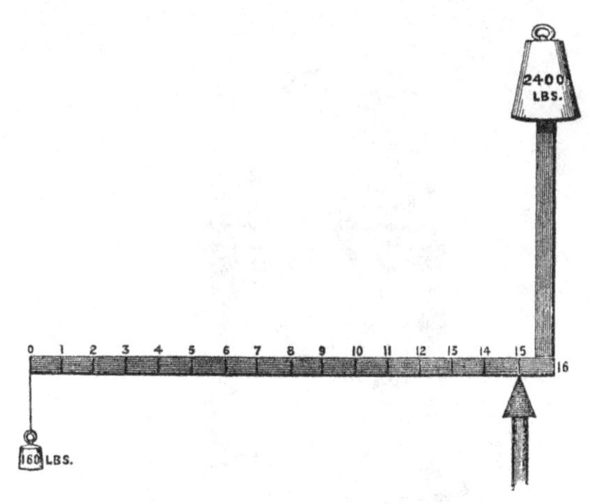

weight could be raised but a very little way. On this I soon imagined that, by pulling round a wheel, the weight might be raised to any height by tying a rope to the weight and winding the rope round the axle of the wheel, and that the power gained must be just as great as the wheel was broader than the axle was thick, and found it to be exactly so, by hanging one weight to a rope put round the wheel, and another to the rope that coiled round the axle; so that, in these two machines, it appeared very plain that their advantage was as great as the space gone through by the working power exceeded the space gone through by the weight; and this property I also thought must take place in a wedge for cleaving wood, but then I happened

$160 \times 15 = 2400$ lbs., that is, with such a lever, instead of being able to lift the assumed 1200 lbs. of roof, he would have been able to have raised a weight or load of roofing of close upon 2400 lbs. had it been required; hence, the lift of 1200 lbs. would be done with the greatest ease,—the father would "*lift up the ponderous roof as if it had been a small weight;*" as by exerting only one half of his strength at the power end of the lever he would raise a roof of 1200 lbs. weight. (The lever used by Ferguson's father:—In the annexed engraving we have a bar or lever 16 feet long—the prop at 15 feet from the power; hence, the power being assumed at 160 lbs., it will balance 2400 lbs. on the top of "*an upright spar*" as shown, or 2400 lbs. hanging to the point at 16—the weight of lever and spar being nothing). We here give a fanciful drawing of Mr. F.'s father raising his cottage roof, in order to show how it was done. Probably the lever used on this occasion was either a timber plank or strong and straight limb taken from some tree.

not to think of the screw.[9] By means of a turning-lathe which my father had, and sometimes used, and a little knife, I was enabled to make wheels and other things necessary for my purpose.[10]

Wheel and Axle.

1719.—I then wrote a short account of these machines, and sketched out figures of them with a pen, imagining it to be the first treatise of the kind that ever was written; but found my mistake when I afterwards showed it to a gentleman, who told me

[9] Ferguson here explains the action of the *wheel and axle question* in language so simple that it can scarcely be misunderstood; but as some of our young readers may not clearly comprehend it, we will lay before them an illustration:—1st, In the wheel and in the axle an infinite number of lines must be supposed to proceed from the centre to the circumference. These lines are called radii, in short they *represent levers*, and are therefore *perpetual levers*. Suppose a line from the axis of the wheel to its circumference to measure 12 inches, and a line from the centre of the axis to its circumference, 3 inches, then we have *two levers*, one of 12 inches, the other of 3; hence, $12 \div 3 = 4$; that is, a weight of 1 lb. on the circumference of the wheel will balance a weight of 4 lbs. on the axle, as shown in the engraving annexed.

[10] Ferguson's father appears to have had a taste for mechanics. We see that he understood the philosophy of the lever, and no doubt, the other mechanical powers also. Here we find him in possession of a turning-lathe. But what had a small crofter or daily labourer to do with a lathe? Mr. F. merely says that his father "*sometimes used*" it, but does not inform us whether he used it in the way of amusement or for profit. We are inclined to think that the father kept it for profit, in order to enable him to eke out his little income in support of the family. In our 4th note, the reader will find that in the early part of last century, Keith and neighbourhood abounded with Spinning-wheels, Pirn-wheels, Reels, and Looms, and for the frequent repairs of such articles a turning-lathe would be absolutely necessary, and thereby become a source of profit.

that these things were known long before, and showed me a printed book in which they were treated of; and I was much pleased when I found that my account (so far as I have carried it) agreed with the principles of mechanics in the book he showed me. And from that time, my mind preserved a constant tendency to improve in that science.[11]

1720.—But as my father could not afford to maintain me while I was in pursuit of these matters, and I was rather too young and weak for hard labour, he put me out to a neighbour to keep sheep, which I continued to do for some years;[12] and in that time I began to study the stars in the night. In the day-time I amused myself by making models of mills, spinning-wheels, and such other things as I happened to see.[13]

[11] Although Ferguson gives us an account of his lever, &c., experiments, and of his writing a treatise on them in two consecutive pages, as if all had happened within a period of a few weeks or months, yet we must not understand it so. We are inclined to think that the interval of time between the raising of the cottage roof and that of writing his "*Treatise*" embraced a period of at least two years. When the roof of the cottage was raised, Ferguson was in his 7th or 8th year, and when he wrote the "Treatise," he would be about 10 years old. A precocious boy at such an age would be able to write a rude treatise as Mr. F. did, but we scarcely think that a boy 7 or 8 years old could have done so from self-instruction.

[12] Ferguson has not given us the *name* of this "*neighbour*"—(his first employer), to whom he was sent to keep sheep,—and as all our inquiries for it in Banffshire had turned out unsuccessful, we had resolved to pass the matter over without note or comment. Latterly, however, our attention was directed to an American publication:—"A Memoir of Mary L. W. Pichard, Boston, U.S., 1856." At page 9 we find the *name* of Ferguson's first master mentioned. His name was ALEXANDER MIDDLETON. We give the extract in full:—"*In January 1802 Mr. Pichard was called to England on business, and took with him his wife and little Mary, then but three months old. They remained there a year and a half visiting both his and her relations in different parts of the kingdom. Mrs. Pichard being connected, on her mother's side, with Alexander Middleton, a Scotch farmer, in whose family Ferguson the astronomer lived as a shepherd-boy, and of whom, with his wife and three children, there are still existing, likenesses drawn in pencil by that lad, so celebrated as a man.*" Besides being thus furnished with the name of his first master, we also discover that Ferguson drew his likeness, and that of his wife and three children. Ferguson did not begin to take likenesses in his native county, Banffshire, until after his return to it from Edinburgh, in the end of the year 1736; and as he left Scotland for London in May 1743, these portraits of the Middleton family must have been taken between the years 1736 and 1743.

[13] Our memoranda relative to this period of Ferguson's life make it evident that he was "*put out*" to keep Mr. Alexander Middleton's sheep in 1720, when he was 10 years old, and continued in his service 4 years, viz. until 1724; and consequently it was between these years that he "*began to study the stars in the night.*" *In the day-time, he informs us, that he* "*amused himself by making models of mills and such other things that he happened to see.*" At this period there were, and are still, several mills near Keith, and then "spinning on the spinning-wheel" was practised by females in almost every house and cot, not only in and around Keith, but over all Scotland; and as "*every spinning-wheel would have its reel,*" no doubt reels would be among the models he fashioned with his *whittle* or knife on the bleak hill-sides. But there is a model he is reputed to have then

1724.—I then went to serve a considerable farmer in the neighbourhood, whose name was James Glashan. I found him very kind and indulgent; but he soon observed that in the

made, on which he is silent, viz. making a "*horse-head clock.*" Our note on this says, "the story is well known here," i. e. Banffshire.

"It is a ludicrous blunder that the French astronomer, Lalande, makes, in speaking of Ferguson, when he designates him as '*Berger au Roi d'Angleterre en Ecosse,*' i. e. the King of England's shepherd for Scotland. He had no claim to this pompous title; but it is true that he spent some of his early years as a keeper of sheep, though not in the employment of the State, but of a small farmer in the neighbourhood of his native place" (Lib. Entg. Know., vol. i., p. 198), in Banffshire,—"*that when in the 11th or 12th year of his age, when employed in herding, he fitted up a clock within the skull of a horse, and fastened it to a willow tree on the banks of the Isla, to let him know when to drive his sheep and cattle home.*" We can scarcely credit this story;—if he is connected with it at all, we should rather be inclined to think that he would use the horse-skull as a sun-dial, for the oddity of the thing, having the stile in the fore-head and the hours laid down round it. A tree, however, would perhaps have served the purpose much better by using the trunk of a tree as a stile, and having stones at proper distances round it to denote the several hours. It would appear that Ferguson did not attempt making a clock until at least 7 or 8 years after the period under review (vid. note 30). Therefore, probably this story of "the horse-head clock" was one of the many "*false and improbable particulars*" which "*had been related of him*" (vid. note 2).

The late Capel Lofft, Esq., has several allusions to Ferguson in his "EUDOSIA, OR A POEM ON THE UNIVERSE" (12mo, pp. 248). "*Printed and sold by W. Richardson in the Strand, and W. Dilly in the Poultry, London* 1781;" now out of print, and very scarce. The following extracts from "*Eudosia,*" relate to Ferguson's shepherd-boy period, &c.:—

"Nor shall thy guidance not conduct our feet,
O honoured shepherd of our later days!
Whom from the flocks, while thy untutor'd soul,
Mature in childhood, trac'd the starry course. (Book I., v. 15.)
Astronomy, enamour'd, gently led
Through all the splendid labyrinths of Heaven,
And taught thee her stupendous laws, and clothed
In all the light of fair simplicity,
Thy apt expression. Hail! ye friends of men:
Who, while ye lived on earth, ranged Heaven, and brought
The hours, the days, the year, to regulate
The life of mortals."———

(Vide "*Eudosia, or a Poem on the Universe;*"
Book II., p. 32, lines 32 to 43.)

And again, in "EUDOSIA," Book V., we have as follows (subject, Eclipses)—referring to this period:—

"Nor thee, ENDYMION of our later age,
Sage shepherd of the northern Scotia's hill,
Shall grateful memory forget. How blest, (125, Book V.)
If to my verse the favouring muse indulge,
The happy clear expression which she gave
Thy simple prose. O teach me to express,
Familiar, what the causes of ECLIPSE,
Their difference, their limits, and their use,—
Since knowledge taught by thee, O! Ferguson,
Seemed intuition to the learner's ear.
Thy parent Earth in infancy beheld
Thee with untutor'd genius on the stars
Intent; above thy years, and far beyond 135.

evenings, when my work was over, I went into a field with a blanket about me, lay down on my back, and stretched a thread with small beads upon it at arm's length between my eye and the stars, sliding the beads upon it till they hid such and such

> Thy humble lot; nor flowery mead, nor play,
> Nor sleep, beguil'd thee from the beauteous scene;
> Nor poverty, nor health infirm, suppress'd
> The active vigour of thy mind; nor all
> That lowering fortune interpos'd, eclips'd
> The piercing ray of thy superior soul,
> Heaven-taught philosopher! Thy wandering sheep
> Saw thee, with eye sublime, reading the Heavens;
> Nor wanted they a guide—the good old man,
> Whom gratefully thy modest page rewards: 145.
> To thy fidelity and genius just
> Provided hours of leisure, that secure,
> Nor fearful for the charge—thy eager view
> Might range the firmament."
>
> (Vide "*Eudosia, or a Poem on the Universe;*"
> Book V., pp. 115–116, lines 123 to 149.)

NOTE.—Mrs. Anne Casborne, of New House, Pakenham, near Bury St. Edmunds, the daughter of the late Capel Lofft, Esq., author of "*Eudosia,*" in one of her letters to us relative to her parents, says, "*My father, Mr. Capel Lofft,* author of 'Eudosia, a Poem on the Universe,' *was born in London on* 25*th Nov.* 1752, *and died at Montcaliere, near Turin, in Italy, on* 26*th May* 1824, *in the* 72*d year of his age. My mother, Anne Emblin, the pupil and the Eudosia of Ferguson's* 'GENTLEMEN AND LADIES' ASTRONOMY,' *was born at Windsor in February* 1753, *and became the first wife of my father, Capel Lofft, on* 20*th August* 1778; *she died on* 8*th September* 1801. . . . . *I was young when she died, and do not recollect her mentioning Ferguson's name, except her showing me, just before her fatal illness, Ferguson's Moveable Planisphere, saying, she would teach me Astronomy, as she had been taught by him. This Planisphere I have still in my possession. I have also a portrait in chalk of Ferguson, left by him to some lady of title at Windsor, who left it to my grandfather. . . . The portrait is somewhat faint, particularly about the mouth; I can discover no initials or date, either in the corners or on the back. I expect a visit from my half-sister, Laura Capel Lofft, the beginning of next month; she was brought up in Italy, and is quite an artist, and a most beautiful copyist. I will ask her to copy Ferguson's portrait for you in the diminished size you wish. . . The size of the original portrait, not including the gilt frame, is* 22 *inches by* 17. . . . *My aunt, who died about* 22 *years ago, has often told me that my mother went as a pupil to study Astronomy under Mr. Ferguson, Bolt Court, Fleet Street, London, when she was young in her teens, about* 1767, *and often stayed with him. She was then in all her youth and beauty, and when* GEORGE III. *used to call her the flower of Windsor. Her ardent taste for Astronomy astonished her younger sister and parents.*"

Mr. Capel Lofft, in the appendix to his "*Eudosia, or a Poem on the Universe*" (p. 222 to 225), gives a short account of Ferguson, condensed from the *original memoir.* At p. 225 he says, "*Some manuscript, tables, diagrams, and a philosophical correspondence of this heaven-taught philosopher, are in my hands, which were given by him to my* EUDOSIA *before our marriage; for he had no pupil whose genius and disposition he more esteemed than hers; nor can I reprove myself for the pride which I often feel in reading over his letters to Miss Emblin, written to her before our marriage.*" At page 224 of Eudosia, Mr. Lofft informs us, that a letter of Ferguson's, relative to his styling his pupil Eudosia in his Astronomical Dialogues, was the cause of his naming his own poem "EUDOSIA." Mrs. Casborne, in alluding to the "*manuscript, tables, diagrams, and a philosophical correspondence*" of Ferguson's sent to her mother, says, "*that they cannot now be found, being long lost.*"

stars from my eye, in order to take their apparent distances from one another; and then laying the thread down on a paper, I marked the stars thereon by the beads, according to their respective positions, having a candle by me.[14] My master

at first laughed at me, but when I explained my meaning to him, he encouraged me to go on; and that I might make fair

[14] Considering that it would be interesting, if it could be ascertained, as to whether or not Mr. Glashan lived to hear of Ferguson's success and renown, we resolved to make inquiries to clear up the point. The only way which appeared to be open for us, at this distant period, was that in which we might get at the date of *his* death. We accordingly addressed a note to the present minister of Keith, the Reverend Mr. Annand, on the subject of the date of his death and age; who, in reply, informs us that he was interred in the church-yard of Keith, where there is a stone to his memory, and that the inscription on it records that "JAMES GLASHAN WAS BORN ON 11th DECEMBER 1686, AND DIED ON 9th JANUARY 1771." These dates make clear several little matters, which, without them, would have been unknown, viz. 1st, It is evident that Mr. Glashan died at the advanced age of 84; 2d, That he was in his 39th year in 1725, when Ferguson (then 15) entered his service; and 3d, That he died only 5 years and 10 months before Ferguson; consequently *he lived nearly through the whole of Ferguson's life-time*, and would therefore hear of his many Astronomical, Mechanical, and other publications, and of how well they had been received and appreciated by the public. He would hear, and no doubt would rejoice when he heard, that his Majesty, George III., had bestowed on him a pension of £50 a-year; and of his having been elected a free member of the most celebrated scientific body in the world—namely, Fellow of the Royal Society. By being so successful in his writings, and by having had so many distinguished marks of honour and respect conferred on him, would gladden the heart of the kind James Glashan, and no doubt would often bring to his mind the days of 1725, when his farm-laddie used to go out into the adjacent fields and hill-slopes, in the cold clear winter evenings, swathed in a blanket or faker, and armed with a lighted lantern, beaded-thread, paper, and pencil, to take the relative bearings of the stars; and of how *he* was wont to sit beside him in the old barn, in the day-time, "*busy with his compasses, ruler, and pen.*" These and similar other little incidents would often be related by the good old man to his friends, and he would contrast his early doings with the high posi-

copies in the day-time of what I had done in the night, he often worked for me himself.——I shall always have a respect for the memory of that man.[15]

1727.—One day he happened to send me with a message to the Reverend Mr. John Gilchrist, minister at Keith, to whom I had been known from my childhood.[16] I carried my star-papers

tion he had attained. An eminent writer says, "*We pay tribute to thy memory, James Glashan. Thy unprecedented kindness to the young rustic astronomer has been universally admired: it shall never be forgotten; future generations will revere thy name and applaud thy worth.*" (See note 17).

We may here note that John Stuart, Esq. of Aucharnie, Aberdeenshire, Secretary of the Spalding Club and Society of Antiquaries of Scotland, is the great-grandson of Mr. James Glashan. He informs us that there are no letters or memoranda extant regarding his great-grandfather, whose memory he holds in the highest esteem.

Mr. Glashan's farm was called ARDNEADLIE, situated about 1 mile to the SSW of Old Keith (at the foot of the Caird's-hill). The farm-house was demolished about 35 years ago; since then, the name Ardneadlie has become obsolete; the site being now known as *Brae-head.* It was therefore in the fields and hill-slopes in the immediate vicinity of the present Brae-head, between the years 1725 and 1729, that Ferguson, a rustic youth, between 15 and 19 years of age, first began to measure the apparent relative distances of the stars, by means of thread and beads, as mentioned in the text.

Regarding Ferguson's beaded-thread operation, we have always found a difficulty in the way when endeavouring to explain how *he* managed to slide the beads on the thread into position, with any degree of accuracy. If both his hands were engaged in holding the thread on the stretch, as the text seems to imply, he could scarcely have shifted them with his mouth, as by such a process, it could not have been done with precision, and would have entailed on him considerable trouble in bringing ever and anon the beads into line with the stars. Mr. Robert Sim, of Keith, has suggested to us the following explanation of the operation, which we adopt, viz. that "Mr. Ferguson, when he went into the fields in the evenings to take the apparent relative distances of the stars, by means of his beaded-thread, would likely carry a stick with him; this he would thrust firmly into the ground,—the ends of the thread would have loops for slipping over the stick-top—one of the loops having been slid down a little way on the said stick; he would then '*lie down on his back,*' and put himself into position, all the while holding the other loop in his hand '*at arm's length between his eye and the stars*' he wished to take. By this method he could easily keep the thread and beads on the stretch, either in a horizontal or angular direction, to meet the position of those he wanted for his map, and as one of his hands would thus always be disengaged, he would employ it in shifting the beads to their proper places on the thread, '*sliding them upon it till they hid such and such stars from his eye.*'" We give a fanciful drawing, in order to illustrate the star-taking process.

It is somewhat remarkable that the celebrated Tycho Brahé, the Danish Astronomer, about the year 1561, and when also in his 15th or 16th year, also "*took the apparent relative distances of the stars by a method of his own devising,*" viz., "*he took the apparent distances of the stars by a common pair of compasses, the hinge of which he used to put to his eye, while he opened the legs until they pointed to the two stars whose relative position he wished to ascertain.*" But by such a method, two stars only could be taken at a time; whereas, by the beaded-thread process of Ferguson, a great many could be taken at one operation.

[15] Mr. Glashan died about 2½ years before Ferguson wrote his Memoir, prefixed to his Select Mechanical Exercises.

[16] The Reverend Mr. John Gilchrist was the first Established Minister of Keith after the subversion of Episcopacy. He was inducted Minister of Keith (the Rev. Mr. Annand informs us) in the year 1700, and died on 13th January 1754, aged

to show them to him, and found him looking over a large parcel of maps, which I surveyed with great pleasure, as they were the first I had ever seen. He then told me that the earth was round like a ball, and explained the map of it to me. I requested him to lend me *that* map, to take a copy of it in the evenings. He cheerfully consented to this, giving me at the same time a pair of compasses, a ruler, pens, ink, and paper, and dismissed me with an injunction not to neglect my master's business by copying the map, which I might keep as long as I pleased.

For this pleasant employment my master gave me more time than I could reasonably expect, and often took the thrashing-flail out of my hands, and worked himself, whilst I sat by him in the barn, busy with my compasses, ruler, and pen.[17]

about 82 years. Mr. Gilchrist appears to have taken an interest in young Ferguson, and no doubt he would rejoice at all times to hear of his welfare and success. In 1754 Ferguson was comparatively little known. He had, however, published an ASTRONOMICAL ROTULA and CARD SUN-DIAL in 1742; "A DESCRIPTION OF A NEW ORRERY" in 1746; The Path of the Lunar Orbit by the Trajectorium Lunarie, 1746; A Dissertation on the Phenomena of the Harvest-moon, &c., 1747; and "AN IDEA OF THE MATERIAL UNIVERSE," 1752. Copies of these would no doubt be sent as presents to the minister of Keith, which would be all he would live to see of the astronomer's works. He would, however, hear that Ferguson had made his debút as a lecturer on Astronomy in 1748, and was appreciated as such by the public. The period of this interview with the minister of Keith appears to have been in 1727, at which time Mr. Gilchrist would be 55 years old, and Ferguson about 17.

[17] We have already alluded to the amiable James Glashan in note 14, and in noticing him again at this juncture, we cannot do better than to quote the just tribute paid to him in pp. 199–200 of the Lib. Entg. Knowl., Vol. I.:—"*This is*

When I had finished the copy, I asked leave to carry home the map; he told me I was at liberty to do so, and might stay two hours to converse with the minister. On my way thither, I happened to pass by the school at which I had been before, and saw a genteel-looking man (whose name I afterwards learnt was Cantley) painting a sun-dial on the wall.[18] I stopt a while to observe him, and the schoolmaster came out and asked me what parcel it was I had under my arm. I showed him the map, and the copy I had made of it, wherewith he appeared to be very well pleased, and asked me whether I should not like to learn of Mr. Cantley to make sun-dials. Mr. Cantley looked at the copy of the map and commended it much, telling the schoolmaster (Mr. John Skinner) that it was a pity I did not meet with notice and encouragement.[19] I had a good deal of conversation with him, and found him to be quite affable and communicative, which made me think I should be extremely happy if I could be further acquainted with him.

a beautiful we may well say, and even a touching picture—the good man so generously appreciating the worth of knowledge and genius, that, although master, he voluntarily exchanges situations with his servant, and insists upon doing the work that must be done himself, in order that the latter may give his more precious talents to their more appropriate vocation. We know not that there is on record an act of homage to science and learning more honourable to the author."

[18] By Ferguson mentioning that he "*happened* to pass by the school," appears to imply that in doing so, it was accidental, not being in the road to the manse from his master's farm of Ardneadlie. On examining an old and large map of the district, we find our surmise seems to be correct. Had Ferguson gone *direct* to the minister from Ardneadlie, a distance of about ¾ths of a mile from Mr. Gilchrist's—he would have had to pass his house before coming to the school. It is therefore evident that he took another road and went direct to Keith, about a mile to the NE, probably to show to his friends there, the copy he had made of the minister's map before taking the original home. On leaving Keith he would proceed westward, and would immediately reach the SE corner of the kirkyard-wall, where he would descry Mr. Cantley painting the sun-dial on the school wall, at a distance of about 35 yards. To reach the school in the NW corner of the kirkyard, he would enter the gateway in the east wall, a little above the SE corner, and then proceed along the diagonal footway to the school. On leaving it he would go on by the south gate in the SW corner of the south wall of the kirkyard, and from thence to the Rev. Mr. Gilchrist's, 30 yards to the west; then on to Ardneadlie, Mr. Glashan's farm, three quarters of a mile directly south. Mr. Gilchrist's manse, now demolished, was situate about 36 yards to the SW of the old Grammar-school, which, as already noticed, is also removed.—This note is for the use of local readers;—that to which it refers has often been matter of local discussion.

[19] The Reverend Mr. Annand, minister of Keith, informs us, that "Mr. John Skinner entered on his duties of Schoolmaster and Session-Clerk of Keith in 1709;" that "the date of his death is not on record; but from a casual remark in the Session-book, made at a meeting of Session held in April 1747 (he says), I find that he died in March 1747." From another source we learn that Mr. Skinner was about 68 years old when he died, and was interred at Keith; hence, he was born about the year 1679, and was schoolmaster of Keith Grammar-school 38 years, and would be about 39 years of age when Ferguson had his "3 *months'*" tuition under him.

I then proceeded with the map to the Minister, and showed him the copy of it. While we were conversing together, a neighbouring gentleman, Thomas Grant, Esq. of Achoynaney, happened to come in, and the Minister immediately introduced me to him, showing him what I had done. He expressed great satisfaction—asked me some questions about the construction of maps, and told me, that if I would go and live at his house, he would order his butler, Alexander Cantley, to give me a great deal of instruction.[20] Finding that this Cantley was the man whom I had seen painting the sun-dial, and of whom I had already conceived a very high opinion, I told Squire Grant that I should rejoice to be at his house as soon as the time was expired for which I was engaged with my present master. He very politely offered to put one in my place, but this I declined.[21]

1728.—When the term of my servitude was out I left my good master, and went to the gentleman's house,[22] where I quickly

[20] From this it would appear, on this occasion, Mr. Grant and his butler, Alex. Cantley, were accidentally in Keith, the former at the manse, the latter at the school, within a few yards of each other. This was a great day for young Ferguson, having had encouraging interviews with his friend Mr. Skinner, the extraordinary Cantley, the worthy Squire Grant, and the kind Mr. Gilchrist.

[21] There were no registers of deaths kept in Scotland until late in last century; hence, there is no written record to show when Mr. Grant died, and there is no grave-stone in Keith (or in any of the neighbouring churchyards) to his memory; probably his remains were not interred in the district. We have made a great many inquiries regarding him, but without leading to any successful result. In balancing our memoranda, however, we are inclined to conclude that Mr. Grant was born about the year 1673, and died about 1748, in the 75th year of his age; consequently, he would be about 55 years old in 1728, the date of this interview with young Ferguson. Having died about 1748, he did not live to hear of Ferguson's future celebrity. The Rev. Mr. Annand, minister of Keith, has kindly sent us the following extract of Mr. Grant's contract of marriage, from his Session Records:—

"*Keith, March 1st,* 1710.
The much honoured Thomas Grant, of Achoynanie, and Mistress Jean Sutherland, grandchild to the Laird of Kinminity, declared their purpose of marriage, and being contracted and orderly proclaimed, were married Aprile the twelfth 1710. *They had three sons and a daughter, viz. Archibald, born* 5 *Dec.* 1711; *Alexander,* 3 *Nov.* 1712; *Isabel,* 15 *June* 1714; *and Walter,* 18*th Feb.* 1716."

[22] ACHWYNANNIE.—This place has had several spellings. In the Keith Register of Mr. Grant's marriage it is spelt *Achoynanie;* Ferguson writes it down Achoynaney, and now the name is generally written *Achwynannie.* It is situated about 1¾ mile to the ESE of Old Keith, at the foot of the Balloch-hill. Mr. Robert Sim, of Keith, has kindly sent us a fine pen and ink drawing of the old house, of which we annex a woodcut. He also has obliged us with the following details regarding it:—

"*Old mansion house of Achwynannie.—I herewith send you a view of the old house of Achwynannie as it is at present, taken from the south. It has lost much of its original appearance by modern innovations, the old windows having been built up, and larger ones struck out, while it has been repeatedly coated over with lime, whereby the base of the turret on the south-west corner is much obscured; the door and windows are, of course, modern. This is understood to be a very old*

found myself with a most humane good family. Mr. Cantley, the butler, soon became my friend, and continued so till his death. He was the most extraordinary man that I ever was acquainted

House of Achwynannie, from the South, 1861.

with, or perhaps ever shall see, for he was a complete master of arithmetic, a good mathematician, a master of music on every known instrument except the harp, understood Latin, French, and Greek, let blood extremely well, and could even prescribe as a physician upon any urgent occasion. He was what is generally called "*self-taught*," but I think he might with much greater propriety have been termed GOD ALMIGHTY'S scholar.[23]

building. The oldest date I have found and attempted to decipher is on a stone originally in the building, but now seen in the wall of the square of offices. It bears two shields; on the dexter one, three lions', or boars' heads, with the letters I. G. (I. Gordon). I was aware from local tradition that about 1667 *it was in the possession of one of that name. On the sinister one, three holly-leaves in chief, and a hunting horn in base, with the letters E. B. (E. Burnett), these being the armorial bearings of Burnett of Leys (Aberdeenshire), date,* 1690. *The property now belongs to the Earl of Seafield."*

23 Mr. Alexander Cantley.—Regarding this remarkable man we have made a great many inquiries, but with no success. William Cantlie, Esq. of Keithmore (Achendown), a descendant of the illustrious Cantley, writes us, that he has no memoranda of *him;* that he had seen several parties, and written to others inquiring for information for us, but without any result. We greatly regret this. Ferguson in his Memoir brings in Mr. Cantley as "*a star of the first magnitude,*" but makes him "*set too soon.*" We would have wished to have known something more about a man so remarkable. Singular it is, that there does not now exist the slightest incident on record, or by tradition, regarding his after life. Had Ferguson written no Memoir of himself, the accomplished Cantley would perhaps

1730.—He immediately began to teach me decimal arithmetic and algebra, for I had already learnt vulgar arithmetic at my leisure hours, from books. He then proceeded to teach me the elements of geometry; but to my inexpressible grief, just as I was beginning that branch of science, he left Mr. Grant, and went to the late Earl of Fife's, at several miles' distance.[24] The good family I was then with could not prevail with me to stay after he was gone, so I left them and went to my father's.

He had made me a present of Gordon's Geographical Grammar, which at that time was to me a great treasure.[25] There

never have been heard of;—excepting to a very few, he would have "*flourished unknown to fame.*"

> "*Full many a flower is born to blush unseen,*
> *And waste its sweetness on the desert air.*"

The Rev. Mr. Merson, Elgin, writes to us, noting that he has in his possession a book which belonged to Mr. Cantley, viz. "MOXON'S TUTOR TO ASTRONOMIE AND GEOGRAPHIE," on some of the leaves of which are his name, beautifully printed with the pen. Mr. Merson has cut one of the names out of his book and kindly sent it to us,—of which the annexed woodcut is an exact fac-simile.

AL:CANTLY

Fac-simile of Alexander Cantley's printed Signature.

From the above cut it will be seen that "CANTLY" is the spelling of the name, and not "*Cantley,*" as Ferguson has it.

One of our notes mentions that "*there can be no doubt that Mr. Alexander Cantley would be buried in the churchyard of Old Keith, as there was a burial-ground there during the last century which belonged to the Cantlies.*"

Of course it cannot now be ascertained when Mr. Cantley died, but it is evident that he was long dead previous to 1773, when Ferguson wrote his Memoir, as he mentions in it that Mr. Cantley continued his friend "*till his death.*"

Moxon's "*Tutor to Astronomie and Geographie*" was a very popular work in its day; *our* copy is the *second edition.* "LONDON: *Printed by Joseph Moxon, and sold at his shop in Russel Street, at the sign of the Atlas,* 1670."

24 This is an anachronism; there was no Earl of Fife at the period here referred to, nor for nearly 30 years after it. Mr. Cantley, in 1730–31, went to William Duff, Esq. of Braco, who became LORD BRACO in 1735, and EARL OF FIFE in 1759. He died 30th Sept. 1763; and as a matter of course, he was "*the late Earl of Fife*" when Ferguson, in 1773, wrote his Memoir. This Earl of Fife was the son of William Duff, Esq. of Dipple, near Fochabers. In 1730–1, William Duff, of Braco, afterwards Earl of Fife, had seats at DIPPLE, BRACO, DELGETTY, ROTHIEMAY, and BALVENY CASTLE.

25 This appears to have been a very popular and successful work—"*Geography Anatomiz'd;* or, the GEOGRAPHICAL GRAMMAR," by Patrick Gordon, M. A., F.R.S.: octavo. Our copy of it is the *twentieth edition*, published in London in 1754. In some old catalogues of books in our possession, we observe that editions of it were issued in 1693 and in 1722. It is dedicated to "THE MOST REVEREND FATHER IN GOD, THOMAS, LORD ARCHBISHOP OF CANTERBURY." The book consists of "Two Parts;"—Part 1st gives "*a general view of the globe,*" *illustrated by definition, description, and derivation*—"*pleasant problems performable by the*

is no figure of a globe in it, although it contains a tolerable description of the globes, and their use. From this description, I made a globe in three weeks at my father's, having turned the ball thereof out of a piece of wood, which ball I covered with paper, and delineated a map of the world upon it—made the meridian ring and horizon of wood—covered them with paper, and graduated them; and was happy to find that, by my globe (which was the first I ever saw), I could solve the problems.

1731.—But as this was not likely to afford me bread, and I could not think of staying with my father, who I knew full well could not maintain me in that way, as it would be of no service to him; and he had, without my assistance, hands sufficient for all his work.[26]

I then went to a miller, thinking it would be a very easy business to attend the mill, and that I should have a great deal of leisure time to study decimal arithmetic and geometry. But my master, being too fond of tippling at an ale-house, left the whole care of the mill to me, and almost starved me for want of victuals, so that I was glad when I could have a little oat-

terrestrial globe—some paradoxical positions in matters of Geography, which mainly depend on a thorough knowledge of the globe—and lastly, a transient survey of the whole surface of the terraqueous globe." Part 2d gives "*a particular view of the terraqueous globe, in extent, situation, division, sub-division, chief towns, name, air, soil, commodities, rarities, archbishops, bishops, universities, manners, language, government, arms, religion, &c.*" Such is an outline of the contents of Cantley's parting-gift to Ferguson. Our copy is embellished with a great many maps,—pages 416: oct.; it is now a very scarce book, having been about 100 years out of print. Section 4th of part 1st contains the "*Strange Geographical Paradoxes,*" pp. 35 to 40 inclusive. They are very curious, and most excellent for exciting the mind on matters Geographical; and no doubt they would prove both interesting and amusing to Ferguson at this period of his life (1731, age 21 years).

[26] As a matter of course, this has reference to "*the hands of the family.*" Near the beginning of his Memoir Ferguson mentions that his father had "*a large family.*" Wishing to ascertain how many of a family Ferguson's parents had, and also their names, we wrote to several friends in the counties of Elgin, Banff, and Aberdeen, but without receiving any reliable notes. One of our correspondents mentions a Margaret Ferguson, eldest sister of the astronomer, who died before 1770. In our copy of Ferguson's will, dated 15th August 1776 (see Appendix), we find the names of three sisters, and probably in the order of their ages, viz. Elspeth, Elizabeth, and Janet; Margaret being dead before the will was made, is of course not named; but it is singular that his "elder brother" John, who was then alive, is not mentioned in it; therefore, the parents of Ferguson had at least *two sons* and *four daughters*,—viz.

1. John Ferguson, born 12th March 1708, at Core of Mayen,		Died at RELASHES, near Rothiemay, in 1796, aged 88.
2. James, born 25 April 1710, at Core of Mayen,		Died at 4 BOLT COURT, LONDON, 16 Nov. 1776, aged 66½.
3. Margaret,	unknown	unknown.
4. Elspeth,	do.	do.
5. Elizabeth,	do.	do.
6. Janet,	do.	Died at BARNHILLS, near Rothiemay, in 1793.

meal, mixed with cold water, to eat. I was engaged for a year in this man's service, at the end of which I left him, and returned in a very weak state to my father's.[27]

1732.—Soon after I had recovered my former strength, a neighbouring farmer, who practised as a physician in that part of the country, came to my father's, wanting to have me as a labouring servant. My father advised me to go to Doctor Young, telling me that the doctor would instruct me in that part of his business. This he promised to do, which was a temptation to me. But instead of performing his promise, he kept me constantly to very hard labour, and never once showed me one of his books. All his servants complained that he was the hardest master they had ever lived with; and it was my misfortune to be engaged with him for half a year. But at the end of three months I was so much over-wrought, that I was almost disabled, which obliged me to leave him; and he was so unjust as to give me nothing at all for the time I had been with him, because I did not complete my half-year's service, though he knew that I was not able, and had seen me working for the last fortnight, as much as possible, with one hand and arm, when I could not lift the other from my side. And what I thought was particularly hard, he never once tried to give me the least relief, further than once bleeding me, which rather did me hurt than good, as I was very weak and much emaciated. I then went to my father's, where I was confined for two months on account of my hurt, and despaired of ever recovering the use of my left arm; and during all that time, the doctor never once came to see me, although the distance was not quite two

27 As Ferguson does not give the name of his "tippling miller," nor that of the mill, we cannot fill up the blank, although to have done so would have been interesting items at this distant period. Mr. Robert Sim, of Keith, in a letter to us, mentions the sites of several old mills, and mills still in operation in and around Keith, viz. 1st, The Mill-o-wood, 1 mile NE of Achoynaney; the present mill is modern. "The old mill," says Mr. Sim, "yet well remembered, stood some distance below the present one, and was an old and romantic-looking thing, from its situation below an immense rock, over and through a crevice of which the water fell on the wheel." (We find that about the beginning of the present century this mill was the one generally understood as that in which Ferguson "*served and starved.*") 2d, Tarnach mill, above 1 mile SW of Achoynaney; 3d, The Earl's mill, on the Isla, a little to the east of Old Keith; 4th, At old Newmills, about 1½ mile NE of Keith, close to which there was a blacksmith who kept an ale-house (many in consequence of this *link* incline to think that it was here where Ferguson starved and the tippling miller enjoyed himself); 5th, Nether mills, about 3 miles down the Isla from Keith, and about 1 mile to the south of Braco house, Grange; 6th, Crook's mill, about 2 miles NW of Keith, &c.

miles.[28] But my friend, Mr. Cantley, hearing of my misfortune, at twelve miles' distance, sent me proper medicines and ap-

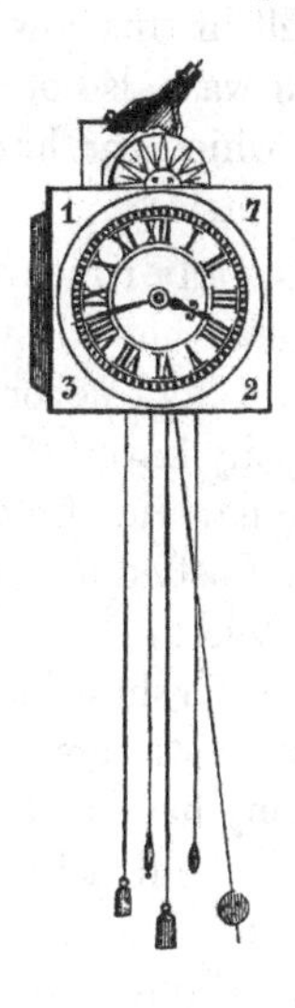

plications, by means of which I recovered the use of my arm;[29] but found myself too weak to think of going into ser-

28 It is evident that Ferguson had not been long in Dr. Young's service when he found out that "*his lines*" had not "*fallen in pleasant places.*" A writer says, "for the sake of humanity, it is to be hoped that very few *Dr. Youngs* have had a resting-place on this nether sphere. It is scarcely possible to conceive of any one so utterly devoid of all the finer feelings of our nature as this Surgeon-farmer." The bad treatment Ferguson here experienced appears never to have been forgotten by him. In 1773, when he wrote his *Memoir*, a date 42 years after he had left this service, the rememberance of his treatment came up in strong relief, and thus made him throw the *picture* into the story of his life. We leave him in Ferguson's hands; what he says of him is to the point, and "will always show the doctor up as a very unpleasant specimen of the human species." We have made a great many inquiries touching the name of the farm occupied by Dr. Young, in order to ascertain the whereabouts of the residence of Ferguson's parents, (Dr. Y. lived within 2 miles of his father's cot), but we have met with no success in this matter—even the time of his death and where buried are unknown. It would have been interesting to have known if he lived long enough to hear of Ferguson's world-wide fame. We hope he did, and regretted his treatment of the embryo astronomer. One of our correspondents says that "his father, who died in the beginning of the present century, understood that Dr. Young was a *queer doctor.* He was also a bit of a farmer; but, a *so-and-so* sort of one." He adds, "*it is likely that this worthy was interred in Old Keith kirkyard.*" Dr. Young's farm was within 2 miles of the now unknown residence of Ferguson's parents. It is supposed by some that his farm was at Pitlurg, a few miles to the SSE of Keith.

29 Ferguson here gives us one of the few distances mentioned in his Memoir. He speaks of his friend Mr. Cantley as being at the Earl of Fife's, 12 mlies distant from his father's cot; probably these 12 miles are *old miles*, if so,

vice again, and had entirely lost my appetite, so that I could take nothing but a draught of milk once a-day, for many weeks.

In order to amuse myself in this low state, I made a wooden clock, the frame of which was also of wood; and it kept time pretty well. The bell on which the hammer struck the hours, was the neck of a broken bottle.[30]

Having then no idea how any timekeeper could go but by a weight and a line, I wondered how a watch could go in all positions, and was sorry that I had never thought of asking Mr. Cantley, who could very easily have informed me. But happening one day to see a gentleman ride by my father's house (which was close by a public road), I asked him what o'clock it then was; he looked at his watch, and told me.[31] As he did that with so much good nature, I begged of him to show me the inside of his watch; and though he was an entire stranger, he immediately opened the watch and put it into my hands. I saw the spring-box with part of the chain round it, and asked him what it was that made the box turn round; he told me that it was turned round by a steel spring within it. Having never seen any other spring than that of my father's gun-lock, I asked how a spring within a box could turn the box so often round as to wind all the chain upon it. He answered, that the spring was long and thin; that one end of it was fastened to the axis of the box, and the other end to the inside of the box; that the axis was fixed, and the

then they are equal to 15 modern ones. If Mr. Cantley was then at Balveny Castle, then a line of 15 miles, in the direction of Keith from it, by the road, terminates at a place called FLOORS, 4½ miles NE of Old Keith, in the parish of Grange, and this FLOORS is 7 old miles, or nearly 9 modern miles, from Durn house.

30 Ferguson made this clock in the year 1731 (when he was 21 years of age). The neck of a broken bottle was an ingenious make-shift for a bell; if securely fixed, it would give out a clear, sharp, and sweet sound. Mr. Dean Walker, lecturer on Astronomy, London, son of the celebrated Mr. Adam Walker, lecturer on Natural and Experimental Philosophy, the friend of Ferguson, had in his possession a great many of Ferguson's papers, and pen and ink sketches; —we called on him in 1831, and were presented with a number of them and other memoranda. On one of these papers was a drawing in pen and ink of this clock, which we give above on a reduced scale, as an interesting memorial of this early period of Ferguson's life. It is remarkbale that Mr. John Harrison, the celebrated inventor of the marine chronometer, when also unwell and in a low condition, in the 21st year of his age, made two wooden clocks of a peculiar construction.

31 Ferguson here informs us that his "*father's house was close by a public road*," but does not say where. The public roads around Keith are those leading to Mortlach and Balveny Castle on the SW; to Fochabers, Elgin, and Spey side on the NW of Keith; to Cullen and Portsoy on the N and NE; and to Grange, Rothiemay, and Huntly on the E and SE. If FLOORS be assumed as the place alluded to, as being "*near a public road*," then that road, which runs *close* by it, is that leading to Durn and Portsoy.

box was loose upon it. I told him I did not yet thoroughly understand the matter. Well, my lad, says he, take a long thin piece of whalebone, hold one end of it fast between your finger and thumb, and wind it round your finger: it will then endeavour to unwind itself; and if you fix the other end of it to

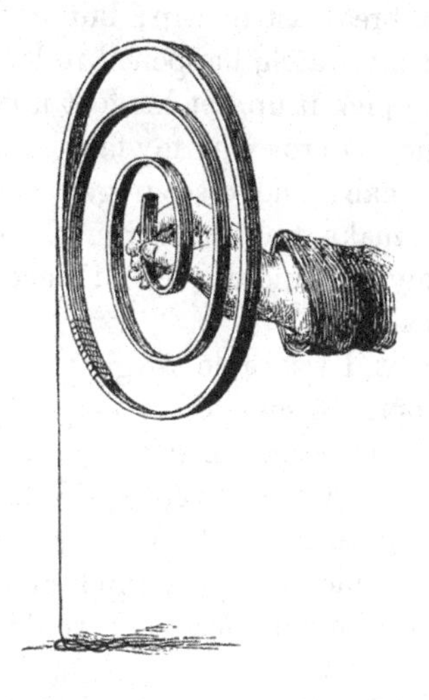

Action of a Watch Main-spring Illustration.

the inside of a small hoop, and leave it to itself, it will turn the hoop round and round, and wind up a thread tied to the outside of the hoop. I thanked the gentleman, and told him that I understood the thing very well.[32] I then tried to make a watch with wooden wheels, and made the spring of whalebone;

[32] This is a happy illustration of the principle and action of the main-spring of a watch; perhaps the simplest that was ever offered or can be given. We give a wood-cut of the arrangement of parts. The interview between Ferguson and this unknown gentleman is taken notice of in the Lib. Ent. Know., vol. I. pp. 203, 204, as follows:—"*Who is there that does not envy the pleasure that must have been felt by the courteous, intelligent stranger, by whom the young mechanician was carried over his first great difficulty, if he ever chanced to learn how greatly his unknown questioner had profited from their brief interview! That stranger might probably have read this narrative, as given to the world by Ferguson, after the talents which this little incident probably contributed to develop, had raised him from his obscurity to a distinguished place among the philosophers of his age; and if he did know this, he must have felt that encouragement in well-doing which a benevolent man may always gather, either from the positive effects of acts of kindness upon others, or their influence upon his own heart.*"

Young Ferguson requesting a sight of the works of the watch, showed that such a piece of mechanism was new to him; and also, that his father was not possessed of one. Indeed, about this period, 1731, watches were only worn by the wealthy.

but found that I could not make the watch go when the balance was put on, because the teeth of the wheels were rather too weak to bear the force of a spring sufficient to move the balance, although the wheels would run fast enough when the balance was taken off.[33] I enclosed the whole in a wooden case, very little bigger than a breakfast teacup; but a clumsy neighbour, one day looking at my watch, happened to let it fall, and turning hastily about to pick it up, set his foot upon it, and crushed it all to pieces, which so provoked my father, that he was almost ready to beat the man; and discouraged me so much, that I never attempted to make such another machine again, especially as I was thoroughly convinced I could never make one that would be of any real use.

1733.—As soon as I was able to go abroad, I carried my globe, clock, and copies of some other maps besides that of the world, to the late Sir James Dunbar, of Durn (about seven miles from where my father lived), as I had heard that Sir James was a very good-natured, friendly, inquisitive gentleman.[34] He received me in a very kind manner, was pleased with what I showed him, and desired I would clean his clocks.

[33] Ferguson, even although he made a minute inspection of the gentleman's watch whilst it was in his hands, would, notwithstanding, experience many untold difficulties in the process of making his wooden one. The sight of the works of the watch by the way-side was but a momentary glance; he could not then have studied the balance—*the balance-spring*—the shape of the verge, and its action on the crown-wheel inside. It is therefore to be presumed that he had subsequent examinations of watches, in the possession of neighbouring gentlemen. He could not, by inspection at the road-side, have seen the balance-spring, or how it was fixed, except by removing part of the works.

[34] After Ferguson had sufficiently recovered from his illness, he informs us that he went to Sir James Dunbar's, at Durn, carrying with him his globe, clock, and maps; and that Durn, the residence of Sir James, was "about seven miles" from where his father then lived. The present Baronet, the Rev. Sir William Dunbar of Kew, informs us, that Sir James Dunbar was born on the 9th January 1668, was served heir to his father, Sir William Dunbar (3d Bart.), on 10th February 1709, and died about the beginning of February 1739, in the 71st year of his age; consequently, Sir James died before Ferguson became known to the world, and would be about 64 years old at the time of Ferguson's visit.—Durn House and estate at one time belonged to a family of the name of Menzies, who sold them to Sir William Dunbar, whose son, Sir James, sold them to John Falconer, Esq., whose relict sold them to Alexander Gordon of Letterfourie, whose son, Sir James, sold them to the Earl of Seafield, the present proprietor. The old house and gateway of which Ferguson speaks were removed about the year 1770. The present Durn House has therefore no connection with Ferguson's history. Conceiving that a view of old Durn House and gateway would be interesting, we addressed a note to Mr. Bryson, factor of the Earl of Seafield, Cullen house, inquiring if there was any view of these extant. In his very obliging reply, he enclosed a reduced drawing of old Durn House and gateway, taken from an old charter in the Charter-chest of the Earl of Seafield, of which we have much pleasure in giving here a copy.

This, for the first time, I attempted, and began to pick up some money in that way about the country, making Sir James's house my home, at his desire.[35]

Durn House Gate-way, 1760.

Two large globular stones stood on the top of his gate. On one of them I painted (with oil colours) a map of the terrestrial globe, and on the other, a map of the celestial, from a planisphere of the stars, which I copied on paper from a celestial globe belonging to a neighbouring gentleman. The poles of the painted globes stood toward the poles of the heavens. On

Durn is about 13 miles NE of Keith, 1 mile E of Fordyce, and 2 miles SSW of Portsoy. Durn is in the parish of Fordyce. In the centre of the parish rise the hills of Durn and Fordyce, extending continuously, and in a crescent form, from NE to SW. The remains of an encampment on the hill of Durn are still visible; there are also other relics, which show that the vicinity had often been the scene of hostile feuds. Quartz rock forms the round-backed eminence of Durn-hill, while boulders of syenite abound at the foot. At Portsoy there is a splendid sea-view, and no doubt our young astronomer journeyed often to this quarter during his stay at Durn House. Portsoy marble has world-wide fame.

Ferguson here gives another of his distances; he says that Durn, the residence of Sir James Dunbar, was "*about* seven miles" from where his father lived. If these 7 miles are old miles, then they are nearly equal to 9 modern miles (see note 29). If we assume *Floors* in the parish of Grange, 4½ miles NE of Keith, to have been the residence of Ferguson's father at this period, it will be found to be *about* 9 miles to the SSW of Durn.

[35] Here Ferguson comes before us as a cleaner and repairer of clocks. About the end of last century the old people in the north of Banffshire used to relate that "*Ferguson cleaned clocks at one shilling each, and oiled them for a penny;*" and that he also "*repaired roasting-jacks, and any other thing that had mechanism about it.*" But it must be recollected that the shilling and the penny of 1732 would purchase more than would the same coins in the present day.

each, the 24 hours were placed around the equinoctial, so as to show the time of the day when the sun shone out, by the boundary where the half of the globe at any time enlightened by the sun was parted from the other half in the shade; the enlightened parts of the terrestrial globe answering to the like enlightened parts of the earth at all times;—so that whenever the sun shone on the globe, one might see to what places the sun was then rising, to what places it was setting, and all the places where it was then day or night throughout the earth.[36]

During the time I was at Sir James's hospitable house, his sister, the Honourable the Lady Dipple, came there on a visit, and Sir James introduced me to her.[37] She asked me whether I

[36] In note 34 we mentioned that this gateway and globular stones were removed about the year 1770. The *globular stones* are, however, still in existence; one of them is built into the west gable of the Church of Ord, about 6 miles SW of Banff, and 4½ SE of Durn; the other one lies on the lawn of the manse of Rothiemay, about 10 miles S of Durn. The Rev. Mr. Moir of Rothiemay has kindly obliged us with particulars of the stone in his possession, and as both stones would be precisely alike, the following note on that of Rothiemay will be equally applicable to the stone at ORD:—Rev. Mr. Moir says, "*the stone in my possession is of freestone, is* 18 *inches in diameter, and weighs* 1 cwt. 3 qrs. 2 lbs., *or* 198 lbs. *No trace of Ferguson's painting is now to be seen on it.*" This is not to be wondered at, as it is now about 132 years since Ferguson "*tried his prentice hand*" on them. The stones, being 18 inches in diameter, have 1018 square inches of globular surface each. Would it not be well to have these stones erected at the Core of Mayen, where he was born, as a memorial of his early genius? The gateway faced the east, and a straight line passing through the centres of the stones lay due north and south. The south globular stone had the terrestrial globe painted on it, and the north one the celestial. We give a sketch of these stones from an old pen and ink drawing done in 1760. In note 23 we mention that the Rev. Mr. Merson of Elgin has in his possession an old copy of Moxon's "*Tutor to Astronomie and Geographie,*" which once belonged to Mr. Alexander Cantley, the friend of Ferguson. In this work there is an engraving of a pillar fountain, on the top of which there is a "*globular stone,*" having on it a map of the world, and the 24 hours of day and night, which was erected at the corner of Leadenhall Street, London, before 1670. The description given of this dial is similar to Ferguson's account of the Durn stone. Mr. Merson therefore thinks it not improbable that Ferguson may have seen Moxon's book while at Mr. Grant's, Achoynaney, in 1730, and that this globular stone-dial suggested to him the painting of the *globular stones* at Durn. We think this very probable; but if Ferguson did see Moxon's book while at Achoynaney, in 1730, he would see in it several representations of the globes, fully mounted; and hence, although in Gordon's Geography there is no figure of the globe, yet, what he had seen in Moxon would enable him to complete his wooden globe without much assistance from Gordon's Grammar.

We may here note that a correspondent of the editor of the Lib. Ent. Know. (vol. II. p. 56), mentions that he "*had seen at Gartness, Stirlingshire, globular stones, with circles of the sphere and constellations*" engraven on them, and concave stones with engravings of a like character, said to have been made by NAPIER (the inventor of logarithms). These are supposed to have been done by him in his youth, when residing at Gartness, about the year 1568.

[37] We are much indebted to Francis Fraser, Esq., of Findrack, Aberdeenshire, and his brother William N. Fraser, Esq., of Tornaveen, (the great-grandsons of Lady Dipple's eldest daughter, Anne Duff, and wife of William Baird, Esq. of

could draw patterns for needle-work on aprons and gowns. On showing me some, I undertook the work, and drew several for her, some of which were copied from her patterns, and the rest I

The Lady Dipple.

did according to my own fancy. On this I was sent for by other ladies in the country, and begun to think myself growing very rich by the money I got for such drawings, out of which I had

Auchmedden), for the following interesting notes on the good and amiable Lady Dipple. "Lady Dipple's maiden name was Jean Dunbar; she was the second wife of William Duff, Esq., of Dipple, near Fochabers, father of the first Earl of Fife; and was the youngest daughter of Sir William Dunbar, Bart., of Durn, by his wife, Janet Brodie, daughter of Dean Brodie of Aldearn. She had one son, Alexander, who died in his infancy in 1722, and four daughters, named Anne, Janet, Mary, and Henrietta. Anne, her eldest daughter, my great-grandmother, was married to William Baird, Esq., of Auchmedden, in the year . . . She had *four brothers* and *three sisters*—viz.: Sir James Dunbar, of Durn; William, who was a physician in Dunse; John, in the army, died in America in 1759; and George, a lawyer in Edinburgh, who also died in 1759. Her sisters were, Anne, married to the fourth Earl of Findlater and Seafield; 2d, ———, married to James Gordon of Letterfourie; and 3d, Grisel, married to Tulloch of Tannachie. . . . I may mention that the original oil-painting of Lady Dipple is at Findrack, is 2 feet 5 inches in length, and 2 feet broad, mounted in a large gilt frame." We also feel ourselves much indebted to William N. Fraser, Esq., for his kindness in sending us a photograph from the oil-painting of Lady Dipple, at Findrack, from which we had a miniature one taken, an engraving from which is annexed, and is pronounced a faithful copy of the original; it cannot fail to be very interesting to every admirer of Ferguson. The same gentleman was also so kind as to send us a tracing from the inscription on the blank leaf of Lady Dipple's Bible, now in the possession of Robert Duff, Esq., at Fetteresso Castle, Kincardineshire,—it is as follows:—

> "Jean Dunbars bibell which she leaves in memorandim to her son-in-law bracow in testimonie of her regard to him whom is in the plais of her Dirly belowed and only son, Alexander Duff who daye'd 1722 aged years, a most hopfowl and beautifowl child."

the pleasure of occasionally supplying the wants of my poor father.[38]

Yet all this while I could not leave off star-gazing in the nights, and taking the places of the planets among the stars by

Lady Dipple's son-in-law became Baron Braco in 1735; then this Bible inscription must have been written after that date. A great many old people in the counties of Elgin, Banff, Aberdeen, &c., who were alive about the end of the last century, who had seen Lady Dipple, represented her as being "a most handsome and beautiful lady." At the time of Ferguson's introduction to her at Durn, in 1732, she would be about 50 years old. Sir James Baird, Bart., Edmonstone, Libberton, near Edinburgh, informs us, that "*Mrs. Duff, wife of William Duff, Esq. of Dipple, was courteously called Lady Dipple, in accordance with the then Scotch fashion of designating the heads of families by the name of the property rather than by the surname.*" It is therefore evident that *Lady* Dipple was merely a title of courtesy, and that *Mrs. Duff*, or *Lady Duff* of Dipple, was the proper designation. Lady Dipple's husband, William Duff of Dipple, died in 1722. We cannot ascertain when Lady Dipple died, or how old she was at the time of her death. Collating and comparing a great many notes we have received on this point, it would appear that she was born at Durn in the year 1681, and died in 1763, aged 82 years. She therefore lived to hear that her protégé had published many useful astronomical works; that he had become a public lecturer on astronomy; that he had received an annual pension of £50 from King George III.; and that he had been elected a Fellow of the Royal Society—honours which would be hailed with joy by the good and pious Lady Dipple.

[38] An aged correspondent informs us, that in his young days (1790) he had been shown (in the north of Banffshire) several pieces of lace, two gowns, and an apron, which were sewed from Ferguson's "*fancy designs*," and that they were very beautiful. We have not been able to ascertain Ferguson's scale of charges for such designs. It is pleasing, however, to observe, that he was so attentive to the wants of his parents when a little money accumulated in his hands.

my above-mentioned thread. By this I could observe how the planets changed their places among the stars, and delineated their paths on the celestial globe.

By observing what constellations the ecliptic passed through in that map, and comparing these with the starry heaven, I was so impressed as sometimes to imagine that I saw the ecliptic in the heavens, among the stars, like a broad circular road for the sun's apparent course, and fancied the paths of the planets to resemble the narrow ruts made by cart-wheels, sometimes on one side of the plain road, and sometimes on the other, crossing the road at small angles, but never going far from either side of it.[39]

Durn House, Banffshire, 1758, from SW.

Sir James's house was full of pictures and prints, several of which I copied with pen and ink; this made him think I might become a painter.

Lady Dipple had been but a few weeks there, when William Baird, Esq. of Auchmedden, came on a visit.[40] He was the

[39] This is an original idea, and a very simple and excellent illustration it is. But had Ferguson used the word zodiac instead of *ecliptic*, it would have been in much better keeping, as the ecliptic is simply an imaginary line, supposed to be described by the centre of the sun in the heavens; whereas, the zodiac is a broad way or belt in which the planets move, and is cut into equal halves by the ecliptic line.

[40] As mentioned in note 37, William Baird, Esq. of Auchmedden was married to Anne Duff, eldest daughter of Lady Dipple. "Squire Baird" appears to have been a most kind-hearted and accomplished gentleman. Sir James Baird, Edmonstone, Libberton, Edinburgh, informs us that "*Mr. William Baird was the last male representative of the family of the House of Auchmedden;*" *and that* "*having been engaged in the rebellion of* 1745, *he was obliged to remain in hiding*

husband of one of that lady's daughters, and I found him to be very ingenious and communicative. He invited me to go to his house and stay some time with him, telling me that I should have free access to his library, which was a very large one, and he would furnish me with all sorts of implements for drawing. I went thither, and stayed about eight months;[41] but was much disappointed in finding no books of Astronomy in his library, except what was in the two volumes of Harris's Lexicon Technicum,[42] although there were many books on geography, and other sciences. Several of these indeed were in Latin, and more in French; which being languages that I did not understand, I had recourse to him for what I wanted to know of these subjects, which he cheerfully read to me; and it was as easy for him, at sight, to read English from a Greek, Latin, or French book, as from an English one.[43] He furnished me with pencils

at Echt, for some years afterwards. His property was not confiscated, but owing to the difficulties he was brought into by having borrowed large sums of money to aid in the cause of the Stuarts, he was obliged to dispose of the family estates, in or about the year 1750. *He had six sons and four daughters, all of whom died without issue, with the exception of Henrietta, the youngest daughter, who married Francis Fraser of Findrack, whose descendant, Francis Fraser, Esq. of Findrack, Aberdeenshire, is in possession of many of his relics; and is, I believe, the possessor of the picture you allude to as painted by Ferguson."* Mr. Humphreys, Aberdeen, informs us that William Baird of Auchmedden was born in 1701, and died in 1777, aged 76 years; and was buried in the churchyard of St. Nicholas in Aberdeen; and that Anne, his wife, was born at Dipple in 1705, and died at Aberdeen in January 1773, aged 68, and was buried beside her husband. The good "Squire Baird" therefore lived through the whole of Ferguson's life-time, and would hear of his world-wide renown, and the honours that had been conferred upon him. At the time of Ferguson's visit to him at Auchmedden he was 32 years old, and Ferguson about 23.

41 Auchmedden is in Aberdeenshire, about 30 miles to the NE of Keith, and 17 miles direct east of Durn House. The estate of Auchmedden is in the parish of Aberdour. The Rev. James Wilson (the present minister of Aberdour), has kindly furnished us with several interesting notes. He says, "*Auchmedden is the name of a considerable estate, with upwards of* 3 *miles of coast line, and running into the interior to a point at least* 6 *miles from the sea. There is now no House of Auchmedden; the ruins of the walls of the old house, however, still remain, and are visible from the road, and the old garden is now turned into a field. The very picturesque fishing village of Pennan, on the property, and within* ½ *a mile of the site of the old house, remains probably not much changed since the days of Ferguson in* 1733, *who must have been astonished at its situation. For six months the sun never shines on the village, which, fronting the north, has a background of very high perpendicular rocks facing the southern shore of the Moray Firth.*"

42 Harris's Lexicon Technicum was a very popular work about the beginning of last century; it has been long out of print. Our copy is in two huge folio volumes, embellished with numerous woodcuts, many of which are very rude and unnecessarily large. It is entitled "LEXICON TECHNICUM, OR AN UNIVERSAL ENGLISH DICTIONARY OF ARTS AND SCIENCES; EXPLAINING NOT ONLY THE TERMS OF ART, BUT THE ARTS THEMSELVES, by John Harris, D.D., and F.R.S.: London, 1725."

43 There are still in existence several MSS. of Mr. Baird, particularly a translation of the Greek of Thucydides. He appears to have had a taste for literary as well as genealogical and antiquarian pursuits, and to have been a gentleman

and Indian ink, showing me how to draw with them; and although he had but an indifferent hand at that work, yet he was a very acute judge, and consequently a very fit person for showing me how to correct my work. He was the first who

William Baird, of Auchmedden.

ever sat to me for a picture, and I found it was much easier to draw from the life than from any picture whatever, as nature was more striking than any imitation of it.[44]

of considerable accomplishments. He wrote a History of his Surname, printed 1857—and some MSS. (by him) of much interest are still extant.

44 This *picture* is still in excellent preservation, in the possession of his great-grandson, Francis Fraser, Esq. of Findrack, Aberdeenshire, who has obliged us with the following particulars regarding it:—"The picture (he says), done by Mr. Ferguson, of my great-grandfather, William Baird of Auchmedden, is in my possession, and has never been out of my family. It is done in Indian-ink, half-length miniature on Bristol card. Its dimensions are 6 inches long, by 4½ inches broad, placed within a small wooden frame rimmed with gilding and glazed. In the lower right-hand corner, in the hand-writing of Ferguson, are as follows—viz. 'J. F. *pinxt*,' and below the portrait

WILLIAM BAIRD OF AUCHMEDDEN
SUMMER 1733.

I may add that the portrait is in excellent condition, apparently as good as it was when first painted." Mr. William N. Fraser, brother of Mr. Fraser of Findrack, has also kindly obliged us with a copy of the original picture, from which our woodcut has been taken, and is in every respect a faithful copy of the original by Ferguson.—Ferguson remarks, that he "*found it was much easier to draw from the life than from any picture whatever, as nature was more striking than any imitation of it.*" The same opinion was held by the Rev. Robert Keith, who, at his first interview with Ferguson, shortly after this, advised him to copy from nature.

Lady Dipple came to his house in about half a year after I went thither; and as they thought I had a genius for painting, they consulted together about what might be the best way to put me forward. Mr. Baird thought it would be no difficult thing to make a collection for me among the neighbouring gentlemen, to put me to a painter at Edinburgh; but he found, upon trial, that nothing worth while could be done among them; and, as to himself, he could not do much that way, because he had but a small estate and a very numerous family.[45]

1734.—Lady Dipple then told me that she was to go to Edinburgh next spring, and that if I would go thither, she would give me a year's bed and board at her house *gratis*, and make all the interest she could for me among her acquaintance there.[46] I thankfully accepted of her kind offer, and instead of giving me one year, she gave me two. I carried with me a letter of recommendation from the Lord Pitsligo (a near neighbour of Squire Baird's),[47] to Mr. John Alexander, a painter at Edinburgh, who allowed me to pass an hour every day at his house to copy from his drawings,[48] and said he would teach me to paint in oil-colours, if I would serve him seven years, and my friends would maintain me all that time; but this was too much for me to desire them to do; nor did I choose to serve so long. I was then recommended to other painters, but they

45 As already mentioned (in note 40), "William Baird, Esq. of Auchmedden, had six sons and four daughters, all of whom died without issue, with the exception of Henrietta, the youngest daughter, who married Francis Fraser of Findrack.

46 Lady Dipple, now aged 53, and Ferguson, aged 24, sailed from Aberdeen for Leith in April 1734. Leith is the seaport of Edinburgh, and is about 2 miles NNE of it. Edinburgh, at this period, was confined within a small space on each side of its principal street, and had a population of about 45,000. It is 395 miles NNW of London, and 110 miles SSW of Aberdeen.

47 Alexander Forbes, fourth Lord Pitsligo, a Scotch Peer.—He was "*out*" in the affair of '45, in the interest of Prince Charles Stuart, the Young Pretender, and was attainted. Pitsligo Castle, formerly the seat of the Lords Pitsligo, is an ancient building, surrounded by extensive gardens. Pitsligo is in the district of Buchan, Aberdeenshire, about 3½ miles W of Fraserburgh, and 7 miles E of Auchmedden. Lord Pitsligo died in 1762.

48 John Alexander was the great-grandson of George Jamesone, the eminent Scottish painter, the "*Vandyke of Scotland.*" It is singular that Ferguson takes no notice of this in his Memoir. Mr. Alexander studied in Italy, residing many years at Florence. He resided in St. John Street, Canongate, Edinburgh, where he died about the year 1752, aged 73. He had an extensive collection of portfolios, filled with the choicest drawings—sketches and prints. "*He was both painter and engraver,*" says one of our correspondents, "and painted the staircase of Gordon Castle 1721 (Rape of Proserpine), and whilst at Rome in 1717, he etched 8 plates after Raphael." (Extract from letter to us from David Laing, Esq., LL.D., Signet Library, Edinburgh).

would do nothing without money. So I was quite at a loss what to do.

In a few days after this, I received a letter of recommendation from my good friend Squire Baird to the Reverend Dr. Robert Keith at Edinburgh,[49] to whom I gave an account of my bad success among the painters there. He told me that if I would copy from nature, I might do without their assistance, as all the rules for drawing signified but very little when one came to draw from the life;[50] and by what he had seen of my

Merchiston Castle, 1754.

drawings brought from the North, he judged I might succeed very well in drawing pictures from the life, in Indian ink, on vellum. He then sat to me for his own picture, and sent me with it and a letter of recommendation to the Right Honourable the Lady Jane Douglas, who lived with her mother the

[49] A slight mistake here; the Rev. Robert Keith was not a Dr.—He was born in 1681, at Uras, in Aberdeenshire, was ordained an Episcopal clergyman in 1710, and consecrated a Bishop of the Scottish Episcopal Church in 1727. He resided in the Canongate, Edinburgh, from 1728 to 1752, when he removed to Bonnington, near Leith (adjacent to Edinburgh), where he died on the 30th January 1757, aged 76 years, and was interred in the Canongate churchyard. He was the author of "*The History of the Affairs of Church and State in Scotland,*" published in 1734, and dedicated to the Lady Jane Douglas, Merchiston Castle. He was related to the Marischal family, and dedicated to Marshal Keith in 1755 his "*Catalogue of Scottish Bishops.*" It is remarkable that Ferguson should have forgotten that he was a Bishop, and to have styled him Dr. Robert Keith.

[50] The Editor of Lib. Ent. Knowl., vol. I. p. 206, remarks, that this was "*certainly a bold counsel to give; but Ferguson having in truth no other resource, followed it, and succeeded beyond his most sanguine expectations.*" *The bold counsel,* however, was but an axiom in art, which Ferguson had adopted on the occasion of his painting, in Indian ink, the portrait of William Baird, Esq., in the year previous, whilst he was sojourning at Auchmedden.

Marchioness of Douglas, at Merchiston House, near Edinburgh.[51] Both the Marchioness[52] and Lady Jane behaved to me in the most friendly manner, on Dr. Keith's account, and sat for their pictures, telling me at the same time, that I was in the very room in which Lord Napier invented and computed the logarithms, and that if I thought it would inspire me, I should always have the same room whenever I came to Merchiston. I stayed there several days, and drew several pictures of Lady Jane, of whom it was hard to say whether the greatness of her beauty,

51 Lady Jane Douglas was the sister of Archibald, Duke of Douglas, and was born on the 17th March 1698. She appears to have been possessed of many singular and extraordinary qualities; her figure and deportment noble, worthy of that race from which she was sprung—of strong natural parts, improved by education, of an insinuating address, and engaging manners. "Upon the 4th of August 1746 (when in her 49th year), Lady Jane was privately married to John Stewart, Esq., commonly called Colonel John Stewart, from his having held that rank in the Swedish service. This gentleman afterwards became Sir John Stewart of Grandtully. As the Duke of Douglas had a prejudice against Mr. Stewart, Lady Jane went abroad, and concealed her marriage . . . Lady Jane had two sons born in Paris. With these, she and her husband returned to Britain; but so hard was their fate, that the Duke of Douglas rejected his nephews as supposititious children, and would not even see Lady Jane nor them when they presented themselves at the gate of his castle. A suit having been raised to prove that the eldest son, Archibald Douglas, Esq., had been a supposititious child, it was given against him in the Court of Session in Scotland by the casting-vote of the Lord President." This finding was afterwards reversed by the House of Lords, and he ultimately succeeded to the dukedom and estates. This cause, "Mr. Archibald Douglas *v.* Duke of Hamilton," is still known as "The great Douglas Cause;" vide "Letters of the Right Honourable Lady Jane Douglas, by Alexander Lockhart, Dean of the Faculty of Advocates." It appears that such was Lady Jane's husband's want of economy, that he was for some time within the Rules of the King's Bench Prison. "Mr. Stewart, her husband, was sunk in debt, prosecuted by his creditors, and thrown into jail. In this destitute condition there was application made for Lady Jane to his late Majesty (George II.), who was graciously pleased to bestow on her a pension of £300 per annum; however, Lady Jane and her husband still continued in very deplorable circumstances; in so much, that when Lady Jane lived at Chelsea with her children, she was at different times reduced to the necessity of selling her clothes and other trifling effects, for the support of her family and her husband." Lady Jane died on 22d November 1753, in the 57th year of her age, and was buried beside her mother, the Marchioness of Douglas, in Holyrood Chapel burying-ground. Ferguson would hear of Lady Jane's misfortunes, and of her residing in Chelsea; and as he was then residing in London, it is likely that he often visited his former noble and kind patroness.

Merchiston House, or Castle, in which Lady Jane and her mother, the Marchioness of Douglas, resided in 1734, appears to have then belonged to a Mr. Lewis, and that they rented it from him. It stands about a mile and a half SW of the Cross of Edinburgh, now in the suburbs of Edinburgh in that district. The view of it, given above, is from a print of 1754, taken from the NE, showing the Pentland hills on the SW in the distance. Allusion is here made to the room in which Lord Napier computed the logarithms. It is understood that this was the top room of the house, directly under the slates.

52 "Lady Mary Kerr, daughter of Robert, first Marquis of Lothian, and Marchioness of Douglas, and mother of Archibald, first Duke of Douglas, died at Edinburgh, 22d January 1736, in the 58th year of her age, and was interred in Holyrood Chapel burying-ground." The Marchioness therefore died during Ferguson's first residence in Edinburgh, and about a year and a half after his first visit to Merchiston.

or the goodness of her temper and dispositions, was the most predominant.[53] She sent these pictures to ladies of her acquaintance, in order to recommend me to them, by which means I soon had as much business as I could possibly manage, so as not only to put a good deal of money in my own pocket, but also to spare what was sufficient to help to supply my father and mother in their old age.[54] Thus a business was providentially put into my hands, which I followed for six and twenty years.[55]

Lady Dipple being a woman of the strictest piety, kept a watchful eye over me at first, and made me give her an exact account at night of what families I had been in throughout the day, and of the money I had received. She took the money each night, desiring I would keep an account of what I had put into her hands, telling me that I should duly have out of it what I wanted for clothes, and to send to my father; but in less than half a year, she told me that she would thenceforth trust me with being my own banker; for she had made a good deal of private inquiry how I had behaved when I was out of her sight through the day, and was satisfied with my conduct.

1736.—During my two years' stay at Edinburgh, I somehow took a violent inclination to study anatomy, surgery, and physic, all from reading of books and conversing with gentlemen on these subjects; which, for that time, put all thoughts of astronomy out of my mind, and I had no inclination to become acquainted with any one there who taught either mathematics or astronomy, for nothing would serve me but to be a doctor.[56]

53 This euology on Lady Jane is quite borne out by what we have given at the beginning of note 51.

54 It is pleasing to observe that Ferguson was so attentive to the wants of his parents; indeed he appears to have attended to their little wants before attending to his own.

55 The period to which Ferguson here refers is the summer of the year 1734; if to this date we add 26 years, which he followed in limning, we are brought to the year 1760, when it would appear that he entirely abandoned the profession of limning. (See date 1760).

56 "The Chair of Anatomy in the University of Edinburgh was founded in 1720, and Dr. Alexander Monro, primus, was elected professor." This eminent man is justly considered as the founder of its Medical School. From small beginnings, under his fostering care, and aided by coadjutors of the most splendid talents and varied acquirements, it suddenly not only equalled, but became superior in fame, as a School of Medicine, to all the continental seminaries. Dr. Monro, the first professor, died at the age of 70, in 1767. The Chair for the Practice of Medicine in the University of Edinburgh, was founded upon 9th February 1726. Dr. John Rutherford taught the Practice of Physic in it for 40 years, and was the first who delivered lectures in the Infirmary on Clinical Medicine,—Vide Bower's Edinburgh Students' Guide, pp. 32, 33, 63, 64. Thus it was Professors Monro and Rutherford who had given such an impetus to the studies of Ana-

At the end of the second year, I left Edinburgh, and went to see my father, thinking myself tolerably well qualified to be a physician in that part of the country, and I carried a good deal of medicines, plasters, &c., thither; but to my mortification I soon found that all my medical theories and study were of little use in practice; and then finding that very few paid me for the medicines they had, and that I was far from being so successful as I could wish, I quite left off that business,[57] and began to think of taking to the more sure one of drawing pictures again. For this purpose I went to Inverness, where I had eight months business.[58]

1739.—When I was there I began to think of Astronomy again, and was heartily sorry for having quite neglected it at Edinburgh, where I might have improved my knowledge by conversing with those who were very able to assist me. I began to compare the ecliptic with its twelve signs (through which the sun goes in twelve months) to the circle of 12 hours on the dial-plate of a watch, the hour-hand to the sun, and the minute-hand to the moon, moving in the ecliptic, the one always over-taking the other at a place forwarder than it did at their last conjunction before. On this I contrived and finished a scheme on paper for showing the motions and places of the sun and

tomy and Physic shortly before Ferguson's visit to Edinburgh in 1734; and according to an old memorandum, dated shortly after this period, we observe that "anatomy and physic were all the rage in Edinburgh, and throughout Scotland almost every one wished their sons to become doctors." Ferguson seems to have caught the epidemic.

At note 46 we mention that Ferguson arrived in Edinburgh in April 1734; and as he says he was 2 years in Edinburgh, the period to which he refers his departure *from* Edinburgh must be the year 1736. It would appear, from a memorandum in our possession, that Ferguson "returned to the north," somewhere about the end of September 1736. It is not unlikely that "*the affair of the Porteous Mob, of the evening of September 6th,* 1736," may have hastened his departure from Edinburgh, as it "*quite unsettled that city for a great length of time,*" as a chronicler of the period records.

Ferguson comes here before us in a new profession, that of a Doctor. We think that was a dangerous experiment. Alluding to this, Mr. Capel Lofft, at p. 223 of his "Eudosia," says—viz. Mr. F. "*made some attempts in the medical line, but honestly laid them aside soon, not venturing far in an employment to which experimental knowledge is no less required than genius, and where the consequences of error are often so fatal.*" It appears that Ferguson, after his return to the *north,* in September 1736, "*dispensed medicine, gave out plasters, and attended to all the ills that flesh is heir to,*" until early in 1739 (for upwards of 2 years); now and then all the while attending to the "*taking of pictures*" among the gentry and his friends.

57 As mentioned in last note, it must have been about the end of the year 1738 or early in 1739 that Ferguson abandoned the profession of "*country doctor,*" at which period he would be about 28 or 29 years old.

58 Inverness is 59 miles WSW of Keith, and it was, as he notes, the year 1739 when he went thither. In the month of May of this year, Ferguson married Isabella Wilson, daughter of George Wilson, of the neighbouring parish of Grange.

moon in the ecliptic on each day of the year perpetually, and consequently the days of all the new and full moons.[59]

To this I wanted to add a method for showing the eclipses of the sun and moon, of which I knew the cause long before, by having observed that the moon was for one half of her period on the north side of the ecliptic, and for the other half on the south; but having not observed her course long enough among the stars by my above-mentioned thread, so as to delineate her path upon my celestial map, in order to find the two opposite points of the ecliptic in which her orbit crosses it, I was altogether at a loss how and where in the ecliptic (in my scheme) to place these intersecting points; this was in the year 1739.

At last I recollected that when I was with Squire Grant of Achoynaney, in the year 1730, I had read, that on the 1st of January, 1690, the moon's ascending node was on the 10th minute of the first degree of Aries, and that her nodes moved backward through the whole ecliptic in 18 years and 224 days, which is at the rate of 3 minutes 11 sec. every 24 hours; but as I scarce knew in the year 1730 what the moon's nodes meant, I took no further notice of it at that time.

However, in the year 1739, I set to work at Inverness, and after a tedious calculation of the slow motion of the nodes from Jan. 1690 to Jan. 1740, it appeared to me that (if I had remembered right) the moon's ascending node must be in 23 deg. 25 min. of Cancer, at the beginning of the year 1740;[60] and so

His *doctor business* being a complete failure, he again directed his attention to Limning; and, in pursuit of employment, went to Inverness in July 1739. Ferguson informs us that he had eight months business at Inverness. From "*King's Munimenta Antiqua*" it appears that in 1740 Ferguson was at Castle Downie (near Inverness), and the guest of the famous Simon Lord Lovat of "*the* '45," engaged, no doubt, to paint portraits of the chief, or some of his friends. Ferguson's sojourn at Inverness, therefore, appears to have extended into 1740.

59 This is a very simple and ready-at-hand illustration of the relative motions of the sun and moon, and one which we ourselves adopted in our early days. The relative conjunctions of the hour and minute hands, however, do not correspond with those of the sun and moon during the year. The hour and minute hands of a watch comes only 11 times into conjunction in 12 hours; whereas the sun and moon come into conjunction 12 times in the course of 365 days; or more correctly, in 12 hours there are exactly 11 conjunctions of the hour and minute hands; whereas, in 365·243 d. the moon comes 12 times into conjunction with the sun, and $\frac{368}{1000}$ into the *thirteenth conjunction.*

60 It is not now known what process of calculation Ferguson followed in determining the place of the moon's ascending node, on January 1st, 1740, from its position on January 1st, 1690. He merely informs us that the process was a *tedious* one, and gives us only one of his periods—viz. 18 years 224 days, the period of the nodes, which, he says, is "*at the rate of* 3 *min.* 11 *sec. in* 24 *hours.*" This is too slow; our calculation gives 3 min. 10$\frac{5}{8}$ sec. He does not give us the length of the year he used in this calculation; but we presume it would consist of 365 days 5 hours 48 min. 55 sec., as then generally adopted. Assuming this to be the case, we shall

I added the eclipse part to my scheme, and so called *The Astronomical Rotula.*[61]

go over the same calculation. It is evident that from January 1, 1690, to January 1, 1740, there are exactly 50 years—365d. 5h. 48m. 55s. = 365·2423, dec. × 50 = 18262d. 115dec. In 18 years 224 days there are 6798·5016 days (for 18y. 224d. = 18·613 years × 365·2423 = 6798·5106d.); then, 18262d·115, the days in 50 years ÷ 6798·5016, the period of the nodes, and it will give 2·686, that is, from January 1st, 1690, to January 1st, 1740, a period of 50 years. The moon's ascending node had made two entire revolutions, and was $\frac{686}{1000}$ on its way in a *third* revolution:—2·686 = 2 circles 247° 57 min., and 247° 57 min., reckoned backward from 10th min. of 1st degree of Aries, falls upon Cancer 23° 13′. Ferguson makes it "*Cancer* 23° 25′;" this difference of 12 min. between Ferguson's result and our own may arise from his peculiar mode of calculation, or from our own process being worked out with an assumed period of 365d. 5h. 48m. 57s. or 365·2423 days. The difference between Ferguson's result and our own is 12′ of a degree, nearly the "*quarter* of a degree," as noted by Mr. Maclaurin to Ferguson; but whether the 12′ ought to be plus or minus Ferguson's period, we have no means of ascertaining. Our calculation, however, is sufficiently near, as an illustration, to show how such a matter may be determined; and that too in a very short period.

61 This "ASTRONOMICAL ROTULA" is an extraordinary production. It has been long out of print, and is now unknown. We have a copy of the engraving of the largest plate of this Rotula; in the lower left-hand corner of which, we find "*Jas. Ferguson Invt*," and in the lower right-hand corner, "*R. Cooper, Sculp.*" The large oblong sheet in our possession, and which forms the basis of the Rotula, is, in engraved surface, 17½ by 12¼ inches. At the top of the engraving, in German or Church text, are the words "THE ASTRONOMICAL ROTULA;" and immediately below, in four long running lines, it states that it "*shows the place of the sun, moon, and moon's nodes in the ecliptic, with their distances from one another every day in the year; the true time of all the eclipses of the sun and moon from* 1730 *to* 1800 *inclusive, together with the figure of all those that are visible at Edinburgh, London, and Paris; the day of the month, moon's age and southing, high water at several ports, equation of time and moveable feasts, rising and setting of the sun at Edinburgh; the motions, magnitudes, solar distances; hourly velocities, &c. of the planets; the differences of time in most remarkable places of the earth, with plain and easy tables for calculating the true time of new and full moon.*" Near the centre of the engraving are a series of 19 concentric circles (diameter of largest circle, 11⅞ inches, smallest one, 6¼ inches; on these circles are the signs of the zodiac, degrees, minutes, &c.; days of month and names of the month, dominical letter, and other cycle tables. In the top right-hand corner are tables for ascertaining "first day of moon for 300 years; below, rules to find the moon's coming to south by her age, and the comparative diameters of Saturn and Rings, Mars, and the Earth and Moon." In the right-hand corner are tables for finding "first day of each month for ever;" below, table to "find the time of high water at several ports, with comparative diameters of Venus, Jupiter, Mercury, and a Comet." In the lower left-hand corner are cycles for finding new moon; below, is a circular table showing the mean motion of the moon from the sun; under this are "Sun's rising and setting tables;" and directly under the series of circles before noticed, is a table of the equation of time, new and full moon in hours, minutes, &c.; and in the lower right-hand corner is a dial-plate of 24 hours, tables of the sun's equation, and cycles for finding full moon. Such is a glance detail of this complicated Rotula. It is one of those things which "*require to be seen to be appreciated.*" Mr. Cameron of Inverness informs us that there is a complete Rotula in Culloden House, having on the back of it—"*To the Right Honourable Duncan Forbes of Culloden, Lord President of the College of Justice, this Astronomical Rotula is, with profound respect, inscribed, by his Lordship's most obedient, and most humble servant, James Ferguson.*" This complete Rotula is mounted. Our correspondent, Mr. Cameron, mentions, that "*this Rotula is suspended on the wall of the great hall of Culloden Castle, is glazed, and in a rosewood frame, gilt edged. It is worked at the back by wooden knobs or keys.*" It is remarkable that this Rotula should have been dedicated to Duncan Forbes, Esq., instead of Professor Maclaurin, who had done so much to bring it out.

1740.—When I had finished it I showed it to the Reverend Mr. Alexander MacBean, one of the ministers at Inverness,[62] who told me he had a set of almanacs by him for several years past, and would examine it by the eclipses mentioned in them. We examined it together, and found that it agreed throughout with the days of all the new and full moons and eclipses mentioned in these almanacs, which made me think I had constructed it upon true astronomical principles. On this, Mr. MacBean desired me to write to Mr. Maclaurin, professor of the Mathematics, at Edinburgh, and give him an account of the methods by which I had formed my plan, requesting him to correct it where it was wrong. He returned me a most polite and friendly answer (although I had never seen him during my stay at Edinburgh), and informed me that I had only mistaken the radical mean place of the ascending node by a quarter of a degree, and that if I would send the drawing of my Rotula to him he would examine it, and endeavour to procure for me a subscription to defray the charges of engraving it on copper-plates, if I choosed to publish it. I then made a new and correct drawing of it and sent it to him, who soon got me a very

Possibly Ferguson may have offered to dedicate it to him, and have suggested "The Right Honourable the Lord President of the College of Justice," as a name of greater influence.

On the back of the title-page of his pamphlet, entitled, "*The use of a new Orrery,*" published about August, 1746, we find the following advertisement relative to this ROTULA—"*The* ASTRONOMICAL ROTULA, *showing the place of the sun, moon, and moon's nodes in the ecliptic, with their distances from one another every day in the year; the true times of all the eclipses of the sun and moon, from* 1730 *to* 1800 *inclusive, together with the figures of all those that are visible at London, Edinburgh, and Paris; the motions, magnitudes, solar distances, hourly velocities, &c. of all the planets; plain and easy astronomical tables, never before published, for calculating the true time of new and full moon; the hour of the day or night in most remarkable places of the earth, having the time at any one of them given; with several other problems, as shown by a printed direction belonging to this scheme. Price Five shillings.*"

[62] The Reverend Alexander Macbean died at Inverness on the 2d November 1762, aged 78 years. We are indebted to Mr. James Cameron, Inverness, for the following inscription, copied from his grave-stone in the parish churchyard there:—"*Underneath this stone are deposited the remains of the Reverend Mr. Alexander Macbean, who died on the 2d day of November* 1762, *aged* 78 *years, and in the fiftieth year of his ministry, the last forty-two of which was at Inverness. Here are also deposited the remains of Marjory Macbean, his spouse, who died March* 30*th*, 1766, *aged* 86 *years.*" From this inscription, it is evident that he was inducted one of the ministers of Inverness in 1720, when in the 36th year of his age. At the time of Ferguson's visit to him, in 1739, he would be about 55 years old. He therefore lived through the greater part of Ferguson's life-time, and would hear of his celebrity as an author, and public lecturer on Astronomy and Experimental Philosophy; and of his Majesty, George III., having bestowed on him a yearly pension of £50.

handsome subscription by setting the example himself and sending subscription papers to others.[63]

1741.—I then returned to Edinburgh[64] and had the Rotula plates engraved there by Mr. Cooper.[65] It has gone through several impressions, and always sold very well, till the year 1752, when the style was changed, which rendered it quite useless. Mr. Maclaurin received me with the greatest civility when I first went to see him at Edinburgh; he then became an exceeding good friend to me, and continued so till his death.

One day I requested him to show me his orrery, which he immediately did. I was greatly delighted with the motions of the earth and moon in it, and would gladly have seen the wheel-work, which was concealed in a brass box, and the box and planets above it were surmounted by an armillary sphere; but he told me that he had never opened it, and I could easily perceive that it could not be opened but by the hand of some ingenious clock-maker, and not without a great deal of time and trouble.[66]

[63] Mr. Colin Maclaurin, one of the most accomplished mathematicians of his day, was born at Kilmoddan, in Argyleshire, in 1698 (of which parish his father was the minister). On the recommendation of Sir Isaac Newton, he was, in 1725, appointed Professor of Mathematics and Natural Philosophy in the University of Edinburgh. He had a small Observatory erected on the south side of the old College, Edinburgh, furnished with valuable instruments. He was the author of the following works:—1st, Geometrica Organica, sive Descripto Linearum Curvarum Universalis, 1720; 2d, On the Percussion of Bodies, 1724; 3d, A Treatise on Fluxions, 1740; 4th, An Account of Sir Isaac Newton's Philosophical Discoveries. He was also the author of several philosophical tracts and papers. He died suddenly (of dropsy) on 14th July, 1746, in the 49th year of his age, and was interred in Greyfriars churchyard, Edinburgh. Mr. Maclaurin thus died in the prime of life, and before Ferguson had risen to any celebrity.

[64] By reference to our memoranda, it appears that Ferguson returned to Edinburgh early in the spring of 1742. His age at this period would be 32 years.

[65] David Laing, Esq., LL.D., of the Signet Library, Edinburgh, informs us that "Mr. Richard Cooper was a native of Yorkshire, England, and practised many years in Edinburgh as an engraver and copper-plate printer, &c., and resided in St. John's Street, Canongate, Edinburgh. The time of his death we have not ascertained. He was alive in 1761. He was interred in the Canongate churchyard. Cooper was a very eminent engraver." Ferguson, in a foot-note in his Memoir, informs us, that "*Cooper was master to the justly celebrated Mr. Robert Strange, who was at that time his apprentice*" (Sir Robert Strange). Besides engraving Ferguson's Astronomical Rotula plates, he about the same time appears to have engraved another plate for him, now long out of print, and is very rare; we have an impression of it in our possession; the print on the card is $5\frac{1}{2}$ by $3\frac{3}{8}$ inches; the upper part of the card is taken up with a scale of days of the month, signs of the zodiac, &c.; directly under which is a triangular "scale of latitudes in degrees from the equator." At the foot is a sort of semicircular sun-dial, with hours, minutes, lines, and has scales for rectifications, &c., at the sides. In the left-hand corner is "*James Ferguson, Delin.;*" and in the right-hand corner, "*R. Cooper, Sculp.*"

[66] The Secretary of the University of Edinburgh (Alexander Smith, Esq.), informs us, that this orrery is not now in his University. Probably it was the private property of Mr. Maclaurin, and removed at the time of his death, in 1746.

After a good deal of thinking and calculation I found that I could contrive the wheel-work for turning the planets in such a machine, and giving them their progressive motions; but should be very well satisfied if I could make an orrery to show the motions of the earth and moon, and of the sun round its axis. I then employed a turner to make a sufficient number of wheels and axles, according to patterns which I gave him in drawing; and after having cut the teeth in the wheels by a knife, and put the whole together, I found that it answered all my expectations. It showed the sun's motion round his axis; the diurnal and annual motions of the earth on its inclined axis, which kept its parallelism in its whole course round the sun; the motions and phases of the moon, with the retrograde motion of the nodes of her orbit; and consequently, all the variety of seasons, the different lengths of days and nights; the days of the new and full moons and eclipses.[67]

1742.—When it was all completed, except the box that covers the wheels, I showed it to Mr. Maclaurin, who commended it, in presence of a great many young gentlemen who attended his

From Ferguson's description of it, it would seem that it was something similar to those excellent orreries constructed by Rowley and by Wright, during the early part of last century. Sir John Herschel, in his Astronomy (Cabinet Library), writes unfavourably of orreries, planetaria, &c., apparently, because they could not be made to show the comparative masses and distances of the planets; but surely the great many other excellences in an orrery far outweigh these two objections. We understand that, in consequence of his opinion, the demand for orreries, &c., has very much declined. We regret this, for, with all due deference to Sir John, we think that an orrery is of great use—a great *first aid* to the untutored. It is not every one that is born to be an astronomer; perhaps not one in a thousand can form a proper conception of the motions of our planetary system. With a very great many, the eye requires to be taught before the mind; and we would therefore strongly advise the young and the unlearned in astronomy, to go to an orrery and *see* its several beautiful motions. More may be gained by *the sight* in an hour, than by the study of books in a year. The motions in Maclaurin's orrery gave Ferguson intense delight and satisfaction. The sight of it in motion materially aided him in his after studies.

67 It is to be regretted that Ferguson has given no account of the wheel-work of this, his first orrery. It would have been interesting now to have known by what process of calculation he got his numbers for the teeth of his wheels and pinions. As this is not known, it cannot now be shown with what degree of accuracy the several motions were produced. We are inclined to think that it was a rude piece of work, and made not for correct motions, but merely to *show them;* probably contenting himself with the earth turning on its axis in 24 hours, and completing a revolution round the sun in 365¼ days, the synodic revolution of the moon in 29½ days, and the nodes of her orbit in 19 years. Thus, by discarding *minutes* and *seconds*, he would be able to make a simple orrery at little trouble and expense. We may here note, that to produce the *parallelism* of the earth's axis, and the retrograde motion of the nodes of the moon's orbit, shows that Ferguson, at this early period, had an intimate knowledge of the theory and practice of wheel-work (perhaps receiving some assistance from Derham's "Artificial Clockmaker," 3d edit. Lond., 1714).

lectures. He desired me to read them a lecture on it, which I did without any hesitation, seeing I had no reason to be afraid of speaking before a great and good man who was my friend.[68] Soon after that I sent it in a present to the Reverend and ingenious Mr. Alexander Irvine, one of the ministers at Elgin, in Scotland.[69]

1743.—I then made a smaller and neater Orrery, of which all the wheels were of ivory, and I cut the teeth in them with a file. This was done in the beginning of the year 1743; and in May that year I brought it with me to London, where it was soon after bought by Sir Dudley Rider.[70] I have made six

68 This lecture, on his wooden orrery, appears to have been delivered by him in the old College of Edinburgh, in the spring of the year 1742, when in the 32d year of his age. Notwithstanding the celebrated Maclaurin being his friend, it would require from Ferguson some considerable amount of confidence and self-possession to bear him through with this his first public lecture.

69 The Reverend Alexander Irvine was a learned, ingenious, and most amiable man. For a great length of time he was Presbytery clerk of Elgin district. He wrote a small but clear and beautiful hand, as is shown by his entries in the Elgin Presbytery Records. His favourite study was astronomy. On February 18th, 1736–7, he made preparations for observing the annular eclipse of the sun, but was prevented, by clouds, from seeing the annulus for above half a minute. He, however, sent a memorandum regarding what he did see, to the celebrated Professor Maclaurin, at Edinburgh, who, along with other notices of the eclipse, inserted it in the Philosophical Transactions as follows:—1736–7, February 18th. "At Elgin the eclipse was observed annular at 36h. 29′, the larger part of the ring uppermost, by the Reverend Mr. Irvine, who had a view of it for about 30″, but by reason of intervening clouds could not determine the beginning or end of this appearance;" (vide Phil. Trans. Ab., vol. 8th, p. 147). Mr. Irvine contrived various pieces of astronomical mechanism, and also constructed a wooden clock, which has been long in the possession of the Rev. Peter Merson, at Elgin. No traces of the wooden orrery sent to him by Ferguson can now be found.

The Rev. Dr. Francis Wylie, of Elgin, informs us, that Mr. Irvine was ordained at St. Andrews (Lanbryde), the parish next to Elgin on the east, on the 1st March 1725. From this charge he was translated to that of Auldearn, in the Presbytery of Nairn, on 7th January 1731; from which he was translated to the Collegiate charge of Elgin, on 12th August 1735, and died on the 22d December 1758; and, as it would appear, in the 59th year of his age. From this it is obvious that he was an ordained minister for nearly 34 years, and was minister of Elgin 23 years and 8 months. He corresponded with Ferguson until shortly before his death. Ferguson wrote a long and interesting letter to him from London, dated London, 17th January 1758 (see date 1758), about 11 months before Mr. Irvine's death; the original letter of Ferguson is preserved, and may be seen in the Elgin museum. Thus this amiable and ingenious gentleman died about 2½ years after Ferguson published his great work, "*Astronomy Explained, upon Sir Isaac Newton's Principles*," &c., and just as Ferguson was rising into celebrity.

70 This small orrery is still in existence, is now the property of Sir Dudley Ryder's descendant, the Right Honourable the Earl of Harrowby, and is to be seen at his Lordship's residence, Sandon Hall, Staffordshire. The Honourable Henry D. Ryder (his Lordship's son) informs us, that this orrery was much damaged by the fire which destroyed Sandon Hall, on June 6th, 1848; "that the works of this orrery are enclosed within a mahogany box; that the horizon and other circles on it are papered and filled up by the pen; that the horizon circle is about 16 inches in diameter, and that there are neither initials nor date anywhere on it to be found—it is not known what Sir Dudley Ryder paid for it." Sir

Orreries since that time, and there are not two of them in which the wheel-work is alike; for I could never bear to copy one thing of that kind from another, because I still saw there was great room for improvements.[71]

I had a letter of recommendation from Mr. Baron Edlin,[72] at

Dudley Ryder, Lord Chief Justice of England, died 25th May 1756.——"A patent was preparing to create him a Peer, by the title of Lord Ryder of Harrowby, Lincolnshire; he had not kissed his Majesty's hand as was reported." Vide Benjamin Martin's Miscellaneous Correspondence, vol. 1st, p. 306.

Ferguson here mentions that he arrived in London in May 1743. Referring to our memoranda, we think it is evident that he and his wife sailed from Leith for London, in a Leith smack, on Saturday 21st May, arriving in London on Friday 27th May 1743. (See note 102).

71 We apprehend that Ferguson alludes to the *arrangement* of the wheel-work, rather than to *new calculations* for wheels to produce more accurate periods, as during his lifetime he used 365d. 5h 48′ 55″ average, as his value of the length of a year; and to produce this, he continually adopted Camus's fraction for his annual train of wheels—viz. $\frac{7}{25}\times\frac{7}{69}\times\frac{8}{83}=$365d. 5h. 48′ 58″·78 (error $3\frac{3}{4}$ seconds plus). In the year 1826, we made a set of new calculations for an orrery which we constructed; and by following a peculiar mode in working out our continuous ratios, we obtained the following fractions—viz. $\frac{8}{44}\times\frac{10}{89}\times\frac{13}{97}=$365d. 5h. 48′ 55″·38, (error $\frac{38}{100}$ of a second only). It is surprising that Ferguson adhered so firmly to Camus's fraction, never trying to get more accurate wheel-work. Had he, by calculation, come upon our fractional ratio, and adopted it, he would have got a period, for his solar year, within the *third part of a second* of the then ascertained length of the year; but it is likely that the method of obtaining accurate ratios by "continuous fractions" as applicable to wheel-work, was not known in Ferguson's time.

We have succeeded in tracing out four of Ferguson's orreries; 1st, The orrery sold in 1743 to Sir Dudley Ryder, is now the property of his descendant, the Right Honourable the Earl of Harrowby, Sandon Hall, Staffordshire; 2d, Shortly after Ferguson's death, in 1777, a London publican, an admirer of Ferguson, named his house "*The Ferguson's Head*," where there was long to be seen one of Ferguson's large wooden orreries, which the publican had purchased at Ferguson's sale in 1777. (Our late friend, Mr. Andrew Reid of London, brother of Mr. Thomas Reid, watchmaker, Edinburgh, who had seen this orrery in 1787, was our informant; also, see London Mirror, vol. 29th, No. 822, p. 128, for 1832). Mr. Bartlett, watchmaker, Maidstone, discovered this orrery, in 1836, in an old curiosity shop in Old Compton Street, London, and purchased it for £3. He (Mr. B.) presented it to the Manchester League Bazaar, where it was sold to some person unknown; 3d, At the sale of Mr. Thomas Hawys' Mathematical and Philosophical Instruments, on Tuesday, 13th October 1807, there was sold "*a large orrery with glass shade, made by Ferguson*,"—vide cat. of this sale. This orrery appears to have got into the possession of the late Professor Millington, about the year 1818; 4th, In the University of London there is one of Ferguson's orreries, which was presented to this University on 27th June 1851, by George Walker, of Port Louis,—vide appendix, article "ORRERY."

72 Regarding Mr. Baron Edlin, we are much indebted to Robert Cox, Esq., W.S., Edinburgh, for the following particulars:—"In the MS. '*Register of Privy Seals Sign Manuals*,' kept in the Exchequer office, Edinburgh, vol. 2d, p. 250, the Commission or Letters patent, dated 17th August 1730, appointing Edward Edlin, Esq., a Baron of the Exchequer, in place of Edmund Miller, Esq., deceased; also, a warrant dated 5th October 1730, authorizing the payment to him of £500 a-year, in addition to the £500 payable under the Commission, or Letters Patent, in consideration of his leaving his practice of the law in this part of the United Kingdom (i. e. London), to attend our service, and for other good causes and considerations; this warrant is signed 'R. Walpole, Wm. Yonge, Wm. Clayton.' In vol. 4th, p. 282 of the same register, there is a commission dated 29th

Edinburgh, to the Right Honourable Stephen Poyntz, Esq., at St. James's, who had been preceptor to his Royal Highness the late Duke of Cumberland, and was well known to be possessed of all the good qualities that can adorn a human mind. To me his goodness was really beyond my power of expression; and I had not been a month in London till he informed me that he had wrote to an eminent Professor of Mathematics to take me into his house, and give me board and lodging, with all proper instructions to qualify me for teaching a Mathematical School he (Mr. Poyntz) had in view for me, and would get me settled in it. This I should have liked very well, especially as I began to be tired of drawing pictures, in which, I confess, I never strove to excel, because my mind was still pursuing things more agreeable. He soon after told me that he had just received an answer from the Mathematical Master, desiring I might be sent immediately to him. On hearing this, I told Mr. Poyntz that I did not know how to maintain my wife during the time I

Mrs. Ferguson.—From a Painting on Vellum by Ferguson.

must be under the master's tuition. What, says he, are you a married man? I told him I had been so ever since May in the

May 1761, to George Winn, of Lincolns Inn, Esq., as Baron, in place of Edward Edlin, deceased. From another register it appeared that Baron Edlin's widow petitioned, after his death, for his salary, down to 10th December 1760, which thus appears to have been the date of his decease, as it is not the stated period of paying salaries. He had executed, on 22d November 1758, a '*writing*' by which she had right to receive sums due to him. Baron Edlin was of Lincoln's Inn, London."

year 1739. He said he was sorry for it, because it quite defeated his scheme; as the master of the school he had in view for me must be a bachelor.[73]

He then asked me what business I intended to follow? I answered that I knew of none besides that of drawing pictures. On this he desired me to draw the pictures of his lady and chil-

[73] It appears singular that "*how to maintain his wife during the time he would require to be under the master's tuition,*" did not occur to him at that interview with Mr. Poyntz, when he had been informed of what had been done. Probably in the interval, and before this visit, he had been talking the matter over with his wife, when this difficulty would be likely to suggest itself, and first raised it at this visit. Mathematics was by no means in Ferguson's way, he would have been out of his element in such a school; in short, he would not have been successful in it (see also Appendix). Ferguson alludes to his wife only twice in his writings;—viz. 1st, In his "Electricity, 1770," article 96, p. 130; and 2dly, In his Memoir, as here noted, 1773. Relative to Ferguson's marriage, Mr. Robert Sim, of Keith, has kindly extracted for us, from "*Keith Parochial Record,*" the following:—

> " Keith, 1739, *April* 28. *James Ferguson in this parish, and Isabel Wilson in the parish of Grange, were matrimonially contracted at Grange, as a testimony therefrom bears, and being orderlie proclaimed with us were married at Grange May ultimo* 1739."

The Reverend Mr. James Allan, present minister of Grange, has kindly sent us the following extracts relative to Ferguson's marriage, from his Parish Register:—

> " Grange, *April 27th. James Ferguson in the parish of Keith, and Isabel Wilson in this parish, were contracted in order to marriage, and after the ordinar proclamation of banns were married at Grange the* 31*st May* 1739 *years.*"

The Rev. Mr. Allan, in a letter to us, mentions that "the name James Ferguson," in the marriage entry, is written down in extraordinarily large writing; letters at least a quarter of an inch in size, and that there is nothing like it in the Grange Register, which appears to indicate, that the Grange Session-Clerk of that day had considered him as "*a man of note.*" He also informs us that "the Reverend James Murray was minister of Grange from 1700 to 1741, and that, therefore, there is no doubt he officiated on the occasion." (Letter from Rev. James Allan, Minister of Grange). Mr. Ferguson's wife, Isabella Wilson, was born on 24th December 1719. Rev. Mr. Allan also sends the extract of her birth, from Grange Baptism Record—viz. "*George Wilson in Cantly had a daur be his wife Elspet Grant baptised and called Isabell:—Witnesses, John Sandieson, in Keith,—Robert Bremner in Mains of Grange,—Isobell Murray in Cantly, Isobell Geddes in Nether Haughs, and Isobel Grant, Dec. 24th* 1719." She was therefore about 10 years younger than Ferguson. She was the second daughter of George Wilson of Cantly. We annex an engraving of Mrs. Ferguson, taken from the original miniature one in Indian ink, by Ferguson, her husband, about the year 1749-50, when she was in her 30th year. John Ferguson, Esq., the youngest son of the astronomer, had it in careful keeping until shortly before his death in 1833, when he gave it to the eldest of the two Misses Moir, with whom he had so long resided. Some time before her death, Miss M. gave it to her relative, Mrs. Gordon, who again gave it to her son, James Gordon, Esq., 10 Windmill Street, Edinburgh, in whose possession it now is, and to whose kindness, along with that of James L. Rutherfurd, Esq., 10 Windmill Street, Edinburgh, we are indebted for a photograph from it. The original is finely done in Indian ink—is within a small oblong frame, $4\frac{5}{8}$ by $3\frac{7}{8}$ inches—broad front and roughly gilt. In the middle there is an oval opening $3\frac{1}{2}$ by $2\frac{3}{4}$ inches, with a sloping ogee to the portrait—the portrait is seen in this oval. The frame is $\frac{1}{4}$ inch thick mahogany, but perhaps an outer frame enclosed this small one. We have had several photographs, of various sizes, taken from it, from one of which our engraving of Mrs. Ferguson has been taken. (See page 46).

dren, that he might show them, in order to recommend me to others; and told me, that when I was out of business, I should come to him, and he would find me as much as he could: and I soon found as much as I could execute; but he died in a few years after, to my inexpressible grief.[74]

Soon afterward it appeared to me that although the moon goes round the earth, and that the sun is far on the outside of the moon's orbit, yet, the moon's motion must be in a line that is always concave toward the sun; and upon making a delinea-

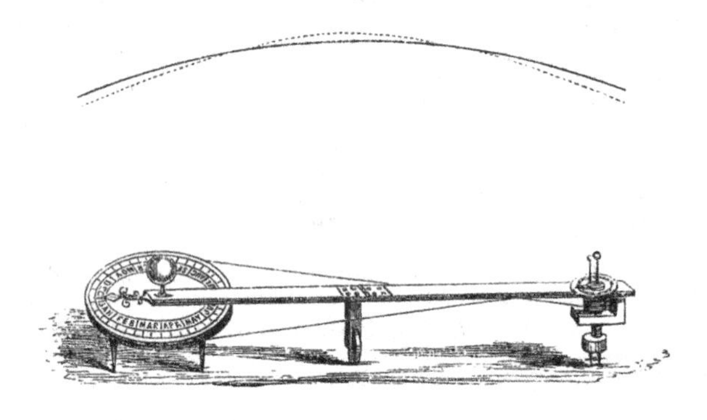

Trajectorium Lunare, 1744.

tion, representing her absolute path in the heavens, I found it to be really so. I then made a simple machine for delineating both her path and the earth's, on a long paper laid on the floor.[75]

[74] The Right Honourable Sir Stephen Poyntz, formerly preceptor to the Duke of Cumberland, and a Privy Counsellor, died on the 17th December 1750, and supposed to have been then about the 65th year of his age (vide Gentleman's Magazine, vol. xx. p. 570). His Royal pupil, William, Duke of Cumberland (of Culloden renown), was born in London, 15th April 1721, and died very suddenly by the bursting of a blood-vessel in the head, at his house in Upper Grosvenor Street, London, on 31st October 1765, aged 45.

[75] At the request of Martin Folkes, Esq., President of the Royal Society, Ferguson published a large copper-plate engraving of the curve generated by his Trajectorium Lunare. This plate has been long out of print, and is exceedingly scarce. We have had an exact copy taken from the one deposited in the British Museum (No. $\frac{552}{23}$ f. 7). The engraving is on a slip of paper, occupying a space of 2 feet $10\frac{3}{8}$ inches in length, by $6\frac{1}{8}$ inches in breadth. The curve in the engraving shows the imaginary curve of the moon for upwards of a lunation, and in two lines, running the whole length of the paper at the top, are as follows:—"*The line delineated which ye moon describes in ye heavens, during ye time of somewhat more than a month, showing that her real path is constantly curved towards ye sun. The dotted line represents a portion of ye earth's annual orbit round ye sun, considered as a circle, ye numerical figures showing its motion from west to east, every day for 34 days, accounted from ye time of any particular new moon. The black line represents ye moon's true path in ye heavens, whilst her apparent orbit about ye earth, expressed by ye several small circles, is itself carried*

I carried the machine and delineation to the late Martin Folkes, Esquire, President of the Royal Society,[76] on a Thursday afternoon. He expressed great satisfaction at seeing it, as it was a new discovery, and took me that evening with him to the Royal Society, where I showed the delineation, and the method of doing it.[77]

When the business of the Society was over, one of the members desired me to dine with him next Saturday, at Hackney, telling me that his name was Ellicott, and that he was a watchmaker.

I accordingly went to Hackney, and was kindly received by Mr. John Ellicott, who then showed me the very same kind of delineation, and part of the machine by which he had done it; telling me that he had thought of it twenty years before. I could easily see by the colour of the paper, and of the ink lines upon it, that it must have been done many years before I saw it. He then told me, what was very certain, that he had neither stolen the thought from me nor had I from him. And from that

by ye earth round ye sun.——Note, that the radius of the earth's orbit in the scheme is of 81 *such parts as ye annexed scale containeth nine, every one of which is supposed to answer to a million of miles in ye heavens; ye radius of ye moon's apparent orbit about ye earth, or her constant distance from ye same is* $\frac{24}{100}$ *of one of those parts, answering to* 240,000 *miles, which is agreeable to ye true distance of ye earth from ye sun, and of ye moon from ye earth.*" The engraving, at foot, has as follows:—"*Martin Folkes, Esq., President of Royal Society, this plate is dedicated by his most humble servant, James Ferguson. Published June 26th,* 1745, *according to act of Parliament.*" On each side of this dedication are explanations of the moon's motion on the curve which occupies the centre part of the engraving. In a letter written by Ferguson to the Rev. Thomas Birch, D.D., Secretary Royal Society, he says, "*My scheme of the moon's path, on large paper, is sold by Mr. Senex, at the Globe, opposite St. Dunstan's Church, Fleet Street: Price* 1s. 6d." After Ferguson published this moon's-path engraving, a great many ingenious men came forward to claim the discovery as theirs; and from one of Ferguson's letters, it is evident that he was charged with having "*pirated*" the *invention* from the "Gentleman's Magazine." Mr. Ellicott, in his interview with Ferguson, at Hackney, mentions, that he "*had thought of it twenty years before,*" or about 1724; and a Mr. Badder published a diagram of the moon's path, in August, 1742, three years before Ferguson published his delineation. As all such discoveries were unknown to Ferguson, *his* had all the merits of an original invention or discovery. We annex a cut of the Trajectorium Lunare, taken from Plate 7th of "*Ferguson's Astronomy.*" (See description of the "Trajectorium Lunare," under date 1744, with Ferguson's letter to Rev. Thomas Birch regarding it, and Badder's lunar curve).

76 Martin Folkes, Esq., President of the Royal Society at this period, in 1745 published "Tables of English Silver Coins, from the Norman Conquest to the present time, with their weights and intrinsic values," illustrated by plates, on which are engraved "*fac-similes of* 1,000 different coins." He died in the year 1754, aged 64.

77 At this period, we are informed, the apartments of the Royal Society were in Crane Court, Fleet Street, London, and that their day of meeting, whilst in session, was on *Thursdays*, at 8 o'clock in the evening.

time, till his death, Mr. Ellicott was one of my best friends.[78] The figure of this machine and delineation is in the 7th Plate of my book of Astronomy.[79]

1744.—Soon after the style was changed I had my Rotula new engraved, but have neglected it too much by not fitting it up and advertising it.[80] After this, I drew out a scheme and had it engraved, for showing all the problems of the Rotula, except the eclipses; and in place of that, it shows the times of rising and setting of the sun, moon, and stars, and the positions of the stars for any time of the night.[81]

1747.—In the year 1747 I published a Dissertation on the Phenomena of the Harvest Moon, with the description of a new orrery, in which there are only four wheels.[82] But having never

78 John Ellicott, F.R.S., the celebrated watch and clock maker, London, was born about the year 1700. The Horological Journal, vol. i. pp. 152—154, mentions, that "he married a Miss Saunderson, in 1726;" that "he had a singularly handsome countenance, conveying the idea of mingled intelligence and firmness; that he was elected a Fellow of the Royal Society in 1738; and that he died suddenly in 1772, aged 72, having dropped from his chair and instantly expired. In addition to his English business, he had considerable dealings in Spain, probably arising from his intimacy with a Spanish Envoy, fond of scientific pursuits." He read several papers before the members of the Royal Society, and was the author of the following works:—*Essays toward discovering the Laws of Electricity*, 1748, 1s. *Description of two methods by which the irregularities in the motion of a Clock, arising from Heat and Cold on the Pendulum, may be prevented, with a Collection of Papers on the same subject, Plates*, 60 *pages quarto*, 3s. 6d.: *London*, 1753. *Account of the influence which two Pendulum Clocks had upon each other, Plates*, 1s. 6d. Mr. Ellicott's shop is said to have been in Cornhill, London. His house was at Hackney, then a village, about 3 miles NE of St. Paul's, now forming part of the great metropolis.

79 As Ferguson's Astronomy is out of print, we give a reduced view of the Trajectorium Lunarie, taken from the 7th Plate. The figure of the Trajectorium on this plate is 12 inches in length.

80 The style was altered in 1752. We have two Almanacs of that date—viz. "*Parker's Ephemeris*," and the Almanac by Tycho Wing. Turning to Parker's Almanac, and to September 3d we find a blank space, in which is inserted, "The old style ends—the new style begins;" and in Wing's Almanac, on same day, he says, "*the old style ceases here;*" and then the 3d September, 1752, in both Almanacs, at one jump, was called the 14th September, and as a matter of course, Ferguson's Rotulas would, in consequence, be "*rendered quite useless*," as he expresses it, as it was put 11 days out of date.

81 These "*Stellar Rotulas*" are now nearly unknown, having been out of print for at least 80 years. We possess one; it is 12 inches in diameter; has 3 moveable circles, on which are engraved the day of the month, signs of the zodiac, hour and minute circles, &c., with an opening in the upper circle, in which is seen the stars visible at London. This appears to have been published in 1753. London: Price 5s. 6d.

82 This work has been out of print for upwards of a hundred years, and is now exceedingly scarce. After many years' search for a copy of it, we lately succeeded in obtaining one. It is a small octavo pamphlet of 72 pages, and is in two sections. Its title-page runs thus:—

"*A Dissertation upon the Phenomena of the Harvest Moon; also, the Description and Use of a New Four-Wheeled Orrery, and an Essay upon the Moon's turning round her own Axis, by James Ferguson. London: Printed for the Author, and*

had grammatical education, nor time to study the rules of just composition, I acknowledge that I was afraid to put it to the press; and for the same cause, I ought to have the same fears still.[83] But having the pleasure to find that this, my first work, was not ill received,[84] I was emboldened to go on in publishing my Astronomy; Mechanical Lectures; Tables and Tracts relative to several Arts and Sciences; The Young Gentleman and Lady's Astronomy; a small Treatise on Electricity, and the following sheets.[85]

1748.—In the year 1748 I ventured to read lectures on the eclipse of the sun that fell on the 14th of July in that year.[86]

sold by J. Nourse, at the Lamb, facing Katherine Street, and by S. Paterson, at Shakespear's Head, opposite to Durham Yard, Booksellers, both in the Strand. MDCCXLVII."

This small work was published in July, 1747. From an advertisement in a London newspaper, of 1747, we find that the price of this work was 1s. 6d. (vide Gentleman's Magazine for July, 1747, vol. 17, p. 348). There are 3 folding-plates in it. The 1st, an exterior view of "The Four-Wheeled Orrery;" but no description of the wheel-work in the letterpress. (Probably Ferguson made these orreries at the time for sale and profit; if so, the arrangement of the wheels, their numbers of teeth, &c., would be esteemed "*private property.*") The other two folding-plates illustrate the rotation of the moon on its axis. (See also note 84.)

Mr. John Nourse was one of Ferguson's earliest London friends. "He was an eminent Mathematician, and well skilled in the Newtonian Philosophy," (vide Memoir of William Emerson, pp. 11 and 12 in Smeaton's Emerson's Mechanics.)

83 To his want of a grammatical or classical education, he probably owes that simplicity of language which Capel Lofft, Esq., so warmly eulogises in the "Eudosia," and which makes the subjects of which Ferguson treats so readily understood—a quality in which many of our modern publications are deficient.

84 Ferguson here gives us to understand that the "Dissertation upon the Phenomena of the Harvest Moon" was his *first work.* It is most singular that Ferguson should have called this his "*first work,*" when, in the previous year, 1746, he published the following octavo work (in 42 pages)—viz.

"*The use of a New Orrery, made and described by James Ferguson. London: Printed for the Author. MDCCXLVI.*"

This was his "*first work,*" and has a large folding-plate of an orrery, apparently the same as is used in the frontispiece of his Astronomy. This, his *first work,* sold at 1s., and is now almost unknown. We had considerable difficulty in procuring a copy of it. What makes it the more remarkable that Ferguson should designate his *Dissertation upon the Phenomena of the Harvest Moon,* &c., his *first work,* is the fact of his alluding in the preface of this work to his *first publication.* In preface to the "*Dissertation,*" p. 5, he says, "*Last year* (1746) *I published a description of an orrery that I made,*" &c.; thereby showing, that his Dissertation was his *second work.*

85 The "*following sheets*" here referred to are his "*Select Mechanical Exercises,*" published in September, 1773, and to which his Memoir is prefixed. (See note 1.)

86 According to our notes, it is evident that CHRIST'S HOSPITAL SCHOOL, NEWGATE STREET, LONDON, was at least one of the places where Ferguson delivered this, his first public lecture on Astronomy, and under the auspices of his friend, Mr. James Hodgson, the then Mathematical Master. This great solar eclipse took place on Thursday, 14th July 1748 (O.S.), and began at London at 9h. 3′ morning, and ended at 12h. 8′ mid-day; the time of greatest obscuration being about 10h. 32′, at which time, about $\frac{11}{12}$ths of the upper part of the sun's disc was obscured, leaving a slender cusp of the sun visible on the right hand

Afterwards, I began to read Astronomical Lectures on an orrery which I made, and of which the figures of all the wheel-work are contained in the 6th and 7th Plates of this book.[87] I next began to make an apparatus for lectures on mechanics, and gradually increased the apparatus for other parts of Experimental Philosophy, buying from others what I could not make for myself, till I brought it to its present state.[88] I then entirely

lower edge. Ferguson, in his "*Gentleman and Lady's Astronomy*," Plate 7th, gives a projection and description of this eclipse. In the Gentleman's Magazine for July, 1748, there is a fine popular projection of this eclipse by Jos. Walker.

87 In "*This book*"—viz. "*Select Mechanical Exercises*," we have a full and minute description of this orrery, which is illustrated by two folding-plates (Plates 6th and 7th). Plate 6th gives a ground plan, or calliper of the wheels, and Plate 7th, a sectional view of them. In note 71 we mention that Ferguson generally adopted $\frac{25}{8} \times \frac{69}{7} \times \frac{83}{7}$ (fraction of Camus) for his annual train, which gives a period of about 4 seconds too slow every year, for the length of the year as adopted in Ferguson's time—viz. 365d. 5h. 48m. 55″, and we there show how such a period might have been measured by another simple train. In this orrery we find the period of the moon's nodes derived from wheels $\frac{56}{59-56} = 59-56 = 3$ or $\frac{3}{56} = 18\frac{2}{3}$ years, or 18 years 244 days, instead of 18 years 224 days nearly, the true period. Instead of the above train of 59—56, had one of $98-93 = 5, = \frac{5}{93} = 18\frac{3}{5}$ been used, then a period of 18y. 219d. would have been produced. The annexed table was compiled by our friend, the late Dr. William Pearson, for Dr. Brewster's Edinburgh Encyclopædia. (Vide Edinburgh Ency., article "Planetary Machines," p. 629).

THE TRAINS OF FERGUSON'S ORRERY.

Motions.	Wheel-work.	Periods.			
Earth's diurnal motion,	$\frac{25}{8} \times \frac{69}{7} \times \frac{83}{7}$ . . .	365d.	5h.	48m.	58·78s.
A lunation, ...	$\frac{30}{64} \times \frac{63}{1}$ . . .	29	12	45	0
Solar rotation, . . .	$\frac{26}{8} \times \frac{69}{7} \times \frac{64}{78}$ . .	25	6	35	36
Revolution of Venus, .	$\frac{25}{8} \times \frac{69}{7} \times \frac{73}{10}$ . .	224	10	47	8
Revolution of Mercury,	$\frac{18}{28}$ of Venus . . .	87	23	47	24
Rotation of Venus, . .	$\frac{8}{74}$ of its revolution	24	22	31	8
Moon's node, ...	$\frac{56}{59-56}$	$18\frac{2}{3}$ years.			
Earth's parallelism, .	$\frac{40}{40} \times \frac{40}{40}$ of a year.				

88 Ferguson was an ingenious and clever practical mechanician. He constructed with his own hands, from his own designs, no less than *eight orreries*, several astronomical clocks, various curious models and apparatus for illustrating and demonstrating the principles of natural and experimental philosophy. He had a room in his house (his sanctum sanctorum) fitted up with lathe, wheel-cutting engine, vice, &c., in short, with all the tools used by clockmakers, mathematical instrument makers, and opticians, &c. He spent much of his leisure time in constructing and repairing models, and making curiosities. In his later life he was assisted by Kenneth M'Culloch, an ingenious Scotch mechanic; and who, in 1801, was engaged on the new Planetarium of the Royal Institution of London (vide Brewster's Edin. Ency., article "*Planetary Machines*," page 636).

left off drawing pictures,[89] and employed myself in the much pleasanter business of reading lectures on Mechanics, Hydrostatics, Hydraulics, Pneumatics, Electricity, and Astronomy;[90] in all which, my encouragement has been greater than I could have expected.[91]

The best machine I ever contrived is the ECLIPSAREON, of which there is a figure in the 13th Plate of my Astronomy.[92] It shows the time, quantity, duration, and progress of solar eclipses, at all parts of the earth.[93] My next best contrivance is the Universal Dialing Cylinder, of which there is a figure in the 8th Plate of the Supplement to my Mechanical Lectures.[94]

It is now thirty years since I came to London,[95] and during

89 When Ferguson notes he *entirely left-off drawing pictures,* he must refer to the year 1760. He commenced taking likenesses as a profession, at Edinburgh in 1734, and he informs us that he *followed this profession for six and twenty years,"* which period, added to 1734, brings out the year 1760 as the year when he abandoned *the profession* of limner.

90 Although it was not until 1760 that Ferguson abandoned the profession of limner, yet he had on a great many occasions, for 12 years before this, delivered Lectures on Astronomy, &c., in London and the provinces. In 1760 he laid aside limning, and adopted and followed the profession of Lecturer on Astronomy and the Sciences, in which profession he continued until within a few months of his death, in 1776.

91 Ferguson devoted himself to the profession of Lecturer on Astronomy, &c., in 1760. Although he mentions that in this profession his "*encouragement had been greater than he could have expected,*" this success, however, is more applicable to the ten or twelve years which preceded his death. He had a severe struggle until 1758, when he sold the copyright of his Astronomy for £300, which gave him a standing. In 1762 his pecuniary circumstances were further improved, when the King granted him an annuity of £50. This made him a man of mark; and in 1763, he was elected a free Fellow of the Royal Society. He ever afterwards maintained a high position, and ultimately amassed a considerable sum of money. (Vide Appendix.)

92 This Eclipsareon Ferguson estimates highly, saying, it was his best contrivance. He appears to have invented it somewhere about the year 1753. It is now only known in his Astronomy, Plate 13. It was sold at the sale of Ferguson's effects, in March 1777, to a Mr. James Ferguson, teacher of Navigation and Astronomy, Hermitage Row, Tower Hill, London, at whose sale, in November 1802, it was again sold. Purchaser not known.

93 A short account of this Eclipsareon was read before the Royal Society in 1754. In 1756, Ferguson published a pamphlet regarding it entitled, "Description of a piece of Mechanism for exhibiting the Time, Duration, and Quantity of Solar Eclipses in all places of the Earth." In the same years, he also sent short descriptions of this machine to the London Magazines.

94 It is not now known when Ferguson contrived "*The Universal Dialing Cylinder.*" It, however, appears to have been invented about the year 1766. In 1767, Ferguson published his Supplement to his Lectures, octavo, pp. 68; with numerous plates. A representation of the Cylinder is given in Plate 8th of the Supplement, and pages 42 to 48 describe its construction and use. This is an ingenious contrivance, which we prefer to the Eclipsareon. (See Gentleman's Magazine for 1769, page 143.)

95 Ferguson wrote his Memoir in August and September, 1773. He came to London end of May 1743; therefore, at the time he wrote his Memoir he had been rather more than 30 years resident in London.

all that time, I have met with the highest instances of friendship from all ranks of people, both in town and country, which I do here acknowledge with the utmost respect and gratitude;[96] and particularly the goodness of our present gracious Sovereign, who, out of his privy purse, allows me fifty pounds a-year, which is regularly paid without any deduction.[97]

96 Our memoranda show, that besides King George III. bestowing a pension on Ferguson, his Majesty also frequently sent him presents, as a mark of esteem; as also did the several branches of the Royal Family. From the nobility and gentry in various parts of the Kingdom he received likewise substantial marks of the estimation in which they held him.

97 On referring to "THE PRIVY PURSE ACCOUNTS," in the State Paper Office, London, we find that this pension of "*fifty pounds a-year*" was granted by King George III., to Mr. Ferguson, in the end of the year 1761, and that he drew the *first half* of his pension—viz. £25, on January 22d, 1762 (the pension being payable half-yearly).

EXTENDED MEMOIR

OF

JAMES FERGUSON.

1743.

FERGUSON LEAVES EDINBURGH FOR LONDON.—It is now not certainly known what made Ferguson resolve to leave Edinburgh. He would, no doubt, see many obstacles in the way to his attaining success in it. On looking around him he could not fail to observe that its trade was in a very depressed state; that there was then a scarcity of the "circulating medium;" that an apathy had settled down on its inhabitants; and that there was a consequent want of enterprise—a state of things, he would find, that had arisen since the Union, owing, in a great measure, to the large numbers of the nobility and gentry abandoning it for London. In short, Edinburgh, between the date of the Union, in 1707, and the year 1750, passed through one of the most depressed periods of its history—a period which has been designated "*the dark age of Edinburgh.*" It is therefore evident that Edinburgh, in 1743, was not the place to give full scope to Ferguson's genius, where he could expect to have success in his self-taught business of limning, or as an occasional writer on his favourite science of astronomy. We may also add, that Ferguson, about a year previous to his leaving Edinburgh, had set afloat in the London literary market there, two speculations—viz. "*The Astronomical Rotula,*" and an *Astronomical Card-Dial.* Of the success of the latter publication, nothing now is known;[98] but the former—the Rotula—was successful. It sold

[98] We refer to this Dial in note 65; but as it now seems to be utterly unknown, we will here give a more full description of it. The top part of it contains "*Scales of the signs,*" and "*Scales of the days of the month;*" and under these, there is an inverted triangular "*Scale of Latitudes, from the Equator to* 60 *Degrees;*" and directly under is the dial for the hours of the day. The top compartment has engraved on it, within a sweeping concave-curve which bounds the top of this dial, the words "*Forenoon hours,*" "*Morning hours;*" and in a similar curve at the bottom of it there are engraved, "*Afternoon hours,*" "*Evening hours.*" Along

well in Edinburgh and other towns, but nowhere so well as in London. Thus, the Rotula became to Ferguson an "*avant couriere.*" His name being thus advertised in the great metropolis, would be to him a matter of great importance, and, no doubt, would in due time become one of his strongest inducements to remove to that city. Besides, he had then no strong tie to bind him to Scotland. His infirm parents had *then* recently died at advanced ages; and therefore, in leaving, he would be spared the pain of bidding them a last farewell.[99] It would appear from our notes, that Ferguson, early in the spring of 1743, after carefully considering all matters, and no doubt receiving the approbation of his friends, finally made up his mind, and determined to leave Edinburgh for London, the great seat of wealth, patronage, and power—the most proper place for him; as *there* he would have the greatest possible chance of meeting with success in his contemplated pursuits.

Having now resolved to leave Edinburgh for London, he began making preparations for his departure. He at once set about finishing a beautiful little orrery with ivory wheels, of his own invention, and which had been in hand for some time previous. After completing its wheel-work, he enclosed it in a mahogany box of twelve sides, having a short pillar rising from each angle of the box, supporting a thin, broad, wooden circle,—the ecliptic circle,—the surface of which he covered with paper, and then neatly inscribed on it, with a pen, the names and days of the month, the signs and degrees of the ecliptic, &c. The wheel-work was concealed from view by a thin circular plate, above which, in the centre, was placed a gilded sun; the earth and moon were ivory balls, which, along with the sun, were set in motion by a winch communicating with the internal wheels. When it was all completed, it had a very handsome appear-

the margin-border there is a "*Scale of Rectifying Signs for the Bead;*" and along the margin-border on the right hand, there is a "*Scale of Altitudes, from the horizon to the zenith.*" In the lower left-hand corner there is engraved, "*J. Ferguson, Delin.;*" and in the lower corner on the right, "*R. Cooper, Sculp.*" After Ferguson's death, in 1776, this Dial, among other things, came into the hands of Ferguson's youngest son, John, who, early in the present century, presented it to his relative, the late Miss Wilson, of Keith, who, in the year 1842, kindly gifted it to us. This copy of the Dial in our possession is perhaps now the only one in existence.

99 Ferguson's parents were dead before the beginning of the year 1743. Their ages at the time of their death cannot now be ascertained; but a traditionary account makes them "*to be at least* 70:"—it follows that they were born somewhere between the years 1670 and 1673.

ance;[100] he also finished, in pen and ink, a beautiful diagram of the solar system, and several mechanical drawings, maps, &c.

After having got in his few outstanding debts arising from his business as a limner, and the sums due to him on sales of his Rotula, he found, on balancing his little accounts, that he was in much better pecuniary circumstances than he had ever been, which was principally owing to the sums he had received from sales of the Rotula; and with the expectation that the Rotula would still continue to sell well, especially in London, he would leave Edinburgh in good spirits.

According to memoranda in our possession, Ferguson and his wife took their departure from Leith (the seaport of Edinburgh)[101] in a smack, for London, on Saturday, 21st May, 1743, having on board, as luggage, a chest, and a few boxes, containing their little wardrobe, the orrery with ivory wheel-work, just alluded to, and a large collection of Astronomical and Geographical Diagrams, Maps, Mechanical Plans, Rotulas, and Card-Dials. The sailing distance between Leith and London is about 500 miles. After a somewhat tedious voyage of seven days, they arrived safe in London, on Friday, May 27th.[102] As Ferguson, on arriving in London, makes a new start in life, we may note that his age was then 33, that of his wife 23, and that they had been married four years (without family.)

FERGUSON IN LONDON.—Shortly after his arrival in London, Ferguson went to the Right Honourable Sir Stephen Poyntz, at St. James's, and delivered into his hands the "*letter of recommendation*" he had received for him, from Mr. Baron Edlin, at Edinburgh.[103] Sir Stephen received him with the greatest

100 As mentioned in note 70, this small elegant orrery is still to be seen at Sandon Hall, Staffordshire, but now damaged by the fire which destroyed Sandon Hall, on June 6th, 1848.

101 LEITH is situated on the Firth of Forth, at the distance of nearly 2 miles NNE of Edinburgh. In Ferguson's day it was a very small place. At the present time, it is a large and flourishing sea-port, with a population of about 30,000.

102 During last century a passage between Leith and London was a very tedious affair,—the Leith smacks, though built with a view to despatch, having, in some extreme cases, occupied about three weeks in sailing from Leith to London when southerly gales prevailed. Indeed, the *voyage* was considered by landsmen an *event*, and so hazardous as to suggest to prudent travellers the propriety of making their wills before it was undertaken. The particulars as to date of Ferguson and his wife's departure from Leith, luggage, &c., were communicated to us in 1831 by his youngest son, John Ferguson, Esq., Edinburgh.

103 We have not been able to ascertain whether or not the Right Honourable Stephen Poyntz was a Knight or Baronet in 1743, at the period of Ferguson's first visit; but in recording his death, the Gentleman's Magazine, vol. xx. p. 570, says

kindness. It was fortunate for Ferguson that, on his arrival in London, he had been introduced to so good a man. Baron Edlin's letter, as a matter of course, would refer Sir Stephen to the Astronomical Rotula, and the Card-Dial, then recently published by Ferguson at Edinburgh, as also to his elegant little Orrery with ivory wheels, then lately made by him; all which would indicate a mind strongly inclined to astronomical pursuits. A conversation would ensue as to his prospects, and how they could be promoted. Ferguson would mention to Sir Stephen that he had followed his self-taught business of limning for the last ten years, that he had got tired of it, and had a strong wish to be employed in some vocation connected with his favourite sciences of Astronomy and Geography, such as that of a public lecturer on Astronomy, and private teacher of Geography and the Use of the Globes. In such professions, there were already in London by far too many—the supply greatly exceeded the demand; and consequently, such gave little promise of being successful.[104] Sir Stephen being well aware of this, would suggest to Ferguson a situation in which he could teach the sciences of Astronomy and Geography, and thus in receipt of a fixed salary, and where his mind would be set at rest regarding the future. On talking over such matters, mathematics would be alluded to, when Ferguson would have to admit that he had never studied this important branch. This would tend to disconcert Sir Stephen, because mathematics being the basis on which those sciences rested, it would be necessary for Ferguson at once to learn it, as, without a competent knowledge of the mathematics, a situation in no scientific academy would be open to him. Sir Stephen would probably say so to Ferguson at parting; and also, that he would think the matter over, and do what he could to get over this difficulty.

Sir Stephen accordingly made immediate inquiries regarding mathematical tutors, and was not long in finding out an academy which would soon be in want of a mathematical master, and secur-

"*The Right Honourable Sir Stephen Poyntz, formerly preceptor to the Duke of Cumberland, and a Privy Counsellor, died on December* 17*th*, 1750."

104 Even down to 1758, London appears to have been well supplied with professors of such branches. In a letter written by Ferguson to the Rev. Alexander Irvine, one of the ministers of Elgin, in the north of Scotland, and of date, London, 17th January, 1758, he informs his reverend friend that "*as to astronomy, there are at present more than double the number that might serve the place, people's taste lying but very little in that way.*" (See Letter by Ferguson, under date 1758.)

ing the situation for Ferguson. He next wrote "*to an eminent Professor of Mathematics to take Ferguson into his house, to give him board and lodging, with all proper instructions to qualify him for teaching a Mathematical School.*" The Mathematical Professor, in his reply to Sir Stephen, requests that Ferguson be sent immediately to him. Sir Stephen sends for Ferguson, and informs him, that he had just received a letter from his friend, the Professor of Mathematics, and that he had agreed to take him into his house, give him board and lodgings, and proper instruction to qualify him for the Mathematical Mastership of the School, which would soon be vacant. Ferguson is very grateful to Sir Stephen for his great kindness, but feels much perplexed, and is now to give utterance to a few words that will frustrate all Sir Stephen's kind endeavours in his behalf. Ferguson told Sir Stephen that he "*did not know how he was to maintain his wife during the time he would require to be under the master's tuition.*" Wife! "*are you a married man?*" asked Sir Stephen in astonishment. Ferguson told him "*he had been so ever since May* 1739." Sir Stephen said he was sorry for it, because it quite defeated his scheme, as the master of the school he had in view for him must be a bachelor.[105] Thus ended "*the mathematical master scheme,*" and with it, all hopes of Ferguson obtaining a permanent situation in that direction.

We are inclined to think that Ferguson never would have gained a *name*, or succeeded as a mathematical tutor. His was not a mathematical but a purely mechanical mind, developing itself in mechanical inventions, and in the construction of orreries, cometariums, astronomical clocks, sun-dials, &c. To his latest day he did not understand Euclid; his constant method to satisfy himself of the truth of any problem, was by measurement with a scale and pair of compasses.[106] Being thus disappointed

105 It has before been remarked as singular, that Ferguson did not take his wife's maintenance into consideration at the first interview, and thus have saved Sir Stephen much trouble. Probably at his first interview he might have formed an idea from what was then said, that a few weeks would likely suffice for being under the professor's tuition; or, at all events, to enable him afterwards to instruct himself; and in that case, his funds on hand would enable him to tide over the time required. If, on his second visit to Sir Stephen, he was told that he would require to be a *few months* under the professor's tuition, he would feel uneasy, and think of his finances. (See note 73).

106 Professor Dugald Stewart observes of Ferguson :—"*I remember distinctly to have heard him say that he more than once attempted to study the elements of Euclid, but found himself quite unable to entertain that species of reasoning. The second pro-*

of becoming a mathematical tutor, he would likely account it a misfortune; but there can be no doubt, that ultimately, the disappointment was to him a great gain, as in process of time, he became a Lecturer on Natural and Experimental Philosophy, of world-wide renown; also, the author of a great many popular books, tracts, and papers on Astronomy, Mechanics, Electricity,

position of the first book he mentioned particularly as one of his stumbling blocks at the very outset; the circuitous process by which Euclid set about an operation, which never could puzzle for a single moment any man who had seen a pair of compasses, appearing to him altogether capricious and ludicrous. He added, at the same time, that as there were various geometrical theorems, of which he had daily occasion to make use, he had satisfied himself of their truth, either by means of his COMPASSES *and* SCALE, *or by some mechanical contrivances of his own invention. Of one of these, I have still a perfect recollection—his mechanical or experimental demonstration of the 47th proposition of Euclid's first book, by cutting a card so as to afford an ocular proof that the squares of the two sides actually filled the same space with the square of the hypothenuse."* Lond. Mech. Mag., vol. 6th, p. 110.

The annexed figure represents Ferguson's dissected card for a mechanical demonstration of the 47th prop. Euclid, 1st book—viz. "*In any right-angled triangle, the square described upon the side subtending the right angle, is equal to the squares described upon the two sides containing the right angle.*"

If "*the squares described upon the two sides*" are cut as in the annexed figure, then placed upon "*the square subtending the right angle,*" they will exactly cover the surface of this square. The dissections are numbered 1, 2, 3, 4, and 5 in the two squares above. If these were cut out in card, wood, or metal, and placed on the corresponding figures in the large square below, they would exactly fill it up—be equal to it; and of course, were the dissections in the two squares above weighed in scales, they would be equal in weight to the large square below.

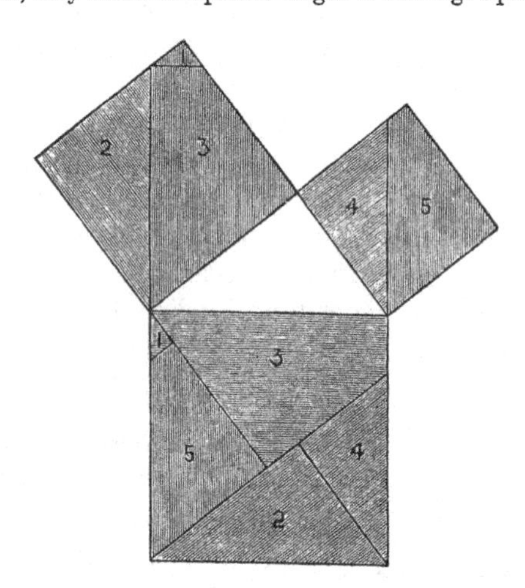

Card used by Ferguson to demonstrate 47 Prop. Euclid, Book 1st.

This figure of Euclid's, prop. 47, Book 1st, is to be found in "HARRIS'S LEXICON TECHNICUM," a work, it will be remembered, that Ferguson perused in the library of William Baird, Esq. of Auchmedden, in the summer of the year 1733.

Perspective, &c;[107] and, at his death, left a considerable funded property.[108]

Ferguson being thus disappointed, and again thrown on his own resources, Sir Stephen asked him what business he intended to follow. Ferguson answered "*that he knew of none besides that of drawing pictures.*[109] On this, Sir Stephen requested him to draw the pictures of his lady and children, that he might show them, in order to recommend him to others; and told him, that when he was out of business to come to him, and he would get as much as he could for him. Ferguson adds, "*and I soon found as much as I could execute, but he died in a few years after, to my inexpressible grief*[110]

Although Ferguson had resumed his business of "*drawing pictures,*" his heart was not in it. His mind, he says, "*was still pursuing things more agreeable,*" that is, "*things astronomical,*" and wished very much to abandon, if possible, the profession. This he could not do all at once, but resolved to do so by degrees. As a first step towards this desirable object, he determined on getting up, during his leisure hours, an astronomical apparatus for public lectures, and private tuition on Astronomy; and thus, by adding to his business of limner, that of lecturer, &c. on Astronomy, he would anticipate a considerable increase to his income, and leave time to determine which of the two professions would ultimately be most conducive to his pecuniary benefit. With these two professions, that of limner and lecturer on Astronomy, &c., Ferguson had a somewhat severe struggle for a living in London, for nearly 17 years—viz. from 1743 to 1760, in which latter year he entirely relinquished his old business of limner, and depended on that of lecturer on Astronomy, &c., which he successfully followed from 1760 till near the period of his death, in 1776.

About the end of the year 1743, Ferguson sold his orrery

107 During the 33 years of Ferguson's residence in London (1743—1776), he, as far as is now known, was the author and publisher of 7 volumes, 15 tracts, and 23 papers, on Astronomy, Mechanics, &c.

108 Shortly after Ferguson's death, in November 1776, several London newspapers had paragraphs, mentioning that he had left about £6,000; but the money and property he left, as mentioned in his WILL, falls considerably short of this sum. (See Ferguson's WILL in the *Appendix*).

109 "Drawing pictures" is the Scotticism for *drawing likenesses*—more properly, *taking likenesses.*

110 Sir Stephen Poyntz lived 7½ years after the time here referred to—viz. June 1743. He died 17th December 1750.

with ivory wheel-work, to Sir Dudley Ryder (afterwards Lord Chancellor of England). It has been conjectured that Ferguson was introduced to Sir Dudley by the Right Honourable Sir Stephen Poyntz, his friend and patron. We have not been able to ascertain the price at which this orrery was sold. As mentioned in note 70, this orrery is now in the possession of the Right Honourable the Earl of Harrowby, at Sandon Hall, in Staffordshire, the descendant of Sir Dudley Ryder. It was much damaged by the fire which destroyed Sandon Hall, in June, 1848, as already mentioned.

1744.

THE ORRERY.—Early in the year 1744, Ferguson commenced getting up an apparatus for public lectures on Astronomy, and for private tuition in Geography, agreeably to his resolve of the previous year. Being now without an orrery, he determined to make a large one, much larger than either of the two he had made, as the first and most desirable piece of machinery for his purpose to begin with. With this view, he laid down on paper, a plan and section of wheel-work for a wooden orrery, to show the motion of the Sun on its inclined axis; the revolution of Mercury round the sun; the revolution of Venus, and rotation on an inclined axis; the revolution of the Earth round the sun, the diurnal rotation on its inclined axis, the parallelism of its axis, and consequently, the alternate succession and different lengths of days and nights throughout the year, and the varied change of seasons; the synodic and periodic revolutions of the Moon round the earth; the retrograde motion of the nodes of its orbit, and therefore, all the eclipses of the sun and moon. At this point in the solar system, Ferguson made a stop,—all the then known superior planets, Mars, Jupiter, and Saturn, being omitted in his arrangements. After getting his orrery plans, and wheel-work calculations completed to his satisfaction, it would appear that he took them to an ingenious turner in wood and metals, with whom he was intimate, and employed him to make it,[111] Ferguson promising to attend to the making

[111] This orrery was the basis of all Ferguson's future orreries, both as regards the prime movers of his wheel-work, and external appearance. In his Memoir, he informs us that he had made *six orreries* since the *two* he had made in Scotland, and that "*there are not any two of them in which the wheel-work is alike.*" This seems to apply to modifications of the wheel-work only.—It would appear from our memoranda, as well as from a notice in Ferguson's "Tables and Tracts,"

of it, and lend a hand when necessary. The making of this orrery appears to have been a tedious affair, partly, no doubt, occasioned by Ferguson being compelled to attend to his limning engagements for a livelihood during the time of its construction, which appears to have occupied little intervals of time, during a period of eighteen months. This being the case, we shall here make a digression, and refer to it again when it is finished—viz. end of the year 1745, and fill up the interval with notices of incidents as they occurred.

THE MOON'S CONCAVE PATH AND TRAJECTORIUM LUNARE.—According to our memoranda, it was towards the end of the year 1744, when to Ferguson "*it appeared, that although the moon goes round the earth, and that the sun is far on the outside of the moon's orbit, yet the moon's motion must be in a line that is always concave toward the sun, and upon making a delineation representing her absolute path in the heavens, found it to be really so.*" Ferguson, in his "ESSAY UPON THE MOON'S TURNING ROUND HER AXIS," page 62 (published in 1747), informs us, that "in discoursing formerly with some gentlemen, I found that they imagined the moon's progressive path made knots or loops once in every lunation, as shown at A and B, in fig. 1 of the annexed diagram; in which case, as seen

Fig. 1.

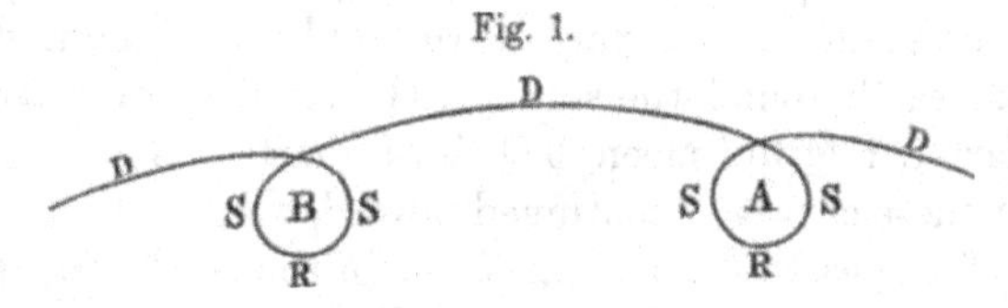

Form of the Moon's Path, as supposed by many Scientific men previous to 1744.

from the sun, she would appear stationary at S, retrograde at R, and in all the rest of her path direct. "This," says Ferguson, "excited my curiosity to draw a representation of her true absolute path, by means of a simple machine," which he contrived

p. 167, whilst treating of this orrery, that he "*got a good workman to make an orrery under my inspection*" (the present orrery):—he adds, that "*the man who made this orrery had never made anything of the kind before, and he is now dead. An ingenious and worthy lady has it now in her possession, whose father was one of my first and best friends in London.*" After this orrery of 1744-45, Ferguson either made or caused to be made under his inspection, five other orreries, which will be found noticed under the dates when they were made. The "*ingenious and worthy lady*" here alluded to by Ferguson is understood to be a daughter of the Right Honourable Sir Stephen Poyntz.

and made — viz. " THE TRAJECTORIUM LUNARE." This machine produced a curve, as is shown in fig. 2: — this curve " shows plainly that the moon can no more appear either stationary or retrograde as seen from the sun, than as seen

Fig. 2.

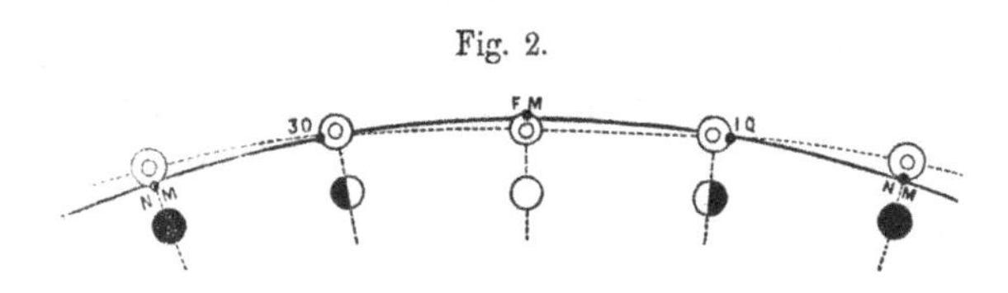

Ferguson's Trajectory of the Moon's Path.—Published 26th June, 1745. Scale, $\frac{1}{12}$th of the original.

from the earth; for her progress is every day at least 880,000 miles more than the whole diameter of her orbit; and as her absolute motion is always concave to the sun, as her relative motion is to the earth, the broadest curve-line from N M to N M represents as much of the ecliptic as the earth describes from any particular new moon to the next following; and the narrowest curve-line, where bounded by the same letters, represents the moon's absolute path during that time, which is concave to the sun, even at the time of new moon. The small dotted circles is intended to represent the moon's imaginary orbit in proportion to the earth's annual path, as the said orbit is carried along with the earth round the sun. N M signifies new moon, 1 Q first quarter, F M full moon, 3 Q third quarter. The right lines crossing these curves, if continued inward, would all meet in the centre of the earth's annual path, or in the sun." In the diagram, the proportion of the curves and orbits are too large—such cannot be accurately delineated on a small scale. For a true projection, we must refer the reader to Ferguson's " ESSAY UPON THE MOON TURNING ROUND HER OWN AXIS," Plate 2d, p. 62; and also to Plate 7th of Ferguson's Astronomy.[112]

[112] Let us suppose, as an illustration, that our earth has no revolution round the sun. In that case, the moon's path round the earth (if it could then revolve) would be in a circular orbit nearly; but as our earth *revolves round the sun*, and is followed by the moon, a compound motion is given to the moon that destroys a circular revolution, and in its stead, produces a curve which sweeps alternately within and without the earth's orbit (see Lunar Orbit as generated by the Trajectorium Lunare, in the annexed diagram, and also in Ferguson's Astronomy, Plate 7th). Suppose a coach wheel to be placed close to a wall, and having a nail projecting from its rim near the circumference; lift the wheel free from the ground, let the nail touch the wall, then turn the wheel round, and the nail

Thus we find that this, the first of Ferguson's important discoveries in Astronomy (and which added much to his reputation), was suggested to his mind during a conversation he had with some gentlemen on the form of the moon's orbit, they maintaining that the absolute path, if projected, would form a series of loop-curves, while Ferguson suspected, what afterwards turned out to be correct, viz. that "*the moon's path was always concave toward the sun.*"

The following is the description of this the Trajectorium Lunare, from Ferguson's Astronomy.

"This machine is for delineating the paths of the earth and moon, showing what sort of curves they make in the ethereal regions. S is the sun, and E the earth, whose centres are 81 inches distant from each other,[113] every inch answering to a million of miles. M is the moon, whose centre is $\frac{24}{100}$ parts of an inch from the earth's, in this machine, this being in just proportion to the moon's distance from the earth.[114] AA is a bar of wood to be moved by hand round the axis *g*, which is fixed in the wheel Y. The circumference of this wheel is to the circumference of the small wheel L (below the other end of the

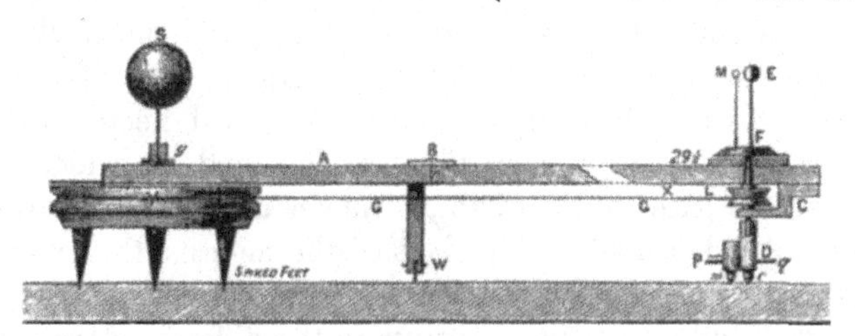

Ferguson's Trajectorium Lunare.
(Section.)

will describe a *circle* on the wall; this done, let the wheel rest again on the ground with the nail touching the wall, roll the wheel forward, and the nail, instead of describing a circle, will now describe a cycloid. In the first case, the wheel had *one* motion only, in the second it had *two*, one round its axis, and another in a rectilinear direction, and hence the difference of the curves generated on the wall; If in the last case, the wheel had gone forward in the arc of a circle, instead of in a straight line, the nail would have generated a curve similar to that of the true path of the moon in absolute space.

113 At the time Ferguson made this machine, and wrote an account of it, the distance of the sun from the earth was estimated at 81,000,000 miles. In constructing a similar machine for the present day, it must be recollected that by the transits of Venus over the sun's disc, in years 1761 and 1769, it is found that the mean distance of the sun from the earth is 95,000,000 miles (in round numbers); and therefore, the centres of the sun and earth in the machine must be 95 inches apart.

114 The $\frac{24}{100}$ parts of an inch, the distance of the centres of the earth and moon in the machine may still be taken, as it will produce no sensible error.

bar), as $365\frac{1}{4}$ days is to $29\frac{1}{2}$, or as a year is to a lunation. The wheels are grooved round their edges, and in the grooves is the cat-gut string G G crossing between the wheels at X. On the axis of the wheel L is the index F, in which is the moon's axis M for carrying her round the earth E (fixed on the axis of the wheel L) in the time that the index goes round a circle of $29\frac{1}{2}$ equal parts, which are the days of the moon's age. The wheel Y has the months and days of the year all round its limb; and the bar A A is fixed to the index I, which points out the days of the months answering to the days of the moon's age, shown by the index F in the circle of $29\frac{1}{2}$ equal parts at the other end of the bar. On the axis of the wheel L is put the piece D, below the cock *c*, in which this axis turns round; and in D are put the pencils *e* and *m*, directly under the earth E and moon M, so that *m* is carried round *e*, as M is round E.

"Lay the machine on an even floor, pressing gently on the wheel Y to cause its spiked feet to enter a little into the floor to secure the wheel from turning. Then lay a paper about four feet long, and the pencils *e* and *m*, cross-wise to the bar; which done, move the bar slowly round the axis *g* of the wheel Y, and as the earth E goes round the sun S, the moon M will go round the earth with a duly proportioned velocity; and the friction-wheel W running on the floor will keep the bar from bearing too heavily on the pencils *e* and *m*, which will delineate the paths of the earth and moon, as already described. As the index I points out the days of the months, the index F shows the moon's age on these days, in the circle of $29\frac{1}{2}$ equal parts; and as this last index points to the different days in its circle, the like numerical figures may be set to those parts of the curves of the earth's path and moon's, where the pencils *e* and *m* are at those times respectively, to show the places of the earth and moon. If the pencil *e* be pushed a very little off, as if from the pencil *m*, to about $\frac{1}{40}$ part of the distance, and the pencil *m* pushed as much towards *e*, to bring them to the same distances again, though not to the same points of space; then as *m* goes round *e*, *e* will go as it were round the centre of gravity, between the earth *e* and moon *m*; but this motion will not sensibly alter the figure of the earth's path or the moon's.[115]

[115] In note 75, we made a few remarks on the "Trajectorium Lunare," which show that the merit of the invention was disputed, and that Ferguson had been

"If a pin as *p* be put through the pencil *m* with its head towards that of the pin *q*, in the pencil *e*, its head will always keep thereto as *m* goes round *e*, or as the same side of the moon is still obverted to the earth. But the pin *p*, which may be considered as an equatorial diameter of the moon, will turn quite round the point *m*, making all possible angles with the line of its progress, or line of the moon's path. This is an ocular proof of the moon's turning round her axis." (Vide Ferguson's Astronomy, article 403).

SEASONS' ILLUSTRATOR.—It would appear from our memoranda, that it was towards the end of this year, 1744, that Ferguson invented and first used his celebrated simple Season Illustrator, and which has ever since been used by lecturers on

charged with having "*pirated*" it from the "Gentleman's Magazine," as the following letter from Ferguson, to the Rev. Mr. Birch, will show.

"LONDON, 12*th Dec*r. 1752.

"REV. SIR,

"The instrument (the Trajectorium Lunare) was shown to the Royal Society in the year 1745, and a painted scheme of the moon's path, taken from a drawing made by it, was presented to the said Society the same year, being published at the desire of the late worthy President, Mr. Folkes. Last summer I gave a draught of the wheel-work of my orrery (in which Venus moves as described by Signior Bianchini), together with a scheme of the Moon's concave path, and a black-lead sketch of the instrument above described, for drawing her path, to Mr. Hawkes, of Norwich, tho in his account published in the Gent. Mag. for this month (Decr. 1752), he has not thought proper to mention it. I am of opinion that altho' that instrument performs well enough by itself, yet it will not be found, however varied from the original plan, to answer so well when put to an orrery; besides, it seems plain to me that he never tried it before he published his account, for if it should be done according to the figure he has given, the moon must go the wrong way round the earth in his orrery to make the pencil which describes her path in the instrument added to it, go the right way round the pencil which describes the annual path of the earth.

"I know it has been insinuated that I pirated this machine (the Trajectorium Lunarie) from a Magazine published some years before my scheme of the moon's path was printed. But I am ready and willing to declare, in the most solemn manner, that I never saw that Magazine, nor heard of any such thing being in it until last summer, when it was handed about by a certain person, with an intent to do me all the prejudice that lay in his power. But the figure of the moon's path in that Magazine, instead of being always concave to the sun, turns off from it in a sharp angle at every new moon,—the paths in different lunations appearing like so many segments of lesser circles joined together at their ends, with their angular points all turned towards one common centre where the sun is supposed to be placed.

"My Scheme of the Moon's Path, on large paper, is sold by Mr. Senex, at the Globe, opposite St. Dunstan's Church, Fleet Street, Price 1s. 6d.

Your faithful servant,

JAMES FERGUSON."

Copied from the Birch collection of Letters in the British Museum, by Mr. A. Burt, copyist.

Astronomy. We annex a view of it. It will be seen that the apparatus consists of two large hoops, the one fitting within the other, and moveable on two pivot-pins, at points directly opposite to each other. These hoops are either supported on two pillars, or held in hands, while a second party, by means of

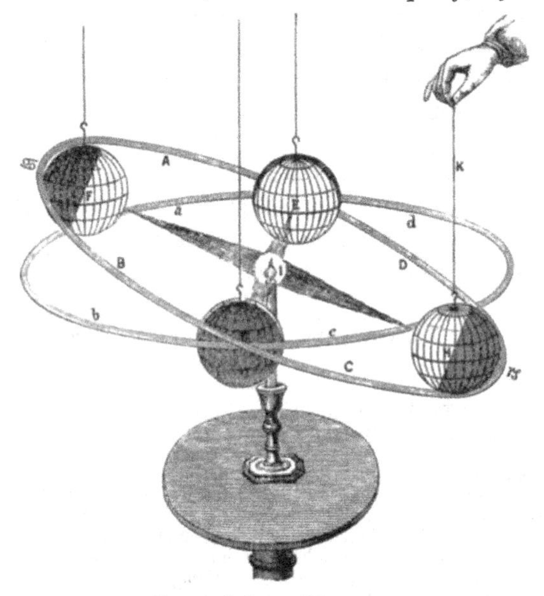

Ferguson's Season Illustrator.

a strong thread, leads a ball or globe gradually round the inside hoop, and if the ball or globe has the thread by which it is suspended twisted in a direction from east to west, it will untwist, and make the globe rotate from west to east, showing thereby the cause of day and night; and by elevating the interior hoop to an angle of 23½ degrees, with the exterior hoop, which represents the ecliptic, and is kept in a horizontal position, the hoop so elevated will show the path or orbit of the earth. These hoops being taken to a table on which a lighted candle is standing, place them around the candle—as in the figure—so that the flame may be in the centre of the hoops, and on a level with the horizontal hoop representing the ecliptic; all this done, twist the thread by which the small globe is suspended in a direction from east to west, and allow it to untwist itself while you lead it and the globe slowly round the elevated hoop, in a direction from west to east, and you will see illustrated, in a very familiar and agreeable manner, the rotation of our earth on its axis in the proper direction, from *west* to *east;* and if the globe, whilst thus rotating, be gradually carried round the inside of the inclined hoop, the alternate succession and different lengths of days and nights throughout the year will be made evident to sight, as also the varied change of the seasons, Spring, Summer, Autumn, and Winter. Referring to the annexed engraving, the letter A indicates the position of the earth at the *vernal equinox*, 21st March. The centre of globe being on a level with the candle flame (which represents the sun), will be enlightened from pole to pole, same as the globe is when in the opposite position, thereby causing equal day and night all over the world. The letter B shows the position of the earth at the *summer solstice*, 21st June. Here it will be observed that the small globe is considerably under the flame of the candle, so much so, that its light proceeds to a point 23½ degrees beyond the north pole, enlightening the whole arctic circle and thereby causing any given point north of the equator of the globe to continue longer in the light than in the darkness, thus giving the *longest day*, mid-summer, to the northern surface of the earth. The lower part of the globe in this position will be found in darkness. The light from the candle falls short of the south pole by 23½ degrees, quite the reverse at the north pole; thus causing mid-winter to all the surface of the earth to the south of the equator. The letter C shows the place of the earth at the *autumnal equinox*, 23d September. Again, as in the opposite position on 21st March, the candle flame enlightens the globe from pole to pole, causing equal day and night over the world; and lastly, the letter D indicates the earth's position at the *winter solstice*, 23d December. Here the light from the candle does not reach the north pole, but falls 23½ degrees short of it, just as many degrees as the light went beyond this pole in the opposite position; but whilst the light falls short of the north pole, it penetrates 23½ degrees beyond the south pole; thus causing mid-winter to the northern parts of the earth, and mid-summer to southern hemisphere. Thus at a cost of a few pence, a simple apparatus may be constructed to illustrate the two-fold motion of the earth—the alternate succession of day and night—with their various lengths, and the change of the seasons, in a way little inferior to that shown in complicated orreries, or other astronomical machinery.

An engraving, similar to the one we here give, will be found in Ferguson's Astronomy, Plate 4th, figure 3d.[116]

1745.

THE TRAJECTORIUM LUNARE — MARTIN FOLKES, ESQ., and the ROYAL SOCIETY.—According to our memoranda, some time during the month of February 1745, Ferguson took his recently invented "*Trajectorium Lunare, and the delineation it had made, to Martin Folkes, Esq., the then President of the Royal Society, on a Thursday afternoon.*"[117] who *expressed great satisfaction at seeing it,* and was convinced *that it was a new discovery.* He took Ferguson with him, the same evening, to a meeting of the Royal Society, and got him to exhibit his machine to the members, and to show how the delineation was projected. At the close of the meeting, one of the members invited Ferguson to dine with him on next Saturday, at Hackney, mentioning that his name was Ellicott, and that he was a watchmaker.[118] Vide Ferguson's own Memoir, p. 49.

116 Bonnycastle, in his Astronomy, has an engraving of this simple apparatus, *improved,* he says, but we can discern no difference in his engraving of the figure from that of Ferguson, excepting in this, the bottom of the candlestick is round and stands on a square table instead of a round one. (See-Bonnycastle's Astronomy.)

117 It may to many appear remarkable that Ferguson, in 1773, when he wrote his Memoir, should have been able to remember the very day of the week that he went to the Royal Society meeting with this instrument. He went to the meeting on a *Thursday afternoon,* he says; but as the Royal Society meetings were then always held, and we believe, still continue to be held on a *Thursday,* it would therefore, in this case, require no effort of the memory to recollect the day. (Vide also note 77).

118 The writer of the short Memoir of "JOHN ELLICOTT, F.R.S." in "*The Horological Journal,*" mentions, that "*Mr. Ellicott frequently introduced friends to the Society, among others, Dolland, Smeaton, and Ferguson.*" This, as regards Ferguson, is a mistake. Ferguson in his Memoir tells us that *he went* to Martin Folkes, Esq., the President of the Society, and that the President introduced him to the Society, in order that he might show and explain his Trajectorium Lunare, and the delineation it had made, to the members; on which occasion, Ferguson and Ellicott appear to have been utter strangers, for Mr. Ellicott, at the close of the said meeting, had to mention to Ferguson that his "*name was Ellicott, and that he was a watchmaker.*" This writer makes another mistake —he says, "*Ellicott who was present,*" when Ferguson showed his machine to the Society, "*stated that he had invented and constructed a similar machine many years before. Ferguson doubting the truth of this assertion, Ellicott invited him to dine at his house at Hackney, and there produced evidence which confirmed his statement,*" &c. It will be observed that this writer does not inform us of the source whence he learned that Ferguson "*doubted the truth of Ellicott's assertion.*" If Ellicott made "*the statement,*" as this writer *says* he did, it is singular Ferguson does not tell us of it in his Memoir. Again, supposing "*the statement*" to have been made, we scarcely think that Ferguson would have doubted the truth of "*the statement,*"—we can imagine in *such a case,* that Ferguson would express his *surprise,* a very different thing from doubting a man's

"Accordingly," says Ferguson in his own Memoir, "I went to Hackney, and was kindly received by Mr. John Ellicott, who then *showed me the very same kind of delineation, and part of the machine by which he had done it, telling me that he had thought of it twenty years before*—viz. in 1725. *I could easily see by the colour of the paper, and of the ink-lines upon it, that it must have been done many years before I saw it. He then told me, what was very certain, that he had neither stolen the thought from me nor had I from him.*" Although Ferguson was therefore not the first who had discovered that "*the path of the moon was always concave to the sun*," yet he had undoubtedly all the merit of an original discoverer.

THE MOON'S CONCAVE PATH, BY FERGUSON, ENGRAVED AND PUBLISHED.—As mentioned in note 75, Ferguson, at the suggestion of Martin Folkes, Esq., and other members of the Royal Society, resolved to publish his delineation of the moon's concave path.[119] He made out a new projection on a sheet of paper $34\frac{5}{8}$ inches in length, by $6\frac{1}{8}$ inches in breadth, giving the absolute motion for a space of 32 days, with a neatly-written explanatory description above and below the delineation, with a dedication to Martin Folkes, Esq., thus:—"TO MARTIN FOLKES, ESQ., PRESIDENT OF THE ROYAL SOCIETY, THIS PLATE IS DEDICATED, BY HIS MOST HUMBLE SERVANT, JAMES FERGUSON. PUBLISHED JUNE 26TH, 1745, ACCORDING TO ACT OF PARLIAMENT." This plate was engraved by J. BICKHAM, and was sold by Mr. Senex, at the sign of the Globe, opposite to St. Dunstan's Church, Fleet Street, London. Price 1s. 6d. (See note 75.)[120] It is now not known how this

word; in short, since Ferguson makes no mention in his Memoir that Ellicott made such "*a statement*," we are inclined to conclude that *no such statement was made.* Then, to give Ferguson a surprise, he invites him to dine at his house, and then and there shows him, no doubt to his very great surprise, the same sort of delineation as his own, of the Lunar Orbit, and part of the machine by which it had been projected. As previously mentioned (note 78), Mr. Ellicott, in 1772 dropped suddenly from his chair, and instantly expired. (See Horological Journal, vol. I., pp. 152–154, Memoir of John Ellicott, F.R.S.

119 Ferguson, in a letter to the Rev. Dr. Birch, London, of date 12th December 1752, says, "*The instrument was shown to the Royal Society in the year* 1745, *and a printed scheme of the moon's path, taken from a drawing made by it, was presented to the said Society the same year, being published at the desire of the late worthy President, Mr. Folkes.*" (See note 115).

120 In same letter to the Rev. Dr. Birch, Secretary, Royal Society, of date 12th December 1752, he says, "*My scheme of the moon's path, on a large sheet of paper, is sold by Mr. Senex, at the sign of the Globe, opposite St. Dunstan's Church, Fleet Street, Price* 1s. 6d." (See letter in the Appendix).

sheet sold. It has been out of print for upwards of a century. Our fac-simile of the engraving was taken from the copy in the British Museum, by Mr. Augustus Burt, London; the Museum number and mark on which, is, No. $^{5}\frac{52}{28}$ f. 7.—On the back of the title-page of Ferguson's first work, a tract, entitled, "THE USE OF A NEW ORRERY" (published about August 1746), we find the following advertisement of this engraved sheet of the Moon's Path:—

> "A Delineation of the Moon's Real Path in the Heavens, showing that her progressive motion is always concave, both to the Earth within her orbit, and to the Sun on the outside thereof. Price One Shilling and Sixpence." Vide Ferguson's Tract, entitled, "The Use of a New Orrery," 1746—fly leaf.

AGNES FERGUSON born.—To Ferguson, the great event of the year 1745 occurred on August 29th. On this day, his first child was born—a daughter—and shortly afterwards was baptized *Agnes*. On one of the fly leaves of a small pocket Bible which belonged to the Ferguson family, we find the following entry in Ferguson's autograph: [121]

"AGNES, born, Thursday, 29th Augt. 1745."

To keep events in harmony with the order of time in which they occurred, we must refer the reader to date 1763, where will be found an account of the "*mysterious disappearance*" and subsequent life of this unfortunate lady, which, previous to our recent researches regarding her, had been recorded as "*a mystery which never would in time be unravelled.*"

THE ORRERY.—About the end of this year (1745), Ferguson got the large wooden orrery finished, which, early in the preceding year, he, and a turner in wood, &c., had commenced to make. In the Tract which Ferguson shortly afterwards published regarding it, he says nothing about the wheel-work *numbers* which produced its various motions; but from the account he gives of it in said tract, it showed, with considerable exactness, the following motions and general phenomena, viz.:—The mo-

121 This Bible is still in good condition, and is at present the property of Doctor James George, of Keith, in Banffshire, who kindly allowed our esteemed friend, Mr. Robert Sim (of same place), to take a copy for us of the writings on its several fly-leaves, all which shall be noticed in the order of time to which they refer, as also in our *addenda*. This Bible is $7\frac{1}{4}$ inches long, $1\frac{1}{2}$ inches broad, and $2\frac{1}{2}$ inches thick; and at foot of the title-page is "London: Printed by Thomas Baskett, printer to the King's Most Excellent Majesty, and by the assigns of Robert Baskett, M.DCC.LVII.'

tion of the sun from west to east on an inclined axis, (in the centre of the machine,) and thereby, the motion of the solar spots; the revolution of the planet Mercury; the revolution of Venus, and her diurnal rotation on an inclined axis, which preserves its parallelism during its annual course; thus exhibit-

Ferguson's Orrery.

ing her different length of days and nights; her change of seasons; and her progressive, stationary, and retrogressive aspects as seen from the earth.—Also, the revolution of the earth round the sun, and its diurnal rotation on an inclined axis, which preserved its parallelism during its course, consequently, showed the alternate succession and different lengths of day and night throughout the year, and the change of the seasons. The Synodic and Periodic revolutions of the Moon were also shown, and also the diurnal rotation, with the retrograde revolution of the nodes of her orbit, and therefore, all the Eclipses of the sun and moon. On the exterior plate, covering the wheelwork, were two dial-plates, having indexes, pointing respectively, when in motion, to the hours of the day of mean solar time, and to the day of the moon's age, while a small inclined dial, under the earth, showed sidereal time. Directly opposite to the earth rose a shaft that carried an index which pointed out on the great ecliptic circle (as it moved along with the earth), the names and days of the months, and the signs and degrees of the ecliptic. In the tract previously alluded to, which is entitled, "THE USE OF A NEW ORRERY MADE AND DESCRIBED BY JAMES FERGUSON—London, pub. 1745," at page 38 he says, that,—" *The justness of the planetary motions in this orrery may be seen in the following manner:*—Turn the handle," says he, "till the pointer comes to the beginning of Aries, observing at that time when the meridian of your place is turned toward the sun; then set the index on the hour circle to XII at noon, and the index on the sidereal dial-plate to 24; then turn the handle, and as the earth proceeds forward in the ecliptic, you will see sidereal index on its dial-plate gaining time on the solar index in the hour circle, which will always point to the same XII., when your meridian turns to the sun; but in a quarter of a year, the sidereal index will be six hours before the solar one; in half a year, 12 hours; in three-quarters of a year, 18 hours; and in a whole year, 24 hours or a whole circle, which it will have gained on the solar index in 365 days, or so many turns of the handle. Note upon the ecliptic the day of any new moon, and fix a bit of paper over against her on the wall, as seen from the earth; then turn the handle $27\frac{1}{8}$ [122] times round, which will bring the moon round the earth, so as to point from it to the bit of paper again; but to bring her round from the sun to the sun again, requires $29\frac{1}{2}$ turns, and the pointer on the ecliptic will have passed over so many divisions to the day of the next new moon, because as every turn of the handle brings any meridian semicircle upon the earth quite round from the sun to the sun again, and carries the solar index round its hour circle, so it advances the pointer on the ecliptic one day forward among the months. Turn on till twelve lunations are accomplished, which will happen eleven days before the pointer comes again to the same day on the ecliptic from which you began to compute; and it is commonly known that twelve lunations come eleven days short of a solar year, which is the foundation of the epact. Turn the handle till the line of the nodes, if produced, would pass through the sun's centre; then note the place of the pointer among the degrees of the ecliptic, and turn the handle till the same node, in the line of nodes, comes between the

[122] These $27\frac{1}{8}$ turns of the handle of the orrery appears to be a typographical error, as a periodical revolution of the moon contains 27 days 7 hours 43 min. 5 sec., or in round numbers, 27 days 8 hours $= 27\frac{1}{3}$; therefore, instead of $27\frac{1}{8}$, read "$27\frac{1}{3}$ *times round.*" We are inclined to think that Ferguson, in this description of his early orrery gives the periods and results in *round numbers.*

centre of the earth and of the sun again; stop there, and you will see the pointer cut the ecliptic almost 19 degrees short of what it did before; then turn the handle till the pointer is carried forward *that* 19 degrees (in which time the nodes are still moving backward) and the line of the nodes will be gone $19\frac{1}{3}$ degrees backward, which is their retrogradation every year. This shifting backward, or contrary to the order of the signs, is the reason why the eclipses happen every year sooner than they did the year before, whereby they are gradually removed from the consequent toward the antecedent signs. If you put a bit of paper or a patch on the sun, over against any part of the ecliptic, so as to be just coming in sight of the earth, and then turn the handle $25\frac{1}{2}$ times round, the sun will have carried the paper or patch, so as to point at the same place of the ecliptic again, but it will require two turns more to bring it in view from the earth again, because the earth has been going forward in the ecliptic while the sun was turning round its axis.[123]

If you observe any meridian semicircle of Venus that looks toward the sun, and turn the handle $24\frac{1}{3}$ times round, the same semicircle will again be turned toward the sun, but he will not be vertical to the same place as before. If you first set Venus and Mercury by hand between the earth and the sun, and note their places in the ecliptic, by laying a thread from it over them to the sun, and then turn the handle 88 times round, Mercury will be gone quite round so as to point from the sun toward the same place of the ecliptic again. But to bring him in a right line between the earth and the sun, or to his next inferior conjunction, will require 28 turns more, in all, 116 days.[124] All this time Venus has gone little more than half round the eclip-

[123] Early in last century some astronomical writers gave the period of the sun's rotation as 25 days 14 ho., 25 days 12 ho., 25 days 6 hours, &c. Ferguson here adopts 25 days 12 ho. $= 25\frac{1}{2}$ days for his period. About the middle of last century, a sidereal rotation of the sun on his axis was supposed to be done in 25 d. 7 h. 59 min. The true period, however, appears to be 25 d. 10 h. 0 min. 2·88 sec.

[124] The conjunctive period of any given planet may be found by multiplying the sidereal periods of revolution into each other, and dividing by a divisor arising from the difference of the periods thus:—Required the time of the conjunctive periods of Mercury and the Earth. The sidereal period of Mercury is 87d. 23h. 15m. 36·12s., or decimally expressed, 87·969168 days. The sidereal period of the Earth is 365 d. 6 h. 9 m. 10·11 sec., or decimally, 365·256367 days, and $365{\cdot}256367 \times 87{\cdot}969168 = 32131298711692656$; and from $365{\cdot}256367 - 87{\cdot}969168 = 277{\cdot}287199$ for a *divisor*. Therefore, $32131298711692656 \div 277{\cdot}287199 = 115{\cdot}8773$ days, which, when reduced, is 115 d. 21 h. 3 m. 18·72 s., or in round numbers, nearly 116 days, as produced by Ferguson's machine.

tic, and therefore you must turn the handle 109 times more round, which will complete her revolution, equal to 225 days in round numbers. But to bring her to her next inferior conjunction, you must turn the handle 258 times more round, which, added to the former 225, makes 583 turns, equal to a year and 218 days.[125]

The motions of Venus and Mercury come nearer the truth in this machine than what is here mentioned, but if they did not, they would still be near enough, when they are so quickly shown as by turning the handle, which may be 365 times done in less than a quarter of an hour, and in this quick way of instruction, the fractional parts of hours and minutes in the annual revolutions cannot be observed, and consequently must be lost as to sense.

I almost believe it is in vain for man to pretend to make the planetary motions so exact in a machine as for ever to agree with their originals in the heavens; but if I was to fit an orrery to the true motions of a well-going clock, I would at a small additional charge make it so, as when moved by the clock, to perform the earth's annual motion in 365 days 5 hours 48 minutes 57 seconds;[126] its motion round its axis (or sidereal day) in 23 hours 56 minutes 4 seconds 6 thirds;[127] its solar or natural day in 24 hours; the moon's motion round the earth in her orbit in 27 days 7 hours and 43 minutes;[128] from new moon to new moon again in 29 days 12 hours 45 minutes; Venus's annual revolution in 224 days 17 hours;[129] her diurnal

125 To ascertain the conjunctive period of Venus with the Earth, see rule in last note. The sidereal revolution of Venus round the Sun is accomplished in 224d. 16h. 49m. 10·93s., which, when decimally expressed, is 224·700821 days; and the sidereal revolution of the earth round the sun is done in 365d. 6h. 9m. 10·11s. decimally expressed, 365·256367 days; therefore, 365·256367 × 224·700821 = 82073·405540377307, and 224·700821—365·256367 = 140·555546 for a *divisor*. Hence, 82073·405540377307 ÷ 140·555546 = 583·9215 days, reduced is 583d. 22h. 6m. 57·6 sec., a conjunctive period close upon that produced by this orrery.

126 It is quite impossible to produce a period of 365d. 5h. 48m. 57s. by means of wheel-work; the nearest approach to such a period is by the ratio or wheel-work Ferguson uses—viz. $\frac{8}{25} \times \frac{7}{69} \times \frac{7}{83} = \frac{392}{143175}$ — 143175 ÷ 392 = 365 d. 5 h. 48 m. 58·78 s. How Ferguson may have obtained his ratio of $\frac{392}{143175}$ will be found in note 133.

127 This period of 23h. 56m. 4s. 6thds. may be ascertained thus:—The sun makes 365 apparent revolutions round the earth in a year, the stars 366 revolutions. The solar year is 365d. 5h. 48m. 51·6, or decimally, 365·242264 days, to which add *one* day, and it will give 366·242264, that is, the sun makes 365·242264 apparent revolutions round the earth in a solar year, and the stars 366·242264 revolutions. Therefore, 365·242264 ÷ 366·242264 = ·99726957 of a solar day; $\frac{99726957}{100000000}$ of a day is 23h. 56m. 4s. 5·45thds., being the true period, and is nearly what is shown by the orrery.

motion in 24 days 8 hours; Mercury's annual motion in 87 days 23 hours;[130] and the sun's motion in 25 days 6 hours. How near the truth these are, I leave to the judgment of those who have read astronomical accounts of the celestial motions."—(Vide Ferguson's "*Use of a New Orrery. London*, 1746," pp. 38—41); also, see notices of his future orreries under their respective dates.

We now annex a plan and section of the wheel-work of this orrery, accompanied by a full description of the machine, extracted from Ferguson's "SELECT MECHANICAL EXERCISES," a careful perusal of which will enable any practical mechanic to construct one for himself.[131]—"In the annexed engraving, Fig. 1 is the ground-plan of the wheel-work of the orrery. The numeral at each wheel shows the number of teeth in that wheel, and the shaded parts show where the teeth of any one wheel take into the teeth of another, as the one turns the other. Fig. 2 is a section or side view of all the wheel-work that can be brought into sight. But in this, some few wheels cannot be shown, for in the orrery itself, take a view of the wheels on any given side, some of them will be unavoidably hid from sight by others that are between them and the eye. Those that come in sight in Fig. 1 have the same numeral figures set to them as the like ones have in Fig. 2, and also the same letters of reference;

128 The true mean periodic revolution of the moon is accomplished in 27d. 7h. 43m. 4·67s., and a mean synodic revolution in 29d. 12h. 44m. 2·88s.

129 Venus's mean tropical revolution round the sun is performed in 224d. 16h. 41m. 31·05s., rotation on axis *not known*. Mercury's mean tropical revolution round the sun is done in 87d. 23h. 14m. 35·98s., and the sun's rotation, 25d. 10h. 0m. 2·88s. During last century the solar rotation was variously estimated by astronomical authors at 25d. 6h.,—25d. 7h. 58m.,—25d. 12h., and at 25d., 14h., &c. Ferguson has adopted the minimum of these periods, all which periods correspond in round numbers to the periods shown by this orrery.

130 The tropical revolution of Mercury round the sun is accomplished in 87d. 23h. 14m. 35·98s. This orrery period of Mercury is therefore 14m. 36s. minus the true mean period of the planet.

131 Ferguson, in his "*Select Mechanical Exercises*," does not give the diameter of the wheels and pinions of this orrery, they are only to be found in his "*Tables and Tracts*," from which they have been taken, and are thrown into this description in order to render it more practically useful. (Vide Ferguson's "Tables and Tracts," pp. 160—167, 1st Edit.).

After a lengthened search for a copy of Ferguson's "*Select Mechanical Exercises*," we were at last successful in procuring one at an old book-stall in Edinburgh, in July 1827. After a careful study of the arrangements of the several wheels given in Plates 6 and 7 of this book, and the description of them there given, we commenced making one in October of that year (all of brass), and finished it in May 1828, and enclosed the wheel-work in a case of 12 sides, as is shown in the print of the external view of the orrery. The wheels were only ⅓d part of the diameters here given.

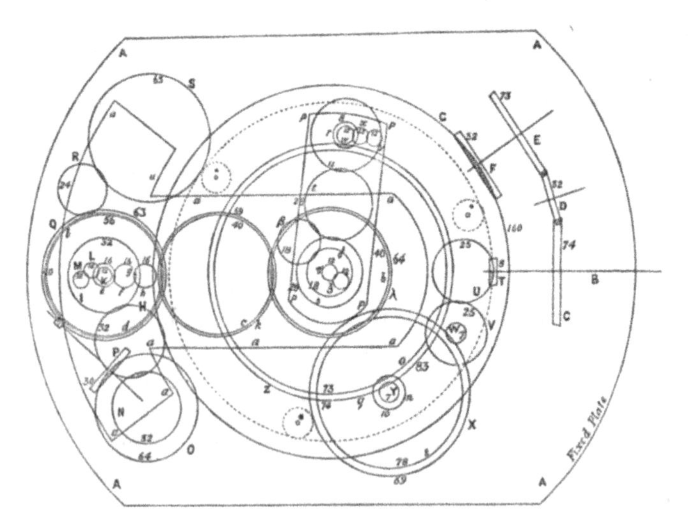

Plan of the Wheel-work of Ferguson's large Orrery.

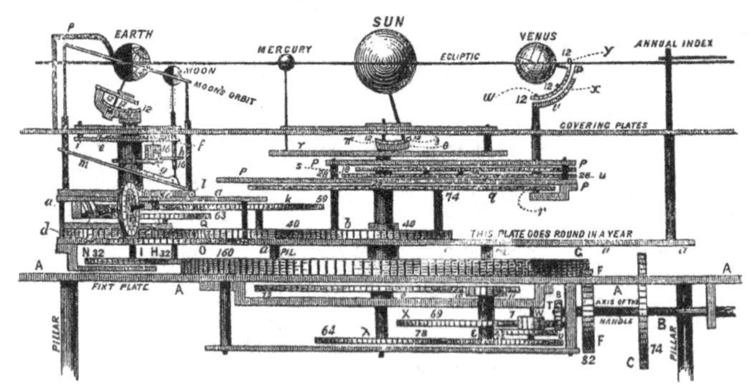

Section of the Wheel-work.

and therefore, in reading the description, it will be necessary to look first at it, and then at Fig. 2, by which means the reader will see the position of these wheels in respect to each other, as they are placed higher or lower in the frames which contain them.

A A A A is a round immovable plate supported by four pillars; some of the wheels are below it, but the greatest number of them are above it. It supports and bears the weight of them all.

B is the axis of the handle or winch by which all the wheels are turned; on its axis is a wheel C of 74 teeth, and 6·12 inches in diameter, which turns a wheel D of 32 teeth, 1·80 inch in diameter, and D turns E, a wheel of 73 teeth, and 6·11 inches in diameter, on whose axis is a wheel F of 32 teeth, 1·80 inch in diameter, turning a wheel G of 160 teeth, and 8·97 inches in diameter, which turns a wheel H of 32 teeth, and 1·80 inch in diameter, and H turns a wheel I of the same number and diameter, on the top of whose axis is a small wheel K of 12 teeth (just under the Earth), which turns a wheel L of the same number and size, and L turns such another wheel M of the same number. The axis of M inclines 23½ degrees, and the earth at the top of it is turned round by it. The wheel H of 32 teeth turns a wheel N of the same number and size as H, on the top of whose axis is an index which goes round a circle of 24 hours (on the plate that covers the wheel-work), in the time the earth turns round its axis. The wheels D and E could not be shown in Fig. 2, because the wheel C of 74 teeth hides them from sight.

On the axis of the wheel N is a wheel O of 64 teeth and 3·60 inches in diameter, turning a contrate wheel P of 30 teeth, 1·6 inch in diameter, on whose axis is an endless screw of a single thread I turning a wheel Q of 63 teeth, 3 inches in diameter, which carries the moon round the earth in her orbit, from change to change, in 29 days 12 hours and 45 minutes. This wheel of 63 teeth turns a wheel R of 24 teeth, 1·23 inch in diameter, which turns a wheel S of 63 teeth, same in diameter as Q, round in 29 days 12 hours and 45 minutes, on whose axis is an index that shows the days of the moon's age, on a circle of 29½ equal parts, on the plate that covers the wheel-work.[132]

[132] We do not know by what process Ferguson obtained the numbers of his wheel-teeth for his lunar train, as the application of continuous fractions to the determining ratios for wheel-work was unknown until within a few years of his death, and it could not therefore be by such a process. It is probable that he followed or modified the roundabout way laid down by the Rev. Dr. W. Derham, in his "*Artificial Clockmaker*," thus:—"Determine on the number of pinions you would like to use, and also on the number of leaves in each. Multiply the number of leaves and this will give the numerator of the ratio, and with which the period of the lunation must be multiplied, and the result will give the denominator of the ratio sought." This is a tedious process, because the denominator must come to a *whole number*, or at least as nearly so as possible. We will give an example. Suppose Ferguson determined on two pinions of 8 leaves each, then $8 \times 8 = 64$, the numerator of the ratio sought, and 29d. 12h. 45m. decimally expressed, is $= 29{\cdot}53125$; therefore, $29{\cdot}53125 \times 64 = 1890{\cdot}00000$, which gives the denominator of the ratio; hence, $\frac{64}{1890}$ is the ratio sought. Were a wheel of 64 teeth to revolve in 24 hours and drive a wheel of 1890, such a wheel would revolve in 29d. 12h. 45m.; but as 1890 teeth cannot be used in one wheel, it requires to be broken down as follows—viz. $\frac{64}{30} \times \frac{1}{63}$ as adopted by Ferguson in this and his other orreries. By adopting the process of *continuous fractions*, to determine such ratios, we find the following ratios :—A lunation of 29d. 12h. 44m. 2·887s. =

On B, the axis of the handle is a pinion T of 8 leaves, $\frac{71}{100}$ inch in diameter, and turns a wheel U of 25 teeth, 2·22

29·5305887 decimally expressed; hence, the prime solid or fraction of $\frac{10000000}{29.5305887}$, which, on being reduced by this continuous fraction process (to be fully explained in the next note, in finding numbers for an annual train of wheels), we arrive at the following ratios:—$\frac{1}{29}$, $\frac{1}{30}$, $\frac{2}{59}$, $\frac{15}{443}$, $\frac{17}{502}$, $\frac{49}{1447}$, $\frac{850}{25101}$, $\frac{1749}{51649}$, &c., although $\frac{64}{1890}$ is not to be found in this accurate process, yet by an engrafting or adding of the *fourth* and the *sixth ratio* together, we obtain Ferguson's numbers, viz. $\frac{15}{443}+\frac{49}{1447}=\frac{64}{1890}$. Ferguson seems never to have been able to get wheel-work to produce a lunation more exact than 29 d. 12 h. 45 m. (a period about 57 seconds too slow in a lunation) excepting in one instance in which he uses a large wheel of 235 teeth, and a pinion of 19 leaves, which will be noticed in our description of that orrery. Fig. 1 in the annexed engraving shows Ferguson's arrangement for the wheel-work of his lunation; if such be adopted, we would suggest *contrate* wheel 30 to be made a bevel wheel, and the upper half of the

Fig. 1.

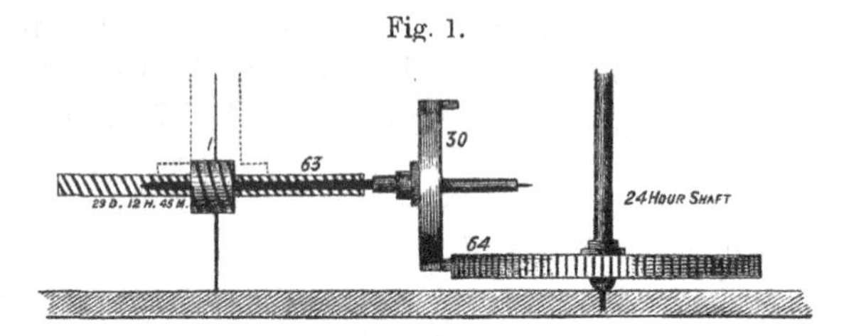

whole of wheel 64 bevelled off to correspond to it,—the lower half being left plain, would work same as shown. We would recommend the arrangement shown in Fig. 2 in preference to that in Fig. 1, thus doing away with contrate and bevel wheels altogether, as also the screw, and giving a more simple and much better

Fig. 2.

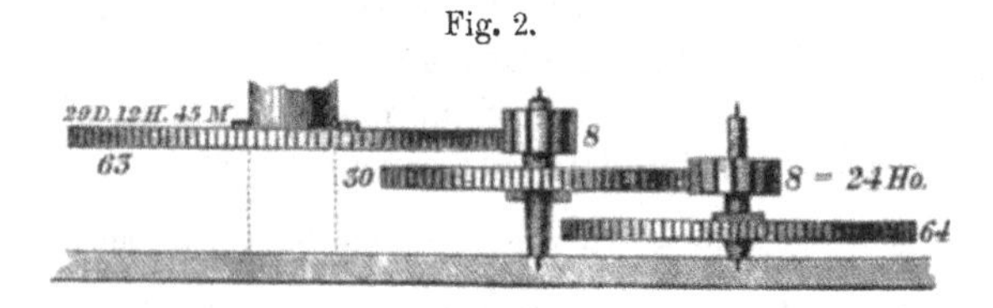

action for these wheels:—thus, wheel 64 becomes useless, and on its axis a pinion of 8 leaves turns in 24 h. and drives wheel 30, and its pinion of 8 drives 63 round in 29 d. 12 h. 45 m. We may here note that the last lunar ratio but one we here give, $\frac{850}{25101}$ cannot be adopted, because 25101 when divided by 9 leave 2789, which is a *prime number*, and therefore cannot be further reduced; the next ratio above this is $\frac{1749}{51649}$, and which can be arranged thus:—$\frac{12}{52}\times\frac{22}{58}\times\frac{53}{137}$ pinion 12 turns round in 24 h. and drives a wheel of 52 teeth, having on its axis a wheel of 22 teeth that turns one of 58 teeth, on whose axis is a wheel of 53 teeth, which turns round a wheel of 137 teeth in 29 d. 12 h. 44 m. 2·881 s. The true mean period of a lunation being 29 d. 12 h. 44 m. 2·887 s.; therefore, this last train is the most perfect that can be found clear of prime numbers. A very simple and tolerably accurate train may be obtained by adding the following lunar ratios together,—viz. $\begin{array}{r}1447\\502\\443\\\hline 2392\end{array}$ and $\begin{array}{r}49\\17\\15\\\hline 81\end{array}$ If we take the fraction $\frac{81}{2392}$ it will produce a period of 29 d. 12 h. 44 m. 26 s. as $2392 \div 81 =$ 29·530964 = 29 d. 12 h. 44 m. 26 s., being a period within 24 s. of the true mean revolution.—2392 may be broken down into 52 and 46, and 81 into 9 and 9, and

inches in diameter, which turns another wheel V of the same number and size, on whose axis is a pinion W of 7 leaves, $\frac{62}{100}$ inch in diameter, which turns a wheel X of 69 teeth, 4·12 inches in diameter, on whose axis is a pinion Y of 7 leaves, $\frac{51}{100}$ inch in diameter, which turns a wheel Z of 83 teeth, 6·12 inches in diameter, once round in 365 days 5 hours 48 minutes 57 seconds, and carrying the earth round the sun in that time,[133] for in

the train will become $\frac{9}{46} \times \frac{9}{52}$; pinion 9 revolves in 24 h. and turns wheel 46, on whose axis is also a pinion of 9 leaves, which turns wheel 52 once round in 29 d.

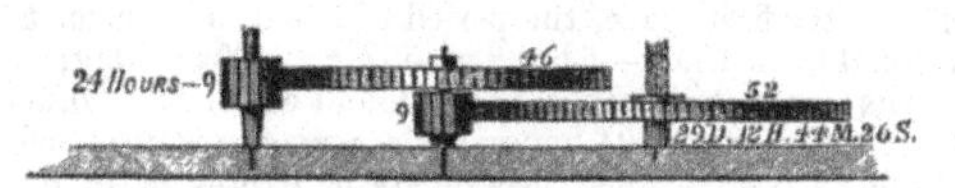

12 h. 44 m. 26 s.; for a train of two wheels and two pinions, this is perhaps the most simple and accurate train, free of prime numbers, which can be found. See the annexed section for arrangement of the wheel-work.

[133] As we have already remarked, the art of computing ratios, and trains of wheel-work, by continuous fractions, was unknown until the year 1771, five years before Ferguson's death. Camus, in his "*Cours de Mathematiques*," Paris, 1767, gives rules for determining trains for planetary revolutions, but in a way so very tedious, that few would be found to trouble themselves with it; *several pages* of his "*Cours*" are taken up with the problem of a motion derived from 12 hours to produce a period of 365 d. 5 h. 49 m. Probably it was by such a process as Camus's, that M. Passeman took 20 years to calculate trains of wheel-work for the solar system, which could now be done in as many days by the application of continuous fractions (or doctrine of exhaustions). Benjamin Martin of London introduced the method of wheel-work calculations, by continuous fractions, in 1771, and about the same period it was successfully carried out by Antide Janvier of Paris, in the computation of the wheel-work of his great Planetarium. Perhaps no one has done so much in applying continuous fractions to wheel-work, as our late friend the Rev. Dr. William Pearson, who, between the years 1800 and 1820, made a great many calculations by this new method for his large Orrery—Planetarium—and Satellite machines.

Referring to Camus's problem, viz. given a *prime mover* of 12 hours; query, the wheels and pinions necessary to produce a motion of 365 d. 5 h. 49 m.? after a lengthened and very tedious process, he arrives at the solid fraction of $\frac{196}{143175}$, which he breaks down into wheels and pinions, thus:—wheels 83, 69, 25, and pinions 7, 7, 4, that is, $83 \times 69 \times 25 = 143175$, and $7 \times 7 \times 4 = 196$. Instead of a mover of 12 hours as in clocks, let us suppose one of 24 hours, as used in orreries, and the solid fraction will become $\frac{392}{143175}$, and the wheels and pinions as follow: —83, 69, 25 for the wheels, and 7, 7, 8 for the pinions; that is, suppose a pinion of 8 leaves turns round in 24 hours and drives a wheel of 25 teeth, which has on its axis a pinion of 7, which drives a wheel having 69 teeth, on whose axis is a pinion of 7 leaves, which turns wheel 83 once round its axis in 365 d. 5 h. 48 m. 58·77 s., a very close approximation to the period proposed:—the solid fraction $\frac{392}{143175} = \frac{8}{25} \times \frac{7}{69} \times \frac{7}{83}$, the exact numbers used by Ferguson for the annual train in his best orreries and astronomical clocks.

We shall now give an example of continuous fractions as applied to finding out a ratio for a period of 365 d. 5 h. 48 m. 55 s., being the mean length of a year as estimated in Ferguson's time. Before going into this problem, it will be necessary to give the rule by which the process is to be conducted, thus:—Any ratio between two given numbers being proposed, consisting of many places of figures, to find a set of integral numbers that shall be the nearest that given ratio.——RULE. —Divide the consequent by the antecedent, and the divisor by the first remainder, and the last divisor by the last remainder, till nothing shall remain, in the same way as finding the greatest common measure for any two numbers; then,

this wheel are four short pillars, whose upper ends are fixed into the lower plate of a moveable frame *a a a a a a a* (Fig. 1), that

for the terms of the first ratio, unity, will always be the first antecedent, and the first quotient the first consequent; for the second ratio, multiply the first antecedent and consequent by the second quotient, and to the product of the antecedent add 0, and to the consequent, add 1, and the sum shall be the terms of the second ratio. For the following ratios, multiply the last antecedent and consequent by the next quotient, and to the product, add the last antecedent and consequent but one, and the sum shall be the present antecedent and consequent. It may be remarked, that if the terms of the given ratio are not *prime to each other*, they must be made so before the proposed operation is begun.

Attending to these rules, we shall proceed to the solution of the question of the period proposed. In the first place, the period of 365 d. 5 h. 48 m. 55 s. must be thrown into a decimal value thus:—5 h. 48 m. 55 s. = 2423 of a day; consequently, 365·2423 days is the period decimally expressed, and as the axis which is assumed to turn once round on its axis in 24 hours is the prime unit from which the annual motion is to be derived, it must necessarily be expressed by *a unit*, and as many cyphers as there are figures in the expression of the fraction of a day, which in the present case are four, then 1·0000 is the antecedent, and 365·2423 the consequent; therefore, the solid fraction becomes $\frac{1 \cdot 0000}{365 \cdot 2423}$, and as they are prime to each other, the operation may be proceeded with.

Operation :—	Therefore		Ratio.
1·0000)365·2423(365.4.7.1.6.1.1 quotients, 365·0000	$0 \times 365 + \frac{1}{0}$	$= \frac{1}{365}$	1st.
2423)10000(4 9692	$\frac{1}{365} \times 4 + \frac{0}{1}$	$= \frac{4}{1461}$	2d.
308)2423(7 2156	$\frac{4}{1461} \times 7 + \frac{1}{365}$	$= \frac{29}{10592}$	3d.
267)308(1 267	$\frac{29}{10592} \times 1 + \frac{4}{1461}$	$= \frac{33}{12053}$	4th.
41)267(6 246	$\frac{33}{12053} \times 6 + \frac{29}{10592}$	$= \frac{227}{82910}$	5th.
21)41(1 21	$\frac{227}{82910} \times 1 + \frac{33}{12053}$	$= \frac{260}{94963}$	6th.
20)21(1 20 1	$\frac{260}{94963} \times 1 + \frac{227}{82910}$	$= \frac{487}{177873}$	7th.

The last ratio $\frac{487}{177873}$ comes very near upon the period proposed, for 177873 ÷ 487 = 365·2422998 = 365 d. 5 h. 48 m. 54 s. 42 thds., being only 18 thirds, *minus* the period required; but such a fraction becomes useless in practice, as 487 is a prime number, and therefore cannot be reduced. The ratio next the last one is $\frac{260}{94963}$, and 94963 ÷ 260 = 365·2423077, or 365 d. 5 h. 48 m. 55 s. 23 thds., a period making an yearly error of 23 thirds too slow; but as this solid fraction is not *prime*, it is capable of being reduced into fractions suitable for a train of wheels, and becomes a very desirable ratio;—thus,—

11)94963
89)8633
97
hence 11 × 89 × 97 = 94963, and

2)260
2)130
5)65
13
or 2 × 10 × 13

This fractional ratio will therefore be $\frac{2}{11} \times \frac{10}{89} \times \frac{13}{97}$, but as the first part of the fraction $\frac{2}{11}$ cannot be used, it must be multiplied by some number so as to raise it. Suppose we multiply it by 5, then $\frac{2}{11} \times 5 = \frac{10}{55}$, which is precisely of the same value as $\frac{2}{11}$; the working or practical fraction therefore becomes $\frac{10}{55} \times \frac{10}{87} \times \frac{13}{97} = \frac{260}{94963}$; Therefore, let a pinion of 10 revolve in 24 hours and turn a wheel of 55 teeth, on whose axis is another pinion of 10, which turns wheel 87, which has a pinion of

turns round on a fixed upright pin in the centre of the plate A A A A, and contains the above mentioned wheels belonging to the Earth and Moon, so that the whole frame goes round the centre pin in the same time with the wheel Z.

This last wheel cannot be seen in Fig. 2, because it lies within the thick wheel G, which is only a thick ring, having 160 teeth on its outside. Its innermost side is represented by a dotted circle in Fig. 1, and it is kept in its place by three rollers marked * * * which turn upon pins fixed in the great immovable plate A A A A.

As the uppermost edge of the contrate wheel F (Fig. 2) must come a little way through the plate A A A A, in order to turn the ring wheel G that lies on the upper side of this plate, and as this wheel turns the wheel H of 32 teeth that belongs to the Earth's diurnal motion, it is plain, that as the wheel H must go round G, in a year, by the annual part of the work, G must be thick enough to turn H at such a distance from or above the

13 that turns 97 once round in the period of 365 d. 5 h. 48 m. 55 s. 23 thds. (See the annexed section of the wheel-work). Had Ferguson been acquainted with this

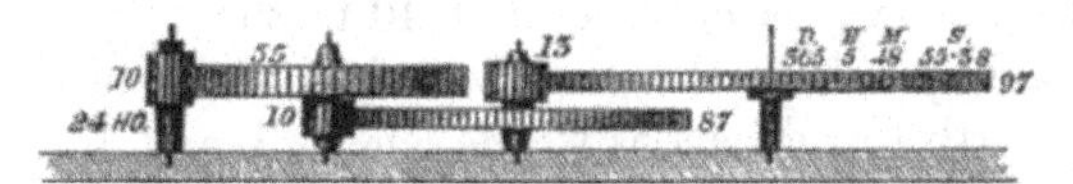

method of finding trains to produce accurate periods, he would undoubtedly have had recourse to it. Instead of always keeping fast to the train of $\frac{8}{25}\times\frac{7}{69}\times\frac{7}{83}$ producing 365 d. 5 h. 48 m. 58·77 s., being an yearly error of 3·77 s. slow, he would have discovered our numbers $\frac{10}{55}\times\frac{10}{87}\times\frac{13}{97}$ producing 365 d. 5 h. 48 m. 55·38 s., which is only 23 thirds too slow,—a period much nearer than the 3·77 seconds per tropical year.

Ferguson's numbers, $\frac{392}{143175}$ is not to be found in the above ratio in their natural fluxion, but nevertheless they may be produced by an ingrafted process, as follows:—Take the 4th ratio $\frac{33}{12053}$ and multiply it by 11 (instead of 6), and then add to it the 3d ratio thus:—

$$\frac{33}{12053}\times 11+\frac{29}{10592}=\frac{392}{143175}=\text{Ferguson's fraction.}$$

But as Ferguson appears to have known nothing of this method when he made his calculation, his fraction could not be produced by such a method. We are inclined to think that he arrived at the solid fraction $\frac{392}{143175}$ by the same process we have given in our last note, viz. by determining on the number of pinions and the leaves each should have, so that their united value, employed as a multiplier on the year period, might produce a *whole number*, or one very near to it; after a great many trials with pinions of 6, 7, and 8, diversified in every manner of way, he would discover that the only arrangement with three sets of pinions multiplied into each other, and again used as a multiplier for the period of the year, were pinions of 8.7.7. thus:—$8\times 7\times 7=392$ and $365{\cdot}2423\times 392=143174{\cdot}9816$, a decimal, nearly equal to 143175, a number capable of being reduced into $25\times 69\times 83$. Ferguson therefore adopted the $\frac{392}{143175}=\frac{8}{25}\times\frac{7}{69}\times\frac{7}{83}$ as the most accurate ratio that he could find for producing the period of his solar year.—We have thus been particular with Ferguson's period of a solar year in order to show how he probably found his ratio, and also to show how such calculations are performed with greater ease and accuracy.

plate A A A A, that wheel H may go over the top of F without touching it; otherwise, when H came round to F, it could not pass by, and would stop the annual motion.

In the centre, just above the upper surface of the moveable frame, plate *a a a a*, is a fixed wheel *b* of 40 teeth, 2·95 inches in diameter, taking into the teeth of the wheel *c*, which has also the same number of teeth and diameter; and these take into the teeth of a third similar wheel, *d*. The axis of this last wheel is hollow, and the top of it is fixed tight at K (see Fig. 2) in the piece K L M that carries the Earth. This part of the work keeps the Earth's inclined axis in a constant parallelism in its annual course round the sun; for as *d* is connected with the fixed wheel *b* by means of the intermediate wheel *c*, and *c* rolls or goes round *b* by the annual work, and as *b*, *c*, and *d* have equal numbers of teeth, *d* must always preserve its parallelism throughout its annual motion.[134] The axis of *b* is fixed into the immoveable plate A A A A, and it is hollow, to let the axis of some wheels below that plate turn within it.

The solid spindle or axis of the wheel I of 32 teeth, turns within the hollow axis of wheel *d* of 40 teeth; and on the top of this solid spindle is the small wheel K of 12 teeth, which turns the earth round its axis by the wheels L and M, of equal number and size with K, as already mentioned.

The hollow axis of the parallelism wheel *d* is within an upright socket, whose lowermost end is fixed into the top plate (marked 56 in Fig. 2) of the moveable frame *a a a a*, and on the top of this socket is fixed a small wheel *e* of 16 teeth, which take into the teeth of another wheel *f* of the same number and size, on the axis of which is a long pinion *g* of 16 leaves, which take into the wheel *h* of 16 teeth, whose axis is hollow, and has

134 The earth's axis does not maintain a constant parallelism. The axis of the earth has a kind of conical motion round the poles of the ecliptic, and at the rate it is moving, it will accomplish a circuit round the poles in 25868 sidereal years. To secure such a motion in an orrery which we made in 1828, instead of "*a fixed wheel of* 40, *d*." we made this wheel to have a slow motion—caused it to accomplish a turn round its axis in 25868 sidereal years, and in doing so, communicated the same motion to the other parallelism wheels, and thus caused the axis of the earth to make a revolution round the poles of the ecliptic of the orrery in 25868 of its sidereal years. The second parallelism, wheel *c*, of 40, although it rolls round in the teeth of the fixed wheel of 40 in the centre, it makes two rotations on its axis in every revolution of the annual frame within which it moves, and thus so operates on the third parallelism wheel *d*, of 40, as to make it turn *no way at all;* any given point on the surface of this wheel *d* always keeps directly in a line with some fixed point at a great distance. In nature, the parallelism is not a constant, but a variable quantity, moving at the rate of 50·1 seconds annually, or through all the signs and degrees of the ecliptic in 25868 sidereal years, as stated.

a black cap on the top of it covering just one half of the Moon. Now, as the socket, on whose top the wheel *e* is placed, is fixed into the annual moving frame, it is plain, that which ever side or tooth of the wheel *e* is once toward the Sun will always be so; and therefore, as the wheel *f*, the pinion *g*, and the wheel *h*, go all round the wheel *e* by the work that carries the Moon round the Earth, and all these have equal numbers of teeth, the wheel *h* will always keep the front part of the Moon's cap facing towards the Sun, and show her to be always full as seen from the Sun, but continually changing her phases as seen from the Earth in her going round it, for when the Moon is between the Earth and the Sun (as represented in Fig. 2), her cap will hide the whole of her from the Earth, but when she is opposite to the Sun, all the half or side of her next the Earth will then appear like a full Moon before the circular edge of the cap, and when she is mid-way between these positions, or in either of her quadratures, she will appear just half enlightened as seen from the Earth.

The axis of the wheel Q of 63 teeth, which carries the Moon round the Earth, is hollow, and turns round upon the before-mentioned fixed socket; to the top of this axis (just under the wheel *e* of 16 teeth, in Fig. 2) is fixed the bar *i*, *f*, which carries the Moon round the Earth by the motion of the wheel Q.

On the top of the axis of the wheel *c* of 40 teeth, is a wheel *k* of 59 teeth, which turns a wheel *l* of 56 teeth, which causes the nodes of the Moon's orbit to go once round, with a retrograde motion, through all the signs and degrees of the ecliptic in $18\frac{2}{3}$ years.[135] The axis of *l* is hollow, and turns upon the hollow

[135] The nodes of the moon's orbit perform a retrograde revolution through all the signs and degrees of the ecliptic in 6798 d. 6 h. 46 m., or in 6798·2819 days = 18 years 223 d. 22 h. 43 m. Ferguson, in all his large orreries, assumes 223 d. to be $\frac{2}{3}$ds. of a year; $\frac{223}{365}$ is far from being $\frac{2}{3}$ds. of a year—it is $\frac{22}{36}$ of a year nearly, a fraction much under $\frac{2}{3}$ds. of a year,—for $\frac{2}{3}$ds. of a year is 243·49654 d. or 243 d. 11 h. 52 m., and if the true period is subtracted from that of Ferguson's it will show an error of fully $19\frac{1}{2}$ days, plus, in the period of the revolution of the nodes, thus:—223 d. 22 h. 43 m.—243 d. 11 h. 52 m. = 19 d. 13 h. 9 m. error too slow in each nodal revolution. Ferguson having assumed $18\frac{2}{3}$ years as his period of the nodes, he would reduce his fraction thus:—$18 \times 3 + 2 = 56$, and as the nodes are retrogressive the 3 must be added to 56 for next wheel thus:—$\frac{56}{59}$ The 56 being on the middle wheel of parallelism of his orrery, takes into and drives a wheel of 59 once round, backward, in $18\frac{2}{3}$ years, or 18 y. 243 d. 11 h. 52 m., instead of 18 y. 223 d. 22 h. 43 m.

The period of the nodes derived from a solar year is found by the rule given in our last note thus:—The mean tropical period of the nodes is 6798 d. 6 h. 46 m. = 6798·2819 days, and 365·2423 that of the year, which, when the two periods are treated, as shewn in our last, give the quotients 18.1.1.1.1.6, &c., and as per rule, work as follows:—

axis of Q, and on the axis of l is a circular plate m (see Fig. 2) fixed obliquely on that axis, and parallel to the Moon's orbit. The work that carries the Moon round the Earth carries also the piece g round upon this oblique plate, and as the lower end of the Moon's axis (which turns within the hollow axis of her cap) is fixed into the piece g, it causes the Moon to rise and fall in her oblique orbit according to her north or south latitude or declination from the ecliptic. As the nodes of her orbit are even with the plane of the ecliptic, one half of her orbit is on the north side, and the other half on the south side of the ecliptic.

On the axis of the wheel X, which has 69 teeth, is a pinion n of 10 leaves $\frac{56}{100}$ inch in diameter, turning a wheel o of 73 teeth 5·82 inches in diameter, which carries Venus round about the Sun in 224 days 17 hours.[136] The axis of the wheel o is

$0 \times 18 + \frac{1}{0} = \frac{1}{18}$ 1st ratio.
$\frac{1}{18} \times 1 + \frac{0}{1} = \frac{1}{19}$ 2d ,,
$\frac{1}{19} \times 1 + \frac{1}{18} = \frac{2}{37}$ 3d ,,
$\frac{2}{37} \times 1 + \frac{1}{19} = \frac{3}{56}$ 4th ,,
$\frac{3}{56} \times 1 + \frac{2}{37} = \frac{5}{93}$ 5th ,,
$\frac{5}{93} \times 6 + \frac{3}{56} = \frac{33}{614}$ 6th ,,

In this case the antecedents and consequents, or numerators and denominators, must be added together to produce the second node wheel; in the 4th ratio will be found Ferguson's numbers $56 + 3 = 59$, and thus is produced his fraction of $\frac{56}{59} = 18\frac{2}{3}$ years. Had Ferguson adopted $\frac{3}{5}$ths. instead of $\frac{2}{3}$ds. of a year, he would have come within 5 days of the truth, instead of $19\frac{1}{2}$ days. The 5th ratio $\frac{5}{93}$ or $93 + 5 = 98$, $\frac{93}{98}$ for the wheels; if a wheel of 93 be used in place of Ferguson's 56, and a wheel of 98 instead of his 59, such a train will produce a period of $18\frac{3}{5}$ years $=$ 18 years 219 d. 3 h. 29 m., a period 4 d. 18 h. 35 m. minus the true period. The 6th or last ratio of $614 + 33 = 647$, and $\frac{614}{647}$, requires compound wheel-work, and as 647 is a prime number, and therefore not reducible, recourse must be had to higher ratios.

136 The true mean tropical revolution of Venus round the Sun is performed in 224 d. 16 h. 41 m. 31 s. Ferguson here gives 224 d. 17 h. as the nett produce of his Venus train, whereas it ought to be 224 d. 20 h. 47 m. 8 s., if we take the wheels he gives for this train, and work them out, they produce that period. Thus, the value of wheel 69, of his annual train, is 30·8035 d. (see last note on the Sun) $=$ 30 d. 19 h. 17 m. 2 s.; therefore, as Ferguson uses $\frac{10}{73}$ for his approximate value, we have the question, If 10 : 30·8035 : : 73? thus, $308035 \times 73 = 22486555 \div 10 = 224{\cdot}86555 =$ 224 d. 20 h. 47 m.; or, $\frac{8}{25} \times \frac{7}{69} \times \frac{10}{73} = \frac{560}{125925}$, and $125925 \div 560 = 224{\cdot}8660714$ d. or 224 d. 20 h. 47 m. 8 s., and not 224 d. 17 h. as given by Ferguson. On referring to Ferguson's Astronomy, to his "Tables and Tracts," and to his "Select Mechanical Exercises," we find the same 224 d. 17 h. written down, instead of 224 d. 20 h. 47 m. 8 s.—the real wheel-work period. This is singular, and suggests that his pen had made a slip when first working out the result of his period for Venus, and that he had never afterwards tested it.

To the orrery maker it therefore becomes necessary to have a nearer ratio for his wheel-work than $\frac{10}{73}$. As by former rule, take 30·80357 and 224·6955 (the period of Venus decimally expressed) and we have the fraction $\frac{3080357}{22469550}$, which, when reduced, according to the rule formerly given, produces the following ratios:— $\frac{1}{7}$, $\frac{3}{22}$, $\frac{7}{51}$, $\frac{10}{73}$, $\frac{17}{124}$. The *fourth ratio* gives Ferguson's numbers. If we adopt $\frac{17}{124}$ instead of them, we shall obtain a period of 224·68486 days $=$ 224 d. 16 h. 26 m. 12 s., which is within 16 minutes of the true mean period; whereas, $\frac{10}{73}$

hollow (because another axis turns within it), and on the top of it is fixed the lower plate of the frame *pppp* which carries Venus round the Sun, and has wheels within it belonging to Venus and Mercury.

Under the lowest plate of this frame is a fixed wheel *q* of 74 teeth 6·12 inches in diameter, being of the same diameter as wheel Z of 83 teeth, which gives the Earth its annual motion, so that in Fig. 1, one and the same circle represents both these wheels. A pinion *r* of 8 leaves $\frac{66}{100}$ inch in diameter takes into the teeth of the fixed wheel *q* of 74 teeth, and is carried round it by the motion of the frame *pppp* that carries Venus round the Sun; consequently, in the time this pinion is carried round the wheel, it will turn $9\frac{1}{4}$ times round its axis, equal to the number of Venus's days and nights, in the time she goes round the Sun.[137]

The wheel *q* of 74 teeth is fixed on the same (before-mentioned) socket, on which the wheel *b* of 40 teeth is fixed. The top of this socket goes through the lower plate of the frame *pppp*, and a wheel *s* of 28 teeth, 1·74 inch in diameter, is fixed upon the top of this socket, just above the same plate. Another wheel *t* of 28 teeth takes into the teeth of *s*, and is carried round it by the motion of the frame, and a third wheel *u* of 28 teeth (which is also carried round by the frame) takes into the teeth of *t;* the axis of *u* is hollow, it turns upon the solid spindle or axis of the pinion *r* of 8 leaves, and on its top is fixed the curved piece *v* (see Fig. 2) that carries Venus on her inclined axis, which, by means of the three last-mentioned wheels of 28 teeth, is kept in a constant parallelism in going round the Sun.[138]

gives an error of 4 h. 20 m. 56 s. Therefore, instead of a pinion of 10 on the shaft of wheel 69, have a pinion of 17 leaves, and let it drive a wheel of 124 teeth, and this will produce the period as noted ($308035 \times 124 = 381964268 \div 17 =$ 224·68486 as given).

137 During the latter half of last century, Bianchini's period of Venus's diurnal motion appears to have been universally adopted—viz. $9\frac{1}{4}$ days in its year, each of its days being equal to $24\frac{1}{3}$ of ours. It has been found, however, that Bianchini was wrong in his measure. Venus's rotation is performed (according to CASSINI) in 23 hours 20 minutes 55 seconds.

138 These three wheels of 28 teeth are subject to the same laws and give the same results as the three wheels of 40 teeth each, which preserve the parallelism of the earth's axis—viz. the first wheel of 28 teeth being *fixed*, and the other two so connected as to roll round it, produces the following results; the second wheel of 28 teeth, rolling round in the teeth of the fixed wheel of 28, performs *two complete rotations on its axis*, while the third wheel of 28, driven by it, moves "*no way at all;*" however paradoxical this may seem, it is nevertheless strictly

On the top of the axis of the pinion r of 8 leaves, and just above the curved piece v (Fig. 2), is a small wheel ω of 12 teeth, which turns another wheel x of the same number and size, and this last wheel turns a third wheel y of the same number, which is fixed on the axis of Venus, and turns her $9\frac{1}{4}$ times round her axis in the time she goes round the Sun, which is just as often as the pinion r turns round in the time it is carried round the fixed wheel q of 74 teeth.[139]

On the top of the axis of the middle wheel t of 28 teeth, is another wheel of the same number and size, which turns a wheel β of 18 teeth, 1·12 inch in diameter, and this wheel turns another wheel δ of the same number and size, whose axis is a hollow socket, on which a bar γ (Fig. 2) is fixed, and this bar carries Mercury round the Sun in 87 days 23 hours.[140]

On the axis of the wheel X (already mentioned) of 69 teeth is a wheel ε of 78 teeth, which turns a wheel λ of 64 teeth once round its axis in 25 days 6 hours. The axis of this wheel turns within the before-mentioned hollow arbours in the centre, and on its top is the small wheel π of 12 teeth, which turns another wheel ξ of the same number and size; this last wheel

correct. The reader will find this point ably discussed by Ferguson in his "*Select Mechanical Exercises*," article "Mechanical Paradox," pages 44—71. (See also Ferguson's Mechanical Paradox in the present work.

[139] Ferguson adopts Bianchini's period of rotation, and it is now universally admitted by astronomers, that "*Bianchini greatly erred.*" The axis of Venus is nearly perpendicular to the plane of the ecliptic, and she performs her rotation in about 23 h. 20 m. 55 s., and not on an axis inclined 75° to the ecliptic, and in $24\frac{1}{3}$ days as given out by Bianchini. (See note 137).

[140] The period of Mercury here given, "87 days 23 hours," is not the real period produced by the wheel-work. It will presently be shown that Ferguson's wheel-work for the period of Mercury produces a period of 87 d. 23 h. 47 m. 8 s. Ferguson here uses one of the parallelism wheels of Venus as the unit of his motion—viz. he has on the axis of the middle wheel of parallelism 28, another wheel of 28 teeth which drives a wheel of 18 teeth round in 87 d. 23 h. 47 m. 8 s. It must be recollected that the middle wheel of parallelism, in rolling round in the teeth of the fixed wheel 28 on the centre, *turns twice round* on its axis, and that circumstance alters the value of the prime fraction of its period; $\frac{18}{28}$, the fraction here used, is $28 \div 18 = 1·55555$, &c., which of course is not the relative ratio of 87 d. 23 h. 47 m. &c., to 224 d. 20 h. 47 m. &c., but as this wheel of parallelism turns *twice* on its axis in one revolution of Venus, the fraction $\frac{18}{28}$ must be treated as follow:—viz. $28 + 18 = 46 \div 18 = 2·55555$, which is the number of times that Mercury's period is contained in that of Venus,—thus—$\frac{224·86606}{2·55555} = 87$ d. 23 h. 57 m., &c., which is a period too much by about 12 minutes. In reducing the ratios for this train to a more accurate period, we find the ratios of $\frac{1}{3}$, $\frac{2}{3}$, $\frac{9}{14}$, &c. a near approach to Mercury's period, and may be obtained by multiplying the last ratio by 7, and adding to it the preceding one, $\frac{9}{14} \times 7 + \frac{2}{3} = \frac{65}{101}$; the $\frac{9}{14}$ is exactly the half of Ferguson's fraction $\frac{18}{28}$ If we adopt the corrected train we have given of Venus—viz. 224·68486 = 224 d. 16 h. 26 m. 12 s., and instead of a wheel of 28 on the middle parallelism wheel, we have one of 101 teeth to drive a

is fixed on the Sun's axis, it turns in the piece θ (Fig. 2), and turns the Sun round his axis in 25 days 6 hours.[141]

The Sun's axis inclines $7\frac{1}{2}$ degrees from a perpendicular to the ecliptic; Venus's axis, 75 degrees; and the Earth's, $23\frac{1}{2}$.

The Earth turns round within a black cap, that always covers the half of it, which at any instant of time is turned quite away from the Sun; the edge of the cap represents the *solar horizon* or circle bounding light and darkness; it is supported by a crooked wire P whose lower end is fixed into the plate that covers the wheels, and is carried round by the annual motion-work. An index (called the *annual index*) goes round the ecliptic by the same work, keeping always opposite to the Sun, and showing the days of the months and the Sun's apparent place in the ecliptic as seen from the Earth.

On looking at Fig. 1, it may appear to many, that the wheels C, D, and E are superfluous, and that the wheel F, which gives motion to the toothed ring G, might have been upon the axis B of the handle. For as F has 32 teeth, and H, that is turned by the teeth of G (and turns the Earth round its axis), has also 32 teeth, F and H would turn round in equal times; and consequently, a turn of the handle would have answered to a turn of the Earth on its axis. This indeed would have been the case if the Earth had no annual motion, but as H goes round G in a year, in the same way that G turns round, H loses five turns in

wheel of 65, then this wheel of 65 will revolve in 87·97849 days, or 87 d. 23 h. 29 m. 1 s., which is a period of only 7 minutes fast of the true mean revolution of Mercury, whose period is 87 d. 23 h. 36 m. See also calculation annexed to the tabular view of the wheel-work of this orrery at the conclusion of the description.

141 The rotation of the Sun is derived from the annual train of the Earth by the addition of two wheels to it,—viz. a wheel of 78 and one of 64; hence, the train for the Sun's rotation becomes $\frac{8}{25} \times \frac{7}{69} \times \frac{78}{64} = \frac{4368}{110400}$, which, when reduced, is 25·274725 days = 25 d. 6 h. 35 m. 36 s., as given under "*the Tabular view of the Wheel-work*" at the conclusion of the description of this orrery. Ferguson, it will be observed, gives the rotation by his wheel-work as "25 *days* 6 *hours*," and has omitted the 35 m. 36 s. which his wheels really produce. The true mean solar rotation is done in 25 d. 10 h. 0 m. 1 s., or 25·4168 days nearly. To obtain this period we may take the value of $\frac{8}{25} \times \frac{7}{69}$ the first part of the train, and belonging to the annual wheels $\frac{8}{25} \times \frac{7}{69} = \frac{56}{1725}$, and 1725 ÷ 56 = 30·8035 days = the value of rotation of wheel 69. Therefore, as the solar rotation is performed in 25·4168 days, their relative values become as the fraction $\frac{25}{30}\!:\!\frac{4168}{8035}$, and this reduced according to the rule already given, leaves the following ratios—viz. $\frac{1}{1}$, $\frac{4}{5}$, $\frac{5}{6}$, $\frac{14}{17}$, $\frac{19}{23}$, $\frac{33}{40}$, $\frac{151}{183}$, &c.; a rotation very near the true period may be had by multiplying the last ratio but one by 1, and adding the preceding one to it,—thus, $\frac{33}{40} \times 1 \times \frac{19}{23} = \frac{52}{63}$—30·8035 × 52 = 16017820 ÷ 63 = 25·4251 days, or 25 d. 10 h. 12 m. 8 s. Therefore, instead of making a wheel of 78 to drive one of 64, cause a wheel of 63 teeth to turn a wheel of 52 teeth, which will produce a period within 13 minutes of the true value, 25 d. 6 h. 35 m. 36 s.; the period of last century is 3 h. 24 m. 24 s. minus the modern value given to the rotation.

going round G (for 5 times 32 is 160, the number of teeth in G), and then the handle would have turned 370 times round in the time the Earth made 365 rotations. To prevent this, and so make the turns of the Earth and handle agree together, C has 74 teeth and E only 73, so that the wheel E will turn five times oftener round than the handle does in 365 turns thereof; and consequently, make the Earth's daily rotation equal to a turn of the handle, or to 24 hours mean solar time.[142] (See Fer-

[142] Ferguson describes this motion correctly. It is evident that the wheel H in going round the ring-wheel G, in the same direction as that wheel turns, must lose 5 rotations on its axis, and the handle would in that case have had to make 370 turns in 365 days, and the wheels 74 and 73 he employs make the turns of the handle and the axis of the earth to agree; for 73 is $\frac{1}{5}$ of 365, and 74 is $\frac{1}{5}$ of 370; and hence, 74 : 370 :: 73 ?—$370 \times 73 = 27010 \div 74 = 365$ = the rectification sought.

As there are several errors in the "*periods*" of Ferguson's orrery in the Edinburgh Encyclopædia (vide note 87, p. 52), the following corrected table of the periods is given:—

TABULAR VIEW OF THE MOTIONS, WHEEL-WORK, AND PERIODS, OF FERGUSON'S ORRERY.

Motions.	Wheel-work.	Periods.
Earth's Diurnal motion,	$\frac{25}{8} \times \frac{69}{7} \times \frac{83}{7}$	365 d. 5 h. 48 m. 58·77 s.
A Lunation, ...	$\frac{30}{64} \times \frac{63}{1}$	29 d. 12 h. 45 m.
Solar rotation, . . .	$\frac{25}{8} \times \frac{69}{7} \times \frac{64}{78}$	25 d. 6 h. 35 m. 36·24 s.
Revolution of Venus, .	$\frac{25}{8} \times \frac{69}{7} \times \frac{73}{10}$	224 d. 20 h. 47 m. 8·57 s.
Revolution of Mercury,	$\frac{18}{28}$ of Venus + 1	87 d. 23 h. 47 m. 8·57 s.
Rotation of Venus, .	$\frac{8}{74}$ of its Revol.	24 d. 7 h. 26 m. 10·65 s.
Moon's Node, . . .	$5\frac{5}{9}\,\frac{6}{56}$	$18\frac{2}{3}$ years.
Earth's Parallelism, .	$40 \times 40 \times 40$ years.	0 deg. 0 min. 0 sec.

The following figures exhibit a practical solution of the periods derived from the wheel-work in the preceding table:—

The Earth's diurnal motion, $\frac{25}{8} \times \frac{69}{7} \times \frac{83}{7} = \frac{143175}{392}$ and $143175 \div 392 = 365·2423469$ days; reduced, it is 365 d. 5 h. 48 m. 58·77 s. as above.

A Lunation, ... $\frac{30}{64} \times \frac{63}{1} = \frac{1890}{64}$ and $1890 \div 64 = 29·53125$ days; reduced, it is 29 d. 12 h. 45 m. 0 s. as above.

Solar Rotation, . . . $\frac{25}{8} \times \frac{69}{7} \times \frac{64}{78} = \frac{110400}{4368}$ and $110400 \div 4368 = 25·274725$ days; reduced, it is 25 d. 6 h. 35 m. 36·24 s. as above.

Revolution of Venus, . $\frac{25}{8} \times \frac{69}{7} \times \frac{73}{10} = \frac{125925}{560}$ and $125925 \div 560 = 224·8660714$ days; reduced, it is 224 d. 20 h. 47 m. 8·57 s. as above.

Revolution of Mercury, . $\frac{18}{28}$ of Venus $= 28 + 18 = 46$, and $46 \div 18 = 2·555555556$ and $224·8660714 \div 2·555555556 = 87·9910714$ = 87d. 23h. 47 min. 8·57 sec., or $224·8660714 \times 18 = 40475892852 \div 46 = 87·9910714$; reduced, = 87 d. 23 h. 47 m. 8·57 s.

Rotation of Venus, . . $\frac{8}{74}$ of a Revolution, $224·8660714 \times 8 = 1798·9285712 \div 74 = 24·3098455$ = 24 d. 7 h. 26 m. 10·65 s.

Moon's Node, ... from $59 - 56 = 3$, and therefore becomes $\frac{3}{56}$ and $56 \div 3 = 18\frac{2}{3}$ years as above.

$18·666666 \times 365·2423469 = 6817·8568986384354$d., or 6817 d. 20 h. 33 m. 56 s., or 18 y. 243 d. 11 h. 52 m.

Earth's Parallelism, . . Wheel 40 *fixed* = 0; the second wheel 40 = 2; the third wheel 40 = 0; hence, the third wheel 40 preserves its parallelism.

guson's "*Select Mechanical Exercises*," 2d Edit. pp. 73—87;—and Ferguson's "Tables and Tracts," 1st Edit. pp. 160—167;—also, Brewster's Edinburgh Encyclopædia, article, "Planetary Machines.")

The following Table exhibits the arrangement and resulting *periods* of the *new wheel-work*, calculated by us, for insuring greater accuracy in the motions of the Moon, the Sun, Mercury, Venus, and the Moon's nodes, in Ferguson's orrery.—The explanations in previous notes show how the new wheels are to be connected with the old train used by Ferguson.

Motions.	Wheel-work.	Periods.
Earth's Diurnal motion,	$\frac{25}{8}\times\frac{69}{7}\times\frac{83}{7}$	365 d. 5 h. 48 m. 58·77 s.
A Lunation, ...	$\frac{52}{9}\times\frac{46}{9}$	29 d. 12 h. 44 m. 26 s.
Solar Rotation, . . .	$\frac{25}{8}\times\frac{69}{7}\times\frac{63}{52}$	25 d. 10 h. 12 m. 8 s.
Revolution of Venus, .	$\frac{25}{8}\times\frac{69}{7}\times\frac{124}{17}$	224 d. 16 h. 26 m. 12 s.
Revolution of Mercury,	$\frac{65}{101}$ of Venus+1.	87 d. 23 h. 29 m. 1 s.
Rotation of Venus, . .	$\frac{8}{74}$ of its Revol.	24 d. 22 h. 31 m. 8 s.
Moon's Node, . .	$\frac{9}{98}, \frac{3}{98}$	$18\frac{3}{5}$ years.
Earth's Parallelism, .	40×40×40.	0 deg. 0 min. 0 sec.

According to recent observations, it has been ascertained that the mean length of a solar or tropical year consists of 365 d. 5 h. 48 m. 51·6 s., and not of 365 d. 5 h. 48 m. 55 s. as adopted during the greater part of last century—afterwards, 365 d. 5 h. 48 m. 48 or 49 s. The period 365 d. 5 h. 48 m. 51·6 s. (a mean between them) has been found to be the more correct period, and is that at present adopted. If it is required to ascertain a series of fractions for wheel-work to produce such a motion, we have to follow the rule and process given in note 133,—viz. the period of 365 d. 5 h. 48 m. 51·6 s., decimally reduced, is 365·242264 days; hence the fraction becomes $\frac{100{\cdot}0000}{365{\cdot}242264}$, and, according to said rule, as the nominator and denominator of this large fraction do not stand prime to each other, they must be made so, otherwise we would be dealing with the multiple of a prime $\frac{100{\cdot}0000}{365{\cdot}242264}\div 2\div 2\div 2=\frac{125000}{45655283}=$ in value to $\frac{1000000}{365242264}$; therefore, $\frac{125000}{45655283}$ being prime to each other, they are now to be reduced as per rule, and the fractions arising from the several quotients obtained will be $\frac{1}{365}$, $\frac{4}{1461}$, $\frac{29}{10592}$, $\frac{33}{12053}$, $\frac{161}{58804}$, $\frac{194}{70857}$, $\frac{1131}{413089}$, &c.,—the fraction $\frac{29}{10592}=$365d. 5h. 47m. 35·171s.: —$\frac{33}{12053}=$ 365 d. 5 h. 49 m. 5·43 s.—$\frac{161}{58804}=$ 365 d. 5 h. 48 m. 49·19 s.—$\frac{194}{70857}=$ 365 d. 5 h. 48 m. 51·95 s.;—$\frac{1131}{413089}=$ 365 d. 5 h. 48 m. 51·564 s. In reducing the two last solids into fractions, we get on prime numbers unsuitable for wheel-work; we shall therefore have to adopt a process discovered by us, about ten years ago; which we have used since, and obtained the greatest possible accuracy in numbers for planetary revolutions. If we, for example, take the fractional ratio $\frac{194}{70857}\times 32+\frac{29}{10592}=\frac{6237}{2278016}$, we here adopt, by random, 32 as a multiple, and take $\frac{29}{10592}$ as the additive fraction, we obtain $\frac{6237}{2278016}$, a ratio which is capable of being reduced into smaller fractions suitable for wheel-work, and at same time, produces a period so close to the tropical period of the earth, as almost to be identical with it. We believe this reducible ratio cannot be surpassed in point of accuracy, for 2278016÷6237 = 365 d. 5 h. 48 m. 51·6017 s., the true period being 365 d. 5 h. 48 m. 51·6000 s. We may here remark that $\frac{194}{70857}$ may be multiplied by any given number, and any of the ratios given may be added, and the result will be a ratio close on the original period, but prime numbers will be met with on reducing them, so high, as to render them unsuitable for wheel-work. After a great many trials with numerous multipliers and additive ratios, we have found none reducible to produce a period so near as the ratio $\frac{6237}{2278016}$,—of being reduced into fractions for wheel-work,—from

THE MOON'S ROTATION DISCUSSION, and THREE-WHEELED ORRERY.—The rotation of the Moon on its axis was a subject which was much discussed about the end of this year (1745), and it continued to be a leading topic of debate for many years afterwards, some admitting that the Moon did *rotate*, whilst the many denied its existence. This was an opportune question of discussion for Ferguson, so soon after his arrival in London, and

this ratio $\frac{6237}{2278016} = \frac{9}{140} \times \frac{21}{148} \times \frac{33}{148} =$ 365 d. 5 h. 48 m. 51·6017 s. We annex a sectional plan of the wheel-work, and shall here merely note, that in the section, pinion 9 must turn once round on its axis in 24 hours, and drive a contrate wheel of 104 teeth, with its teeth looking *downwards*, (in order that the last wheel in the train may turn round in the proper direction). Wheel

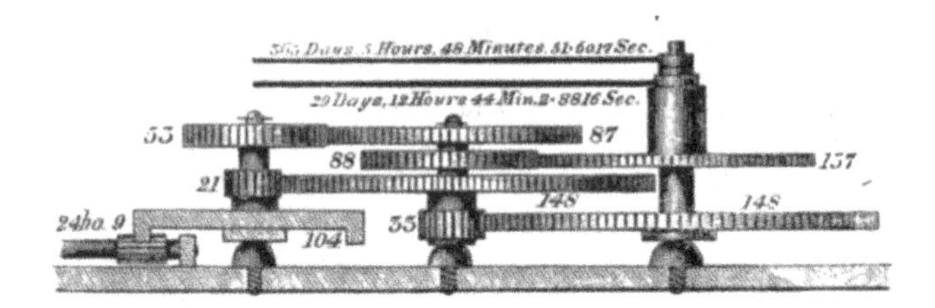

104 has above it a small wheel or pinion of 21 made fast to it, which turns a wheel of 148 teeth, and has on its axis a small wheel of 33 teeth, that turns wheel 148 once round in 365 d. 5 h. 48 m. 51·6017 s., and a bar on its axis will be carried round along with it in this period. In giving this extremely accurate ratio to the public, we are induced to append to it a train of wheels equally accurate for the period of a mean lunation, which consists of 29 d. 12 h. 44 m. 2·887 s. Immediately above the small wheel 21 (on the axis of wheel 104, of the annual train), is made fast a wheel having 53 teeth, which drives a wheel of 87 teeth, and on the *same socket*, under it, is riveted a wheel of 88 teeth ; these two wheels are here supposed to move round on the shaft of the middle wheel 148, but they may be placed on a stud by themselves in any convenient position ; but in the present case, assuming they turn round on the axis of the middle wheel of 148 teeth, it must be understood that they do so, not fixed on said axis, but both turning round together loose upon it, and have therefore no connection with the motion of the axis of said wheel 148. The last-named wheel 88 being fast on the socket of wheel 87, turns round with it in the same time, and turns a wheel of 137 teeth—(which turns *loose* on the axis of the last-mentioned wheel of 148 teeth) in 29 days 12 h. 44 m. 2·881 s.—For wheel 53 of the Moon's train is made fast to the axis of wheel 104 of the annual train, which turns on its axis in 11·5555555555 days:—the lunation, decimally expressed $=$ 29·5305889673 d.—therefore, $\frac{11\cdot5555555555}{29\cdot5305889673}$ days is the fractional expression of the ratio, and on being reduced according to rule in note 133, it will be found that $\frac{4664}{11919}$ is the nearest reducible ratio,—11·5555555555 $\times$ 11919 $=$ 137730666660045 $\div$ 4664 $=$ 29·5305889078 days, or 29 d. 12 h. 44 m. 2·88677472 s., the true mean period being 29 d. 12 h. 44 m. 2·887 s., and $\frac{4664}{11919} = \frac{88}{137} \times \frac{53}{87}$, or $\frac{53}{87} \times \frac{88}{134}$, the wheels used and as added in the sectional plan. Or the wheel-work may be demonstrated thus :—$\frac{9}{104} \times \frac{53}{87} \times \frac{88}{137} = \frac{41976}{1239576}$, and 1239576 $\div$ 41976 $=$ 29·5305889, &c. $=$ 29 d. 12 h. 44 m. 2·881, &c. s., as in former result. Thus, we have given to the modern orrery-maker numbers for a mean solar revolution of the Earth round the Sun, and a mean synodic revolution of the Moon round the Earth, so accurate, that we venture to say that it is not probable they will be surpassed.

offered an excellent opportunity to him for the display of his astronomical talents and ingenuity. Ferguson took the proper side of the question, and ably and successfully demonstrated that "*the Moon did rotate on its axis during its periodical revolution round the Earth.*" [143] To the many, however, an ocular demonstration was necessary, and Ferguson observing this, invented and constructed a simple, but very ingenious and effective machine, which he afterwards called a THREE-WHEELED ORRERY, and which satisfactorily exhibited the existence of a rotation. This machine will be found described, with an engraving, in Ferguson's "*Dissertation upon the Phenomena of the Harvest Moon,*" &c., pp. 49—72 (published in London in 1747).

We annex a cut of this machine, reduced from the engraving in said book, and shall describe its several parts and its use, as it is now scarcely known.

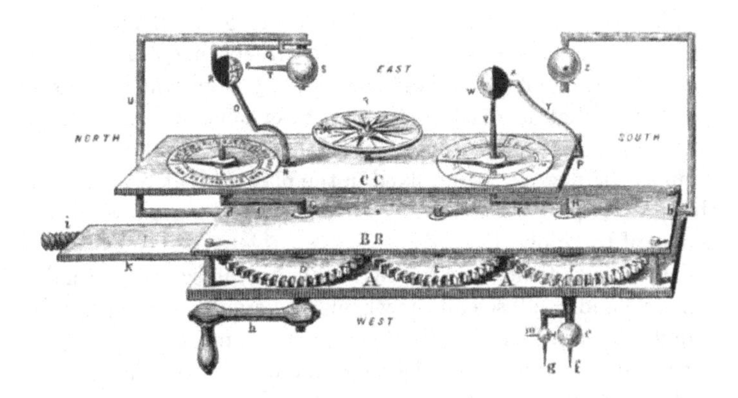

Ferguson's Three-Wheeled Orrery.

Description of "*the Three-Wheeled Orrery.*"—"A A and B B are two fixed plates kept together by pillars, as in the figure.

143 The Moon's rotation question was revived a few years ago, and was discussed by parties who did not appear to be well read on the subject of compound motion. As it would be difficult to give a mathematical demonstration of the Moon's rotation without the aid of a diagram, we shall confine ourselves to a *popular illustration* of the question which will be understood by all.

In the first place we may mention that the Moon, in its revolution round the earth, *always* keeps the *same side* or hemisphere towards our earth—it does this by virtue of its rotation.

Illustration.—Place a table in the centre of a room, and on the middle of it set a terrestrial globe—let this globe be the representative of our earth. Next,

D, E, and F are three wheels of equal numbers of teeth, biting into one another; so that if any one of them be turned, the other two turn equally round with it, D and F turning the same way, and E the contrary. On the axis of D and of F (above the plate B B), at G and H, is fixed the crank I, and the crank K, whose perpendicular shanks come through the plate C C at L and M, within which plate they easily turn. The bar *k* has one of its ends fixed into the plate A A, and has a screw, as *i*, on its other end, which, being screwed into a table or wall, will keep the machine from shaking, while the moveable parts of it are wrought by the handle *h*. In the plate B B at *b* is fixed the wire *a*, having the globe Z on its other end hanging over the centre of the wheel F. On the shank V, as an axis, is fixed the globe W and the index M, and on the other shank the index L is fixed. The crooked wire Y has its lower end fixed in the plate C C at *p*, and its upper end into the cap X, within which the ball W turns freely round. In the other end of the plate B B is fixed the wire U at *d*, whereon the globe S hangs over the centre of the wheel G, so as it may freely turn round on the wire. To the globe S is fixed the piece Q, from which comes a small wire that is fixed in the cap R, of such a length as to allow the cap to turn freely round the globe P, whose axis O inclines 23½ degrees, being fixed in the plate C C at N; and from the globe S proceeds a wire T which always points toward the centre of the globe P.

Having fixed the machine to a table by the screw *i*, turn the handle *h* and you'll see all the wheels and cranks turning equally round with the handle. By the cranks, the moveable plate C C is carried over and over the fixed plates, in a parallel motion, without turning round any axis whatsoever; for if a seaman's compass, as *q*, be hung on this plate, the same points thereof

suppose your head to represent the moon, then walk slowly round the table, keeping your face all the while directed to the globe, (as the Moon does to the Earth), and you will find that during your circuit round the globe, with your face always directed to it, *all the sides of the room come successively and directly before your face.* Now stand still, and make an effort to bring in succession, all the four sides of the room directly before your face, as they came when walking round the globe, and you will find that there is only *one way* in which it can be done,—viz. turning round upon yourself, that is, by performing a motion of *rotation.* In walking round the said globe as directed, *every side* of the room came into *view*, and *directly before your face.* Standing still, and rotating upon yourself, also brings *all the sides of the room* into *view*, and *directly before your face,*—the effects being similar, they must arise from the same cause,—viz. in walking round the globe you performed a rotation exactly as is done by the Moon during the period of her revolution round the Earth.

will constantly look to the same points of the horizon, and will always stand perpendicularly over the same parts of this moving plate which never changes its parallelism from the fixed plates; but the shank or axis V turns the index M quite round a graduated circle on the plate C C, marked 1, 2, 3, 4, in every revolution of the handle. And at the same time the globe W, on which this axis is carried round about Z, to which it still keeps the same side, but shows itself quite round to all the points of the horizon, and turns round within the cap X, which still faces towards one and the same point thereof; because it is fixed by the crooked wire Y into the plate C C. In one revolution W will see Z all round, because Z turns not; but if the cap X be taken away, and these globes connected by a wire, they will turn equally round their axes, and will constantly keep the same sides to one another; in which case, either viewed from the other, would appear to be at rest on its own axis; but viewed by an eye at rest, and at a distance from them both, they turn equally round their axes.

Now as the index M and the globe W are both fixed on the shank or axis V they must turn equally with it; and as in every revolution of the handle, they both show themselves round to all the fixed points of the horizon, the index turning round a circle which keeps parallel thereto, and the globe turning round within a cap which still looks to one point thereof. It is plain that both the globe W and the index Y do each turn round an axis or mathematical line within itself, as the shank V necessarily does; and this shank might be fixed on any place of the crank K (provided the other shank was fixed on the like place of its crank, without which the machine would not perform), for it would still turn the index and globe round the same way as it now does by being fixed to the end thereof whereon it stands. Therefore, while the whole crank is carried round the axis H of the wheel F, every atom of the crank must turn round an axis or mathematical line, within itself; and the whole being carried round through different points of space, in the time that every part or atom turns round its own axis, their cohesion cannot be destroyed, nor the crank reduced to dust." (Ferguson's Essay on the Moon's turning round her own Axis, pp. 65, 67, and 70). At page 70 of the same work, Ferguson says, "A few wheels, by a multiplied motion, from the axis of the index L to the Earth's axis, would give the Earth its diurnal motion, which any skilful mechanic might easily contrive." Ferguson kept this hint in mind, for shortly afterwards he calculated and arranged the "*few wheels*" necessary, and introduced them into a very simple machine which showed the Earth's rotation. Vide "THE CRANK ORRERY," pp. 127—130.

1746.

PROJECTION OF THE GREAT SOLAR ECLIPSE OF 14TH JULY, 1748.—Early in the year 1746, Ferguson calculated the times of beginning, middle, and end of the great eclipse of the Sun, and the digits obscured, which was predicted to happen on 14th July 1748. After having finished his calculations, he made a careful projection of the eclipse, had it engraved, and shortly afterwards published, with a description of it, engraved on the plate. This is one of those early literary productions of Ferguson's now forgotten and unknown. The following is a copy of Ferguson's advertisement of the projection of this eclipse, taken from page 2d of that now *very scarce* Tract of his, the *first* of his published works, entitled, "The Use of a New Orrery," published in London in July 1746.)

> "A Projection of the Sun's Eclipse, as it will happen on the 14th of July 1748, in the forenoon, at London, Edinburgh, and Rome, by James Ferguson. Price One Shilling."

Having published his *projection* of this eclipse early in 1746, and as the eclipse did not take place until July 1748, the projection was therefore on sale *two years* before it took place. Ferguson probably published his projection so long before the event happened for the purpose of making known to the public his powers as an original calculator, and thus to prepare the way for his lectures on this eclipse, which were delivered before considerable audiences in several places in London during the year 1748. (See date 1748—*Solar Eclipse*).

A TRACT ENTITLED "THE USE OF A NEW ORRERY," PUBLISHED BY FERGUSON.—About the month of July 1746, Ferguson published his first work, a small 12mo tract of 42 pages, entitled "*The Use of a New Orrery made and described by James Ferguson: London,* 1746." This tract by Ferguson appears to have been written and published by him for the purpose of making the public aware of the merits and accuracy of the large wooden orrery, which he had got completed in the previous year. In his preface he gives an explanation of the term orrery, and alludes to several early astronomical machines which exhibited the motions of the celestial orbs; after which, he gives a minute description of the appearance, the motions, and the general phenomena arising therefrom, in his own recently constructed orrery, and which we have described (see the orrery, date 1745). The frontispiece of this tract has a large and elegant folding engraving of the external appearance of his orrery, which is the same engraving he afterwards prefixed to the several editions of his celebrated work on Astronomy, from which we are inclined to conclude that although Ferguson did not construct the wheel-work of any two of his orreries alike, he appears to have, at least in his large orreries, preserved the same external appearance. This frontispiece (reduced) will be found at the beginning of the description of the wheel-work of this orrery, under date 1745, when it was finished. This tract is now exceedingly scarce. We only obtained our copy, after a search of several years.

THE PHENOMENA OF VENUS REPRESENTED IN AN ORRERY. —Towards the end of this year (1746) Ferguson made a delineation of the apparent path of Venus as seen from the Earth, deduced from their motions and positions during the period of their conjunctive revolution (a period of 583 days). After finishing his delineation, Ferguson wrote out a concise account of it, and presented it to a member of the Royal Society, who shortly afterwards showed the delineation at a meeting of that body, and read the paper descriptive of it. The members, after the said paper was read, applauded it, and ordered it to be recorded in their Transactions, which was duly done. The following is the title of the paper in the abridged Transactions of the Royal Society.

> "THE PHENOMENA OF VENUS AS REPRESENTED IN AN ORRERY, by James Ferguson." Vide Phil. Trans. Abrid. IX. p. 226, date, 1746.

In delineating the apparent path of Venus as seen from the Earth, Ferguson, in the 138th section of his Astronomy, mentions, that he made use of the orrery represented on the frontispiece, facing the title-page of that work; consequently, the orrery he finished in 1745 is the same one he refers to in his Astronomy, 1st Edition, 1756. In using this orrery for the delineation, he removed from it the balls representing the Sun, Mercury, Venus, the Earth, and the Moon, and put black lead pencils in their places with the points turned upward; this done, he then fixed a circular sheet of pasteboard so that the Earth kept constantly under its centre in going round the Sun, and also kept its parallelism; then by pressing gently with one hand upon the pasteboard to make it touch the three pencils, with the other hand he turned the winch that put in motion the machinery of the orrery. After having performed the operation, he inspected the tracings made on the under side of the pasteboard by the orrery pencils, and found that the Sun had described a *circle*, whilst the pencils of Mercury and Venus described looped curves, Mercury making the greatest number of loops, Venus the least, in accordance with the relative times of their revolutions. He afterwards confined his views to an exact tracing of the apparent path of Venus seen from the Earth as deduced from the orrery, and made it the subject of the paper already referred to, and which was read by one of the members of the Royal Society at one of the meetings of that body about the end of the year 1746. Of this orrery, see our *remarks* on its wheel-work under date 1745; also, Ferguson's Astronomy, Section 138; and for an engraving of the apparent paths of the Sun, Mercury, and Venus, see Plate III. Fig. 1 of same work.

THE FOUR-WHEELED ORRERY.—In the summer of this year (1746), Ferguson invented and constructed an orrery having only *four wheels*, and without either pinion or screw. As formerly noticed, this very simple orrery nevertheless exhibited the Earth's annual motion and the parallelism of its axis; the alternate succession, and different lengths of day and night throughout the year, and the beautiful change of seasons; the Sun's place in the *ecliptic;* the Sun's *right ascension, amplitude,* and *declination;* as also *solar* and *sidereal* time; the *periodic, synodic,* and *diurnal* motions of the Moon; its age and phases; the retrograde motion of the *nodes* of its orbit; and consequently, all the eclipses of the Sun and Moon.

In his "Dissertation upon the Phenomena of the Harvest

Moon, and Description and Use of a New Four-Wheeled Orrery," Ferguson gives a large folding engraving of the exterior of this orrery, and with great minuteness describes its external parts, its motions, phenomena, &c., but says nothing about the internal mechanism by which so many astronomical particulars were exhibited; his not having done so was the cause which induced the public of that day to consider this orrery "*a mechanical puzzle and a paradox.*" Few could account for so many astronomical motions and phenomena being produced by *four-wheels only*, and that too without the aid of a pinion or a screw.

The internal mechanism of this orrery was known to but few until the year 1756, when Ferguson published his great work, "*Astronomy explained upon Sir Isaac Newton's Principles.*" In chapter XII. of this work he describes his astronomical machinery. In section 399 of this chapter he gives an account of one of his astronomical machines, which he designates THE CALCULATOR, alluding to which, he says, "*In the year* 1746 *I contrived a very simple machine, and described its performance in a small Treatise upon the Phenomena of the Harvest Moon, published in* 1747. *I improved it soon after by adding another wheel, and called it* THE CALCULATOR." [144] From this it is evident that the internal mechanism of the Calculator, with the exception of an additional wheel, was precisely the same as that of the four-wheeled orrery. Having in our possession a copy of Ferguson's "*Dissertation,*" we have looked carefully over it and compared the engraving and description of the orrery with the engravings and description of his Calculator in his Astronomy, and out of them have made a *sectional view,* which we shall presently describe. In the meantime, we give an engraving of the exterior of the four-wheeled orrery, reduced from the original in Ferguson's book, and also extracts taken out of his description of it. He says,

"In this machine, S represents the Sun, E the Earth, and M the Moon. W, a wire is fixed in the Sun and points toward the Earth's centre, but does not touch its surface, because of the graduated arc G, which is fixed in the south point of the horizon

144 The additional wheel here referred to was an *apogee wheel*, or rather pulley, introduced for the purpose of showing the mean place of the Moon's apogee in the heavens, and regarding which the reader will find details in the description of the CALCULATOR, under date 1749.

H, and must go round with the motion of the Earth, so as not to be hindered by the said wire. The horizon has all the points of the compass drawn upon it, and likewise a graduated circle showing the degrees of amplitude both of the Sun and Moon.

The Four-Wheeled Orrery.

The graduated arc has its degrees numbered from the horizon to the zenith upon one side, for showing the altitude of the Sun and Moon as they pass over the meridian; and on the other side it is graduated from above the equator to the tropics and a little beyond them, for showing the declination of the Sun and Moon. The before-mentioned wire, which, in the following parts of this description I shall call a solar-ray, shows the Sun's amplitude at rising and setting, every day of the year, as it cuts the horizon on the eastern or western side thereof, while the Earth is turned round its axis; and in passing over the graduated arc, it shows both the declination and meridian altitude of the Sun, which is different every day, as the amplitude likewise is; for the Earth's axis is inclined 23½ degrees from a line supposed to be drawn perpendicular to the ecliptic; and as the Earth is carried round about the Sun, its axis still continues parallel to itself; that is, if in any part of the Earth's annual path, a line be drawn parallel to its axis, the said axis will always keep parallel to that line.

While the Moon goes round the Earth, she rises above and falls below the plane of the ecliptic, or annual path of the Earth's centre, in her orbit, which is $5\frac{1}{3}$ degrees inclined thereto; consequently, as one half of the Moon's orbit is above the ecliptic, and the other half below it, her orbit will cut the ecliptic in two points diametrically opposite to one another, and these points are called the *Moon's nodes.* They are marked upon the inclined plane P, to which plane the Moon always keeps parallel, because the lower end of her axis lies upon it; and because of its inclination, the axis will rise and fall in the socket Q, which is fixed in the bar T, that carries the Moon round the Earth. To this bar a small index U is fixed, which constantly shows the Moon's place in her orbit, among the signs on the plate V. These signs always keep parallel to themselves, as they go round the Sun; but the inclined plane P with its nodes goes backward, so that each node recedes through all the above signs in almost 19 years. On this inclined plane there are two graduated circles, the outermost whereof shows the degrees and parts of a degree of the Moon's latitude, as it is north or south, and the innermost circle shows her distance from either of her nodes.

That node from which the Moon ascends northward above the plane of the ecliptic, is called the ascending or north node; because from it her latitude (or declination from the ecliptic) is called *north latitude ascending,* until she comes to the highest part of her orbit, or to her greatest north latitude. From thence she has *north latitude descending,* until she comes to the south node, so called, because from it the Moon descends southward below the ecliptic, until she comes to her greatest south latitude, in the lowest part of her orbit; whence she ascends in south latitude, until she comes again to the north node; and then ascends from it in north latitude, as before.

Two indexes are put upon the Earth's axis, and by the Earth's diurnal motion they are carried round the dial-plate X, which is divided into 24 hours. The shortest index shows the sidereal, and the longest, the solar time.

The axes of the before-mentioned plates turn concentrically within one another, and come from the wheel-work that is contained in the box B,[145] wherein the Sun's axis is fixed, and by that means the solar-ray W constantly points toward the Earth's centre as it is carried round the Sun.

The large plate A is immoveable, because on it is drawn the ecliptic (which is the standard of the whole instrument) with its twelve signs, each containing *thirty degrees*, within which is a circle of the *twelve months* divided into their respective days in such a manner, that the degrees of the ecliptic show the Sun's place, as they stand over these days throughout the whole year. The next or innermost circle contains 24 hours, which begin with Aries at the 9th day of March; their use is to show the Sun's right ascension in time every day of the year. All these are shown by the annual index Y fixed to the box, and so goes round the above-mentioned circles, as the box is by hand turned round upon a pin fixed in the centre of the large immoveable plate.

If we imagine this circle of signs to be raised up from the immoveable plate, as high as the centres of the Earth and Sun, it will represent the *situation* of the ecliptic, in whose plane the Earth always moves about the Sun, which remains in the centre; but will appear, as seen from the Earth, to go through all the twelve signs in a year, because he is seen by the inhabitants of the moving Earth, who cannot perceive their own motion as they are carried along therewith.

While the Earth is carried once round the Sun by hand, it shows all the beautiful variety of the four seasons, and the Moon goes 12 *times* and $\frac{1}{3}$ more round the Earth, from the Sun to the Sun again; showing her lunations and phases; also, her latitudes and distance from her nodes, as she rises above and falls below the ecliptic in her inclined orbit, whose *nodes* go backward $19\frac{1}{3}$ degrees every year; and consequently, in $18\frac{2}{3}$ years, or little more, they go once backward through all the signs and degrees of her orbit without ever disturbing the position of her signs; which, as above said, still keep their parallelism as they go round the Sun, by looking toward their corresponding signs in

[145] Ferguson, in his description of the four-wheeled orrery, does not explain the wheel-work in the inside of the box. Probably he made such simple orreries for sale, and thus kept its mechanism secret; be this as it may, we shall annex a sectional view of the contents of the box, composed by us, from the sectional view of *his* "Calculator," which, he says, it resembled, excepting in the addition of one wheel. (See article *Calculator*, date 1749).

the ecliptic.[146] And in this small circle of signs the Moon's *place* is shown at any time, as her age is on a circle of $29\frac{1}{2}$ equal parts upon the box, round the lowermost edge of her inclined plane or orbit.

On the box are several *tables* for showing the place of the *north node,* with the *moon's age, week day,* and *day of the month,* for any time past or to come. By these tables the machine may be rectified in two minutes to any given time, so as to show the days of the weeks and months whereon the *new* and *full moons,* and *eclipses* of any year, did or will happen.

The Sun is 13 inches in circumference, the Earth 9, and the Moon $2\frac{1}{2}$; the box is 18 inches long, and the large immoveable plate is 27 inches in diameter. The Earth in this machine has no diurnal motion, but such a motion may be given to it by means of a string wrapped round its axis."

One side of the Moon keeps always toward the Earth, and is engraven, to distinguish it from the other. She turns round her axis within a black cap, which shows her phases to be different every day, as seen from the Earth; but to the Sun she always appears full, and turns herself once round to him in every lunation, having only one day and one night in that time, equal in length to $29\frac{1}{2}$ natural days on our Earth." (Ferguson's Dissertation on the Harvest Moon, and Use of a New Four-Wheeled Orrery, pp. 19—22). This description is followed by 26 pages of print, showing how to rectify the orrery for exhibiting the change of the seasons—the different lengths of day and night—sidereal time—the Sun's place in the ecliptic—right ascension—declination and amplitude—the phenomena arising out of the Moon's motions—eclipses—precession of the equinoxes—the Moon's rotation, &c., much too long for insertion here.

As mentioned in our note, Ferguson does not describe the internal mechanism of his orrery-box. This omission, no doubt, was intentional; done, to prevent any one from making it, and for securing the invention entirely to himself. That the reader may be made acquainted with the arrangement of the works inside said box, we have designed a sectional plan of it, from the pulley-work of his "*Calculator,*" which, along with the

146 Instead of "in $18\frac{2}{3}$ years or *little more,*" it should have been, "in $18\frac{2}{3}$ years or *little less,*" as $18\frac{2}{3}$ years is equal to 18 years 243 days nearly; whereas the *nodal period* is one of about 18 years 224 days, or $18\frac{3}{5}$ years nearly.

short description we shall give, we trust, that its several simple parts will be readily understood, and easily made by any ingenious mechanic.

In the annexed figure, A A represents, in section, the large circular ecliptic-plate, 27 inches in diameter, and B B, that of the moveable box which contains the works. A strong wire $\frac{1}{4}$ inch in diameter, and square in the middle at *c*, is driven down through the centre of the ecliptic-plate, and down a little way into the pedestal K. Two holes $\frac{1}{4}$ inch in diameter are made

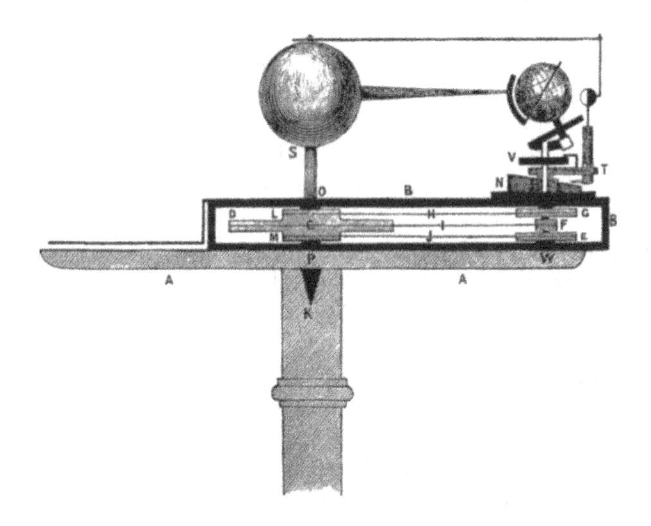

Section of the Four-Wheeled Orrery, 1746.

in the top and bottom of the box at O and P, which fits on the axis S K in the centre, and thus the box moves freely on the axis or shaft as a pivot. The pulley *c* is turned out of one piece of wood, as shown in the section, and fits tightly on the square part of the thick wire in the centre at *c;* and thus this large pulley is immoveable, and the box may be carried round the ecliptic-plate without in the slightest degree altering its position. The under part of the large fixed central pulley is grooved at M, and this groove must be sharp at the bottom and 2 inches in diameter; the central part D is also sharp grooved (as shown by the dark line across its centre), and $6\frac{18}{100}$ inches in diameter. The upper part has likewise a sharp cut groove at L, 2 inches in diameter. Thus the large fixed central pulley is actually a combination of *three pulleys*, but, being in one mass, Ferguson, singularly enough reckons it as *one wheel*, or rather as *one pulley* only, and therefore has designated it "*the Four-Wheeled Orrery.*"

Near the right hand end of the bottom of the box there rises out of it, at W, a perpendicular spindle $\frac{1}{10}$th of an inch in diameter, which rises to a point nearly touching the lower part of the oblique dial-plate under the Earth. On this *fixed* spindle the pulley E turns, having a long hollow axis which turns on the before-mentioned spindle and reaches to near its top. Near the upper end of the hollow axis of this pulley there is fixed a small ecliptic circle, and a little above it and on the top of this hollow axis is made fast the oblique-piece which carries the Earth and its sidereal dial-plate. This pulley must, like those in the centre, have a sharp-bottomed groove, and be exactly 2 inches in diameter, taken from the bottom of the groove. This pulley E being 2 inches in diameter, is exactly of the same diameter as the pulley L in the centre. These two pulleys are connected together by means of a cat-gut string J (*not crossed*); being thus connected, pulley E will maintain its parallelism when the box is carried round the large ecliptic-circle A A; and consequently, as formerly mentioned, cause the Earth to show the varied changes of the seasons.

Above pulley E there is a small pulley F having a sharp groove cut into it; the diameter, taken from the bottom of the groove, must be exactly $\frac{1}{2}$ inch in diameter; a cat-gut string I (*crossed*) connects this $\frac{1}{2}$-inch pulley with the large one C in the centre, and when the box is moved round the ecliptic-circle, the $\frac{1}{2}$-inch pulley will make $12\frac{36}{100}$ rotations, because 6·18 inches being the diameter of the bottom of the groove in the large pulley C, and ·5 of an inch that of the small pulley, the fraction thus becomes $\frac{5}{6\ 18} = 6{\cdot}18 \div {\cdot}5 = 12{\cdot}36$, and these 12·36 rotations is very nearly equal to the number of revolutions which the Moon makes round the Earth in a year. This $\frac{1}{2}$-inch pulley has a hollow axis which turns on the hollow axis of the pulley E under it, and ascends to near the under surface of the small ecliptic plate V, and carries on its top a horizontal bar T, having a short tube made fast to its extreme end; the Moon's shaft works up and down into this tube, according to its contact with the inclined orbit below, as formerly described. This horizontal bar being fixed to the top of the tube of the hollow axis of the $\frac{1}{2}$-inch pulley, must with it move round in the same time, and with it the Moon, causing it to revolve round the Earth in 29 days 12 hours 45 minutes, as indicated by the passage of the annual index over the *days* in the large ecliptic-circle, or $12\frac{36}{100}$ revolutions round the Earth in the time taken by the annual index in traversing all the signs and degrees of the ecliptic. The Moon is half covered with a black cap, which, by means of a bent wire, is connected with a swivel-pin on the top of the Sun; this pin has a small hole in it, through which the end of the long wire has a lateral motion to accommodate the variable distances of the Moon from the Sun in its revolution round the Earth; the other end of this wire being *fixed* to the back of the Moon's cap, causes the front of it to look always towards the Sun during its circuit round the Earth, thereby showing the varied phases of the Moon as seen from the Earth—and always a *full Moon* as seen from the Sun.

Above the $\frac{1}{2}$-inch pulley there is another one, G, having a sharp-bottomed groove, the diameter of which, measured from the bottom of the groove, must be $1\frac{9}{10}$ inch nearly; this pulley is, by means of an *uncrossed* cat-gut string, connected with the upper central pulley L, which, as before noted, is exactly 2 inches in diameter, as measured from the bottom of its groove. This pulley being $\frac{1}{10}$th of an inch less than that in the centre, will not keep its parallelism as the lower pulley E does, but will gain a little yearly, giving a retrograde motion to it in $18\frac{2}{3}$ years, which is the period of the *Moon's nodes.* Instead of a pulley $1\frac{9}{10}$ inch in diameter at the bottom of its groove, one of $1\frac{894}{1000}$ will give a more correct nodal revolution; but as it is impossible to make a groove to answer this fraction, the bottom of the groove ought to be made $1\frac{9}{10}$ inch in diameter, and then by trials, nicely reduced a little, by means of a triangular-shaped file. This pulley G has a hollow axis which turns on that of the Moon, and ascends about 1 inch above the surface of the box, and carries on its top the oblique orbit of the Moon N, and causes it to make a retrograde revolution through all the signs and degrees of the ecliptic in $18\frac{2}{3}$ years.

In concluding our description of the internal arrangements of this very simple orrery, we may note that the grooves in all the pulleys ought to be shaped like the letter V, sharp at the bottom, in order that the cat-gut strings may not slip during the motion of the box. That a greater degree of security may be given to these cat-gut strings to prevent them slipping, we would recommend that each of the strings, at the letters H I J, be furnished with a pulley moving on a joint, and by means of a spring, cause the pulley to press the several cat-gut strings *inwards;* this would keep the strings tight on the pulleys in all weather and under all circumstances.

As noticed at the beginning of the description, Ferguson mentions that he improved this machine soon after it was made "*by adding another wheel*" to it, and then called it "*the Calculator.*" Whether Ferguson then made an entirely new machine by adding an extra wheel to it, or simply altered the four-wheeled orrery into the Calculator, is now not known. If he altered the four-wheeled orrery into the Calculator, he would do so by placing the pulleys into a new frame and bringing them into closer contact, as shown in the pulley-work of the Calculator; and in having the works concealed from view by a circular cover-plate, moving with the annual motion. (See view of the Calculator, under date 1749). It is singular that Ferguson should call this machine a *four-wheeled orrery;* a *four-pulleyed,* or rather a *six-pulleyed orrery* would have been the more correct designation; because, the *central mass* does not act as *one,* but as *three distinct pulleys.* Ferguson calls these pulleys, *wheels.* Wheels are usually understood to have *teeth* or *cogs,* while pulleys are *wheels without teeth,* simply *grooved* or *plain* round the circumference.

1747.

"THE IMPROVED CELESTIAL GLOBE."—About the beginning of the year 1747, Ferguson improved the Celestial Globe by adding a piece of apparatus to it which enabled him to show the apparent diurnal revolutions of the Sun and Moon, with the times of their rising, culminating, and setting, the age of the Moon, &c. This improved globe was exhibited before the members of the Royal Society, and at same time, there was read a paper, written by Ferguson, descriptive of the new improvement and its use. The paper and the improved globe were approved of by the members, and the description was ordered to be printed, which was done in due course. The paper stands indexed in their Transactions as,

> "An Improvement of the Celestial Globe, by James Ferguson, read before the Royal Society, on 14th May, 1747." See Phil. Trans. Vol. 44, page 535.

We annex an engraving of the Celestial Globe, as improved by Ferguson, along with a full description of it.

Improved Celestial Globe.

"On the north pole of the axis, above the hour circle, is fixed an arc M K H of 23½ degrees, and at the end H is fixed an upright pin H G, which stands directly over the north pole of the ecliptic, and perpendicular to that part of the surface of the globe. On this pin are two moveable collets at D and H, to which are fixed the quadrantal wires N and O, having two little balls on their ends for the Sun and Moon, as in the figure. The collet D is fixed to the circular plate F, on which the 29½ days of the Moon's age are engraven, beginning just under the Sun's wire N; and as this wire is moved round the globe the plate F turns round with it. These wires are easily turned if the screw G be slackened; and when they are set to their proper places, the screw serves to fix them there, so that when the globe is turned, the wires with the Sun and Moon may go round with it; and these two little balls rise and set at the same times and on the same points of the horizon, for the day to which they are rectified, as the Sun and Moon do in the heavens.

Because the Moon keeps not in her course in the ecliptic (as the Sun appears to do), but has a declination of 5⅓ degrees on each side from it in every lunation, her ball may be screwed as many degrees to either side of the ecliptic as her latitude or declination from the ecliptic amounts to at any given time; and for this purpose, S is a small piece of pasteboard, of which the curved edge at S is to be set upon the globe at right angles to the ecliptic, and the dark line over S to stand upright upon it.

From this line, on the convex edge, are drawn 5⅓ degrees of the Moon's latitude on both sides of the ecliptic; and when this piece is set upright on the globe, its graduated edge reaches to the Moon on the wire O, by which means she is easily adjusted to her latitude, found by an ephemeris.

The horizon is supported by two semicircular arcs, because pillars would stop the progress of the balls when they go below the horizon in an oblique sphere.[147]

To rectify this Globe.—Elevate the pole to the latitude of the place; then bring the Sun's place to the ecliptic for the given day to the brass meridian, and set the hour-index to XII. at noon, that is, to the upper XII. on the hour-circle, keeping the globe in that situation; slacken the screw G and set the Sun directly over his place on the meridian, which being done, set the Moon's wire under the number that expresses her age for that day on the plate F, and she will then stand over her place in the ecliptic and show what constellation she is in; lastly, fasten the screw G, and laying the curved edge of the pasteboard, S, over the ecliptic, below the Moon, adjust the Moon to her latitude over the graduated edge of the pasteboard, and the globe will be rectified.

Its use.—Having thus rectified the globe, turn it round, and observe on what points of the horizon the Sun and Moon balls

[147] This simple auxiliary apparatus of Ferguson's may be used with effect by those who have their Celestial Globe mounted on a pillar. We had the apparatus made for our globe, and the cost was only 30s.; and we believe that the same mechanician continues to supply them at same price.

rise and set, for these agree with the points of the compass on which the Sun and Moon rise and set in the heavens on the given day; and the hour-index shows the times of their rising and setting, and likewise the time of the Moon's passage over the meridian.

This simple apparatus shows all the varieties that can happen in the rising of the Sun and Moon, and makes the phenomena of the Harvest Moon plain to the eye. It is also very useful in reading lectures on the globes, because a large company can see this Sun and Moon go round, rising above and setting below the horizon at different times, according to the seasons of the year, and making their appulses to different fixed stars. But in the usual way, where there is only the places of the Sun and Moon in the ecliptic to keep the eye upon, they are easily lost sight of, unless they be covered with patches." (See Ferguson's Astronomy, Chapter XII. Astronomical Machinery, article 401, and Plate III., figure 3; also, The Universal Magazine, Vol. 7, with Plate, &c.)

"A DISSERTATION UPON THE PHENOMENA OF THE HARVEST MOON," &c., Published.—Ferguson published this his second work in July, 1747, being a small 8vo of 72 pages, and illustrated by *three* folding engravings. As formerly mentioned, Ferguson, in his own memoir, notices this publication, and calls it his *first work;* but it was really his *second.* At page 5 of his preface to this "*Dissertation,*" he *refers* to his *first work*—apologises for an error in it, and also for the printer, for his faulty printing. We do not find "*the Dissertation*" advertised in any of the London newspapers or magazines *before* July, 1747; and as it comes first into notice this month, we conclude it was then just published. The following is a copy of an advertisement of it in the Gentleman's Magazine:—"*Dissertation on the Phenomena of the Harvest Moon; 3 Copper-plates; by James Ferguson.*" (Gentleman's Magazine, Vol. 17, p. 348, July, 1747; see also note 82.)

THE PLANETARY GLOBES INSTRUMENT.—During the summer of 1747, Ferguson invented and made, for the illustration of his future lectures, a curious and simple apparatus, which he afterwards designated "THE PLANETARY GLOBES." The annexed wood-cut shows its appearance, and the following is a description of its several parts, its use, &c.

In the engraving the letter T represents a terrestrial globe fixed on its axis, standing upright on the pedestal C D E, on which is an hour circle, having its index fixed on the axis, which turns somewhat tightly in the pedestal so that the globe may

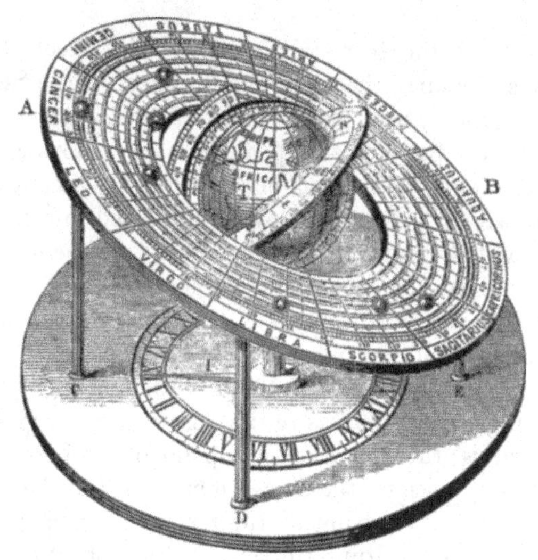

Planetary Globes.

not shake, to prevent which, the pedestal is about two inches thick, and the axis goes quite through it, bearing on a shoulder. The globe is hung in a graduated brazen meridian, much in the usual way, and the thin plate N, NE, E, is a moveable horizon, graduated round the outer edge, for showing the *bearings* and *amplitudes* of the Sun, Moon and planets. The brazen meridian is grooved round the outer edge, and in this *groove* is a slender semicircle of brass, the ends of which are fixed to the horizon in its north and south points. This semicircle slides in the groove as the horizon is moved in rectifying it for different latitudes. To the middle of the semicircle is fixed a *pin* which always keeps in the zenith of the horizon, and on this pin the quadrant of altitude *q* turns, the lower end of which, in all positions, touches the horizon as it is moved round the same. This quadrant is divided into 90 degrees from the horizon to the zenithal pin, on which it is turned at 90. The great flat circle or plate, A B, is the ecliptic, on the outer edge of which the signs and degrees are laid down, and every fifth degree is drawn through the rest of the surface of this plate towards its centre. On this plate are *seven grooves,* to which seven little balls are adjusted by *sliding wires,* so that they are easily moved in the grooves without danger of starting out of them.

The ball next the terrestrial globe is the Moon, the next without it is Mercury, the next Venus, the next the Sun, then Mars, then Jupiter, and lastly, Saturn; and in order to know them, they are separately stamped with the following characters:—●, ☿, ♀, ☉, ♂, ♃, ♄. This plate or ecliptic is supported by four strong wires, having their lower ends fixed into the pedestal at C D and E, the fourth wire being hid by the globe. The ecliptic is inclined 23½ degrees to the pedestal, and it is therefore properly inclined to the axis of the globe which stands upright on the pedestal.[148]

To rectify this machine, set the Sun and all the planetary balls to their geocentric places in the ecliptic, for any given time, by an ephemeris; then set the north point of the horizon to the latitude of your place on the brazen meridian, and the quadrant of altitude to the south point of the horizon; which done, turn the globe with its furniture till the quadrant of altitude comes right against the Sun, viz. to his place in the ecliptic; and keeping it there, set the hour index to the XII. next the letter C, and the machine will be rectified not only for the following problems, but for several others, which the ingenious reader may easily find out:—

Problem I. To find the amplitudes, meridian altitudes, and times of rising, culminating, and setting of the Sun, Moon, and planets. Problem II. To find the altitude and azimuthy of the Sun, Moon, and planets, at any time of their being above the horizon. Problem III. The Sun's altitude being given at any time, either before or after noon, to find the hour of the day and the variation of the compass in any known latitude.

[148] From an old sketch of this machine which was shown to us in the winter of 1831, by our friend Mr. Deane F. Walker, lecturer on Astronomy in London, it would appear (from the measurements there given), that the terrestrial globe of this machine was 9 inches in diameter; the large ecliptic, 21 inches in diameter; the pedestal or base, 2 feet diameter; the height of the lowest point of the ecliptic from the base, 9 inches, and the highest point about 10 inches. The balls representing the Moon, Mercury, Venus, the Sun, Mars, Jupiter, and Saturn, were of ivory, and about three-fourths of an inch in diameter each.

(Vide Ferguson's Astronomy, Chap. XII., article "*The Planetary Globes;*" and for engraving of same, see Plate III. and fig. 3 of same work.)

Astronomical Clock.—About the end of the year 1747, Ferguson invented and made a large wooden model of an Astronomical Clock, for the illustration of his future lectures, which

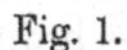

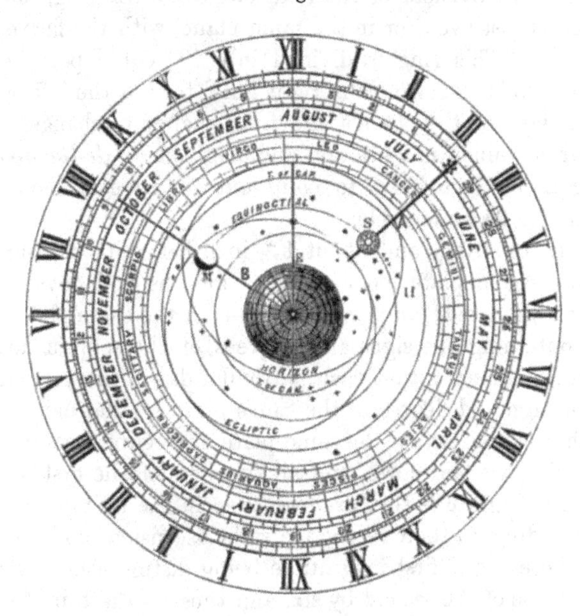

Dial Face of Astronomical Clock.

showed, with considerable accuracy, the apparent diurnal motion of the Sun, Moon, and Stars; with the times of their rising, southing, and setting; the places of the Sun and Moon in the ecliptic; and the age and phases of the Moon throughout the year. Ferguson, in his "*Tables and Tracts,*" London, 1767, describes this clock at pages 111—116, and informs us that it was made "*about twenty years ago,*" thereby showing that it was made in 1747; and on referring to our notes, we find that it appears to have been made about the close of this year. This clock is also described in Ferguson's "*Select Mechanical Exercises,*" pages 19—31, illustrated by a folding copper-plate engraving of the wheel-work and dial-plate, which we have reduced and here annexed. The following is a description of this curious clock in all its details:—

The dial-plate of this clock is represented by the annexed Fig. 1. It contains all the 24 hours of the day and night, and each hour is divided into 12 equal parts, so that each part answers to 5 minutes of time.

Within the divisions of the hour-circle is a flat ring, the face of which is just even (or in the same plane) with the face of the hour-circle. This ring is divided into 29½ equal parts, numbered 1, 2, 3, 4, &c., from the right hand toward the left, which are the days of the Moon's age from change to change. The ring turns round in 24 hours, and has a *fleur-de-lis* upon it, serving as an hour-index to point out the time of the day or night in the 24 hour circle.

Within the ring, and about $1\frac{4}{10}$ inch below its flat surface, is a flat circular plate, on which the months and days of the year are engraved; and within these, on the same plate, is a circle containing the signs and degrees of the ecliptic, divided in such a manner as that each particular day of the year stands over the sign and degree of the Sun's place on that day.

Within this circle, on the same plate, the ecliptic, equinoctial, and tropics, are laid down, and all the stars of the first, second, and third magnitude that are ever seen in the latitude of London, according to their respective right ascensions and declinations; those of the first magnitude being distinguished by eight points, those of the second by six, and those of the third by five. This plate turns round in 23 hours 56 minutes 4 seconds and 6 thirds of time, which is the length of a sidereal day;[149] and consequently, it makes 366 revolutions (as the stars do in the heavens) in the time the Sun makes 365, the number of sidereal days in a year exceeding the number of solar days by one.

Over the middle of this plate, and about four-tenths of an inch from it, is a fixed plate E to represent the Earth, round which the Sun, Moon, and stars move in their proper times,—

149 The length of a sidereal day may be determined as follows:—The mean length of a solar year is 365 d. 5 h. 48 m. 51·6 s.; or decimally, 365·242264 days, in which time the stars make 366·242264 revolutions; therefore, 365·242264÷366·242264 = ·99726957 of a solar day; ·99726957 reduced is 23 h. 56 m. 4 s. 5·148 thirds. Those who do not understand decimals may state the question thus:—If 366·242264 : 24 h. : : 365·242264 ?

viz. the Sun in 24 hours, the Moon in 24 hours 50½ minutes, and the stars in 23 hours 56 minutes 4 seconds 6 thirds. The Sun S is carried round by a wire A which is fixed into the inside of the Moon's age-ring, even with the *fleur-de-lis;* the Moon, M, is carried round by a wire B which is fixed to the axis of a wheel below the Earth, E, and the star plate is turned round by a wheel at the back of the dial-plate.

Over the dial-plate is a glass, as is in common clocks. On this glass is an ellipsis H, drawn with a diamond, to represent the horizon of the place for which the clock is to serve, and across this horizon is a straight line *e* E *d* (even with the two XII's.) to represent the meridian. All the stars that are seen at any time within this ellipsis are above the horizon at that time, and all those that are without it, are then below the horizon. When any star on the plate comes to the left-hand side of the horizon (the stars moving from left to right), the like star in the heavens is rising; when it comes under the meridian line *e* E, the like star in the heavens is on the meridian of the place; and when it comes to the right-hand side of the horizon, the like star in the heavens is setting.

When the point of the ecliptic that the Sun's wire A intersects comes to the left-hand side of the horizon on any day of the year cut by that wire, the *fleur-de-lis* will be, at the time of the Sun's rising, in the 24 hour circle; and at the time of his setting when the intersection of the Sun's wire and ecliptic comes to the western side of the horizon. The like is to be understood with regard to the rising and setting of the Moon, when the point of the ecliptic, which her wire B intersects, comes to the left and right hand sides of the horizon.

Every 24 hours, the Moon's wire shifts over one day of her age in the circle of 29½ equal parts on the flat ring before mentioned. Each of these day spaces is divided into four equal parts, for showing the Moon's age to every sixth hour thereof. Thus, as the Moon's wire B stands in the engraving, it shows the Moon to be 8 days and 18 hours old. It shifts quite round the ring, and carries the Moon round from the Sun to the Sun again in 29 days 12 hours and 45 minutes.

The Sun, on its wire A, goes 365 times round in 365 days; and in that time, the star-plate, with the months and days of the year upon it, goes 366 times round; so that for every revolution of the Sun, the star-plate advances forward, under the Sun's wire, through the space of one day in the circle of months; and by this means, the Sun's wire shows the day of the month throughout the whole year; and at the same time, for each particular day in the year, it shows the Sun's place in the ecliptic in the circle of signs. The Moon's wire B cuts the Moon's place in the circle of signs for every day of her age throughout the year.

The whole circle of signs shifts round, under the Sun's wire A, in 365 days 5 hours 48 minutes 58 seconds, which is the time the Sun takes in going quite round the ecliptic; and the Moon's wire shifts over all the circle of signs in 27 days 7 hours and 43 minutes, which is the time of the Moon's going round the ecliptic; and thus, by these different motions of the Sun and Moon, there are always 29 days 12 hours and 45 minutes between any conjunction of the Sun and Moon and the next succeeding one.

The Moon M is a round ball, half black, half white; it turns round its axis, or wire B, in 29 days 12 hours 45 minutes, and so shows all the different phases of the Moon for every day of her age, from change to change. When the Sun and Moon are together, or in conjunction, the white side of the Moon-ball, M, is toward the Sun, and the black side toward the eye of a spectator looking at the clock, who then can see no part of the white (or apparently enlightened) side of the Moon. When the Moon is full, or opposite to the Sun, all the white side of the ball M is turned toward the spectator's eye; and when the Moon is in either of her quarters, or 90 degrees from the Sun, the spectator sees half the black and half the white side of the ball, so that the white part of it then appears like the Moon in her first or last quarter.

The annexed engraving (Fig. 2) represents the dial-work of this clock, in ground plan, and in section; showing the wheels at the back of the dial-plate, between it and the wheels of the common movement, which are contained between two fixed plates, wherein the pivots of their axes turn in holes.

A long pin or spindle is fixed into the movement-plate, next the back of the dial-plate, perpendicular to both these plates. This spindle goes through the centre of the dial-plate, and has the Earth-plate E (Fig. 1) fixed on the end of it.

On this spindle, and close to the movement-plate in which it is fixed, is a fixed pinion A of 8 leaves (Fig. 2) which takes into the teeth of a wheel B, and of a wheel C. The number of teeth in B is 35, and of C is 50.[150]

Fig. 2.

Ground Plan.

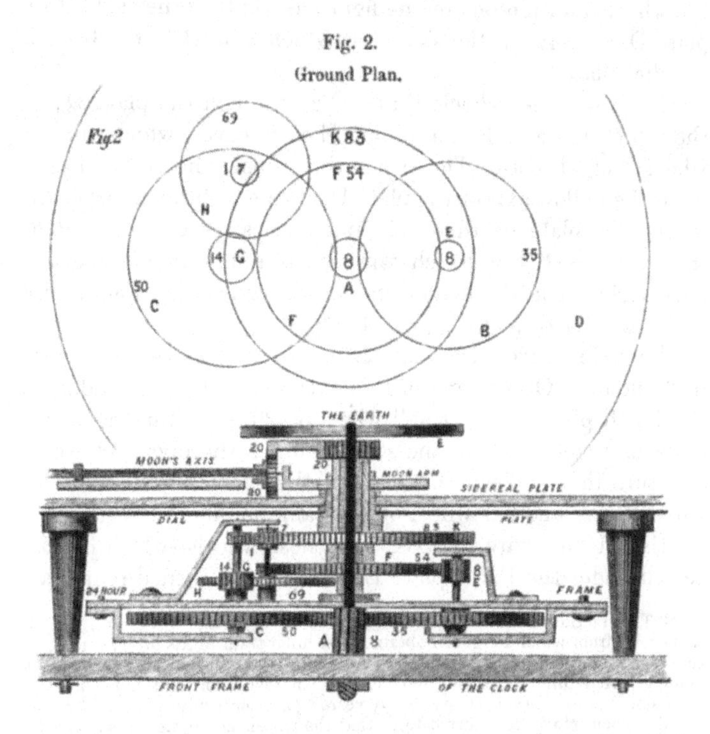

These two wheels hang and turn upon a large plate D, which, by the common clock movement, is turned round upon the axis of the fixed pinion A in 24 hours; and consequently, as these two wheels are carried round the fixed pinion, and its leaves take into their teeth, each of them will be so far turned round its axis, in 24 hours, as is equal to 8 of their teeth. The axis of the plate D is hollow, and turns upon the solid fixed spindle or axis of the pinion A; and the side of the plate D that is nearest to the clock movement is almost contiguous to the end

150 Ferguson, in his "*Tables and Tracts,*" in his description of this clock, makes these wheels 70 and 100; being thus doubled, it became necessary to double the fixed central pinion also; and hence he has it 16. $\frac{16}{70} = \frac{8}{35}$, and $\frac{16}{100} = \frac{8}{50}$ as here given ("*Tables and Tracts,*" pages 115, 116).

of the pinion A, next to the dial-plate of the clock; hence, it is plain, that the two wheels B and C, which are carried round the pinion A, and are turned by it, must be on that side of the plate D which is almost contiguous to the pinion. All the rest of the wheels and pinions in the figure are on the other side of the plate D, namely, on the side of it which is next to the back of the dial-plate.

The axes of the wheels B and C go through the plate D; on the top of the axis B is a pinion E of 8 leaves, which turns a wheel F of 54 teeth. The axis of this wheel is hollow, and turns upon the hollow axis of the plate D; it comes through the centre of the dial-plate of the clock, and carries the Moon round by the wire B in Fig. 1, which wire turns round in a piece that goes tight upon the hollow end of the axis of F (Fig. 2) just under the Earth-plate E (Fig. 1.)[151]

The hollow axis of the plate D turns round, as the plate does, in 24 hours. On the end of this axis, just under the middle of the Earth-plate E, is a small wheel of 20 teeth turning a contrate, or bevel wheel, of the same number, the pivots of whose axis turn in the piece that carries the Moon's wire, and this wheel turns another wheel of the same number of teeth fixed on the Moon's wire or axis. By these wheels (which lie concealed under the Earth-plate E) the Moon is turned round her

[151] The fractions or wheel-numbers for a lunation being $\frac{8}{35} \times \frac{8}{54} = \frac{64}{1890}$, the same solid from which Ferguson obtained the lunar train in his orrery (see orrery *lunation* under date 1745), produces the imperfect lunation of 29 d. 12 h. 45 m. being a period fully 57 seconds too slow in each revolution of the Moon, which soon amounts to a serious error. Many years ago, when studying the wheel-work of this curious clock, it occurred to us that the wheels and pinion producing the lunation might be dispensed with, and a more accurate train substituted, as there was no necessity of having *two separate motions* from the fixed pinion of 8 leaves in the centre,—wheel 50, the first wheel in the sidereal train, being quite sufficient to convey any number of motions on the upper surface of plate D. We therefore made a calculation for a new lunar train, and made Ferguson's second wheel of 69, in his sidereal train, the unit or basis of our calculation. It is evident that $\frac{8}{50} \times \frac{14}{69} = \frac{112}{3450}$ or $\frac{56}{1725}$, and $1725 \div 56 = 30{\cdot}803571$ days for the period of wheel 69, and as a mean lunation, is 29 d. 12 h. 44 m. 2·8 s., or decimally expressed, 29·530588 days, we have the fraction $\frac{29}{30}\,\frac{530588}{803571}$, which, on being reduced by continuous fractions (as in the example shown in the description of the orrery under date 1745), we obtain the fractions $\frac{1}{1}$, $\frac{23}{24}$, $\frac{93}{97}$, $\frac{116}{121}$, &c. If we take the two last fractions $\frac{116}{121}$, we obtain a train that closely approximates the true lunation,—viz. $30{\cdot}803571 \times 116 = 3573{\cdot}214236 \div 121 = 29{\cdot}530696$ d., or 29 d. 12 h. 44 m. 12 s., which is a period within $9\frac{1}{4}$ seconds of the true mean lunation, whereas by the numbers used in the clock, the error is 57 seconds. To any clockmaker who may make this clock we would suggest that a wheel of 121 teeth be made fast to the shaft of pinion 7 of wheel 69, and let it drive a wheel of 116 (in the place where wheel 54 now is), and this wheel of 116 teeth, which is a much nearer approach to the true period, will turn round in 29 d. 12 h. 44 m. 12 s.

axis in 29 days 12 hours 45 minutes, as before mentioned, and shows her different phases. These three last-mentioned wheels are not represented in Fig. 2, because they could not be put in without confusing it. (In our new section, however, these wheels, as also the position of all the rest, are distinctly shown).

On the axis of the wheel C, of 50 teeth, is a pinion G of 14 leaves, which turns a wheel H of 69 teeth, on whose axis is a pinion I of 7 leaves, turning a wheel K of 83 teeth. This wheel is pinned fast to the back of the star-plate, which, together with the wheel, turns round in a sidereal day, or in 23 hours 56 minutes 4 seconds 6 thirds of mean solar time;[152] and the before-mentioned wheels, which belong to the Moon, will carry her round from the meridian to the meridian again in 24 hours 50½ minutes,[153] if the plate D, on which all the wheels hang, be turned round in 24 mean solar hours.

The plate D carries the ring of 29½ equal parts round, within the fixed 24 hour-circle on the dial-plate, in 24 hours, by means of four pillars, whose opposite ends are fixed into the plate D, and into the ring. By this the Sun is carried round in 24 hours (its wire A being fixed into the ring), and also the *fleur-de-lis* that shows the time, like an index, in the 24 hour circle."

Ferguson concludes the description of his clock by saying, "I contrived this clock above twenty years ago, and made a model of it in wood, which I have still in my custody, and since that time." (See Ferguson's "*Select Mechanical Exercises*," pp. 19—31).

This is the same clock which Ferguson describes in his "*Tables and Tracts*," pp. 111—116, the only difference in the two descriptions being in the numbers of the teeth in two of the wheels, and in the number of leaves in one of the pinions. In

152 This pinion of 14 leaves, and wheel of 50 teeth, is equal in value to 7 leaves and 25 teeth, as is used by Ferguson in his orrery. This train is as follows:—$\frac{8}{50} \times \frac{14}{69} \times \frac{7}{83} = \frac{784}{286350}$, and 286350÷784 = 365 d. 5 h. 48 m. 58·78 s.; but as 50 and $\frac{14}{69}$, $\frac{7}{83}$ are carried round the fixed pinion 8 every 24 hours, the effect produced on the last wheel of the train, viz. wheel 83, is this; although this wheel is carried round with the 24-hour frame, yet any given point on its surface will make a revolution round a fixed point—any fixed point on the dial-plate, in 23 h. 56 m. 4 s. 6 thirds; this wheel of 83 therefore gains *one* revolution in 365 d. 5 h. 48 m. 58·78 s. The true length of a sidereal day is 23 h. 56 m. 4 s. 5½ thirds; and Ferguson's train for producing the period he here gives is simply the yearly train of his orrery, $\frac{8}{25} \times \frac{7}{69} \times \frac{7}{63}$, made to revolve inversely, and in 24 hours. (See orrery, under date 1745).

153 This period of 24 h. 50½ m., the period of a mean apparent diurnal revolution of the Moon, is here expressed in round numbers only; that produced by the wheel-work is 24 h. 50 m. 31½ s.

his "*Select Mechanical Exercises,*" from which we have extracted the description of the clock, he gives 8 as the number of leaves in the fixed pinion. In his "*Tables and Tracts,*" he doubles this number, and makes 16 to be the number of leaves in the fixed pinion; and consequently, he must double the numbers of the respective wheels which act in it. Thus, instead of B having 35 teeth, and C 50, he makes these wheels 70 and 100; but in both cases the value and result of the fraction is the same, for $\frac{8}{35}\times\frac{8}{50}=\frac{16}{70}\times\frac{16}{100}$

In his "*Tables and Tracts,*" first published in 1767, in beginning his short description of this clock, he mentions that he made it "*about twenty years ago,*" that is, about 1747. In his "*Select Mechanical Exercises,*" first published in 1773, he notes that he "contrived this clock *above* twenty years ago;" therefore, the descriptions of it given in these two works, no doubt refer to the same clock.

1748.

SIMPLE LUNAR MOTION.—According to our memoranda, as also from the date on an old sketch of this Lunar Motion work, it appears to have been invented by Ferguson early in the year 1748. This simple train of wheel-work was often used by him in his astronomical machines that exhibited the apparent diurnal revolution of the Moon; and since his time, it has often been adopted by others, and introduced into several pieces of clockwork. Annexed is a section of this train of wheels, taken from

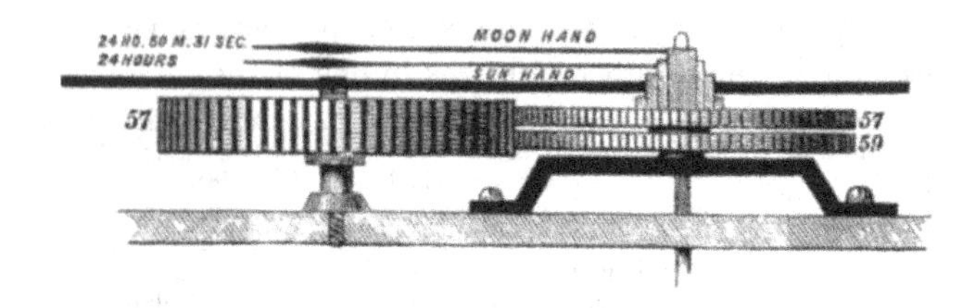

Front Frame of the Clock.

the old sketch referred to, with the description, as given in "Tables and Tracts," pp. 126, 127.

"Let a thick wheel of 57 teeth be turned round in 24 hours,

and take into the teeth of two wheels of equal diameters, one of which has 57 teeth, and the other 59; these wheels lying close upon one another, and the axis of the one turning within the axis of the other. A wire fixed on the axis of the wheel 57 will carry a Sun round in 24 hours, and a wire fixed on the axis of the wheel of 59 will carry a Moon round in 24 hours $50\frac{1}{2}$ minutes.[154]

If the Sun carries a plate round with him in 24 hours, and the limb of the plate be divided into $29\frac{1}{2}$ equal parts for the days of the Moon's age, the Moon will show her age in the

[154] This is the nearest approximation to the mean apparent diurnal revolution of the Moon that can be obtained by so simple wheel-work. Ferguson gives, as the product of his train, the period of 24 h. $50\frac{1}{2}$ m.; the exact period arising from the wheels is, however, 24 h. 50 m. $31\frac{33}{57}$ s., for $57 : 24 : : 59 = 24\cdot8421 = 24$ h. 50 m. $31\frac{33}{57}$ s. as given. The true mean apparent diurnal revolution of the Moon is performed in 24 h. 50 m. 28 s. 22 thirds 48 fourths, or when decimally expressed, 24·8412166667 hours: and if we take 24 hours as our *unit*, the fraction of the period will become $\frac{24\cdot0000000000}{24\cdot8412166667}$, which, on being reduced by the process of *continuous fractions*, as shown in the example for obtaining a solar revolution of the Earth in the description of the orrery finished in 1745, which process we need not repeat here, but shall give a few of the approximate fractional results, which are as follows:—$\frac{1}{1}$, $\frac{28}{29}$, $\frac{57}{59}$, $\frac{428}{443}$ $\frac{485}{502}$, $\frac{2368}{2451}$, $\frac{24108}{24953}$. &c. It will be observed that Ferguson's numbers are in the third ratio, the next fraction is $\frac{428}{443}$, but as its denominator is a prime number, it becomes useless in wheel-work; the next or fifth fraction $\frac{485}{502}$ is also in a manner inapplicable, because its denominator, $502 \div 2$, gives 251, which is a prime number, and inconveniently large for wheels; but if we add the fractions $\frac{428}{443} + \frac{485}{502} = \frac{913}{945}$, which is capable of being reduced into the following fractions suitable for wheel-work:—$913 = 11 \times 83$, and $945 = 15 \times 63$; the reduced fraction thereby becomes $\frac{83}{15} \times \frac{11}{63}$, that is, if a wheel of 83 teeth turn round in an hour, and drive a pinion of 15 leaves, having fixed to it a pinion of 11 leaves, which is made to drive a wheel of 63 teeth concentrically with wheel 83, and this wheel of 63 teeth will revolve in 24 h. 50 m. 28 s. 15 thirds, which is an error of only 7 *thirds* 48 *fourths*, or about an hour too fast at the end of 81 years. If we take the fraction $\frac{2368}{2451}$, we will obtain a very near approach to the mean apparent diurnal revolution of the Moon; this fraction is capable of being reduced as follows:—$2368 = 37 \times 64$, and $2451 = 43 \times 57$; therefore, let a wheel of 64 teeth turn round on its axis in 24 hours, and drive a wheel of 43 teeth, to which must be made fast the wheel of 37 teeth, and this wheel of 37 be made to turn a wheel of 57 teeth concentrically with wheel 64, the result will be, that wheel 57 will turn round on its axis in 24 h. 50 m. 28 s. 22 thds. 42·1619 fourths, being an error of only 5·8381 fourths.

The last fraction is prime and therefore unsuitable for wheel-work, but still a very near approximate ratio may be thus obtained,—thus, take $\frac{24108}{24953} + \frac{57}{59} = \frac{24165}{25012} = \frac{148}{135} \times \frac{169}{179}$; therefore, let a wheel having 179 teeth turn round on its axis in 24 h. and drive a wheel of 148 teeth, on whose axis is made fast a wheel of 135 teeth, which is made to drive a wheel of 169 teeth concentrically with wheel 179, this wheel of 169 teeth will, by this arrangement, turn once round its axis in 24 h. 50 m. 28 s. 22 thds. 47·5977648 fourths, which is perhaps *the nearest* wheel-work approximation to the apparent diurnal revolution of the Moon *ever obtained.* These are the numbers we thus obtained, some years ago, for a very accurate astronomical machine which we had then made; it is so near to nature, that a thousand years may pass away before any error worth mentioning would occur. The ratio $\frac{57}{59}$, as before shown, produces a period of 24 h. 50 m. 31 s. 34 thirds, = 3 s. 12 thirds too much in every *lunar day*, which amounts to an error of nearly one hour in 3 years and 2 months.

divisions of that plate, and may be made to turn round her axis, and show her phases, by a wheel of any number of teeth on her axis, and taking into the teeth of a contrate wheel of the same number, fixed on the axis of the wheel of 57 teeth, which carries the Sun." [155]

FERGUSON'S LONDON RESIDENCE.—Ferguson's residences in London, for the first few years after his arrival, from 1743 to 1748, is now not known. In our endeavours to discover the localities of his several residences in the great metropolis during said period, we instituted a tedious search through a great many London newspapers and magazines of that day, thinking it probable that he would either advertise himself as a limner, or affix his address to some astronomical communication to the editors. In our search, we were unsuccessful; and it is not unlikely that his residences in London, from 1743 to 1748, will now for ever be a blank.

In the Library of the British Museum, we found a series of that exceedingly rare work, "KENT'S LONDON DIRECTORY,"

[155] Ferguson here alludes to two wheels to produce a rotation of the Moon on her axis, and as the description may not be readily understood, we annex a section of the solar and lunar wheels, with the position of the two extra wheels mentioned by him. On the top of the hollow axis of the Sun wheel, of 57 teeth, is made

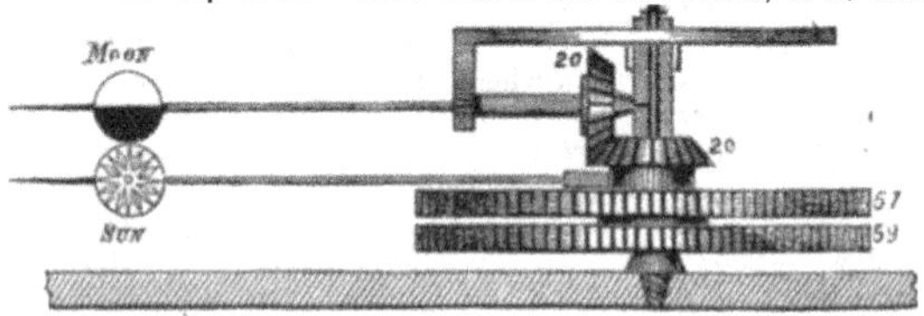

fast a small bevel wheel B of 20 teeth ; another bevel wheel A, having also 20 teeth, is fixed on the end of the Moon's wire or shaft, takes into that of the other bevel wheel, and is made to revolve by the difference of their velocities, thus, 57 turns round in 24 hours ; and 59, to the top of whose axis is made fast the bar which carries the Moon's shaft and bevel wheel A, turns round in 24 h. 50 m. 31 s.; therefore, suppose a hole to be drilled through both wheels, 57 and 59, then place the thick wheel of 57 teeth, formerly mentioned, into action with them; this done, turn the thick wheel, and it will be found that the two loose wheels, 57 and 59, will operate thus:—before the two drilled holes can come *directly* opposite to each other again, wheel 57 will have made $29\frac{1}{2}$ revolutions, and wheel 59 only $28\frac{1}{2}$ revolutions; hence, the difference being *one*, a revolution is communicated to bevel wheels A and B, causing B to make one revolution on its axis during the period of a mean synodic revolution of the Moon; and if the Moon be a ball half black, half white, by thus turning on its axis, it will show the phases of the Moon in that time. C is a bar of brass, fixed to the top of the hollow axis of the Moon wheel 57 ; D is a short piece with a hole in it to sustain the axis of the Moon, and through which it passes ; E is the socket of the bar C, into the side of which the short pivot of bevel wheel A turns.

commencing with the year 1748; looking into it, we discovered the name and address of Ferguson, which is certainly the earliest that has been found. The following is extracted from said directory:—

1748.
"*James Ferguson, Astronomical and Mathematical Teacher, Surrey Street, Strand.*"

Whether the domestic residence of Ferguson in 1748 was in Surrey Street, or whether he had there only an apartment for the purpose of teaching and delivering lectures on Astronomy, Geography, &c., is not now known. As Surrey Street in the Strand would be then central and convenient for tuition and lectures, it is probable that the above address in 1748 was that of his lecture-room.

LECTURES DELIVERED ON THE GREAT SOLAR ECLIPSE OF JULY 14TH, 1748. It would appear from our memoranda, that Ferguson commenced delivering popular lectures on the Great Solar Eclipse of 14th July 1748, somewhere about middle of April, that year (or about three months before the eclipse occurred). Ferguson's illustrative apparatus at this early period, consisted of an *orrery*, a *pair of globes*, an *Improved Celestial Globe*, the Four-Wheeled Orrery, Planetary Globes, wooden model of an Astronomical Clock, and several large diagrams of *the solar system, eclipses*, &c.

It is now not known what success Ferguson had with these his *first public lectures*, neither is it now known at what places in London they were delivered; after considerable research, we have ascertained at least *one* of the many places,—viz. Christ's Church Hospital School-room, Newgate Street, under the auspices of *his* friend, Mr. James Hodgson, the then mathematical master of that establishment.[156] (Vide Universal Magazine,

[156] It has been supposed that the folding engraving in Hinton's Universal Magazine, vol. 2d, p. 284, represents Ferguson delivering his first lecture. We have no means of testing the accuracy of this, but we may give a description of it. The engraving is entitled "*The First Lecture on the Sciences of Geography and Astronomy.*" The view represents a hall, in the middle of which there is a long table covered with a cloth, and on which stands, near the right hand end of it, *an armillary sphere*, and half hanging over the side near to it is *a map of the world*; on the floor there are *three terrestrial globes*; on the back wall there are *two large diagrams*, the one on the left has on it the "*Tychonick system*," on the other on the right there is a large "*mariner's compass*;" At the extreme end of the table stand an audience of *four gentlemen*, dressed in the costume of the period, one of whom rests his right hand on a large diagram of the

Vol. 2d, page 284, Aug. 1748; as also our note 86). It is evident from our memoranda, that Ferguson did not confine his lectures to London, but delivered lectures on this eclipse, &c., at a great many places in the provinces,—Bristol being one of his stations, where he delivered lectures to considerable audiences in the large room of the Bell Inn, Broad Street. (See also Ferguson's Mechanics, sixth Edit., p. 89, where he mentions being in Bristol 12 years ago, that is 12 years before 1760, when his Mechanics were first published).

THE GREAT SOLAR ECLIPSE OF THURSDAY, 14TH JULY, 1748, excited general interest throughout the kingdom. The following is extracted from a lecture on this eclipse, in the "*Universal Magazine*" of 1748, from the lecture by Ferguson, delivered in Christ's Hospital School. "The Sun on this occasion is first centrally eclipsed in North America, near Quebec, in Canada (lat. 46° 50′ N, and 75° 30′ W, from London), to whose inhabitants the body of the Sun appears centrally eclipsed at 4 hours 40 minutes, at the time of his rising. From thence, the central shade passes by the river St. Lawrence, and leaving Nova Scotia to the south, it crosses the north part of Newfoundland and the Atlantic Ocean; lands on the western Isles of Scotland; then passing by Perth, Dundee, Cupar, and St. Andrews, it takes the German Ocean in lat. 56° 30′ N, and coasting the south of Denmark, it leaves Norway and Sweden to the north, and enters Germany in lat. 54° 15′ N, by the mouth of the Elbe; whence, proceeding to Hamburgh, near the south part of the Baltic Sea, Brandenburg, Berlin, Frankfort, and Breslau, it enters Poland, passes near Cracow; and after passing part of Hungary and Turkey in Europe, it takes the Euxine, or Black Sea, between Constantinople and the mouth of the Danube; then it enters Turkey

"*Ptolemean system.*" At the other end of the table, on the extreme right, stands *the lecturer*, supposed to be *Ferguson*, robed in a loose flowing gown, and cap on head, a la academia, having in his right hand a long pointing rod with which he points to the armillary sphere on the table, while his lecture is directly before him in a small frame on the end of the table. Behind the lecturer is an open door looking out into a terrace and garden; opposite to which, and immediately at the lecturer's back, there is an old chair with an immensely tall back and extended arms, in which the lecturer had evidently been sitting before commencing his lecture. We have a decayed copy of this scarce engraving, and would have reduced it for the present work had we been certain that the lecturer represented was Ferguson.

in Asia; passes by the Euphrates and the Persian Gulf; then it entered the Arabian Sea, and there leaves the Sun centrally eclipsed at his setting on the coast of Coromandel, in the East Indies, lat. 12° 0′ N, and 80° 0′ E, from London."—Hinton's Universal Magazine, Vol. 2d, pp. 234—237.

In London, this Solar Eclipse was not total, only partial, as London lay about 420 miles to the south of the central track of the shadow. At the time of the greatest obscuration at London, which happened 10 hours 30 minutes forenoon, only about $\frac{10}{12}$ths of the northern limb of the Sun was hid from view. We annex a representation of this eclipse, showing the appearance of the

View of the Solar Eclipse of July 14th, 1748.

Sun at the time of his greatest obscuration at London, as also the curved track of the Moon, taken from the Gentleman's Magazine for July, 1748; as also the following note regarding it:—[157]

"Thursday, 14th July, 1748, there was a great eclipse of the Sun, beginning at London about 3 minutes after 9, and ending about 8 minutes after 12. During the eclipse, Venus made a beautiful appearance through the telescope, in the form of a

157 As already mentioned, after delivering a great many lectures on this eclipse in London and vicinity, Ferguson went to Bath and Bristol, and delivered the same lectures in these cities; with what success is now not known.—Vide also Ferguson's Lectures on Select Subjects, p. 89, where he alludes to having been in Bristol in 1748.

crescent. The darkness was scarce perceived, though the markets at Covent Garden and St. James's were almost destitute of gardeners, who, being terrified by false accounts, were afraid to come, lest they should have to go home in the dark."—Gent. Mag. Vol. 18th, p. 330, July, 1748.

For a Geometrical Projection of this Solar Eclipse, we refer the reader to Ferguson's "*Young Gentlemen and Ladies' Astronomy*," Dialogue X., "On the Projection of Solar and Lunar Eclipses," where will be found a minute description of the construction and track of this eclipse, illustrated by a fine folding engraving, which most likely is a fac-simile of the "*Projection of the Sun's Eclipse*" published by Ferguson early in 1746, as already noticed. (See also note 86.)

"An Answer to Ferguson's Essay on the Moon's turning on her Axis."—In the midst of the bustle and excitement that attended the first series of his Astronomical Lectures, in the summer of this year, Ferguson was startled by the publication of an anonymous pamphlet in London, entitled, "*An Answer to Mr. Ferguson's Essay upon the Moon's turning round its own Axis, subjoined to his Dissertation upon the Phenomena of the Harvest Moon, &c.*" (8vo, 52 pages, and 3 folding engravings). "*London: Printed for J. Roberts at the Oxford Arms, in Warwick Lane, MDCCXLVIII.*"

The arguments offered by this anonymous writer, in his "Answer" to Ferguson's "Dissertation," are weak—of no force whatever; consequently, they made no impression on those who understood the question in dispute. Ferguson published no special rejoinder, probably thinking that it sufficiently refuted itself. It would seem, however, from our memoranda, that a discussion on the "*moon's rotation*" was carried on in the newspapers, &c., of the time, between the writer of the "*Answer*," Mr. Ferguson, and others, and that the author of the "*Answer*" had the worst of it, "and somewhat lost his temper." During the discussion, Ferguson, finding that a Mr. Grove, at Richmond, in Surrey, was the author of the "*Answer*," resolved to have a personal interview with him on the subject of his *anti-lunar rotation* theories. Ferguson accordingly went to Richmond, but Mr. Grove refused to see him. A writer of last century, referring to this rudeness of Mr. Grove, says, "*most generous Ferguson; most unhandsome and narrow-minded Grove, you ought to have been polite.*" The following are copies of notes written on the cover of Mr. Grove's "*Answer*" to Ferguson's "*Dissertation*," in our possession :—

1st note says, "*This Answer to Mr. Ferguson's Essay is by Mr. Grove of Richmond, Surrey.*"
2d do. "*A composition of ignorance and petulance.*"
3d do. "*This man was so annoyed at last with Ferguson, that he refused to see him when he went to visit him ; so disagreeable is it to be convicted of error, when an ignorant man sets up for learned.*"

In the year 1751, Mr. Grove again went into print with "*A Supplement*" to his "*Answer on Ferguson's Essay.*" (See date 1751.)

JAMES FERGUSON *born.*—Referring to our extracts from the fly-leaves of the small pocket Bible which belonged to the Ferguson family, and noticed under date 1745, we find that Ferguson's eldest son, baptized *James*, was born in October 1748. The following is the copy of the entry, (the original on the fly-leaf in the Bible is in Ferguson's autograph) :—

"JAMES, born Tuesday 11th October, 1748."

This son, of whom little is known, died of consumption in 1772, to which, and also to date 1763, the reader is referred for other notices.

THE CRANK ORRERY.—According to our memoranda, as also from a date on an old drawing of this orrery, it is evident that Ferguson invented and made it about the close of the year 1748. The principal parts of this orrery so strongly resemble those of *the three-wheeled* one which Ferguson made at the end of the year 1745 to demonstrate the Moon's rotation, that we are inclined to assume, that in designing and constructing this one, he had *that* orrery in view. The annexed engraving represents a view of its several parts, which will be clearly understood by the following description :—

In his description of this orrery, Ferguson says, "In this machine, which is the simplest I ever saw for showing the diurnal and annual motions of the Earth, together with the

Moon and her nodes, A and B are two oblong square plates, held together by four upright pillars, of which three appear at *f*, *g*, and *g* 2. Under the plate A is an endless screw on the axis of the handle *h*, which works in a wheel fixed on the same axis with the double-grooved wheel E; and on the top of this axis is fixed

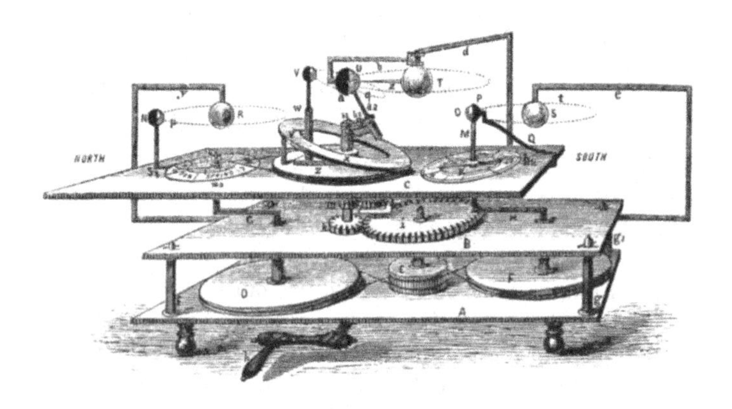

The Crank Orrery.

the toothed wheel *i*, which turns the wheel *k*, on the top of whose axis is the small wheel *k* 2, which turns another, *b* 2, and that turns a third, which, being fixed on *a* 2, the axis of the Earth U turns it round and the Earth with it. This last axis inclines at an angle of 23½ degrees. The supporter, X 2, in which the axis of the Earth turns, is fixed to the moveable plate C.

"In the fixed plate B, beyond H, is fixed the strong wire *d*, on which hangs the Sun T, so as it may turn round the wire. To this Sun is fixed the wire or solar ray Z, which (as the Earth U turns round its axis) points to all the places that the Sun passes vertically over every day of the year. The Earth is half covered with a black cap *a*, as in the former orrery (of 1745), for dividing the day from the night; and as the different places come out from below the age of the cap, or go in below it, they show the times of Sun-rising and setting every day of the year. This cap is fixed on the wire *b*, which has a forked piece C turning round the wire *d*; and as the Earth goes round the Sun, it carries the cap, wire, and solar ray, round him, so that the solar ray constantly points toward the Earth's centre.

"On the axis of the wheel *k* is the wheel *m*, which turns a wheel on the cock or supporter *n*, and on the axis of this wheel, nearest *n*, is a pinion (hid from view) under the plate C, which pinion turns a wheel that carries the Moon V round the Earth U; the Moon's axis rising and falling in the socket W, which is fixed to the triangular piece above Z, and this piece is fixed to the top of the axis of the last-mentioned wheel. The socket W is slit on the outermost side, and in this slit the two pins near Y, fixed in the Moon's axis, move up and down, one of them being above the inclined plane Y X, and the other below it. By this mechanism, the Moon V moves round the Earth U in the inclined orbit *q*, parallel to the plane of the ring Y X, of which the descending node is at X, and the ascending node opposite to it, but hid by the supporter X 2.

"The small wheel E turns the large wheels D and F, of equal diameters, by cat-gut strings crossing between them; and the axes of these two wheels are cranked at G and H above the plate B, with a motion which carries the Earth U round the Sun T, keeping the Earth's axis always parallel to itself, or still inclining toward the left hand of the plate, and showing the vicissitudes of seasons. As the Earth goes round the Sun, the wheel *k* goes round the wheel *i*, for the axis of *k* never touches the fixed plate B, but turns on a wire fixed into the plate C.

"On the top of the crank G is an index L, which goes round the circle *m* 2 in the same time that the Earth goes round the Sun, and points to the days of the months, which, together with the names of the seasons, are marked on this circle.

"This index has a small grooved wheel L fixed upon it, round which, and the plate Z, goes a cat-gut string, crossing between them; and by this means the Moon's inclined plane, Y X, with its nodes, is turned backward, for showing the times and returns of eclipses.

"The following parts of this machine must be considered as distinct from those already described.

"Toward the right hand let S be the Earth hung on the wire *e*, which is fixed into the plate B, and let O be the Moon fixed on the axis M, and turning round within the cap P, in which, and in the plate C, the crooked wire Q is fixed. On the axis M is also fixed the index K, which goes round a circle *h* 2 divided into 29½ equal parts, which are the days of the Moon's age; but to avoid confusion in the scheme, it is only marked with the numeral figures 1, 2, 3, 4, for the *quarters.* As the crank H carries this Moon round the Earth S in the orbit *t*, she shows all her phases by means of the cap P, for the different days of her age, which are shown by the index K; this index, turning just as the Moon O does, demonstrates her turning round her axis, as she still keeps the same side toward the Earth.

" At the other end of the plate C, a Moon N goes round an Earth R in the orbit P. But this Moon's axis is stuck fast into the plate C at S 2, so that neither Moon nor axis can turn round; and as this Moon goes round her Earth, she shows herself all round to it, which proves, that if the Moon was seen all round from the Earth, in a lunation, she could not turn round her axis.

N.B.—" If there were only the two wheels D and F, with a cat-gut string over them, but not crossing between them, the axis of the Earth U would keep its parallelism round the Sun T, and show all the seasons—as I sometimes make these machines; and the Moon O would go round the Earth S, showing her phases, as likewise would the Moon N round the Earth R; but then, neither could the diurnal motion of the Earth, U, on its axis, be shown, nor the motion of the Moon V round the Earth." [158] (Ferguson's Astronomy, Chapter XXII., Art. 398, and Plate VI., Fig. 1st.)

1749.

THE CALCULATOR.—The beginning of the year 1749 found Ferguson busy with the construction of an astronomical machine, which, when finished, he called *The Calculator.* He, in Chapter XII. and Section 399 of his Astronomy, describes the Calculator, and informs us that its internal mechanism was similar to that of "The Four-Wheeled Orrery," which he made in 1746, with the exception of an additional wheel, and that in consequence of this additional wheel, he discarded the original name of " *The Four-Wheeled Orrery,*" and adopted that of " *The Calculator.*" The mechanism of The Four-Wheeled Orrery of 1746 being identical with that of the *Calculator*, excepting that the latter had *an additional wheel*—we have only to refer the

158 Ferguson, although he minutely describes the positions of the several wheels in this Crank Orrery, does not mention how many teeth each wheel has; this is to be regretted, because the numbers of teeth in the wheels are as important to be known as are their positions.

reader to the description and section of the former, under date 1746, in order that the interior works of the *Calculator* may be understood; and shall here only describe the position of "*the additional wheel,*" and what it accomplished.

Annexed is a small section of the pulley-work of the Calculator (Fig. 1); on comparing it with that of "*The Four-Wheeled Orrery,*" it will be found that they are alike, with the exception of *an additional pulley* in the former, on the right-hand side, and the pulleys of the same being brought into closer contact. In the section, which we here give, of the Calculator, it will be observed that the large central pulley is made to act as *four pulleys,* whilst in "The Four-Wheeled Orrery" it acts as *three;* on the right of the section of the Calculator there are *four pulleys,* in the Four-Wheeled Orrery there are *three* only.

Fig. 1.

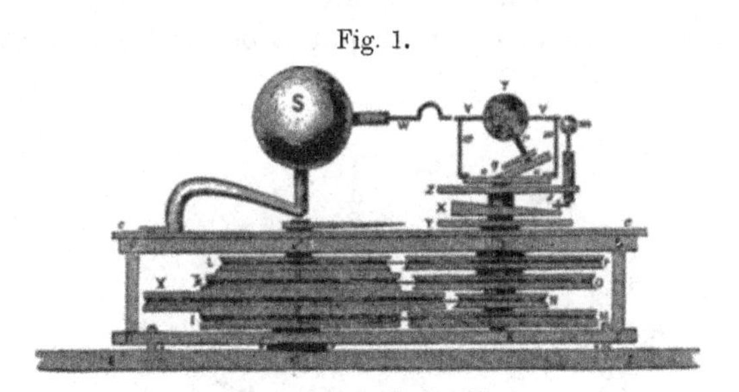

Section of the Pulley-Work.

The position of the additional pulley, or *wheel,* as Ferguson calls it, will therefore be readily discovered in the top pulley of the system of moveable pulleys on the right-hand side of the annexed section, and which is connected with the new cut groove opposite to it in the top of the large *fixed* pulley in the centre of the machine; therefore, the Calculator having *an additional wheel,* it has, by such an arrangement, *two wheels* or pulleys *added* to it. The new pulley has been introduced for the purpose of carrying a circular plate to show the place of the Moon's apogee, perigee, &c., and must therefore have a progressive motion through all the signs and degrees of the ecliptic in 8 years and 309 days; and in order that such a motion should be given to this new pulley, it must bear the same relation to the centre pulley as 62 days to 55; that is, if 62 represents the diameter of the groove of the new pulley, then 55 will represent that of the groove of the *fixed* pulley in the centre. In order to throw the numbers 62 and 55 into a workable form, we require to find lesser numbers to represent their relative values; by an easy process, we find that when they are reduced to *inches* they will stand thus:—2 inches, and $1\frac{77}{100}$ inch; therefore, let the new moveable pulley, on the right in the section, have the diameter of the bottom of its groove exactly 2 inches, and that of the groove on the top of the large *fixed* pulley in the centre, and opposite to it, be $1\frac{77}{100}$ inches, or $1\frac{3}{4}$ inch fully; and on this being done, and the two pulleys connected by a cat-gut string, it will be found that on the frame containing the pulley-work being moved $8\frac{6}{7}$ times completely round the central stem, the pulley of 2 inches will have made one revolution exactly, which is the period of revolution of the Moon's apogee.[159] The new or additional pulley is furnished with a short hollow axis, which turns round upon the hollow axis of the pulley below it, which carries

Fig. 2.

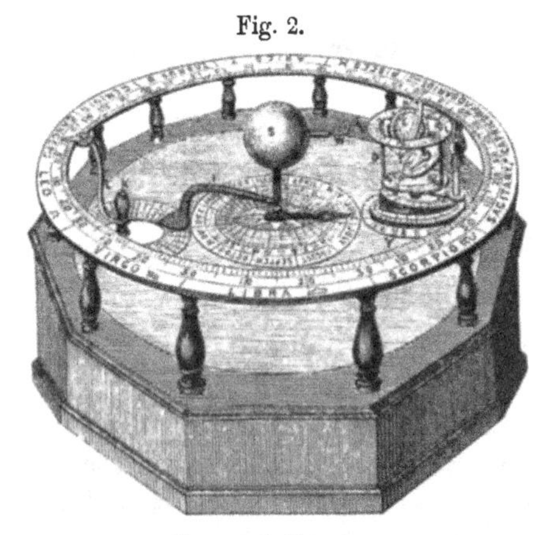

Ferguson's Calculator.

round the Moon's orbit, and ascends about a quarter of an inch above the upper surface of the circular plate which covers the works, and on which is tightly fixed a plate, having on it the places of the Moon's apogee, perigee, &c., which moves round in $8\frac{6}{7}$ years, and thus shows when the Moon is in apogee or perigee.

The pulley-works of the CALCULATOR having been brought

[159] The mean periodic revolution of the Moon's apogee is accomplished in 3231 days 11 hours 25 minutes, or 8 years and $309\frac{1}{2}$ days, equal to $8\frac{6}{7}$ years nearly; hence, $8\times7+6=62$, and $62-7=55$, or $\frac{62}{55}$, the numbers given.

into closer contact, allows them to be put into a box of a much more convenient size and form than that of its original. The Four-Wheeled Orrery (Fig. 2) shows the exterior appearance of the CALCULATOR after it was finished; it will be seen that Ferguson has put the works into a polygonal box, very much resembling that of his orrery, which gives the CALCULATOR a handsome appearance, and is altogether a great improvement on "The Four-Wheeled Orrery."

For further particulars, see Ferguson's Astronomy, Chapter XII., Section 399, article "*The Calculator*." Somewhere about the year 1765, Ferguson further improved his Calculator by substituting wheel-work for pulleys and cat-gut strings, thus insuring a greater degree of accuracy in the motion of its several parts. (See Orrery under date 1765.)

FERGUSON RESOLVES to Lecture on MORE SCIENCES than ASTRONOMY.—Ferguson had by this time (1749) been a public lecturer on Popular Astronomy for rather more than a year, during which period he had delivered several courses of lectures in London, as also in the provinces. Although his lectures were not altogether a failure, yet he had gained considerable experience, and had seen sufficient to convince him that it would not be beneficial for his pecuniary interest to confine his lectures entirely to Astronomy, as he had hitherto done, but to extend his course to other subjects, such as Mechanics, Pneumatics, Hydraulics, Hydrostatics, Electricity, Optics, and Dialling; thus, by diversifying his course, he rightly concluded that he would likely draw larger audiences, and consequently have a much better chance of success.

Having thus resolved on extending his course of lectures, the question of the extent and cost of apparatus next engaged his attention. In order properly to illustrate and demonstrate so many subjects as he had chosen, would require a large and complex apparatus, attended with considerable expense. The purchasing and collecting this apparatus would be to him a question of time; because, regulated by the state of his finances, and as his means were but very slender, he resolved to make, with his own hands, the greater portion of his instruments, and occasionally to purchase, from opticians, and others, those only that he could not conveniently make himself.

He therefore set earnestly to work, and got up a set of large and accurately delineated astronomical diagrams for the illustration of the theories of Ptolemy, Copernicus, and Tycho Brahé; as also, diagrams for the demonstration of the cause of the Ebb and Flow of the Tides—the Phenomena of Eclipses, &c.—He also at this time invented and made an ingenious tide-dial, and also made a centrifugal machine for the illustration of central forces. The following are sketches and descriptions of these two pieces of apparatus.

THE TIDE-DIAL.—"The outside parts of this machine consist of, 1st, an eight-sided box, on the top of which, at the corners, is shown the phases of the Moon, at the octants, quarters,

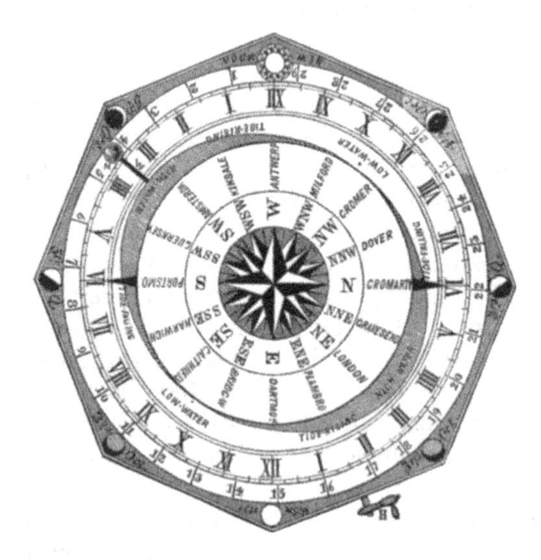

The Tide-Dial.

and full; within these is a circle of $29\frac{1}{2}$ equal parts, which are the days of the Moon's age, counted from the Sun at new Moon round to the Sun again. Within this circle is one of 24 hours, divided into their respective halves and quarters.—2d, A moving elliptical plate, painted blue, to represent the rising of the tides, under and opposite to the Moon, which has the words, *high water, tide falling, low water, tide rising*, marked upon it. To one end of this plate is fixed the Moon M, by the wire W, and goes along with it.—3d, Above this elliptical plate is a circle, with the points of the compass upon it, and also the names of above 200 places in the large machine (but only 32 in the figure, to avoid confusion), set over those points on which the Moon bears when she raises the tides to the greatest heights at these places twice in every lunar day; and to the north and south points of this plate are fixed two indexes I and K, which show the times of high water, in the hour circle, at all these places.—4th, Below the elliptical plate are four small plates, two of which project out from below its ends at new and full Moon; and so, by lengthening the ellipse, shows the spring tides, which are then raised to the greatest heights by the united attractions of the Sun and Moon. The other two of these small plates appear at low water when the Moon is in her quadratures, or at the sides of the elliptic plate, to show the neap tides; the Sun and Moon then acting crosswise to each other. When any two of these small plates appear, the other two are hid; and when the Moon is in her octants, they all disappear, there being neither spring nor neap tides at those times. Within the box are a few wheels for performing these motions by the handle or winch H.

"Turn the handle until the Moon M comes to any given day of her age in the circle of 29½ equal parts, and the Moon's wire W will cut the time of her coming to the meridian on that day in the hour circle; the XII. under the Sun being mid-day, and the opposite XII. mid-night; then looking for the name of any given place on the round plate (which makes 29½ rotations, whilst the Moon M makes one revolution from the Sun to the Sun again), turn the handle till *that* place comes to the words *high water*, under the Moon, and the index, which falls among the forenoon hours, will show the time of high water in the forenoon of the given day. Then turn the plate half round till the same places come to the opposite high-water mark, and the index will show the time of high water in the afternoon of that place. And thus, as all the different places come successively under and opposite to the Moon, the indexes show the times of high water at them in both parts of the day; and when the same places come to the low-water marks, the indexes show the times of low water. For about three days before and after the times of new and full Moon, the two small plates come out a little way from below the high-water marks on the elliptical plate, to show that the tides rise still higher about these times; and about the quarters, the other two plates come out a little from under the low-water marks, towards the Sun, and on the opposite side, showing that the tides of flood rise not then so high, nor do the tides of ebb fall so low, as at other times.

"By pulling the handle a little way outward, it is disengaged from the wheel-work, and then the upper plate may be turned round quickly by hand so, as the Moon may be brought to any given day of her age in about a quarter of a minute; and by pushing in the handle, it takes hold of the wheel-work again.

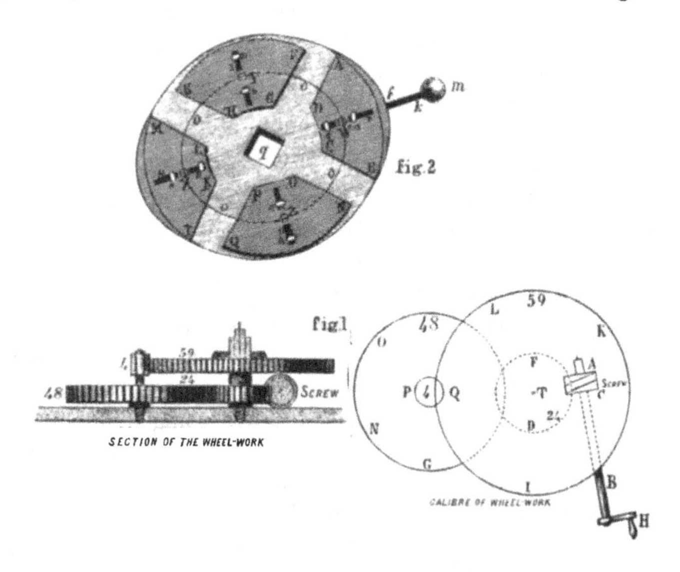

The following is a description of the wheel-work inside the box :—" On A B, the axis of the handle H (Fig. 1), is an endless screw C, which turns the wheel F E D of 24 teeth round in 24 revolutions of the handle; this wheel turns another, O N G, of 48 teeth, and on its axis is the pinion P Q of 4 leaves, which turns the wheel L K I, of 59 teeth round in 29½ turnings or rotations of the wheel F E D, or in 708 revolutions of the handle, which is the number of hours in a synodical revolution of the Moon. The round plate, with the names of places upon it, is fixed on the axis of the wheel F E D, and the elliptical or tide-plate, with the Moon fixed to it, is upon the axis of the wheel L K I; consequently, the former makes 29½ revolutions in the time that the latter makes one. The whole wheel F E D, with the endless screw C, and dotted part of the axis of the handle A B, together with the dotted part of the wheel O N G, lie hid below the large wheel L K I.

"Fig. 2, represents the under side of the elliptical or tide-plate *a, b, c, d,* with the four small plates A B C D, E F G H, I K L M, N O P Q, upon it; each of which has two slits, as T T, S S, R R, U U, sliding on two pins, as *n n,* fixed in the elliptical plate. In the four small plates are fixed four pins at W, X, Y, and Z, all of which work in an elliptic groove *o o o o,* on the cover of the box below the elliptical plate, the longest axis of this groove being in a right line with the Sun and full Moon; consequently, when the Moon is in conjunction or opposition, the pins W and X thrust out the plates A B C D and I K L M a little beyond the ends of the elliptic plate at *d* and *b,* to *f* and *e*; whilst the pins Y and Z draw in the plates E F G H and N O P Q quite under the elliptic plate to *g* and *h.* But, when the Moon comes to her first or third quarter, the elliptic plate lies across the fixed elliptic groove in which the pins work; and therefore, the end plates A B C D and I K L M are drawn in below the great plate, and the other two plates E F G H and N O P Q, are thrust out beyond it to *a* and *c.* When the Moon is in her octants, the pins W, X, Y, Z, are in the parts *o o o o* of the elliptic groove, which parts are at a mean between the greatest and least distances from the centre *q,* and then all the four small plates disappear below the great one." (Vide Ferguson's Astronomy,—3d Edit. 4to, 1764, pp. 296, 297, 298.)[160]

Centrifugal Machine.—Ferguson, after finishing his ingenious Tide-Dial, turned his attention to the making of a machine for demonstration of the law of centrifugal forces, especially as applied to Astronomy.

It is well known, he observes, that "all globular bodies, whose parts can yield, and which do not turn on their axes, must be

160 A description, with reference plates, of a Tide Rotula, with accurate motions, invented by us, will be found in the London Mechanics' Magazine, Vol. 16. No. 425, pp. 1, 2, 3, and 4. We have in our possession a very neatly-written explanation of the theory of the tides, with a diagram for reference, in the autograph of Ferguson.

perfect spheres, because all parts of their surfaces are equally attracted toward their centres. But all such globes as do turn on their axes will be oblate spheroids; that is, their surfaces will be higher, or farther from the centre, in the equatorial than in the polar regions; for as the equatorial parts must move quickest, they must have the greatest centrifugal force, and will therefore recede farthest from the axis of motion."

The following is an engraving and description of the Centrifugal Machine constructed by him at this period, to illustrate this law:—"Thus, if two circular hoops A B and C D, made thin and flexible, and crossing each other at right angles, be

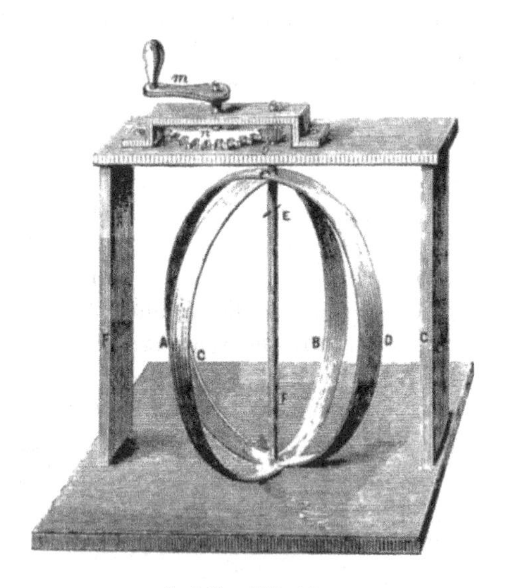

Centrifugal Machine.

turned round their axis E F by means of the winch *m*, the wheel *n*, and pinion *o*, and the axis be loose in the pole or intersection *e*, the middle parts, A, B, C, D, will swell out so as to strike against the sides of the frame at F and G, if the pole *e*, in sinking to the pin E, be not stopped by it from sinking farther; so that the whole will appear of an oval figure, the equatorial diameter being considerably longer than the polar. That our Earth is of this figure, is demonstrable from actual measurement of some degrees on its surface, which are found to be longer in the frigid zones than in the torrid; and the difference is found to be such as proves the Earth's equatorial diameter to be thirty-six miles more than its axis. Seeing then, that the Earth is higher at the equator than at the poles, the sea, which like all other fluids naturally runs downward (or toward the places which are nearest the Earth's centre), would run towards the polar regions, and leave the equatorial parts dry, if the centrifugal force of the water, which carried it to those parts, and so raised them, did not detain and keep it from running back again towards the poles of the Earth." These observations are from Ferguson's description of this machine in his "Lectures on Select Subjects in Mechanics, Hydrostatics, Pneumatics, Optics, and Astronomy."

FERGUSON'S RESIDENCE.—According to an autograph letter of Ferguson in the collection of the late Dawson Turner, Esq., of Yarmouth, Ferguson, about 1749, resided in lodgings in Great Pulteney Street.

1750.

THE MECHANICAL PARADOX.—It would appear from our memoranda, as also from other references, that it was during the year 1750 that Ferguson invented and made his celebrated machine, called "The Mechanical Paradox." In several of his works he mentions that he made it "on a very particular occasion," but without informing us of anything regarding this "particular occasion." But it is now certain that he made this curious machine for the purpose of silencing a London watchmaker who did not believe in the doctrine of the Trinity, as will be shown by a very interesting letter, written by Ferguson a few months before his death, to a clerical friend of his in the north of Scotland, a copy of which the reader will find at the conclusion of the description of the Paradox. The following is from Ferguson's own description of the machine:—[161]

[161] In three of Ferguson's works the Mechanical Paradox is described, but they all differ as to the date when it was first made. In a Tract, published by him in 1764, entitled, "The Description and Use of a new machine, called The Mechanical Paradox, invented by James Ferguson, F.R.S," p. 4, we find him saying that it was "*on a very particular occasion, about fourteen years ago,*" when he contrived the Paradox, that is, 14 years before 1764, that he wrote this Tract, which refers to the year 1750, the year we have adopted. This Tract was

"On a very particular occasion, about fourteen years ago, I contrived the Mechanical Paradox, which has been shown and explained to many, and which I shall here describe.

"It is represented in the annexed figure, in which A is called *the immoveable* plate, because it lies on a table whilst the machine is at work; B C is a moveable frame to be turned round an upright axis *a* (fixt into the centre of the immoveable plate), by taking hold of the knob *n*.

"On the said axis is fixt the immoveable wheel D, whose teeth take into the teeth of the thick moveable wheel E, and turns it round its own axis as the frame is turned round the fixt axis of the immoveable wheel D, and in the same direction that the frame is moved.

Fig. 1.

Mechanical Paradox.

"The teeth of the thick wheel E take equally deep into the teeth of the three wheels F, G, and H, but operate on these

specially written to describe the Paradox, and nothing else, and we must admit the "14 *years ago*," or 1750, as the correct date of the invention of this curious machine. Ferguson, in his "Tables and Tracts," London, 1767, p. 171, says "*The Mechanical Paradox is a small kind of orrery which I contrived and made about fifteen years ago.*" If we subtract 15 years from 1767 it will give 1752 as the date when Ferguson made it; and again, in his "Select Mechanical Exercises," 3d Edit., Lond. 1770, p. 46, he says, "*On a very particular* occasion, about eighteen years ago, I contrived a small machine called the Mechanical Paradox," and 18 years subtracted from 1773, gives 1755 as the date when it was made; thus, Ferguson gives 1750, 1752, and 1755 as the dates of his invention of the Paradox, and we find it difficult to reconcile them; but we incline to favour the earliest date, 1750, as given in the Tract he wrote purposely to explain the machine.

wheels in such a manner, that whilst the frame is turned round, the wheel H turns *the same way* that the wheel E does, the wheel G turns *the contrary way*, and the wheel F *no way at all.*

"Before we explain the principles on which these three different effects depend, it will not be improper to fix some certain *criteria* for bodies turning or not turning round their own axis or centres, and to make a distinction between absolute and relative motion.

"1*st*, If a body show all its sides progressively round toward a certain fixed point in the heavens, the body turns round its own axis or centre, whether it remains still in the same place, or has a progressive motion in any orbit whatever; for unless it does turn round its own centre, it cannot possibly have one of its sides toward the west at one time, toward the south at another, toward the east at a third time, and toward the north at a fourth. This is the case with the Moon, which always keeps one side toward the Earth, but shows the same side to every fixed point of the starry heaven, in the plane of her orbit, in the time she goes once round her orbit; because in the time that she goes round her orbit, she turns once round her own axis or centre;—on the contrary, if a body still keeps one of its sides toward a fixed point of the heaven, the body does not turn round its own axis or centre, whether it keeps in one and the same place, or has a progressive motion in any orbit or direction whatever. This is the case with the card of the compass of a ship, which still keeps one of its points towards the magnetic north, let the ship be at rest or sail round a circle many miles in diameter.

"Both of these cases may be exemplified either by a cube or a globe having a pin fixt into either of its sides to hold it by. We shall suppose a cube, because its sides are flat.—Sit down by a table, and hold the cube by the pin, which may be called its axis, and keep one of its sides toward any side of the room. Whilst you do this, you do not turn the cube round its axis, whether you still keep it in the same place or carry it round any other fixed body on the table; but if you try to keep any side of the cube toward the fixed body whilst you are carrying it round the same, you will find that you cannot do so without turning the pin round (which is fixt into the cube) betwixt the finger and thumb whereby you hold it, unless you rise and walk round the table, keeping your face always toward the fixed body on the table, and then both yourself and the cube will have turned once round; for the cube will have shown the same side progressively round to all sides of the room, and your face will have been turned toward every side of the room, and every fixed point of the horizon.[162]

"2*d*, If a ship turns round, and at the same time, a man stands on the deck without moving his feet, he is turned absolutely round by the motion of the ship, tho' he has no relative motion with respect to the ship; but if, whilst the ship is turning round, he endeavours to turn himself round the contrary way, he thereby only undoes the effect that the turning of the ship would otherwise have had upon himself, and is, in fact, so far from turning absolutely round, that he keeps himself from turning at all, and the ship turns round him as round a fixed axis, although, with respect to the ship, he has a relative motion.

"Fig. 2 is a small plan or flat view of the machine in which the same letters of reference are put to the wheels in it as to those in Fig. 1, for the conveniency of looking at both the

Fig. 2.

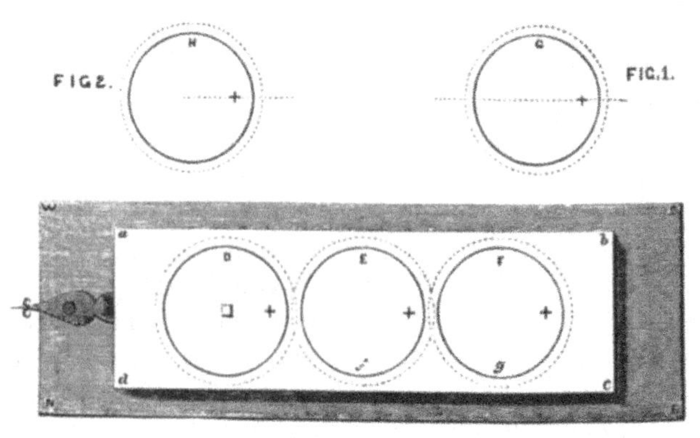

Plan of the Wheels, Frames, &c., of the Mechanical Paradox.

figures, in reading the description of them.—W S E N is the immoveable plate; D, the immoveable wheel on the fixt axis in the centre of that plate; E, the thick moveable wheel, whose

[162] We give a familiar illustration of the Moon's rotation in note 143, to which the reader is referred.

teeth take into the teeth of the wheel D; and F is one of the thin wheels, over which G and H may be put; and then F, G, and H will make a thickness equal to the thickness of the wheel E, and its teeth will take equally deep into the teeth of them all. The frame that holds these wheels is represented by the parallelogram *a b c d;* and if it be turned round, it can give no motion to the wheel D, because that wheel is fixt on an axis which is fixt into the immoveable plate.

"Take away the thick wheel E, and leave the wheel F where it lies, on the lower plate of the frame. Then turn the frame round the axis of the immoveable plate W S E N (denoted by A in Fig. 1), and it will carry the wheel F round with it. In doing this, F will still keep one and the same side toward the fixt central wheel D, as the Moon still keeps the same side to the Earth; and although F will then have no relative motion with respect to the moving frame, it will be absolutely turned round its own centre *g* (like the man on the ship whilst he stood without moving his feet on the deck), for the cross mark on its opposite side will be progressively turned toward all the sides of the room.

"But if we would keep the wheel F from turning round its own centre, and so cause the cross mark upon it to keep always toward one side of the room, or like the magnetic needle, to keep the same point still toward one fixed point in the horizon, we must produce an effect upon F resembling what the man on the ship did by endeavouring to turn himself the contrary way to that which the ship turned, so as he might keep from turning at all, and by that means keep his face still toward one and the same point of the horizon. And this is done by making the numbers of teeth equal in the wheels D and F (suppose 20 in each), and putting the thick wheel E between them so as to take into the teeth of them both; for then, as the frame is turned round the axis of the fixed wheel D, by means of the knob *n*, the wheel E is turned round its axis by the wheel D; and for every space of a tooth that the frame would turn the wheel F, in direction of the motion of the frame, the wheel E will counteract that motion by turning the wheel F just as far backward with respect to the motion of the frame, and so will keep F from turning any way round its own centre, and the cross mark near its edge will be always directed towards one side of the room. Whether the wheel E has the same number of teeth as D and F have, or any different number, its effect on F will be still the same.

"If F had one tooth less in number than D has, the effect produced on F, by the turning of the frame, would be as much more than counteracted by the intermediate wheel E, as is equal to the space of one tooth in F; and therefore, while the frame was turned once round in direction of the letters W S E N on the immovable plate, the wheel F would be turned the contrary way, as much as is equal to the space taken up by one of its teeth; but if F had one tooth more than D has, the effect of the motion of the frame (which is turned F round in the same direction with it), would not be fully counteracted by means of the intermediate wheel E, for as much of that effect would remain as is equal to the space of one tooth in F; and therefore, in the time the frame was turned once round, the wheel F would turn on its own centre, in direction of the motion of the frame, as much as is equal to the space taken up by one of its teeth; and here note, that the wheel E (which turns F) always turns in direction of the motion of the frame.

"And therefore, if an upright pin be fixed into the lower plate of the frame, under the centre of the wheel F, and if the wheel F has the same number of teeth that the fixt wheel D has, the wheel G one tooth less, and the wheel H one tooth more; and if these three wheels are put loosely upon this pin, so as to be at liberty to turn either way, and the thick wheel E takes into the teeth of them all, and also into the teeth of the fixt wheel D, then, whichever way the frame is turned the wheel H will turn *the same way*, the wheel G *the contrary way*, and the wheel F *no way at all*. The less number of teeth G has, with respect to those in D, the faster it will turn backward; and the greater number of teeth H has, with respect to those in D, the faster it will turn forward, reckoning *that* motion to be backward which is contrary both to the motion of the frame and of the thick wheel E, and *that* motion to be forward which is in the same direction with the motion of the frame and of the wheel E; so that the turning or not turning of the three wheels, F, G, H, or the direction and velocity of the motions of those that do turn round, depends entirely on the relation between their numbers of teeth, and the number of teeth in the fixt wheel D, without any regard to the number of teeth in the moveable wheel E."[163] (Vide "The Description and Use of a new Machine called the Mechanical Paradox," invented by James Ferguson, F.R.S., London, 1764, pp. 4—9; also, Ferguson's "Tables and Tracts relative to several Arts and Sciences," London, 1767, pp. 171, 172; Ferguson's "Select Mechanical Exercises," London, 1773, pp. 46—57.

We now subjoin the long and interesting letter already referred to, written by Ferguson to his friend the Rev. Mr. Cooper, of Glass, regarding the origin of the Mechanical Paradox.[164]

FERGUSON'S INTERESTING AUTOGRAPH LETTER on the ORIGIN of the MECHANICAL PARADOX.—

"REVEREND AND DEAR SIR,—I am glad that my last letter came safe to your hands, and do return you my sincere thanks for delivering the one enclosed in it to my sister, begging that you will now repeat the same favour, as it is exactly on a similar occasion, and she may still be in need of a small supply. I thank God that I am now much recovered of my gravel, and in hopes of getting quite well again.

"I herewith send you an account of my *Mechanical Paradox*, and my three letters to parson Kennedy, who is now very quiet. He has been most sadly trimmed by all the monthly reviewers. My interview with the watchmaker was as follows:— One evening I went to a weekly club with a friend, and on our entering the room (or very soon after), the watchmaker began to hold forth violently against a Trinity of persons in the Godhead, wondering at the impudence of the person who broached such an absurd doctrine, and at the weakness and folly of every one who believed it. I happened to sit just opposite to him, with the table between us, and (you may believe) plenty of wine and punch upon it. I gave him a severe frowning look, on which he asked my opinion concerning the Trinity. I told him that all my belief thereof depended upon the opinion I had of the sure knowledge and veracity of the revealer, but that I did

163 We may mention that such results from wheel-work combinations were known long before the year 1750, when Ferguson made the *Paradox*, as we find similar combinations of wheels used in the orreries and planetary machines of Rowley, Wright, and others.

164 Glass is a parish in the south-western district of Banffshire, the Church and Manse of which are situated about 10 miles to the south of Keith.

not think it was a proper subject to be talked of over our bottles, bowls, and glasses, and should therefore be desirous of talking to him about his own business. Very well, said he, let us talk about it. Sir, said I, I believe you know very well how one wheel must turn another, or how a pinion must turn a wheel, or a wheel turn a pinion. I hope I do, said he. Then, said I, suppose you make one wheel as thick as other three, and cut teeth in them all, and then put the three thin wheels all loose upon one axis, and set the thick wheel to them, so that its teeth may take into those of the three thin ones; now turn the thick wheel round; how must it turn the others? Says he, your question is almost an affront to common sense; for every one who knows anything of the matter must know that, turn the thick wheel which way you will, all the other three must be turned the contrary way by it. Sir, says I, I believe you think so. Think! says he, it is beyond a thought—it is demonstration that they must. Sir, said I, I would not have you be too sure, lest you possibly be mistaken; and now what would you say if I should say that, turn the thick wheel whichever way you will, it shall turn *one* of the thin wheels *the same way*, the other *the contrary way*, and the third *no way at all*. Says he, I would say that there never was anything proposed that could be more absurd, as being not only above our reason, but contrary thereto, and also to plain fact. Very well, says I. Now, Sir, is there anything in your ideas more absurd about the received doctrine of the Trinity than in this proposition of mine? There is not, said he; and if I could believe the one, I should believe the other too. Gentlemen, said I (looking at the company), you hear this, bear witness to it. The watchmaker asked me whether I had ever made or seen such a machine. I told him I had not, but I believed that I could make it, although I had never thought of it till that instant.[165] By G—d, says he, your head

[165] Ferguson, we have no doubt, never made a machine before this time expressly for the purpose of demonstrating such a proposition, but he was at the time familiar enough with the principle of such a machine, as he had previously made orreries, in which was wheel-work, for preserving the parallelism of the Earth's axis in its revolution round the Sun, and *the wheel*, which in such a set of wheels supports the oblique axis of the Earth, may be said to have *no motion at all;* because any given tooth in it preserves its position to a given fixed point at a great distance; he would also know that the wheel which directed the Moon's nodes *went backward*, with respect to the wheel which maintained the parallelism of the Earth's axis, while the wheel which directed the apogee motion *went forward*, in respect to it. Therefore, all that Ferguson would have to do in this

must be wrong, for no man on earth could do such a thing. Sir, said I, be my head wrong or right, I believe I can not only do it, but even be able to show the machine, if I may be admitted into this company, on this day se'ennight. The company who, with serious faces, were very attentive to all this, requested that I would come.

"So I made the machine all of wood, and carried it (under my coat) to the same room on the day appointed; and there was the watchmaker. Well, old friend, says he, have you made your machine? Yes Sir, said I, there it is—let us take it in pieces. Are these wheels fairly toothed and fairly pitched into the thick wheel? Yes they are, said he. I then turned round the great wheel, whose teeth took into those of the three thin wheels, and asked him whether the uppermost thin wheel did not turn *the same way* as the one did that turned it; whether the next wheel below did not turn *the contrary way;* and the lowermost thin wheel *no way at all?* They do, said he, but there is a fallacy in the machine. Sir, said I, do you detect the fallacy, and expose it to the company. He looked a long while at it, took it several times in pieces, and put it together again. Sir, said I, is there any fallacy in the machine? I confess, said he, I see none. There is none, said I. How the devil is it then, said he, that the three thin wheels should be so differently affected? the thing is not only above all reason, but is even contrary to all mechanical principles. For shame, Sir, said I, ask me not how it is, for it is a simpler machine than any clock or watch that you ever made or mended; and if you may be so easily nonplus'd by so simple a thing in your own way of business, no wonder you should be so about the Trinity; but learn from this not for the future to reckon *every* thing absurd and impossible that you cannot comprehend. But now I hope you remember what you said at our last meeting here; namely, that if you could believe such a thing as this, you would then believe the doctrine of the Trinity—you own the truth of the machine, what do you say to your promise? He humm'd and ha'd, and asked me whether I would let him take it home to consider it.

matter was to get one thick wheel to drive the parallelism, nodes, and apogee wheels, and this again with a fixed central wheel of same number of teeth to give motion to the thick wheel, and consequently, to the other three, by being carried round it, in the way shown by the machine.

I told him he might, but desired he would bring it to me to-morrow morning. He promised he would, and did so; but gave it me with some hearty curses, telling me he saw it was true, but did not understand it, and wanted me to explain it to him, which I refused. I kept it for six years without finding any person who could explain the principles on which it acted, and then put the Sun and Earth, with the ecliptic and Moon's orbit, to it, seeing it would then be a kind of orrery, and published the description, which I send you, in order to save myself the trouble of explaining it any longer. As it is now finished, it makes a good orrery for showing the causes of the different seasons, and the times of eclipses, &c.[166]

Reverend Sir,
Your most obliged, humble servant,
JAMES FERGUSON."

No. 4 in Bolt Court, Fleet Street,
LONDON, *April 10th*, 1776.

SIMPLE LUNAR WHEEL-WORK.—The annexed drawings are taken from one of Ferguson's old MS. papers, of date 1750. Although the wheels and pinions in this very simple train do not produce an accurate synodic period of the Moon, they are perhaps sufficiently accurate for all common purposes. According to Ferguson's paper, he used this train for one of his orreries, as also for turning round on its axis, a ball half black, half white, which exhibited the Moon's phases. It was on this train that the celebrated Mudge *operated*, through the medium of an ingenious but complicated arrangement of wheels and endless screws, which he added to it, thereby causing a period of 29 days 12 hours 44 minutes 3 seconds 12 thirds.

In the figures annexed, the pinion of 8 leaves, marked A, is the *prime mover*, and must *turn once round in* 24 *hours*, and drive a wheel of 42 teeth, on the centre of which there is made fast a pinion of 8 leaves, which drives round a wheel of 45 teeth in 29 days 12 hours 45 minutes exactly, a period too slow by

[166] It is not known if the original of this letter is in existence; there are, however, one or two copies still extant, and in Ferguson's autograph,—one of which is in the possession of our friend, Mr. Robert Sim, at Keith, and we observe that he has inserted it in the Banffshire Journal, No. IX., in his "*Walk from Keith to Rothiemay.*" From some other copy, the letter was inserted in No. 4, pp. 49 and 50 of "*The Horological Journal,*" London, December 1858.

nearly 57 seconds; an error which would amount to about a day in 120 years.

Ferguson appears through life to have been satisfied with this imperfect lunation-train, with its monthly error of 57 seconds; we find it, or some modification of the root from which the train

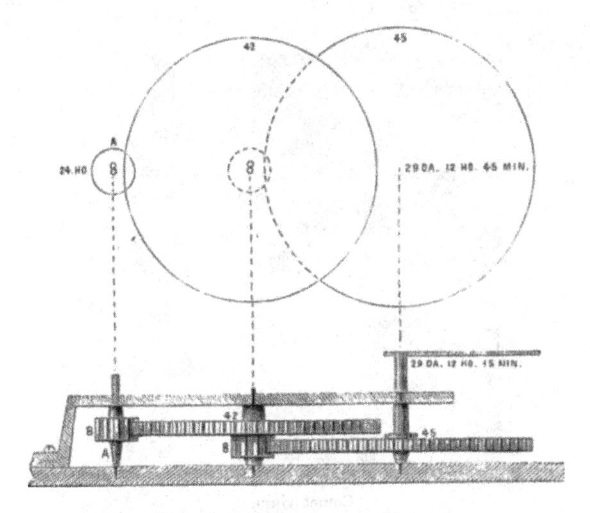

Ferguson's Simple Lunation-Train.

is derived, in all his orreries and other astronomical machinery; this is surprising, when a more accurate period could be obtained, and that too with as simple a train as that one he had adopted. Ferguson not having tried to obtain more accurate numbers for his lunation machinery, is, we think, a proof that the calculation of wheel-trains, by the process of *continuous fractions,* was unknown to him. We have shown in note 173, to which the reader is referred, a train more accurate. The calculation of the train is derived from the fraction $\frac{64}{1890}$. If $64 : 24 :: 1890 = 29$ days 12 hours 45 minutes, and 8×8 the pinion leaves $= 64$, and 42×45, the wheel teeth $= 1890$.

According to our memoranda, "Ferguson, early in the year 1750, commenced to make a great many models,—of levers, wedges, inclined-planes, the wheel and axle, pumps, carts, corn-mills, &c," for the illustration of the extended course of lectures he had determined on. It is very likely that these models are those which he afterwards described in his "Lectures on Select Subjects in Mechanics," &c.

THE COMETARIUM.—From the same source we find that Ferguson, "about the same period, made a very neat astrono-

Fig. 1. Fig. 2.

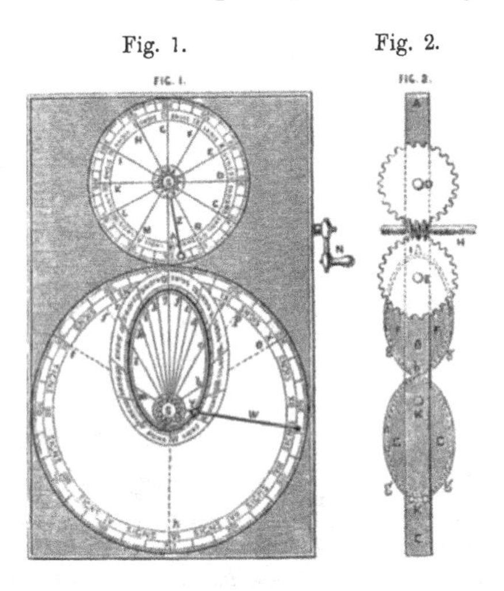

Cometarium.

mical machine, called a Cometarium, similar to the one then lately invented by Dr. Desaguliers." Ferguson gives an engraving and description of this machine in his Astronomy, which we here transcribe. "The Cometarium.—This curious machine shows the motion of a comet or eccentric body moving round the Sun, describing equal areas in equal times, and may be so contrived as to show such a motion for any degree of eccentricity. It was invented by the late Dr. Desaguliers.[167]

The dark elliptical groove round the letters *a*, *b*, *c*, *d*, *e*, *f*, *g*, *h*, *i*, *k*, *l*, *m*, is the orbit of the comet Y; this comet is carried round in the groove, according to the order of the letters, by

[167] Dr. Desaguliers, an eminent lecturer on Natural and Experimental Philosophy in London, between the years 1718 and 1744. He is the author of Mechanics, 1 vol., Plates. London, 1719; "Lectures on Experimental Philosophy," 2 vols., Plates. London, 1734. He died in 1744, aged 61 (the year after Ferguson's arrival in London). We have in our possession an ancient small circular horologe or clock which belonged to Dr. D.; after his death, it came into the possession of Dr. Franklin in 1757; of Mr. Ferguson in 1766; of Kenneth M'Culloch in 1774; and of G. W., as shown on the inside of the brass lid on the back of it.

the wire W fixed in the Sun S, and slides on the wire as it approaches nearer to, or recedes farther from the Sun, being nearest of all in the perihelion *a*, and farthest in the aphelion *g*. The areas *a* S *b*, *b* S *c*, *c* S *d*, &c., or contents of these several triangles, are all equal; and in every turn of the winch N, the comet Y is carried over one of these areas; consequently, in as much time as it moves from *f* to *g*, or from *g* to *h*, it moves from *m* to *a*, or from *a* to *b*; and so of the rest, being quickest of all at *a* and slowest at *g*.

Thus, the comet's velocity in its orbit continually decreases from the perihelion *a* to the aphelion *g*, and increases in the same proportion from *g* to *a*.

The elliptic orbit is divided into 12 equal parts or signs, with their respective degrees, and so is the circle *n o p q r s t n*, which represents a great circle in the heavens, and to which the comet's motion is referred by a small knob on the point of the wire W. Whilst the comet moves from *f* to *g* in its orbit, it appears to move only about five degrees in this circle, as is shown by the small knob on the end of the wire W; but in as short time as the comet moves from *m* to *a*, or from *a* to *b*, and it appears to describe the large space *t n* or *n o* in the heavens, either of which spaces contains 120 degrees or four signs. Were the eccentricity of its orbit greater, the greater still would be the difference of its motion, and *vice versâ*.

A B C D E F G H I K L M A is a circular orbit for showing the equable motion of a body round the Sun S, describing equal areas A S B, B S C, &c., in equal times with those of the body Y in its elliptical orbit, above mentioned, but with this difference, that the circular motion describes the equal arcs A B, B C, &c., in the same equal times that the elliptical motion describes the unequal arcs *a b*, *b c*, &c.

Now, suppose the two bodies Y and I to start from the points *a* and A at the same moment of time, and each having gone round its respective orbit, to arrive at these points again at the same instant, the body Y will be forwarder in its orbit than the body I all the way from *a* to *g*, and from A to G; but I will be forwarder than Y through all the other half of the orbit, and the difference is equal to the equation of the body Y in its orbit. At the points *a*, A, and *g*, G, that is, in the perihelion and aphelion, they will be equal; and then the equation vanishes. This shows why the equation of a body moving in an elliptic orbit, is added to the mean or supposed circular motion from the perihelion to the aphelion, and subtracted from the aphelion to the perihelion, in bodies moving round the Sun, or from the perigee to the apogee, and from the apogee to the perigee in the Moon's motion round the Earth.

This motion is performed in the following manner by the machine:—A B C is a wooden bar (in the box containing the wheel-work) Fig. 2, above which are the wheels D and E; and below it the elliptic plates F F and G G, each plate being fixed on an axis in one of its focuses, at E and K; and the wheel E is fixed on the same axis with the plate F F. These plates have grooves round their edges precisely of equal diameters to one another, and in these grooves is the cat-gut string *g g*, *g g*, crossing between the plates at *h*. On H, the axis of the handle or winch N (Fig. 1), is an endless screw, as shown in Fig. 2, working in the wheels D and E, whose numbers of teeth being equal, and should be equal to the number of lines *a* S, *b* S, *c* S, &c., in Fig. 1, they turn round their axes in equal times to one another, and to the motion of the elliptic plates; for, the wheels D and E having equal numbers of teeth, the plate F F being fixed on the same axis with the wheel E, and the plate F F turning the equally big plate G G by a cat-gut string round them both, they must all go round their axes in as many turns of the handle N as either of the wheels has teeth.

It is easy to see, that the end *h* of the elliptic plate F F, being farther from its axis E than the opposite end *i* is, must describe a circle so much the larger in proportion; and therefore move through so much more space in the same time; and for that reason, the end *h* moves so much faster than the end *i*, although it goes no sooner round the centre E. But then the quick moving end *h*, of the plate F F, leads about the short end *h* K of the plate G G with the same velocity; and the slow moving end *i* of the plate F F coming half round as to B, must then lead the long end *k* of the plate G G, as slowly about, so that the elliptical plate F F and its axis E move uniformly and equally quick in every part of its revolution; but the elliptical plate G G, together with its axis K, must move very unequally in different parts of its revolution; the difference being always inversely, as the distance of any point of the circumference of G G, from its axis at K; or in other words, to instance two points, if the distance K *k* be four, five, or six times as great as the distance K *h*, the point *h* will move in that position four, five, or six times as fast as the point *k* does, when the plate G G has gone half round; and so on for any other eccentricity or difference of the distances K *k* and K *h*. The tooth *i*, on the plate F F, falls in between the two teeth at *k*, on the plate G G, by which means the revolution of the latter is so adjusted to that of the former, that they can never vary from one another.

On the top of the axis of the equally-moving wheel D, in Fig. 1, is the Sun S in Fig. 2; which Sun, by the wire Z fixed to it, carries the ball I round the circle A B C D, &c. with an equable motion, according to the order of the letters; and on the top of the axis K, of the unequally-moving ellipsis G G in Fig. 2, is the Sun S in Fig. 1, carrying the ball Y unequably round in the elliptical groove *a b c d*, &c. N.B.—This elliptical groove must be precisely equal and similar to the verge of the plate G G, which is also equal to that of F F.

In this manner, machines may be made to show the true motions of the Moon about the Earth, or of any planet about the Sun, by making the elliptical plates of the same eccentricities, in proportion to the radius, as the orbits of the planets are, whose motions they represent; and so, their different equations in different parts of their orbits may be made plain to sight, and clearer ideas of these motions and equations acquired in half an hour than could be gained from reading half a day about such motions and equations." (Ferguson's Astronomy, 3d Edit. 4to. London, 1764, pp. 288—291).[168]

We have no other notes regarding Ferguson for 1750, but may remark that this year closed upon him with the death of his earliest London friend and benefactor, the Right Honourable Sir Stephen Poyntz, who died on 17th December, 1750 (understood to have died in the 64th year of his age). Ferguson, in his Memoir, informs us that the death of this amiable gentleman was to him the source of "inexpressible grief." (See note 74.)

168 Cometariums constructed on this plan, and sufficiently large for the lecture room at a cost of about £2 10s.; when made with eccentric wheels (instead of pulleys and cat-gut strings), the price may rise to £4.—In note 66 we express a similar opinion as to the utility of machinery for the illustration of astronomical motions.

1751.

PERPETUAL POCKET ALMANAC, 1751.—In the spring of the year 1751, Ferguson invented and published a curious "*Perpetual Pocket Almanac*," (long out of print, and now very scarce). He sent one of them in a present to his early friend, Mr. Rose of Geddes, enclosed in an interesting letter, of which the following is a copy. It is the earliest of Ferguson's letters known to be in existence; besides referring to this Almanac, it shows that his time was then taken up with his business as a limner; in delivering lectures on his favourite science of Astronomy; in making various mechanical models for the illustration of his lectures; and that his residence was in Margaret Street, Cavendish Square.

"Margaret Street, Cavendish Square,
LONDON, *June 1st*, 1751.

"SIR,—Be pleased herewith to accept of one of my Perpetual Pocket Almanacs, newly published. It will lie in your pocket-book, and with a pin you may easily shift the month and Moon-plates, as directed, provided you hold the instrument edgeways betwixt your thumb and three last fingers, keeping your fore-finger's point lightly at the back, against the openings through which the said plates are shifted. This way of holding keeps it flat, and does not pinch the plates. I am sorry that the table of semi-diurnal arcs, for showing the rising and setting of the Sun and Moon, will not answer in your latitude; but all the rest is universal, and the circles for showing the places of the Sun and Moon are adapted to the new style, as that is so soon to take place.

"I am still going on in the old way of drawing, and lecturing upon my astronomical machines, of which I've now got a good collection, and have lately finished two working models of water-mills, in one of which, the water-wheel moves a train for turning two mill-stones, and for sifting the flour as it is ground.[169] The other is of a mill to go by water, without ever a wheel or trundle, and yet it will grind as well as any, and need no repairs until the water rots out the few simple materials which compose the machine.[170]

169 This appears to be the same sort of mill mentioned by Ferguson in the list he gives of his apparatus in his "Tables and Tracts," viz. "*A model of a water mill for winnowing and grinding corn, drawing up the sacks, and boulting the flour.*" (Ferguson's "*Tables and Tracts.*" London, 1767, p. 320).

170 The second mill, although rather obscurely defined, appears also to be "Barker's mill," noticed in same list, viz. "*A model of Dr. Barker's water mill (for grinding corn), in which mill there is neither wheel nor trundle.*" (Ferguson's "*Tables and Tracts.*" London, 1767, p. 320; also, see a figure and description of this mill in the supplement to Ferguson's "*Lectures on Select Subjects in Mechanics.*")

"I should be exceedingly glad to hear from you, how you and your worthy lady and family are, to all of whom, my wife and I offer sincerely our best wishes; and I am, with great regard,[171]

SIR,

Your most humble servant,

JAMES FERGUSON."

"To Hugh Rose of Geddes, Esq., *at Nairn.*"

Perpetual Almanack.—Invented by James Ferguson, 1751.

Fig. 1. Fig. 2.

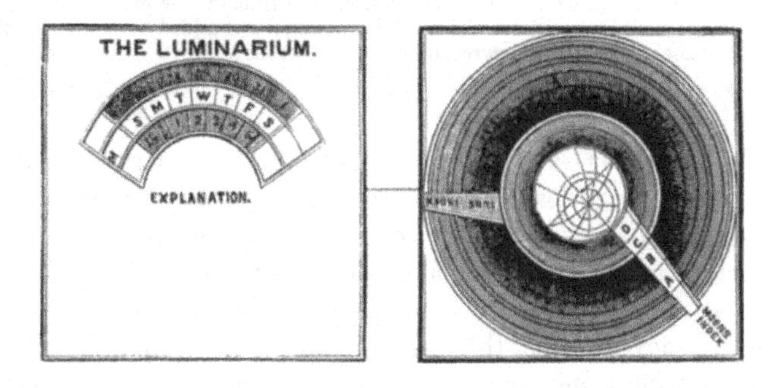

"The Perpetual Pocket Almanac" here referred to (and shown in Fig. 1 and Fig. 2 above), consists of two cards fully three inches square, held together by being stitched round the edges of the four sides with stout white silk thread, thus allowing space enough between them for the working of two moveable concentric paper circles, on the largest of which are the days of the month, seen through the cut out open arc at top (see Fig. 1); on the smaller circle are the days of the Moon's age, seen through the lower open arc. These two circles being concentric, move on a strong thread as their axis, fixed in the centre of the card. The two open arcs have a bridge-piece between them, *left* in cutting out the arcs, on which are, in capitals, the initial letters of the days of the week. Underneath are the following

[171] This letter is extracted from "A Genealogical Deduction of the family of Rose of Kilravock," printed for the Spalding Club, 1848. This is the earliest written letter of Ferguson's that we have seen.

rules or "EXPLANATION" for working both sides of the Almanac-card, which we insert below, as there is not space enough for them in our reduced engraving.

"EXPLANATION."

Fig. 1. Shift towards the left hand all that is visible of the above plates, every Sunday, and they will show the day of the month, and age of the Moon, every day of the week.

Fig. 2. On the other side, set the Sun's index to the day of the month, and it will cut his place in the ecliptic in the circle marked A (on the Moon's index), and half the time of his staying above the horizon in the circle B. Keeping the Sun's index there, set the Moon's index to the day of her age in the circle C, and it will cut the time of her coming to the meridian in the circle D, and the hours and minutes of her semi-diurnal arc in the circle B, which arc, subtracted from the time of the Moon's coming to the meridian, gives the time of her rising; and added thereto, gives the time of her setting, agreeable to her mean motion.

J. FERGUSON, *inv.* B. COLE, *sculp.*

On the other side of the card (Fig. 2) are a series of circles not moveable), described on the card, having in them the names of the month, days, &c.; the names of the signs of the ecliptic, with its degrees; and a circle of the Sun's semi-duration above the horizon. Above, are two small paper concentric circles, moveable on the same thread as the circles in Fig. 1; the lowermost one, or that next the card, has on it the twenty-four hours of the day (divided into half hours and quarters), and round its outer edge are the 29½ days of the Moon's age. This circle has an index marked "*The Sun's Index*," and extends over the whole of the fixed circles on the card; and immediately above this paper circle, and working on the same centre, is a smaller one, having on it some of the circles of the sphere, and has an index marked "*Moon's Index*," also extending over the whole of the fixed circles on the card. On this index are the letters A, B, C, D, which serve as guides to the several circles under them (see Fig. 2). The names, figures, and divisions on Fig. 2 are not given in our engraving, being too numerous for insertion.[172]

LUNAR MOTION.—About same period, Ferguson contrived "An easy way of showing the phases of the Moon in a clock."

[172] A copy of this very scarce Almanac is in the possession of Principal Forbes, D.C.L., LL.D., &c., St. Andrews, to whose kindness, for the loan of it, we are much indebted.

"Let a wheel of 16 teeth be fixed on the axis of a wheel of 15, and the wheel of 16 turn a wheel of 63, on whose axis, let a ball, half black, half white, be fixed, and project half way out through a round hole in the dial-plate.

Then if the wheel of 15 teeth be always moved one tooth in 12 hours, the ball will be turned round in 29 days 12 hours 45 minutes, and show all the varied phases of the Moon." (See Ferguson's "Tables and Tracts." London, 1767, pp. 127, 128.) The description we have just quoted is so very simple that it can scarcely be misunderstood; and Ferguson appears to have

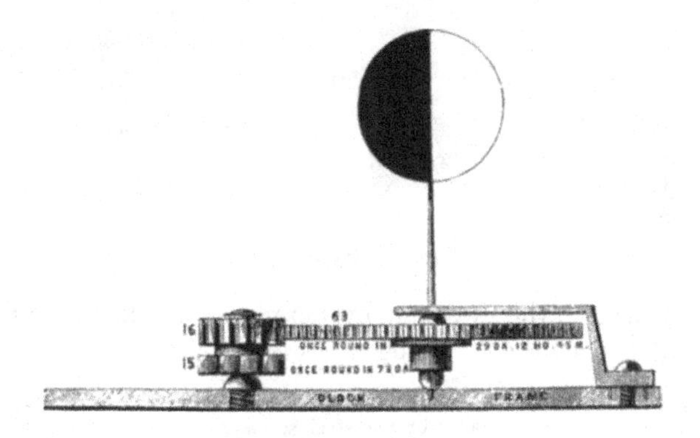

Ferguson's Simple Lunation-Work.

had the same idea of it, as he has no drawing to illustrate the description. We, however, shall annex a drawing of it, in order that the description may be better understood. It is obvious that the Moon, in this wheel-work arrangement, moves by daily *jerks* or starts, although its motion is a simple one—a continuous gradual motion is desirable, and ought to be preferred.[173]

173 We allude to a similar motion at note 151, &c. The present lunar wheel-work is evidently derived from the same root, viz. from the solid $\frac{64}{1890}$; for $64 \div 4 = 16$ for one of the wheels Ferguson here adopts, and $1890 \div 4 = 472\frac{1}{2}$, or decimally, $472{\cdot}5 \div 7\frac{1}{2}$ days $= 7{\cdot}5 = 63$ for the number of teeth in the wheel which carries the Moon; and as wheel 15 turns round in $7\frac{1}{2}$ days, with 16 pinned to it, therefore moving round also in $7\frac{1}{2}$ days, we have $63 \div 16 = 3{\cdot}9375 \times 7{\cdot}5 = 29{\cdot}53125$ d., or 29 d. 12 h. 45 m.; see notes 132—151. In note 132 we give wheel-work of our own for a gradual continuous motion (without jerks or starts), which is also very simple in its arrangement, and easy to be made, besides being about 31 seconds each lunation nearer the true motion than that given above. We made a model of it about 40 years ago, and have found it to answer well, and and to be accurate. Although we have given a sketch of the same train in note 132, we shall here repeat it, with a drawing, to show its application.

HYGROSCOPE.—In the year 1751, Ferguson invented and made a "New Hygroscope," which was noticed in many of the newspapers and magazines of the time. The following descrip-

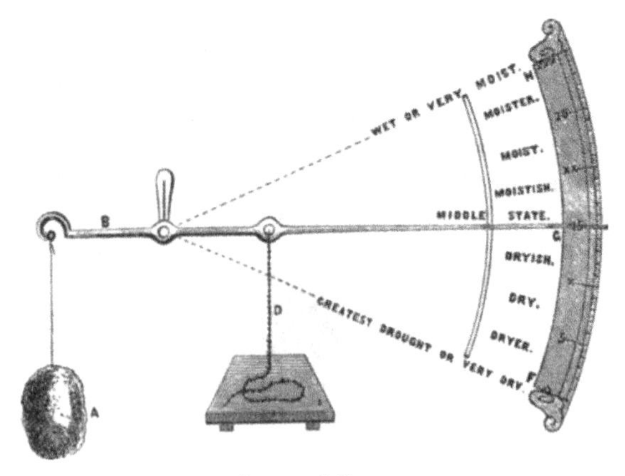

An Improved Hygroscope.

tion of it, as also the drawing, by Ferguson, is taken from the Universal Magazine for October 1751,—viz.

Description:—A A represents part of the front frame of the clock; B a bridge of brass secured to the front frame by two steel screws at *c* and *c*; within this is placed the lunar wheels, thus:—a pinion of 9 leaves make a rotation on its axis

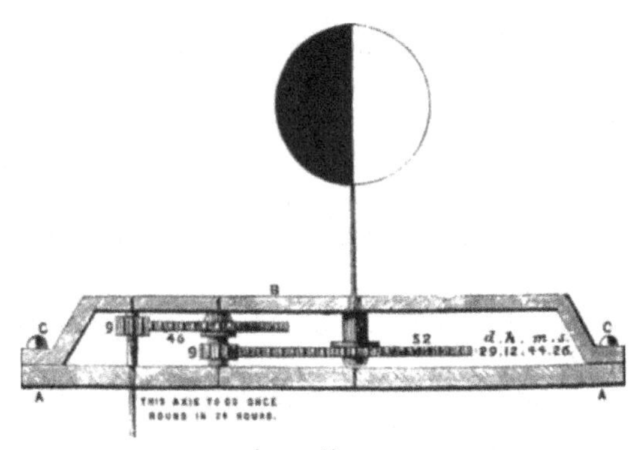

Lunar Motion.

in 24 h., and drives a wheel of 46 teeth having a pinion of 9 leaves, which drives a wheel of 52 once round (along with the ball half black, half white, representing the Moon) in 29 d. 12 h. 44 m. 26 s., for $9 \times 9 = 81$, and $46 \times 52 = 2392 = \frac{81}{2392}$, and $2392 : 81 = 29 \cdot 530861 =$ 29 d. 12 h. 44 m. $26\frac{64}{100}$ s. The number $\frac{81}{2392}$, as stated at note 132, being derived from the addition of the three solids $\frac{49}{1447} + \frac{17}{502} + \frac{15}{443} = \frac{81}{2392}$.

"There have been several hygroscopes invented by different authors for determining the true state of the atmosphere; but that recommended by the great Mr. Boyle, of weighing a piece of sponge in a pair of accurate scales, seems to come nearest to the truth. But the difficulty and time requisite to adjust the weights, and discover the true state of the air, render it very tedious and troublesome. Another method, therefore, is wanting, whereby we may perceive, at all seasons, by inspection only, the most minute alterations with respect to moisture or dryness, and this may be performed by the hygroscope represented in the annexed figure.

"A represents a thin piece of sponge, so cut, as to contain as large a surface as possible. This is fastened by a fine silk thread to the beam B, and is exactly balanced by another thread of silk at D, strung with the smallest lead shot at equal distances, and so adjusted as to cause the index to point at G, the middle of the graduated arc F, G, H, when the air is in a middle state between the greatest moisture and the greatest dryness. I represents a small table or shelf for that part of the shot which is not suspended to rest on.

"By this instrument, the most minute changes in the atmosphere, with regard to moisture and dryness, may be discovered by inspection only, as is abundantly evident from the figure." (Vide Hinton's Universal Magazine, Vol. 9th, p. 159). (Note, —A sponge, well soaked in a solution of tartar, suspended when quite dry, has been recommended). A contributor to the London Mechanics' Magazine, Vol. 1st, p. 360, introduces, apparently as *new*, a hygrometer exactly the above! Ferguson's hygroscope was published seventy-two years before the London Mechanics' Magazine was heard of.

A SUPPLEMENT TO THE ANSWER TO FERGUSON'S ESSAY.— The same author who, in 1748, published the anonymous pamphlet, entitled "An Answer to Mr. Ferguson's Essay upon the Moon's turning round its own Axis," again went to press in 1751, with a second anonymous pamphlet, entitled "A Supplement to the Answer to Mr. Ferguson's Essay on the Moon's turning round its own Axis. Printed for J. Roberts at the Oxford Arms, in Warwick Lane, MDCCLI," (an octavo of 24 pp. and 2 copperplates).

The lunar-rotation dispute, which had raged so fiercely for some years before the appearance of this Supplement, had now subsided into a calm, and ceased to be referred to,—in time, as well as in matter, this publication was out of place.

Why this author (Mr. Grove) should twice attack Ferguson, *masked*, is now not known; it has been suggested "that he probably stood in fear of the newspaper and magazine critics of his day, thinking it possible they might give to his local habitation and his name, *rotation* on an awkward axis."

The London newspapers and magazines of 1751 have been examined, but without finding even allusion made to this Supplement; hence it may be inferred that neither Ferguson nor any of the other writers on the subject of lunar rotation took any notice of it.

In taking leave of this anonymous writer we may add, that although many are of opinion that Ferguson went to Richmond to see him, immediately after the publication of his first anonymous pamphlet in 1748 (see page 126), yet, nevertheless, we think it as probable that Ferguson's visit to Mr. Grove at Richmond would be in 1751, *after* the publication of the Supplement; because, it begins by saying—"*It is now above two years since I published an Answer to Mr. Ferguson's Essay upon the Moon's turning round on an Axis within her. I expected not any reply from him, as I concluded, that what he had ventured abroad to the public, he had set forth what he took to be the strength of his cause, and had thereupon, to his entire satisfaction, received the approbation of those of his own opinion,*" &c. We scarcely think that such remarks as these would have been used had Ferguson's visit to him been *before* they were written. It is indeed probable that, had Ferguson gone to Richmond shortly after the publication of the "ANSWER" in 1748, there would have been no SUPPLEMENT to it in 1751.

FAMILIAR IDEA of the DISTANCES, &c., of the PLANETS.—According to an old memorandum, it was in the year 1751 that Ferguson first promulgated his familiar illustration of the distances and magnitudes of the Sun and Planets; as this "familiar illustration" has often been quoted, and too often without acknowledging the source from whence it was derived, we give it in Ferguson's own words,—

"The dome of St. Paul's is 145 feet in diameter. Suppose a globe of this size to represent the Sun; then a globe of $9\frac{7}{10}$ inches will represent Mercury; one of $17\frac{9}{10}$ inches, Venus; one of 18 inches, the Earth; one of 5 inches diameter, the Moon (whose distance from the Earth is 240,000 miles); one of 10 inches, Mars; one of 15 feet, Jupiter; and one of $11\frac{1}{2}$ feet, Saturn, with his ring four feet broad, and at the same distance from his body all round.

"In this proportion, suppose the Sun to be at St. Paul's; then Mercury might be at the Tower of London; Venus at St. James's palace; the Earth at Marybone; Mars at Kensington; Jupiter at Hampton Court; and Saturn at Cliefden, all moving round the cupola of St. Paul's as their common centre."[174] (Ferguson's "Tables and Tracts," 1st Edit., pp. 153, 154).

SATELLITE MACHINE.—About this period (1751), Ferguson informs us that he constructed a machine "to represent the motions of Jupiter's satellites round Jupiter, in a clock, and show the times of their eclipses in Jupiter's shadow." In his "Tables and Tracts," he describes this machine as follows:—

"On four hollow arbors, let there be four bent wires of different lengths, to carry the satellites round Jupiter, as the arbors are turned round within one another; and let Jupiter be fixed on the top of a solid axis or spindle, on which all the arbors are turned round; the wires being so bent, as that the satellites, on their tops, may be of the same height with Jupiter's ball. The diameters of the satellites should not be above a sixth or seventh part of the diameter of Jupiter, and to be at their proper distances from him; the distance of the nearest satellite should be $5\frac{2}{3}$ semi-diameters of Jupiter distant from his centre; the second satellite 9 semi-diameters of Jupiter distant from his

[174] Ferguson made this calculation when the adopted distance of the Sun from the Earth was taken at 81,000,000 of miles. The results of calculations arising out of the transits of Venus over the Sun's disc in the years 1761 and 1769, make this distance fully $\frac{1}{8}$th more, or 95,000,000 miles in round numbers for the distance of the Sun from the Earth. (See also note 113). Since Ferguson's time, the planets Uranus, Vesta, Juno, Ceres, Pallas, and Neptune have been discovered, besides, upwards of 70 asteroids, which would require a new calculation to make them all correspond to modern measures of planetary distances; this we shall leave as a task for the curious reader.

centre; the third $14\frac{1}{3}$ semi-diameters; and the fourth $25\frac{1}{3}$ of his semi-diameters from his centre.

Let four wheels of different sizes, and different numbers of teeth, be fixed upon the lower end of the above-mentioned arbors, in a conical manner; the wheel on the smallest arbor,

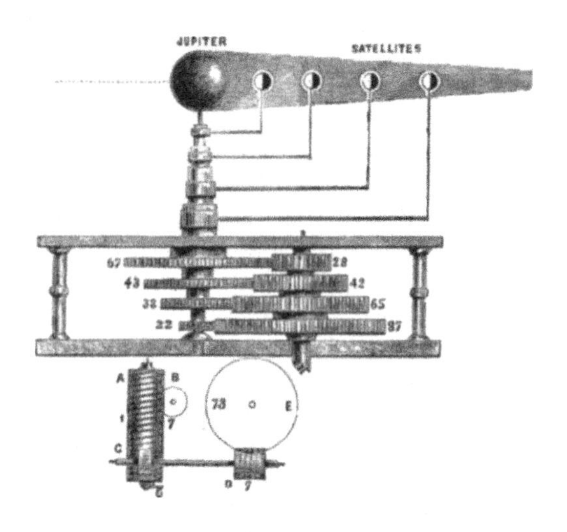

Ferguson's Satellite Machine.—Section of the Wheel-Work.

that carries the first satellite, having 22 teeth; the wheel on the next arbor that carries the second satellite, 33 teeth; the next bigger wheel on the arbor that carries the third satellite, 43 teeth; and the largest wheel of all, on the arbor that carries the fourth or outermost satellite, 67; the biggest wheel being the uppermost, and the smallest the lowermost.

These four wheels must be turned by other four, all fixed on a solid axis, in an inverted conical manner, with respect to the former wheels on the hollow arbors; and then, all the four on the solid axis will be turned round in one and the same time.

The smallest wheel (or uppermost one) on this axis must have 28 teeth, and turn the wheel of 67 teeth, which carries the fourth satellite.

The next wheel on the axis must have 42 teeth, and turn the wheel of 43 teeth which carries the third satellite.

The next bigger wheel below on the axis, must have 65 teeth, and turn the wheel of 33 teeth, which carries the second satellite; and

The lowermost, and biggest wheel on the axis, must have 87 teeth, and turn the wheel of 22, which carries the first satellite. Then,

If the clock turns the solid axis with all its wheels round in 7 days, the first satellite will be carried round Jupiter in 1 day 18 hours 28 minutes 57 seconds; the second satellite in 3 days 13 hours 17 minutes 46 seconds; the third in 7 days 3 hours 59 minutes 54 seconds; and the fourth satellite in 16 days 18 hours 0 minutes 0 seconds; which agrees so nearly with their revolutions in the heavens, as not to differ sensibly, in a long time, from them.

And then, if a piece of black wood be turned, a little conical in its shape, having its thickest end as broad as the diameter of Jupiter is long, and be made hollow to fix on the back of Jupiter, and have notches cut in it for the satellites to pass through, it will represent Jupiter's shadow; and when the satellites are in the notches, it will show them to be eclipsed.

The times of the immersions of the satellites of Jupiter into his shadow, or of their emersions from it, may be had from *White's* Ephemeris every year; and if the satellites are once put just entering the notches for the immersions, or just leaving it for the emersions at the proper times by the clock; they will keep right to the times thereof for more than a year afterward, without needing any new adjustment. And in order that they may be so set, without affecting the wheels that move them, their wires should be fixed into round collars which go moderately tight on the tops of the four hollow arbors, so as they may be carried about Jupiter by the tightness of the collars, and yet at any time may be moved and set right by hand." (See Ferguson's "Tables and Tracts." Lond., 1767, pp. 154—159).

The following is a tabular view of the wheel-work and periods produced by this satellite machine, and a column showing the true revolutions:—[175]

[175] We have compiled the above table from the satellite wheel-work. The last column, entitled "True periods in the Heavens," is taken from a MS. table by Ferguson, in our possession. Such a satellite, with brass wheels, may be purchased for about £5.

Satellites.	Wheel-work.	Periods produced by the Wheel-work.	Ferguson's 'True Periods in the Heavens.'
1st	$\frac{87}{22}$ of 7 days	1 d. 18 h. 28 m. 46 s.	1 d. 18 h. 28 m. 36 s.
2d	$\frac{65}{33}$ do.	3 d. 13 h. 17 m. 32·30 s.	3 d. 13 h. 17 m. 54 s.
3d	$\frac{42}{43}$ do.	7 d. 4 h. 0 m. 0 s.	7 d. 3 h. 59 m. 36 s.
4th	$\frac{28}{67}$ do.	16 d. 18 h. 0 m. 0 s.	16 d. 18 h. 5 m. 6 s.

It is to be observed that Ferguson makes $\frac{87}{22}$ of 7 days = 1 day 18 hours 28 minutes 46 seconds; whereas 87 : 7 d. : : 22 produces 1 day 18 hours 28 minutes 58 seconds nearly; also, he makes $\frac{65}{33}$ of 7 days = 3 days 13 hours 17 minutes 46 seconds; but 65 : 7 : : 33 = 3 days 13 hours 17 minutes 32·3 seconds; lastly; $\frac{42}{43}$ of 7 days = 7 days 3 hours 59 minutes 54 seconds; but 42 : 7 d. : : 43 produces 7 days 4 hours exactly.

Ferguson adds that "all the numbers of teeth in the wheels are here copied from Mr. Roemer's satellite instrument, except those for the second satellite, where Mr. Roemer has a wheel of 63 teeth turning a wheel of 32, instead of which, I make a wheel of 65 turn a wheel of 33, which is much nearer the truth."[176] Ferguson concludes his description by informing us that

[176] By referring to the last column of the above satellite table,—which, as already mentioned, is from a manuscript table by Ferguson, it will be seen that the "true period in the heavens," of the second satellite of Jupiter, is set down at 3 d. 13 h. 17 m. 54 s.; and hence, Ferguson's wheels of $\frac{65}{33}$ was a much nearer approach to the true period than Roemer's wheels of $\frac{63}{32}$. Roemer's wheels of $\frac{63}{32}$ produce a period of 3 d. 13 h. 20 m., while Ferguson's wheels of $\frac{65}{33}$ produce a period of 3 d. 13 h. 17 m. 32·30s., being a period within 21 s. of the true period, as then understood. Sir John Herschell in his Astronomy gives 3 d. 13 h. 14 m. thus throwing Ferguson's wheel-work period about $3\frac{1}{10}$ minutes too slow. If Sir John's period of 3 d. 13 h. 14 m. be adopted, then, by continuous fractions, we find that if we make the wheels for the second satellite $\frac{62}{35}$, it will produce a period of 3 d. 13 h. 13 m. 2·6 s., which is within 57 s. of his period. We may here note that we once had in our possession Ferguson's wheel of 33 which carried the second satellite in his machine; it is of boxwood, and the teeth beautifully cut into it. This wheel is now among the *Ferguson relics* in the Museum at Banff.

Wheels $\frac{87}{22}$, used for the 1st satellite, produce a period *too slow* by about 22 seconds. By the process of continuous fractions, applied to the period of the 1st satellite, from a *prime mover* of 7 days, we find that the fraction $\frac{91}{23}$ produces a period within $5\frac{1}{2}$ seconds of the true time,—thus, let a wheel of 91 teeth turn round in 7 days, and be made to drive a wheel of 23 teeth, this wheel of 23 teeth will turn once round on its axis in 1 d. 18 h. 28 m. 41·5 s.; the true period being 1 d. 18 h. 28 m. 36 s.

There is a rude woodcut figure, and also a full description of Roemer's satellite machine, in "*Harris's Lexicon Technicum,*" a work which Ferguson says in his Life, that he perused while he was on a visit to Squire Baird at Auchmedden, in 1733; it is probable that Ferguson *then* copied the woodcut figure and the description. (See Harris's Lexicon Technicum. London, 1725; article "Satellite Machines.")

"About 16 years ago,[177] I made one of these instruments to be turned with a winch by hand. It had a dial-plate divided into the months and days of the year, within which was a circle divided into twice twelve hours. On this plate there were two indexes, one of which was moved round over all the 365 days of the annual circle in 365 turns of the winch; and the other index was moved round over all the 24 hours in one turn of the winch, by which means I could, in a very short time, show at what times of the day the satellites would be eclipsed throughout the whole year;[178] and, after having the above numbers for

177 Ferguson here mentions that he made his satellite machine "about 16 years ago," that is, 16 years before the year 1767, when his "Tables and Tracts," (from which this extract is taken), were published, and consequently refers to the year 1751.

178 Ferguson made three satellite machines,—the one just described was the first he made, simple in its construction, and being copied from Roemer's instrument, it would, like it, have a *prime mover* of 7 days. Shortly after it was finished, Ferguson discarded the prime mover of 7 days, and added to it one of 24 *hours*. We have in our possession (pasted on a small piece of mahogany) a short description (in MS. by Ferguson), showing its arrangement in its altered state, and of which the following is a copy,—he says, "On the axis of the handle is an endless screw which turns a wheel of 12 teeth, on whose axis is a pinion of 6 leaves, turning a wheel of 16 teeth, which turns a wheel of 42, on whose axis is a pinion of 7 leaves, turning a wheel of 60 teeth, on whose axis is a pinion of 12 leaves, turning a wheel of 73 teeth, on whose axis is the annual index which goes round the circle of months in 365 revolutions of the hour index, which is on the axis of the first wheel and pinion. The above-mentioned wheel of 42 teeth turns a wheel of the same number, on whose axis three other wheels are fixed, one of which has 87 teeth, the next above it 65, the next above which is the wheel 42, and the next, or uppermost, is a wheel of 28 teeth." The several wheels and pinions in this part of the train are to be arranged thus,—$\frac{6}{16} \times \frac{16}{42} \times \frac{7}{60} \times \frac{12}{73} = \frac{8064}{2943360}$, and $2943360 \div 8064 = 365$ exactly; or simply, $\frac{6}{42} \times \frac{7}{60} \times \frac{12}{43} = \frac{504}{183960}$, and $183960 \div 504 = 365$ as before. We annex a sectional view of the wheels and pinions calculated and adopted by Ferguson for his *new prime mover*

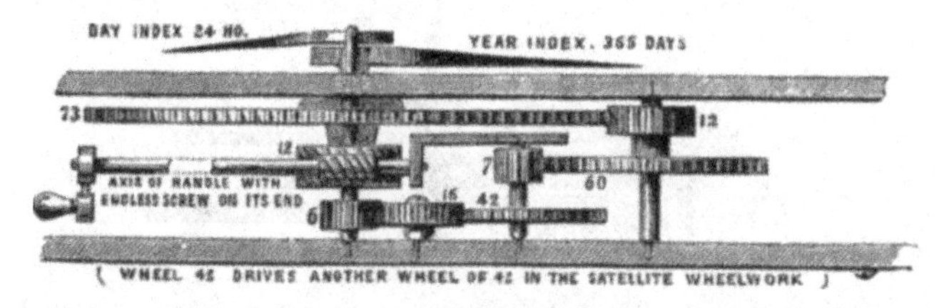

of 24 hours, in order that it may be the more easily understood. Below our sketch of the wheel-work of Ferguson's satellite machine will be found a very simple plan we adopted to accomplish the same motions;—A is a long endless screw, making a revolution in 24 hours, which turns round a small wheel B of 7 teeth in 7 days, which wheel drives the satellite machinery. The long endless screw also drives a small wheel of 5 round in 5 days, which has an endless screw D that drives wheel 73 round in 365 days, or in 365 turns of the endless screw. By this simple arrangement we obtain movers of 1 day (or 24 hours), of 7 days, and of 365 days.

the motions of the satellites, any clockmaker may easily construct a machine of this sort, by which the times of the immersions and emersions of the satellites may be known before-hand, in order to be prepared for observing them in the heavens." ("Tables and Tracts." Lond., 1767, pp. 158, 159).

We have no other memoranda of Ferguson for the year 1751, excepting, that "he was occasionally occupied towards its close, in delivering lectures on astronomy, and the rectification of the Calendar."

1752.

FERGUSON'S *seeming* APPLICATION for some VACANT SITUATION.—Early in the year 1752, his attention appears to have been directed to some vacant situation or office in or about

To adopt 7 days as the *prime mover* of the wheel-work in a satellite machine is rather an inconvenient *mover;* a prime mover of 24 hours is preferable. We give the reader a tabular view and section of the wheel-work of a satellite machine made in 1805 by the late Dr. Wm. Pearson, having a prime mover of 24 hours, that its wheels and resulting periods may be compared with the Ferguson-Roemer machine just described.

Satellites.	Wheels.	Periods by Wheels.	Dr. Pearson's True Synodic Periods.
1st	$\frac{39}{69}$ of 24 h.	1 d. 18 h. 27 m. 41 s.	1 d. 18 h. 28 m. 35·95 s.
2d	$\frac{27}{96}$ do.	3 d. 13 h. 20 m. 0 s.	3 d. 13 h. 17 m. 53·75 s.
3d	$\frac{18}{129}$ do.	7 d. 4 h. 0 m. 0 s.	7 d. 3 h. 59 m. 35·87 s.
4th	$\frac{8}{134}$ do.	16 d. 18 h. 0 m. 0 s.	16 d. 18 h. 5 m. 7·09 s.

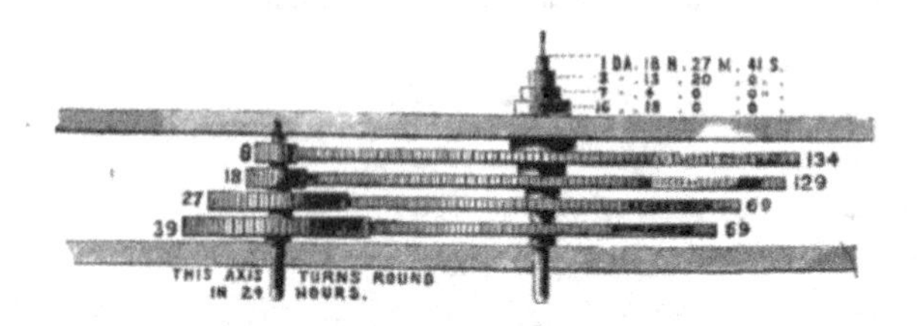

This sectional view of Dr. Pearson's satellite machine will be readily understood from the description of Ferguson's machine. We may just note that wheels 39, 27, 18, and pinion 8, are all fixed on the same axis, and revolve with it in 24 hours, which is the unit or prime mover. Wheel 39, turning round in 24 hours, drives wheel 69 round in 1 d. 18 h. 27 m. 41 s. Wheel 27, turning round in 24 hours, drives wheel 69 round in 3 d. 13 h. 20 m. Wheel 18, turning round in 24 hours, drives wheel 129 round in 7 d. 4 h.; and pinion 8, turning round in 24 hours, drives wheel 134 round in 16 d. 18 h. It is somewhat remarkable that two of the periods in this machine are precisely alike to those in Ferguson's, viz. the periods of the 3d and 4th satellites.

London, and he became a candidate for it; and as the matter is nowhere afterwards mentioned, it is to be presumed that he was one of probably a great many unsuccessful candidates. All our endeavours to ascertain something about this vacant office have proved fruitless.

In reference to it, however, is the following letter from Martin Folkes, Esq., President of the Royal Society, to the Rev. Mr. Birch, its Secretary, at the Royal Society's house in Crane Court:—

"I pray the favour of Mr. Birch that he will please to present this recommendation at the meeting of the Society this afternoon, as it is my humble request that any of the gentlemen present, to whom it may not be disagreeable, to promote the success of Mr. James Ferguson, who is very strongly recommended to me as a gentleman of the greatest merit.

From the Society's most dutiful,
humble servant,

"*Feby.* 20*th*, 1752." M. FOLKES."

Copied from the Birch collection of letters in the British Museum, No. 4308, Vol. 9, p. 282.

During the first months of 1752, we find that "Ferguson was still busy with his lectures on Astronomy, and on the rectification of the Calendar; and it would appear that he had considerable success, principally owing, perhaps, to the curiosity of the public being then aroused on the subject of the Calendar, and the proposal for striking off 11 days from the year 1752."[179]

[179] "The rectification of the Calendar was effected on 3d September, 1752; the 3d of September of this year was called the 14th, thus striking off the 11 days which had been in excess." We have before us "An Almanac for the year of our Lord 1752, and from the World's Creation 5754 years, by Tycho Wing, Philomath." On looking into the September of this year, in it, we find the following notice, which is inserted in a blank space between September 2d and 14th:

Sept. 1752.
"According to an Act of Parliament passed in the 24th year of his Majesty's reign, and in the year of our Lord 1751, the old style ceases here, and the new style takes place; and consequently, the next day, which in the old account would have been the 3d, is now to be called the 14th: so that all the intermediate nominal days, from the 2d to the 14th, are omitted, or rather annihilated this year; and this month contains no more than 19 days."

Although this is not the proper place to enter on a discussion of the Calendar, a few remarks, however, may be made to show the nature of the subject Ferguson had to deal with. In A.D. 325, the vernal equinox occurred on March 21st. About the year 1750, the vernal equinox occurred on March 9, making a differ-

THE LUNARIUM.—About the beginning of the year 1752, Ferguson invented a sort of Rotula, which he called "The Lunarium." A few of these Lunariums are still to be found. It consists of several concentric moveable card-discs, the largest of which is about 15 inches in diameter, the smallest, $5\frac{3}{4}$ inches, on which are engraven the $29\frac{1}{2}$ days of Moon's age; the 24 hours of the day; circles full of dates for 6000 years before and after the year 1800, &c. Within a space on the middle card-disc, we find the following:—

> "A LUNARIUM, showing the Days of all the New and Full Moons, and the Moon's Age every day for 6000 years before or after any year in the 18th century, and the time of the Moon's Southing on each day of her Age. By James Ferguson."

This Lunarium has been long out of print and is now very scarce. Our friend Mr. Robert Sim, in Keith, has a MS. copy of one. We have the MS. middle card-disc of one, from which the above note is extracted.

ence of 11 days; this was occasioned by estimating and using 365 d. 6 h. as the length of the year, taking in the 6 hours every fourth year, as they in that time amounted to 24 hours or 1 day. The true mean length of a year is 365 d. 5 h. 48 m. 52 s., which, on being subtracted from 365 d. 6 h., shows a difference of 11 m. 8 s., that is, the year estimated at 365 d. 6 h. is too long by 11 m. 8 s.; this excess, 11 m. 8 s. $\times 4 = 44$ m. 32 s. that had been added to the Calendar *every* four years since A.D. 325. About the year 1750, these 44 m. 32 s. added every four years since A.D. 325, had amounted to 11 days very nearly; and hence it was necessary to strike off these 11 days in order to restore the vernal equinox again to 21 March. Had the year consisted of 365 d. 6 h. exactly, then, by taking up these 6 hours every fourth year, and making a day of them, would have been perfect, and engendered no error in any time to come; the yearly excess of 11 m. 8 s. amounts to an entire day in about $129\frac{1}{2}$ years.

It will be seen from those remarks that the question of the Calendar is simply one of dealing with the 5 h. 48 m. 52 s. What part of a day is 5 h. 48 m. 52 s.? —By continuous fractions it is shown that the fractional equivalents for this 5 h. 48 m. 52 s. are $365\frac{1}{4}$ days, $365\frac{7}{29}$ days $365\frac{8}{33}$; now, it has been shown that $365\frac{1}{4}$ is a term too great by 11 m. 8 s.; then try $365\frac{7}{29}$, this gives a nearer approximation to the truth, as $365\frac{7}{29}$ days reduced is $=$ 365 d. 5 h. 47 m. 35 s., being a period only 1 m. 17 s. faster than the true period; therefore, by adding 7 entire days to the Calendar every 29 years, the error of a day would not occur in a less period than 992 years. Suppose all the months every 29th year to have 31 days each, this would add these 7 days to the Calendar, and no rectification would be required for near 1000 years. $365\frac{8}{33}$ would give a period still nearer, For a full exposition of the Calendar, and how in future it is to be dealt with by leap centenary years, see any work on Astronomy. We have in our possession a small and very rare volume, entitled, "THE EARL OF MACCLESFIELD'S SPEECH IN THE HOUSE OF PEERS, ON MONDAY THE 18TH DAY OF MARCH, 1750, AT THE SECOND READING OF THE BILL FOR REGULATING THE COMMENCEMENT OF THE YEAR, &c. LONDON: PRINTED BY CHARLES DAVIS, PRINTER TO THE ROYAL SOCIETY, M.DCC.LI.," which gives a very clear description of the Calendar, and how it was to be rectified.

ROTULAS.—The several Astronomical Rotulas which Ferguson had invented and had published previous to 1752, were, by the alteration of the style this year, as before noticed, rendered entirely useless. He therefore directed his attention to correcting them, and had them newly engraved, to suit the altered style, to take effect on 3d September.

LUNAR TABLES and LUNAR NODE TABLES.—During the year 1752, Ferguson calculated and published these two sets of tables; the first one was entitled "A Table, showing the days of all the mean changes of the Moon, from A.D. 1752 to 1800, New Style." The second was "A Table, showing the days of the Sun's conjunctions with the nodes of the Moon's orbit, from A.D. 1752 to 1800, New Style." These tables were printed on large sheets of paper at One Shilling each; they have been long out of print. They were, however, reprinted by Ferguson in his "Tables and Tracts." London, 1767, pp. 30, 31, and 34, where the reader will find them.

GEOMETRICAL CARDS, &c.—Ferguson appears to have transcribed on cards, and in a very careful and neat manner, every curious mathematical or geometrical problem that happened to come in his way. At the time of his death, in 1776, he had a large collection of these cards, and also papers and drawings on various philosophical subjects. Many of these cards and papers came into the possession of the late amiable Capel Lofft, Esq. On making inquiries regarding these, we learn that "it is now a very long time since they were seen, and it is to be feared that they are irretrievably lost." Very few of these curious cards can now be in existence; as we happen to have seven, we shall here give copies and descriptions of five of them, viz. Nos. 1, 2, 3, 4, 5, dates 1752; the other two will be found under dates 1743 and 1770, illustrating questions under discussion in these years.

CARD NO. 1.—The cards on which the geometrical figures are delineated and described are $4\frac{7}{8}$ inches by $3\frac{1}{8}$ inches, with the initials J. F., and date 1752, in the lower right-hand corners; No. 1 is entitled, "To find a Square equal in Area to a given Circle." Although ingenious, it can be received as an approximation only to what is generally and popularly known as "The Squaring of the Circle," that is, the area of a circle being given, to find a square whose area will mathematically be expressed by the same quantity. This cannot be done, because the mathematical ratio of a diameter to a circumference cannot be determined; if this fact were borne in mind by professed squarers of the circle, before entering upon their squaring operations, it would save them both time and trouble.

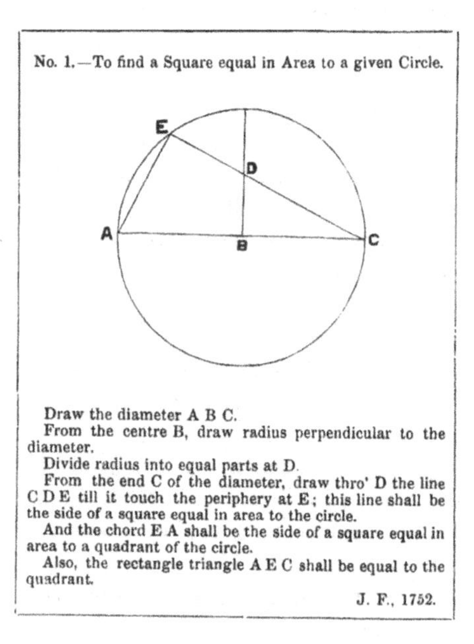

No. 1.—To find a Square equal in Area to a given Circle.

Draw the diameter A B C.
From the centre B, draw radius perpendicular to the diameter.
Divide radius into equal parts at D.
From the end C of the diameter, draw thro' D the line C D E till it touch the periphery at E; this line shall be the side of a square equal in area to the circle.
And the chord E A shall be the side of a square equal in area to a quadrant of the circle.
Also, the rectangle triangle A E C shall be equal to the quadrant.

J. F., 1752.

The Squaring the Circle, like the Trisection of the Angle, and the Duplication of the Cube, are indeterminate questions. On the back of this Geometrical Card, we find the following note:—*Also, the square formed on the chord A E (and equal to that triangle) shall be the biggest that can be inscribed in the semicircle, and equal to the biggest circle that can be inscribed in the semicircle.*

GEOMETRICAL CARD No. 2.—This card is entitled, To make two equal Circles, whose Areas, taken together, shall be equal to the Area of a given Circle, or four equal Crescents, the sum of whose Areas shall be equal to the Area of a given square.

In looking into Ferguson's *Select Mechanical Exercises*, we find, in Plate IX., Fig. 5, a diagram similar to the annexed cut; and at page 125, a more full description than what is given on the card,—we shall therefore give the details, and shall also do the same with Cards 3, 4, and 5.

"Let *c d k i* be the given circle. In this circle, describe the square *e f m l;* and on the middle points of any two sides, as at *e* and *f* as centres, describe the two circles *A a* B E A and B *b* C E B; the areas of these two circles, taken together, shall be equal to the area of the given circle *c d k i*.

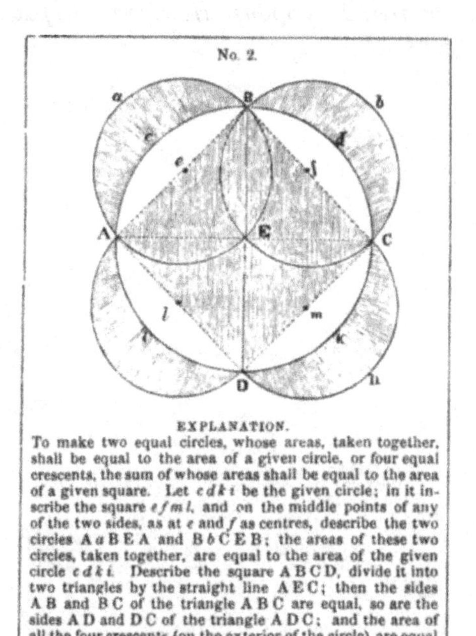

EXPLANATION.

To make two equal circles, whose areas, taken together, shall be equal to the area of a given circle, or four equal crescents, the sum of whose areas shall be equal to the area of a given square. Let *c d k i* be the given circle; in it inscribe the square *e f m l*, and on the middle points of any of the two sides, as at *e* and *f* as centres, describe the two circles A *a* B E A and B *b* C E B; the areas of these two circles, taken together, are equal to the area of the given circle *c d k i*. Describe the square A B C D, divide it into two triangles by the straight line A E C; then the sides A B and B C of the triangle A B C are equal, so are the sides A D and D C of the triangle A D C; and the area of all the four crescents (on the exterior of the circle) are equal to the area of the square, &c. J. F., 1752.

"Draw the diagonal A E C, which will divide the square into two triangles A B C and A D C, right-angled at B and D. Now as the sides A B and B C of the triangle A B C are equal, and so are the sides A D and D C of the triangle A D C, and the areas of circles being as the squares of their diameters, and the hypothenuse A C being squared, is equal to the square of the sides A B and B C or A D and D C; the larger semicircle A *c*

B d C is equal to the two lesser semicircles A a B e A and B d C f B; consequently, if you subtract the two common portions A c B e A and B d C f B, the two remaining crescents A a B c A and B b C d B will be equal to the two triangles AEBA and BCEB, which make one half of the square $e\ f\ m\ l$; and therefore, the sum of the areas of all the four outward crescents is equal to the area of the whole square."

Geometrical Card No. 3 is represented in the annexed woodcut. It is entitled Geometrical Methods for producing *Triangles, Squares, Pentagons, Hexagons, Heptagons, Octagons, Nonagons, &c.*

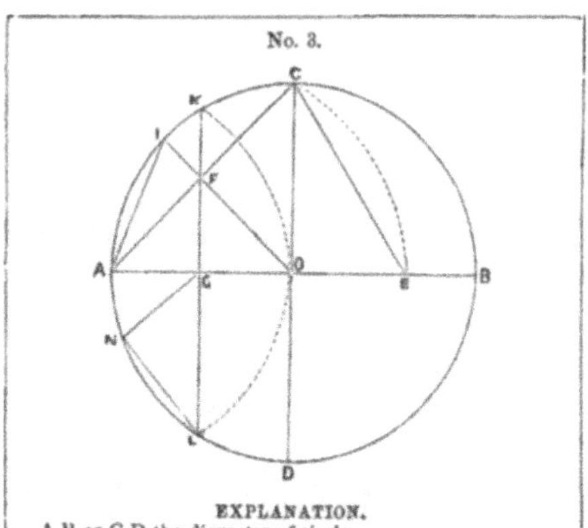

EXPLANATION.

A B or C D the diameter of circle.

Put one foot of the compasses in A, extend the other foot to O, and draw the dotted curve K L, and draw the line K L which will be the side of a *triangle*, which may be inscribed in the circle. A C is the length of a line equal to the side of a *square* inscribed in the circle. Put one foot of the compasses in G and extend the other to C, and draw the dotted arc C E, then draw the line C E, and it will be the side of a *pentagon*, inscribed in the circle. Any of the semi-diameters will be the side of a *hexagon*, inscribed in the circle. Half the line K L or K G L G are the sides of a *heptagon*, inscribed in the circle. Divide the line A C into two equal parts, then draw the line O F I, then the line A I is the side of an *octagon*, inscribed in the circle. Divide the arc L A K into three equal parts, then draw the line L N, and it will be the side of a *nonagon*, inscribed in the circle. The line O E is the side of a *decagon*, inscribed in the circle. The line N G is the side of an *undecagon*, and A G or G O the sides of a duodecagon, inscribed in the circle.

J. F., 1752.

As Ferguson gives a more full description of these methods in another paper, we shall insert it also.

"How to find the side of any regular polygon, from a trigon to a duodecagon, which may be inscribed in any given circle.

Suppose the circle A B C D. *First,* through the centre O draw the straight line A O B, the diameter, dividing the circle into two equal parts. *Second,* take in your compasses half the diameter A O or O B, and putting one foot in A, with the other mark off K and L points, and draw the straight line K L, which line will be the length of the side of *a triangle,* inscribed in the circle. *Third,* draw the line C D through the centre O, cutting the diameter A B at right angles; and then join A C, and this line will be the length of the side of *the square,* inscribed in the circle. *Fourth,* set one foot of the compasses in G; extending the other to C, describe the arc and chord C E, and the length of the chord will be the side of *a pentagon,* inscribed in the circle. *Fifth,* any of the semi-diameters A O, O B, &c., are equal in length to the sides of *a hexagon,* inscribed in the circle. *Sixth,* half the line K L, viz. K G or L G, are the sides of *a heptagon,* inscribed in the circle. *Seventh,* divide the line A C into two equal parts in F; then draw the line O F I, cutting the circumference of the circle in I; then join A I, and the length of this chord or line will be the length of the side of *an octagon,* inscribed in the circle. *Eighth,* divide that part of the circle L A K into three equal parts, one third of which is from L to N; then draw a line from L to N, and it will be the length of the side of *a nonagon,* inscribed in the circle. *Ninth,* the length of the line O E is equal to the side of *a decagon,* inscribed in the circle; and *lastly,* the line N G is the length of the side of *an undecagon* or figure of eleven sides, inscribed in the circle; and the lines A G, G O, are equal to the length of the side of *a duodecagon,* inscribed in said circle."

GEOMETRICAL CARD No. 4.—It will be seen that the diagram in the annexed woodcut is the same as on Plate IX., Fig. 6th, in the *Select Mechanical Exercises;* and at page 130 we find the following full description of it. The Card is entitled *Matter Infinitely Divisible.*

"Let A B be a straight line produced to an infinite length beyond B, and straight throughout. On this line let there be an infinite number of equilateral triangles placed, as A *a b*, *b c d*, *d e f*, *f g h*, &c., whose bases A *b*, *b d*, *d f*, *f h*, &c., touch one another upon the right line A B; and let the side *a b* of the first triangle be of any given length, as suppose an inch, and each side of each triangle be of the same length with *a b*.

"Then, from the point A, draw the straight line A *c* to the top of the second triangle *b c d*, and A *c* shall cut *a b* in the middle point at *m*.

"From the point A draw the straight line A *e* to the top of the third triangle *d e f*, and A *e* shall cut *a b* at *n*, in two-thirds of its length from *a*, leaving only one-third remaining from *n* to *b*.

Matter infinitely Divisible

EXPLANATION.

To show that an angle may be continually diminished, and yet never be reduced to nothing, and consequently, that matter is infinitely divisible.—Let A B be a straight line produced to an infinite length beyond B, and straight throughout. On this line let there be an infinite number of equilateral triangles placed as A *a b*, *b c d*, *d e f*, *f g h*, &c., whose bases A *b*, *b d*, *d f*, *f h*, &c. touch one another upon the right line A B, and let the side *a b* of the first triangle be of any given length, as suppose an inch, and each side of each triangle be of the same length with *a b*.

Then, from the point A, draw the straight line A *c* to the top of the second triangle *b c d*, and A *c* will cut *a b* in the middle point at *m*, &c. Here it is plain that every line drawn from A to the top of the next triangle, beyond that to which the last preceding line was drawn, will make a less angle with the line A B than the last preceding line did. But no right line drawn from the point A to the top of any triangle placed upon A B, even at an infinite distance from A, could ever coincide with the line A B, although every succeeding line will make a less angle with A B than the line last drawn before it did; and therefore, the angle at A will be continually diminishing, but can never come to nothing; consequently, the whole line *a b* will never be exhausted or quite cut off by any line drawn from A to the top of any triangle; and therefore, a part of it will still remain between *a* and *b*, which proves that matter is infinitely divisible.

No. 4. J. F., 1752.

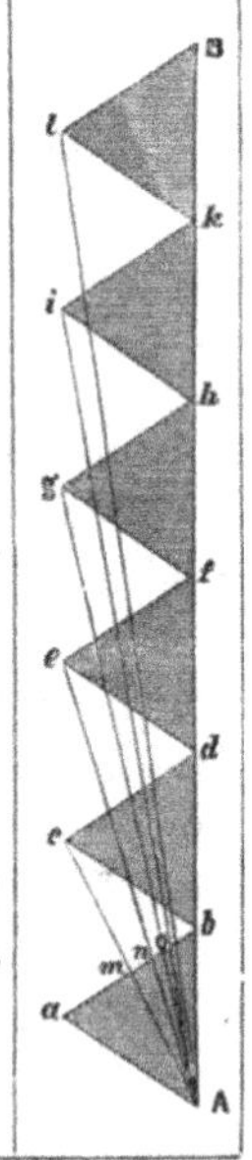

"From the point A draw the right line A *g* to the top of the fourth triangle *f g h*, and A *g* shall cut *a b* at *o* in three-fourths of its length from *a*; and consequently leave one-fourth of it remaining from *o* to *b*.

"Here it is plain that every line drawn from A to the top of the next triangle, beyond that to which the last preceding line was drawn, will make a less angle with the line A B than the last preceding line did. But no right line drawn from the point A to the top of any triangle placed upon A B, even at an infinite distance from A, could ever coincide with the line A B, although every succeeding line will make a less angle with A B than the line last drawn before it did; and therefore, the angle at A will be continually diminishing, but can never come to nothing.

"Consequently, the whole line *a b* will never be exhausted or quite cut off by any line drawn from A to the top of any triangle; and therefore, a part of it will still remain between *a* and *b*, which proves that matter is infinitely divisible."

GEOMETRICAL CARD No. 5.—This card is entitled, *A Card, showing how any given Circle may be cut so as to form two Ovals, without any waste;* and the annexed woodcut, with its short description, will be sufficiently clear for showing how such may be done.

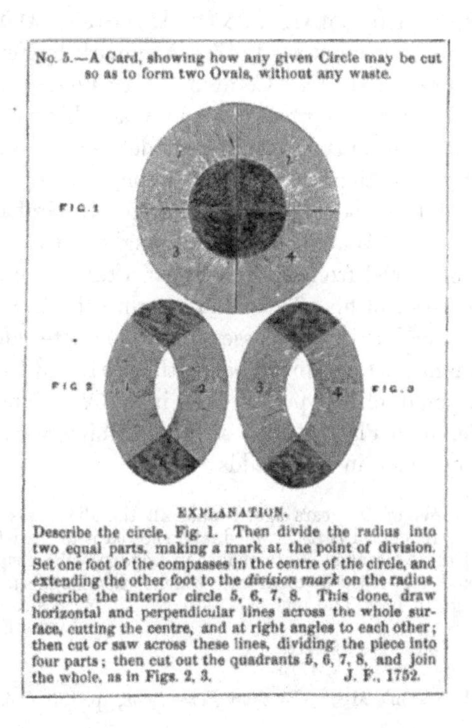

These cards were given to us in 1831 by Deane Walker, Esq., London, Lecturer on Astronomy and the Eidouranion, which had belonged to his father, Adam Walker, Esq., Lecturer on Natural Philosophy, and an old friend of Ferguson.

In looking over Hinton's "Universal Magazine" for 1752, we find that Ferguson, this year, published a pamphlet on the Lunar Eclipse, to take place in 1753, April 17th, entitled,—

> "A Calculus of the Lunar Eclipse of the 17th April next (1753) to every digit of observation as the shadow comes and goes off. By James Ferguson."

Note.—The middle of this Eclipse of the Moon occurred at 7 o'clock, on the evening of 17th April, 1753.

We have not been able to procure a copy of this pamphlet; it is likely that it would go out of print as soon as it served the occasion which brought it forth.

MODEL OF THE SOLAR, LUNAR, AND SIDEREAL PART OF THE ASTRONOMICAL CLOCK OF 1747.—A remark by Ferguson in his recently discovered MS. "Common Place Book," shows, that at or about the time he invented and made his curious clock in 1747, that he also made a wooden model of that part of it which actuated the motions of the Sun, Moon, and Stars; that after he finished the model, he showed it to Mr. Graham, the celebrated London Watch and Clock maker; that Mr. Graham bought the model from him, and that after whose death, he attended the sale of his effects, and bought it back, and ever afterwards retained it in his possession. Our note informs us that Mr. Graham died on November 16th, 1751, and that his effects were disposed of by public sale in 1752. Ferguson, in his MS. "Common Place Book," after describing a clock similar to the one he made in 1747, adds,

> "Upwards of 20 years ago, I made all the above described part (the Astronomical part) in wood, and showed it to the late eminent Mr. George Graham, Watchmaker in London, who bought it from me, and kept it till he died, after which, when his effects were sold, I bought it back, and now (1775), have it in my possession.
>
> JAMES FERGUSON."
>
> Bolt Court, Fleet Street,
> LONDON, *Decr. 27th*, 1775.

(Vide Ferguson's MS. "*Common Place Book*," p. 135, Col. Lib., Edin.)

Ferguson, in many instances, treats dates rather loosely, and he does so here, in saying, "*upwards of* 20 *years ago I made*," &c., as the above note was written by him in 1775, he, in the ordinary way of reckoning, would seem to point to sometime about the years 1753 or 1754. It is evident, however, from what he says, that Mr. Graham had the model for some considerable length of time before his death, which happened in 1751; and as this model is part of the Astronomical part of the Clock of 1747, it is evident that had he said *about* 30 *years ago*, it would have been much nearer the true date, when said model was made and sold.

An "interesting domestic memorandum" for Ferguson, occurred in November of this year, viz., the birth of a son, baptized *Murdoch*. The following is a copy of extract of the birth; the original is in the autograph of Ferguson, on the fly-leaf of his Bible, now in the possession of Dr. George at Keith:

"MURDOCH (born) Friday 3d Novr., 1752, N.S."

We have not been able to ascertain who Murdoch was named after; probably after an old and valued friend of Ferguson's, viz. Mr. Murdoch, who, in 1748, published "An Account of Sir Isaac Newton's Philosophical Discoveries." This is now Ferguson's third child, and it is somewhat singular that none of them are named after either their grandfather or grandmother, as is customary. James, the name given to the first-born son, appears to have been named after his father, instead of being named John, the prænomen of his paternal grandfather. Who Agnes was named after is now not known; but it is certain, it was not after either of her grandmothers, as the name of her paternal grandmother was Elspeth Lobban, and that of her maternal grandmother, Elspeth Grant. Regarding Murdoch Ferguson, see also dates 1768, 1775, and 1776; as also the Appendix.

Our memoranda show that Ferguson, "towards the end of 1752, was busy reading lectures on Astronomy in London and the provinces,—in expounding the nature of the recently altered Calendar,—and on the Eclipse of the Moon, already alluded to.

THE TRAJECTORIUM LUNARE MACHINE AND CURVE CONTROVERSY REVIVED.—Sometime in the year 1752, Ferguson was again annoyed by the revival of the Trajectory Machine and Curve Controversy of 1746. One of his former detractors had, unluckily for himself, got his hands on some old London magazine of date 1742, and found in it, a Trajectory of the Moon's path. In the fulness of his joy, and no doubt with a little malignity against Ferguson, he at once sent it abroad that the scheme of the Moon's path in this magazine must have been known to Ferguson, and that he had pirated the scheme in it, and palmed it upon the world as an original discovery of his own.

It was soon discovered that the Trajectory Curve of Badder's in this magazine was *most ridiculous, most absurd,* and that it had no resemblance to the *true curve* of the lunar path *projected by Ferguson's Machine.* On this becoming known, the person who had brought up Badder's scheme from its oblivion, was, along with his abettors, covered with confusion and shame—they had gratuitously exposed their own ignorance, and shown themselves to be persons who could form no opinion on the merits of the question at issue, in short, incapable of pointing out which was the right or which was the wrong curve.

This second assault, although spiritedly begun, suddenly collapsed, and came to nought, and instead of injuring Ferguson, as intended, it resulted in doing him a great deal of good; it showed that his traducers had not studied Astronomy to any good purpose, and that the curve projected by Ferguson's Trajectorium Lunarie was the true one.

In note 75 we allude to the first attack on Ferguson regarding his machine and the curve projected by it; and at note 115 will be found a copy of the letter written by Ferguson to the Rev. Mr. Birch, Secretary of the Royal Society, respecting this controversy. In this letter it will be seen, that besides referring to the charge of being a "literary pirate," he complains of a Mr. Hawkes of Norwich, who had received from him an orrery projection of the curve of the Moon's path, and who had published it in the Gentleman's Magazine for 1752, without acknowledging from what source he had got it.

1753.

ASTRONOMICAL LECTURES, AND PUBLICATION.—Early in 1753 we find Ferguson at Norwich, reading lectures on Astronomy, and the then forthcoming Eclipse of the Moon, on 17th April of this year. At same time and place, he published a pamphlet of 16 octavo pages, entitled,

> "A Brief Description of the Solar System, to which is subjoined an Astronomical Account of the year of our Saviour's Crucifixion. By James Ferguson. Norwich: Printed by W. Chase, for the author. MDCCLIII. Price Fourpence."

We have in our possession a copy of this very rare pamphlet, and as the title-page informs us, it gives a brief description of the solar system, &c., in particular as regards planetary magnitudes, distances, and motions; concluding with an interesting "Astronomical account of the year of our Saviour's Crucifixion,"—subjoined to which is "A Table of the mean Times of all the Passover Full Moons, adapted to the Meridian of Jerusalem,"—showing the days of the week on which they occurred, "from the 21st year after the common date of our Saviour's Birth, to the 40th Year."

ASTRONOMICAL ROTULA REPUBLISHED AND ALTERED TO NEW STYLE.—The alteration of the Style on 3d September, 1752, having rendered Ferguson's previously published Rotulas useless, he, early in 1753, brought out a new one, calculated for the new Style. The following is the advertisement of it, taken from a fly-leaf of his "Analysis of a Course of Lectures:"—

> "The Astronomical Rotula, showing the Day of the Month, Change and Age of the Moon, the places of the Sun and Moon in the Ecliptic, with the Times of all the Eclipses of the Sun and Moon, from A.D. 1752 to A.D. 1800, is sold at his house for Five Shillings and Sixpence."

Some years ago we had for some time in our possession one of these Rotulas, which consisted of a series of moveable circular cards, the largest being about 15 inches in diameter, and the smallest one about 11 inches. These Rotulas are now exceedingly scarce, and when they are to be had, they command high prices.

THE ECLIPSAREON.—Our memoranda show that Ferguson invented and made a new astronomical machine, which he called an Eclipsareon, for "Exhibiting the Time, Duration, and Quantity of Solar Eclipses at all places of the Earth." In his Memoir he says, "the best machine I ever contrived is the Eclipsareon, of which there is a figure in the 13th Plate of my Astronomy." The annexed cut is from a reduced drawing of this 13th Plate, and the description is also from the same work, article 405.

Eclipsareon.

"The principal parts of this machine (says Ferguson) are, 1st, a terrestrial globe A, turned round its axis B by the handle or winch M; the axis B inclines $23\frac{1}{2}$ degrees, and has an index which goes round the hour-circle D in each rotation of the globe; 2d, A circular plate E, on the limb of which the months and days of the year are inserted. This plate supports the globe, and gives its axis the same position to the Sun, or to a candle properly placed, that the Earth's axis has to the Sun upon any day of the year, by turning the plate till the given day of the month comes to the fixed pointer, or annual index G; 3d, A crooked wire F, which points toward the middle of the Earth's enlightened disc at all times, and shows to what place of the Earth the Sun is vertical at any given time; 4th, A penumbra or thin circular plate of brass I divided into 12 digits by 12 concentric circles, which represent a section of the Moon's penumbra, and is proportioned to the size of the globe; so that the shadow of this plate, formed by the Sun, or a candle placed at a convenient distance, with its rays transmitted through a convex lens to make them fall parallel on the globe, covers exactly all those places upon it that the Moon's shadow and penumbra do on the Earth; so that the phenomena of any solar eclipse may be shown by this machine with candle-light almost as well as by the light of the Sun; 5th, An upright frame H H H H, on the sides of which are scales of the Moon's latitude or declination from the ecliptic. To these scales are fitted two sliders K and K, with indexes for adjusting the penumbra's centre to the Moon's latitude, as it is north or south, ascending or descending; 6th, A solar horizon C, dividing the enlightened hemisphere of the globe from that which is in the dark at any given time, and showing at what places the general eclipse begins and ends with the rising or setting Sun; 7th, A handle M which turns the globe round its axis by wheel-work, and at the same time, moves the penumbra across the frame by threads over the pulleys L, L, L, with a velocity duly proportioned to that of the Moon's shadow over the Earth, as the Earth turns on its axis; And as the Moon's motion is quicker or slower according to her different distances from the Earth, the penumbral motion is easily regulated in the machine by changing one of the pulleys.

To rectify the Machine for use.—The true time of new Moon and her latitude being known, if her latitude exceeds the number of minutes or divisions on the scales (which are on the side of the frame, hid from view in the figure of the machine), there can be no eclipse of the Sun at that conjunction; but if it does not, the Sun will be eclipsed to some places of the Earth; and, to show the times and various appearances of the eclipse at those places, proceed in order as follows:—

To rectify the Machine for performing by the light of the Sun.—1st, Move the slides K K till their indexes point to the Moon's latitude on the scales, as it is north or south ascending or descending, at that time; 2d, Turn the month-plate E till the day of the given new Moon comes to the annual index G;

3d, Unscrew the collar N a little on the axis of the handle, to loosen the contiguous socket on which the threads that move the penumbra are wound, and set the penumbra by hand till its centre comes to the perpendicular thread in the middle of the frame; which thread represents the axis of the ecliptic; 4th, Turn the handle till the meridian of *London* on the globe comes just under the point of the crooked wire F; then stop, and turn the hour-circle D by hand till XII. at noon comes to its index, and set the penumbra's middle to the thread; 5th, Turn the handle till the hour-index points to the time of new Moon in the circle D; and holding it there, screw fast the collar N; lastly, Elevate the machine till the Sun shines through the sight-holes in the small upright plates O, O, on the pedestal; and the whole machine will be rectified.

To rectify the Machine for showing by candle light.—Proceed in every respect as above, except in that part of the last paragraph where the Sun is mentioned; instead of which, place a candle before the machine, about four yards from it, so as the shadow of intersection of the cross threads in the middle of the frame may fall precisely on that part of the globe to which the crooked wire F points; then, with a pair of compasses, take the distance between the penumbra's centre and intersection of the threads; and equal to that distance, set the candle higher or lower, as the penumbra's centre is above or below the said intersection; lastly, place a large convex lens between the machine and candle, so as the candle may be in the focus of the lens, and then the rays will fall parallel, and cast a strong light on the globe.

Its use.—These things done, which may be sooner than expressed, turn the handle backward, until the penumbra almost touches the side H F of the frame; then turning it gradually forward, observe the following phenomena:—1st, Where the eastern edge of the shadow of the penumbral plate I first touches the globe at the solar horizon, those who inhabit the corresponding part of the Earth see the eclipse begin on the uppermost edge of the Sun, just at the time of its rising; 2d, In that place where the penumbra's centre first touches the globe, the inhabitants have the Sun rising upon them centrally eclipsed; 3d, When the whole penumbra just falls upon the globe, its western edge at the solar horizon touches and leaves the place where the eclipse ends at Sun-rise on his lowermost edge—continue turning, and, 4th, the cross-lines in the centre of the penumbra will go over all those places on the globe where the Sun is centrally eclipsed; 5th, When the eastern edge of the shadow touches any place of the globe, the eclipse begins there; when the vertical line in the penumbra comes to any place, then is the greatest obscuration at that place; and when the western edge of the penumbra leaves the place, the eclipse ends there—the times of all which are shown on the hour-circle; and from the beginning to the end, the shadows of the concentric penumbral circles show the number of digits eclipsed at all the intermediate times; 6th, When the eastern edge of the penumbra leaves the globe at the solar horizon C, the inhabitants see the Sun beginning to be eclipsed on his lowermost edge at its setting; 7th, Where the penumbra's centre leaves the globe, the inhabitants see the Sun-set centrally eclipsed; and lastly, where the penumbra is wholly departing from the globe, the inhabitants see the eclipse ending on the uppermost part of the Sun's edge, at the time of its disappearing in the horizon.

N.B.—If any given day of the year on the plate E be set to the annual-index G, and the handle turned till the meridian of any place comes under the point of the crooked wire, and then the hour-circle D set by the hand till XII. comes to its index; in turning the globe round by the handle, when the said place touches the eastern edge of the hoop or solar horizon C, the index shows the time of Sun-setting at that place; and when the place is just coming out from below the other edge of the hoop C, the index shows the time when the evening twilight ends to it. When the place has gone through the dark part A, and comes about so as to touch under the back of the hoop C on the other side, the index shows the time when the morning twilight begins; and when the same place is just coming out from below the edge of the hoop next the frame, the index points out the time of Sun-rising; and thus, the times of Sun-rising and setting are shown at all places in one rotation of the globe, for any given day of the year; and the point of the crooked wire F shows all the places over which the Sun passes vertically on that day." (Ferguson's Astronomy, 4to Edit., London, 1764; article 405, pp. 298—301).

Although Ferguson esteemed "*The Eclipsareon*" machine his best invention, it does not appear to have been made for sale by any of the London Optical, Mathematical, and Philosophical Instrument makers. We, some years ago, called on a great number of them, who, in reply to our inquiries regarding it, assured us, that they never had such a machine on sale, nor ever heard of one being made by London Instrument makers; we also called on a great many dealers in second-hand Astronomical and Philosophical Instruments, and they too never heard of such a machine as an Eclipsareon having been made. Probably the difficulty of adjusting and using the Eclipsareon may have been an obstacle in the way of making them for sale; or if any difficulty ever existed, it may håve arisen from Ferguson himself not giving a description and drawing of the wheel-work which turned the globe, &c. It is difficult to see why an explanation of such an essential part of the machine as the wheel-work has been omitted. This Eclipsareon appears to have been sold at Mr. Ferguson's sale in March 1777, to a James Ferguson, Teacher of Navigation, &c., Hermitage Row, near the Tower, London, at whose sale (according to the Catalogue before us), the Eclipsareon was sold at the end of the first day's sale as "*Lot* 102.—*An Eclipsareon made by J. Ferguson, F.R.S., in a case, to W. Walker*, for **£**3 3s."

We have no other notes in our collection regarding Ferguson for the year 1753.

1754.

Ferguson, early in 1754, had been solicited by several of his friends, members of the ROYAL SOCIETY, to send his Eclipsareon, and a description of it, and of its use, to the SOCIETY meeting, in order that that learned body might see it, and thereby bring it and its inventor prominently into notice. Ferguson complied with the request made to him, and wrote out a description of the Eclipsareon, and its use, and lost no time in sending it to the Royal Society House in Crane Court, Fleet Street. In their Transactions, it is indexed as a

"Description of a piece of Mechanism for exhibiting the Time, Duration, and Quantity of Solar Eclipses at various places of the Earth." Read before the Royal Society, 21st February, 1754. (Phil. Trans., Vol. 48, p. 520, or Abrid. Trans. X., p. 456; also, "The Universal Magazine," Vol. 17, pp. 9, 10).

The Eclipsareon was exhibited, and the description and use of it read to the members of the Royal Society at their meeting, on Thursday, 21st February, 1754, under title as annexed; and an account of it was entered in their Transactions, as above noticed.

FERGUSON'S ADDRESS, 1754.—He appears to have been residing in *Broad Street, Golden Square*, early this year, as we observe in Kent's London Directory for *this year* the following entry:—

1754.—James Ferguson, Astronomer, Broad Street, Golden Square.

According to a remark which we saw many years since, in a letter in the autograph of Ferguson, and which we at the time took a note of, it appeared as certain, that Ferguson, in 1754, projected and began to write and collect materials for his great work, "Astronomy Explained upon Sir Isaac Newton's Principles;" a full account of which will be found under date 1756, when the work was published.

PUBLICATION —Sometime during the year 1754, Ferguson published an 8vo pamphlet of 32 pages with 2 engravings,—entitled,

"An Idea of the Material Universe deduced from a survey of the Solar System. By James Ferguson. London: Printed for the author. MDCCLIV. Price One Shilling."

The first engraving, or plate, represents the usual diagram of the SOLAR SYSTEM; and Plate 2d, SIDEREAL SYSTEMS (intended to illustrate "the architecture of the heavens.")

In its matter it is something similar to his "Brief Description of the Solar System," published at Norwich in 1753, as already noticed. This pamphlet has long been out of print, and it was with the greatest difficulty we secured a copy of it.

1755.

FERGUSON INVENTED AND MADE A LARGE WOODEN ORRERY. —In the year 1755, Ferguson invented and made a large wooden orrery for exhibiting the annual motion of the Earth, the parallelism of its axis; and consequently, the change of the seasons; the Moon's periodic, synodic, anomalistic, nodal, and diurnal motions, with the nodal and apogee motions of her orbit, &c. In this wooden orrery we find a very simple and tolerably accurate lunation period, more accurate than in any of the orreries he had previously made.[180] In his Tables and Tracts, Ferguson says, "About 12 years ago, (that is, in 1755), I made a large wooden orrery, for showing only the motions of the Earth and Moon, with the retrograde motion of her nodes, and the phenomena arising from all their motions. The Earth had not its diurnal and annual motions carried on by means of a winch, but by hand; and as the Earth was moved round the Sun, the Moon was carried round the Earth in her orbit, and her nodes had their retrograde motions. As there is something very particular and simple in the construction of this machine, and as the Moon's motion in it will not vary above one degree from the truth in 304 years; and as it answers as well in leap years as in common years, and has only seven wheels and one pinion in it; I shall here mention its use, but must beg to be excused from describing the position of its wheels and their numbers of teeth, because I intend to instruct my son, if both he and I live till the proper time, how to make it for his own benefit. Besides, it would be very difficult to make it intelligible by a description, without seeing it; especially as some of the wheels are not only divided into very uncommon numbers of teeth, but also that, in some of the wheels, equal numbers of teeth are contained in unequal spaces, for showing the inequality of the Earth's annual motion round the Sun, and of the Moon's motion round her orbit.[181] It shows the following matters very readily.

180 The lunar-trains of all his orreries, made previously, produce 29 d. 12 h. 45 m. exactly, the true mean length of the synodical revolution of the Moon being 29 d. 12 h. 44 m. 2·88 s. In the lunar train of this wooden orrery the period is 29 d. 12 h. 44 m. 25¼ s. (assuming the year to consist of 365¼ days).

181 If the large wheel of 235 teeth had its common central hole filled up, and a fresh hole made in it, about $\frac{1}{60}$ part of the diameter of the wheel out of the true centre, then if the bar with pinion 19 had a sliding motion to accommodate itself to the new centre distances, it would actuate the motion of pinion 19 in a manner similar to that described by Ferguson. Ferguson describes the wheel-work of this orrery in his "Select Mechanical Exercises," from which, it will be observed, that the wheels here alluded to as having "*uncommon numbers of teeth*" refers to wheels 235, 59, 55, 19, containing *odd numbers* of teeth in them. Ferguson here adds that, "*in some of the wheels*, equal numbers of teeth are contained in *unequal spaces*;" in his description of the wheel-work he only mentions *one wheel* being so divided, viz. the large wheel 235; therefore, the word "some," used by Ferguson here, ought to be *one* (*one of the wheels*). Ferguson wrote the above in 1767, and the description of the wheel-work in 1773.

The lengths of days and nights at all places of the Earth, and at all times of the year; with all the vicissitudes of seasons. The Sun's place in the ecliptic on any given day of the year, and time of the day; with his declination, altitude, and azimuth at any time; also, his amplitude, and the time of his rising and setting. The time of the day, by the observed altitude or azimuth of the Sun. The variation of the compass, in any place, whose latitude is known, by a single observation of the Sun's altitude, taken at any time, either in the forenoon or afternoon. The Moon's periodical and synodical revolution, with her rotation on her axis, and different phases. The retrograde motion of her nodes, and direct motion of her apogee. Her mean anomaly and her elliptic equation, by which her true place in her orbit is very nearly found at any time. Her latitude, declination, altitude, and azimuth at any time when she is above the horizon. Her amplitude, and the time of her rising and setting, however affected by her latitude. The times of all the new and full Moons, and of all the solar and lunar eclipses within the limits of 6,000 years before or after the Christian Era; with an easy method of rectifying the machine, in less than two minutes of time, for the beginning of any given year within these limits; and when it is once rectified, it will keep right for 304 years either backwards or forwards; at the end of which time, the Moon must be set one degree forward in her orbit. The small difference in the time of the Moon's rising, in harvest, throughout the week in which she is full; and the great difference in the time of her setting during that week. The recession of the equinoctial points in the ecliptic. The phenomena of the tides, and the causes of many irregularities in their heights, and times of ebbing and flowing." (See Ferguson's Tables and Tracts, 1st Edit., pp. 168—171).

Ferguson, in his "Select Mechanical Exercises," describes the wheel-work of this orrery as follows, but gives no illustrative engraving of it. We therefore annex one from his description, to assist the reader the more clearly to understand it.

In this work he says, "This is the orrery mentioned in my Tables and Tracts (page 169, 2d Edition), which I intended to keep for my son, who was then serving an apprenticeship to a Mathematical Instrument maker.[182] But as it has pleased God

[182] The late Mr. Andrew Reid, watchmaker, London, who died about the year 1834, aged 84 (and who, as formerly mentioned, was on intimate terms of friend-

to call him from this world to a better, I shall now freely communicate it to the public.

"A large wheel of 235 teeth is fixed in the box that contains the work (see the annexed sectional view of the wheel-work) the centre of the wheel being in the centre of the box, directly

Section of the Wheel-work of the Orrery.

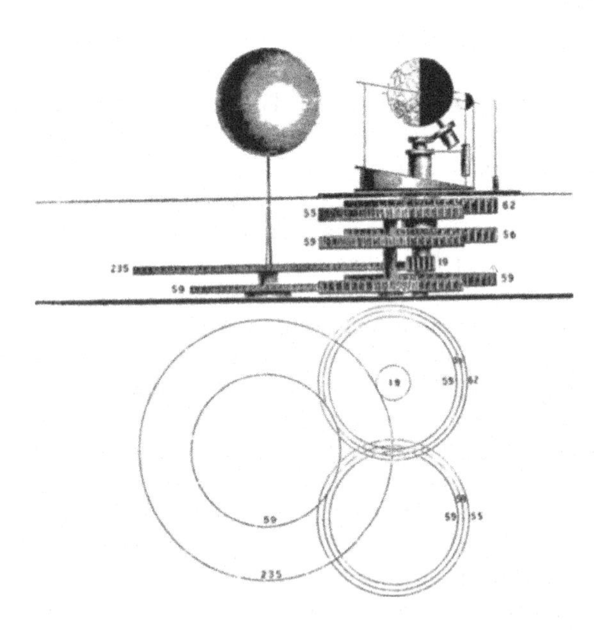

Plan of the Wheel-work of the Orrery.

under the Sun's centre. On this wheel runs a pinion of 19 leaves, carried round the teeth of the wheel by the annual motion of the Earth; and by this means, the pinion is turned round its own axis for every 19 teeth that is carried onward, in going

ship with the Ferguson family), informed us, that James, the son here alluded to, was, about the year 1763, apprenticed to Mr. Nairne, Optician and Mathematical Instrument maker, Cornhill, London, the son being then in his 15th year. In 1767, when Ferguson published his "Tables and Tracts" (from which the first part of this description has been taken), the son would then be in the 4th year of his apprenticeship. He died five years after this, in 1772, the year before "Select Mechanical Exercises" appeared; thus, Ferguson had kept the wheel-work of this orrery private, for his son's benefit, for upwards of 17 years. (See dates 1763 and 1772).

round the wheel.[183] Now, supposing this pinion to be carried round the wheel in $365\frac{1}{4}$ days, the pinion will be turned round its own axis in 29 days 12 hours 44 minutes 25 seconds,[184] and a bar on the axis of the pinion will carry the Moon round the Earth from change to change in that time. This comes so near to the truth, as to vary but one day in the Moon's course in 335 years; and these are the nearest numbers for such simple wheel-work that can possibly be found for mean lunations.[185]

183 Wheel 235 and pinion $19 = \frac{19}{235}$: that is, there are 235 complete synodic revolutions of the Moon in 19 years very nearly.

The following is the method for finding the ratio $\frac{19}{235}$ by "continuous fractions:"—The period of a tropical or solar year is 365 d. 5 h. 48 m. 51·6 s., and a synodic revolution of the Moon round the Earth, 29 d. 12 h. 44 m. 2·8 s. Required, the continuous fractions of these periods? 365 d. 5 h. 48 m. 51·6 s.; decimally expressed, is 365·242264 days, and 29 d. 12 h. 44 m. 2·8 s. = 29·530589 days, and $\frac{29 \cdot 530589}{365 \cdot 242264}$ are prime to each other; therefore,

```
29·530589)365·242264(12
          354367068
          ---------
          10875196)29530589(2
                   21750392
                   --------
                    7780197)10875196(1
                             7780197
                             -------
                             3094999)7780197(2
                                     6189998
                                     -------
                                     1590199)3094999(1
                                             1590199
                                             -------
                                             1504800)1590199(1
                                                     1504800
                                                     -------
                                                       85399)1504800(17
```

Quotients 12·2·1·2·1·1·17, &c.

Then as by former rules already given (see description of the orrery under date 1745).

$$0 \times 12 + \frac{1}{0} = \frac{1}{12}$$
$$\frac{1}{12} \times 2 + \frac{0}{1} = \frac{2}{25}$$
$$\frac{2}{25} \times 1 + \frac{1}{12} = \frac{3}{37}$$
$$\frac{3}{37} \times 2 + \frac{2}{25} = \frac{8}{99}$$
$$\frac{8}{99} \times 1 + \frac{3}{37} = \frac{11}{136}$$
$$\frac{11}{136} \times 1 + \frac{8}{99} = \frac{19}{235}$$ &c. = the fraction used by Ferguson.

We may note that $\frac{19}{235}$ is a remarkable fraction; 19 is "the cycle of the Moon, commonly called the golden number, and is a revolution of 19 years, in which time the conjunctions, oppositions, and other aspects of the Moon are within an hour and a half of being the same as they were on the same days of the months 19 years before." (See Ferguson's Astronomy, article 383). In 19 years there are very nearly 235 complete mean lunations. This cycle of 19 was discovered about the year 430 B.C. by Meton, and is called the "Metonic Cycle." We are of opinion that, on this occasion, Ferguson simply adopted the Metonic Cycle for the lunar ratio in this orrery,—the 19 years = 19 leaves in the Moon pinion, and 235 lunations = 235 teeth in the large fixed wheel; thus, he would find the ratio $\frac{19}{235}$ ready at hand, and for a simple ratio, produces a very correct lunation.

184 This period from the ratio $\frac{19}{235}$ may be found as follows:—Ferguson assumes pinion 19 to be carried round wheel 235 in $365\frac{1}{4}$ days, or decimally expressed, in 365·25 days; therefore, $235 \div 19 = 12 \cdot 368421$, and $365 \cdot 25 \div 12 \cdot 368421 = 29 \cdot 530851$, or 29 d. 12 h. 44 m. 25·526 s. But if he had assumed 365 d. 5 h. 48 m. 58 s. as he has adopted in his other orreries, or 365·242347 days, then we will have a less period, for $365 \cdot 242347 \div 12 \cdot 368421 = 29 \cdot 530232$, or 29 d. 12 h. 43 m. 32·045 s. nearly.

185 Ferguson, referring to this orrery in his "*Tables and Tracts,*" p. 170, mentions, that when the Moon is accurately adjusted "*it will keep right for 304 years either backwards or forwards.*" In his "*Select Mechanical Exercises,*"

"But in nature, the Earth moves unequally round the Sun, so that there are 8 days more between the vernal and autumnal equinox than between the autumnal and vernal. And therefore, in common orreries, where this circumstance is taken no notice of, the Earth's position to the Sun cannot be right at both the equinoctial points.[186]

"In order to avoid this error, I first divided the ecliptic into 360 equal parts or degrees, and then, after having put the names of the signs to it, I laid down the days of the year from an ephemeris against the degrees of the Sun's place in the ecliptic, for each day respectively throughout the year. By this means, the daily spaces answering to the Earth's unequal motion round the Sun, were so divided, as to be continually and gradually lessening from the 30th of December till the first of July; and then as gradually lengthening from the first of July to the 30th of December; as the Earth's progressive annual motion is swiftest of all on the 30th of December, and slowest of all on the 1st of July.

"The days of the months being unequally divided, so as to answer to the Earth's unequal motion round the Sun, I made these divisions a pattern or scale for dividing the 235 teeth of the above-mentioned wheel into such unequal spaces as would agree with the spaces allotted for the days answering to them. But these gradual inequalities of the teeth were so very small, and the difference so little between the widest and narrowest, that the pinion (whose leaves were all equal) run very smoothly through all the teeth of the wheel; as the leaves of the pinion were sized to these teeth,

p. 90, he says, that when so set, it will keep right for 335 years (a difference here of 31 years). In trying the period, we find that the Moon will keep right for about 314 years,—thus, 29 d. 12 h. 44 m. 25 s. 32 thds. = the period of pinion 19, and Ferguson adopts 29 d. 12 h. 44 m. 3 s. 7 thds. as the true period, as estimated in his time; therefore, 29 d. 12 h. 44 m. 3 s. 7 thds.—29 d. 12 h. 44 m. 25 s. 32 thds. = 22 s. 25 thds. of error in each synodic revolution of the Moon (or 22·417 s.) Pinion 19, which carries the Moon, makes 12·368421 revolutions or turns in $365\frac{1}{4}$ days; hence, $12·368421 \times 22·417 = 277·262$. In a day, there are 86400 seconds, which, on being divided by 277·262, will give $311\frac{6}{10}$ years before pinion 19 can make one day of error, either backwards or forwards, after it has been correctly rectified, which nearly agrees with the period he gives in his Astronomy, section 385.

[186] In common orreries, this is never attended to; still, there have been made, both in this and other countries, orreries and planetaria for exhibiting the equated motions of the Moon and all the planets round the Sun. In this country, the name of our late friend the Rev. Dr. Pearson stands prominent; in France, that of Antide Janvier, and in America, that of Mr. Rittenhouse.

which were at a mean rate between the greatest and least distances from one another. By this contrivance, the mean lunation was always 29 days 12 hours 44 minutes 25 seconds throughout the whole year; and the pinion was among the least distant teeth of the wheel when the annual index was at the 1st of July, and among the most distant teeth when the index was at the 30th of December.

"For the parallelism of the Earth's axis a wheel of 59 teeth was fixed in the middle of the work, with its centre directly over the centre of the wheel of 235 teeth, and the teeth of another wheel of 59 took into the teeth of the former; and those into the teeth of a third wheel of the same number, on the top of whose axis the piece that carries the Earth on its oblique axis was fixed. And, as the Earth was moved round the Sun, these three wheels kept the parallelism of the Earth's axis, as described in the *Mechanical Paradox*, and the former orrery.

"Above the middlemost wheel of 59, and on its axis, is fixed a wheel of the same number, which takes into a wheel of 56 teeth; this last wheel is just below the Earth, and turns the nodes of the Moon's orbit quite round backward, in $18\frac{2}{3}$ years.[187]

"Above the last-mentioned wheel of 59 teeth, and on the same axis with it, is a wheel of 55, turning a wheel of 62 teeth below the Earth; and this wheel of 62 moves the Moon's apogee plate quite round forward, in 8 years 312 days.[188] And from this plate a wire rises, and points out the place and motion of the apogee in the Moon's ecliptic."

Ferguson mentions that there were only 7 wheels and 1 pinion in this orrery, but according to his description of it, just quoted, it is evident that there are 8 wheels and 1 pinion in it.[189] In the year 1828 we made an orrery of this kind from

[187] As formerly noted, Ferguson, in all of his large orreries, used this incorrect train of wheels. The period of the Moon's nodes is a retrograde period of 18 years and 224 days nearly; but $18\frac{224}{365}$ years is not $18\frac{2}{3}$ years, but a period of $18\frac{3}{5}$ years *nearly*, as formerly shown; and $18 \times 5 + 3 = 93$, and $93 + 5 = 98$; therefore, by using $18\frac{3}{5}$ as the fraction, we have $\frac{93}{98}$ for the new and more correct set of wheels, instead of $\frac{56}{59}$ See also note 135.

[188] $\frac{55}{62}$ does not appear to have been produced by continuous fractions, but as follows:—Apogee period $=$ 8 years 312 days, and as $\frac{312}{365} = \frac{6}{7}$ nearly, we have $8\frac{6}{7}$ years $=$ period of apogee; hence, $8 \times 7 + 6 = 62$, and $62 - 7 = 55$, wheels of 55 and 62 as given above.

the description, but so contrived it, that it had "*only* 7 *wheels and* 1 *pinion.*"

The annexed sectional view will, when kept in connection with the previous details, sufficiently show the arrangement

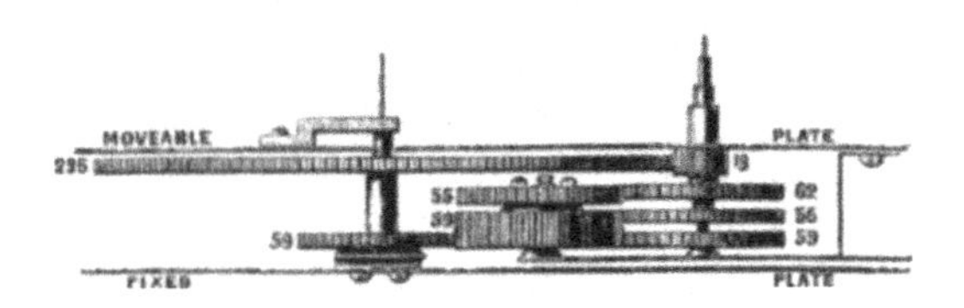

without a description of their actions (the only difference being in having a thick wheel of 59 driving the two wheels of 59 and 56). Such a machine might be made for about £7.

Ferguson adds, that "As the Moon goes round the Earth, she comes to her mean changes, nodes, and apogee, in the proper times; and at all intermediate times, her distance from her apogee and nodes are shown on her ecliptic, orbit, and apogee-plate; on which last her mean anomaly and elliptic equation are shown; by which means her true place in the ecliptic, and her latitude, may be very nearly found for any given time.

"The days of the months throughout the year are laid down in a diagonal manner, in a spiral of four revolutions, marked 0, 1, 2, 3 for leap year, and the first, second, and third years after. The annual index, in these spirals, being at the given day of any month, for either of these years, all the other motions and phenomena will be right for that day; and by means of these diagonals, the lunation is brought still nearer the truth than as above specified.

[189] This orrery has the following wheels,—viz.

A large wheel having 235 teeth (for Moon's motion), . .	1 wheel.
Parallelism of Earth's axis, *three wheels* of 59 teeth each, . . .	3 do.
Moon's nodes, a wheel of 59 and one of 56 teeth, . . .	2 do.
Moon's apogee, a wheel of 55 and one of 62 teeth, . . .	2 do.
Total number,	8 wheels, and one pinion of 19 leaves.

Therefore, there are 8 wheels and 1 pinion in this orrery, and not 7 wheels and 1 pinion as Ferguson has it. It is singular that his pen has made "a slip" in a matter like this.

"Within this set of spirals are tables, which show the places of the Sun, Moon, Ascending Node, and Apogee, for the noon of the first day of January in any year, within the limits of 6,000 years, both before or after the Christian Era; And, by means of these tables, the Orrery may, in less than two minutes of time, be rectified for the beginning of any of these years; and then, all the motions, not only for that year, will be right, but also for 334 years afterward, without needing any rectification.

"By comparing this description with that of the orrery in the 8th Plate of my Astronomy (see Sect. 399), it will be very easily understood, particularly those parts which show the parallelism of the Earth's axis, the motion of the Moon's nodes and apogee." (Ferguson's "Select Mechanical Exercises," 2d Edit., Lond., 1778, pp. 88—95).

1755.

THE MECHANICAL PARADOX (of 1750) CONVERTED INTO AN ORRERY (1755).—Ferguson, near the close of his long letter, given at pp. 145—148, mentions, that he kept the Mechanical Paradox (in its original form) "for six years without finding any person who could explain the principles on which it acted, and then put the Sun and Earth, with the ecliptic and Moon's orbit, to it, seeing it would then be a kind of orrery." From this remark, it would appear that he converted his Mechanical Paradox into the orrery form sometime in the year 1755. The annexed woodcut shows the Paradox in its transformed orrery state; between the years 1750 and 1760:—Ferguson made several of these Paradox Machines and Orreries; but appears to have forgotten when each was made—this may account for the discrepancies in his dates respecting them. We have in our possession one of these orreries made by him about 1759. The following is his description of the Paradox Orrery, taken from his "Select Mechanical Exercises."

"Having solved the Paradox, and described the cause of the different effects which are produced upon the three wheels F, G, and H, we shall now proceed to show some uses that may be made of the machine.

"This machine is (as altered) so much of an orrery, as is sufficient to show the different lengths of days and nights, the vicissitudes of the seasons, the retrograde motion of the nodes of the Moon's orbit, the direct motion of the apogeal point of her orbit, and the months in which the Sun and Moon must be eclipsed.

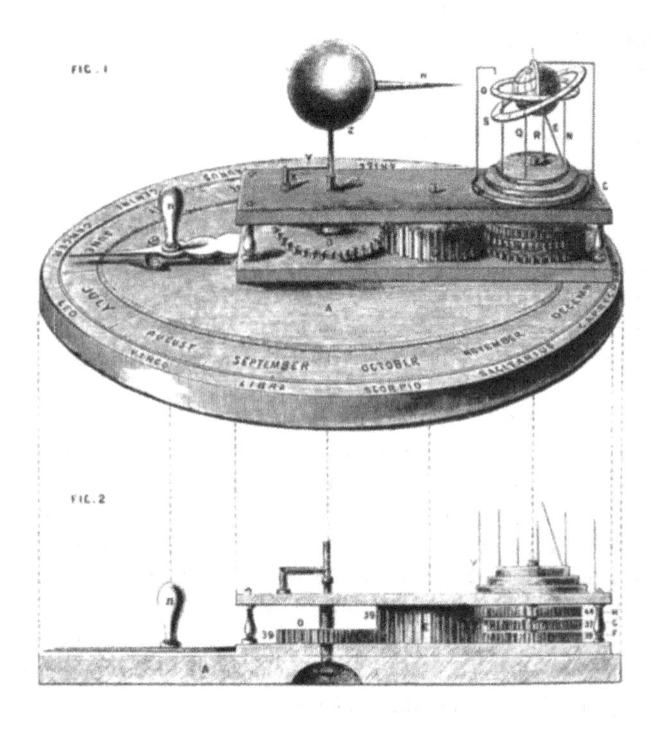

"On the great immoveable plate A (Fig. 1) are the months and days of the year, and the signs and degrees of the zodiac, so placed, that when the annual index h is brought to any given day of the year, it will point to the degree of the sign in which the Sun is on that day. This index is fixed to the moveable frame B C, and is carried round the immoveable plate with it, by means of the knob n. The carrying this frame and index round the immoveable plate, answers to the Earth's annual motion round the Sun, and to the Sun's apparent motion round the ecliptic in a year.

"The central wheel D (being fixed on the axis a, which is fixed in the centre of the immoveable plate) turns the thick wheel E round its own axis by the motion of the frame; and the teeth of the wheel E take into the teeth of the three wheels F, G, H, whose axes turn within one another, like the axes of the hour, minute, and second hands of a clock or watch, where the seconds are shown from the centre of the dial-plate.

"On the upper ends of these axes are the round plates I, K, L; the plate I being on the axis of the wheel F, K on the axis of G, and L on the axis of H; So that, whichever way these wheels are affected, their respective plates, and what they support, must be affected in the same manner; each wheel and plate being independent of the others.

"The two upright wires M and N are fixed into the plate I, and they support the small ecliptic O P, on which, in the machine, the signs and degrees of the ecliptic are marked. This plate also supports the small terrestrial globe *e* on its inclining axis *f*, which is fixed into the plate near the foot of the wire N. This axis inclines 23½ degrees from a right line, supposed to be perpendicular to the surface of the plate I, and 66½ to the plane of the small ecliptic O P, which is parallel to that plate.

"On the Earth *e* is the crescent *g*, which goes more than half way round the Earth, and stands perpendicular to the plane of the small ecliptic O P, directly facing the Sun Z; its use is to divide the enlightened half of the Earth next the Sun from the other half which is then in the dark; so that it represents the boundary of light and darkness, and therefore ought to go quite round the Earth; but cannot, in a machine, because, in some positions, the Earth's axis would fall upon it. The Earth may be freely turned round on its axis by hand, within the crescent, which is supported by the crooked wire *w*, fixed to it, and into the upper plate of the moveable frame B C.

"In the plate K are fixed the two upright wires Q and R; they support the Moon's inclined orbit S T in its nodes, which are the two opposite points of the Moon's orbit where it intersects the ecliptic O P. The ascending node is marked ☊, to which the descending node is opposite, below *e*, but hid from view by the globe *e*. The half ☊ T *e* of this orbit is on the north side of the ecliptic O P, and the other half *e* S ☊ is on the south side of the ecliptic. The Moon is not in this machine: but, when she is in either of the nodes of her orbit in the heavens, she is then in the plane of the ecliptic; when she is at T in her orbit, she is in her greatest north latitude; and when she is at S, she is in her greatest south latitude.

"In the plate L is fixed the crooked wire U U, which points downward to the small ecliptic O P, and shows the motion of the Moon's apogee therein, and its place at any given time.

"The ball Z represents the Sun, which is supported by the crooked wire X Y, fixed into the upper plate of the frame at X. A straight wire W proceeds from the Sun Z, and points always toward the centre of the Earth *e*; but toward different points of its surface at different times of the year, on account of the obliquity of its axis, which keeps its parallelism during the Earth's annual course round the Sun Z; and therefore must incline sometimes toward the Sun, at other times from him, and twice in the year neither toward nor from the Sun, but sideways to him. The wire W is called *the solar ray*.

"As the annual index *h* shows the Sun's place in the ecliptic for every day of the year, by turning the frame round the axis of the immoveable plate A, according to the order of the months and signs, the solar ray does the same in the small ecliptic O P; for, as this ecliptic has no motion on its axis, its signs and degrees still keep parallel to those on the immoveable plate. At the same time, the nodes of the Moon's orbit S T (or points where it intersects the ecliptic O P) are moved backward, or contrary to the order of signs, at the rate of 19⅓ degrees every *Julian* year;[190] and the Moon's apogeal wire U U is moved forward, or according to the order of the signs of the ecliptic, nearly at the rate of 41 degrees every *Julian* year; the year being denoted by a revolution of the Earth *e* round the Sun Z; in which time, the annual index *h* goes round the circles of months and signs on the immoveable plate A.[191]

"Take hold of the knob *n*, and turn the frame round thereby; and in doing this, you will perceive that the north pole of the

[190] The nodes of the Moon's orbit retrograde 18 degrees 53 minutes annually. In this machine, wheels 37 and 39 produce the nodal period, which consists of 6798 days nearly, in 18½ years = 6757 days, showing an error of 41 days, minus the true mean period. (See also note 189).

[191] A mean revolution of the apogee of the Moon's orbit is accomplished in about 3232 days; therefore, if 3232 days : 360° : : 365¼ days ?—365¼, or decimally, 365·25 × 360 = 1314900 ÷ 3232 = 40·683 degrees or 40° 41′; that is, the apogee progresses in the Moon's orbit at the rate of 40° 41′ every 365¼ days. The wheels used in this machine for producing an apogeal motion, are wheels, one of 39 teeth, the other 44; therefore, 39—44 = 5, and 44 ÷ 5 = 8⅘ years for a machine revolution of the apogee, which is a period of 3212 days instead of 3232 days, showing an error of 20 days, *minus* in said true mean period of the apogee, thus showing that the apogee pointer progresses in the Moon's orbit in the machine at the rate of 40 degrees 56 minutes per annum

Earth *e*, is constantly before the crescent *g*, in the enlightened part of the Earth toward the Sun, from the 20th of March to the 23d September; and the south pole all that time behind the crescent in the dark; and, from the 23d of September to the 20th of March, the north pole is constantly in the dark, behind the crescent, and the south pole in the light before it; which shows, that there is but one day and one night at each pole, in the whole year; and that, when it is day at either pole, it is night at the other.

"From the 20th of March to the 23d of September, the days are longer than the nights in all those places of the northern hemisphere of the Earth which revolve through the light and dark, and shorter in those of the southern hemisphere. From the 23d of September to the 20th of March, the reverse.

"There are 24 meridian semicircles drawn on the globe, all meeting in its poles; and as one rotation or turn of the Earth on its axis is performed in 24 hours, each of these meridians is an hour distant from the other, in every parallel of latitude. Therefore, if you bring the annual index *h* to any given day of the year, on the immoveable plate, you may see how long the day then is at any place of the Earth, by counting how many of these meridians are in the light, or before the crescent, in the parallel of latitude of that place; and this number being subtracted from 24 hours, will leave remaining the length of the night. And if you turn the Earth round its axis, all those places will pass directly under the point of the solar ray, which the Sun passes vertically over on that day, because they are just as many degrees north or south of the equator, as the Sun's declination is then from the equinoctial.

"At the two equinoxes, viz. on the 20th of March and 23d of September, the Sun is in the equinoctial, and consequently has no declination. On these days, the solar ray points directly toward the equator, the Earth's poles lie under the inner edge of the crescent, or boundary of light and darkness; and in every parallel of latitude, there are twelve of the meridians or hour-circles before the crescent, and twelve behind it; which shows that the days and nights then are each twelve hours long at all places of the Earth. And if the Earth be turned round its axis, you will see that all places on it go equally through the light and the dark hemispheres.

"On the 21st of June, the whole space within the north polar circle is enlightened, which is 23½ degrees from the pole, all round; because the Earth's axis then inclines 23½ degrees toward the Sun; but the whole space within the south polar circle is in the dark; and the solar ray points toward the tropic of Cancer on the Earth, which is 23½ degrees north from the equator. On the 20th of December the reverse happens, and the solar ray points toward the tropic of Capricorn, which is 23½ degrees south from the equator.

"If you bring the annual index *h* to the beginning of January, and turn the Moon's orbit S T by its supporting wires Q and R till the ascending node (marked ☊) comes to its place in the ecliptic O P, as found by an Ephemeris or by Astronomical Tables, for the beginning of any given year; and then move the annual index by means of the knob *n*, till the index comes to any given day of the year afterward, the nodes will stand against their places in the ecliptic on that day. And if you move the index onward, till either of the nodes comes directly against the point of the solar ray, the index will then be at the day of the year on which the Sun is in conjunction with that node. At the times of those new Moons, which happen within seventeen days of the conjunction of the Sun with either of the nodes, the Sun will be eclipsed; and at the times of those full Moons, which happen within twelve days of either of these conjunctions, the Moon will be eclipsed. Without these limits there can be no eclipses either of the Sun or Moon; because, in nature, the Moon's latitude or declination from the ecliptic, is too great for the Moon's shadow to fall on any part of the Earth, or for the Earth's shadow to touch the Moon.

"Bring the annual index to the beginning of January, and set the Moon's apogeal wire U U to its place in the ecliptic for that time, as found by Astronomical Tables; then move the index forward to any given day of the year, and the wire will point on the small ecliptic to the place of the Moon's apogee for that time.

"The Earth's axis *f* inclines always toward the beginning of the sign Cancer on the small ecliptic O P. And, if you set either of the Moon's nodes, and her apogeal wire, to the beginning of that sign, and turn the plate A about, until the Earth's axis inclines toward any side of the room (suppose the north side), and then move the annual index round and round the immoveable plate A, according to the order of the months and signs upon it, you will see that the Earth's axis and beginning of Cancer will still keep toward the same side of the room without the least deviation from it; but the nodes of the Moon's orbit S T will turn progressively towards all sides of the room, contrary to the order of signs in the small ecliptic O P, or from east, by south, to west, and so on: and the apogeal wire U U will move the contrary way to the motion of the nodes, or according to the order of the signs in the small ecliptic, from west, by south, to east, and so on quite round. A clear proof that the wheel F, which governs the Earth's axis and the small ecliptic, does not turn any way round its own centre; that the wheel G, which governs the Moon's orbit O P, turns round its own centre backwards, or contrary both to the motion of the frame B C and thick wheel E; and the wheel H, which governs the Moon's apogeal wire U U, turns round its own centre, forward, or in direction both of the motion of the frame, and of the thick wheel E, by which the three wheels F, G, and H, are affected.

"The wheels D, E, and F, have each 39 teeth in the machine; the wheel G has 37, and H 44; as shown in Fig. 2.

"The parallelism of the Earth's axis is perfect in the machine;[192] the motion of the apogee very nearly so; the motion of the nodes not quite so near the truth, though they will not vary sensibly therefrom in one year. But they cannot be brought nearer, unless larger wheels, with higher numbers of teeth, are used.[193]

"In nature, the Moon's apogee goes quite round the ecliptic in eight years and 312 days, in direction of the Earth's annual motion; and the nodes go round the ecliptic in a contrary direction, in eighteen years and 225 days. In the machine, the apogee goes round the ecliptic O P in eight years and four-fifths of a year, and the nodes in eighteen years and a half.

192 By this is meant that the parallelism does not alter or deviate from a given fixed point at a great distance, and is therefore a *perfect parallel motion*. The axis of the Earth does not preserve its parallelism in its revolutions round the Sun, but makes a circle round the pole of the ecliptic in 25920 solar, or 25868 sidereal years.

193 The progressive motion of the apogee, and retrograde motion of the nodes, worked by wheels 44 and 37 respectively, is simply a question of "*fast and slow motion*," compared with the wheel 39, whose value is 0; and thus is the cause of the phenomena arising out of the performance of the wheels. For higher numbers of wheel teeth, see note 189.

"Notwithstanding the difference of the numbers of teeth in the wheels F, G, and H, and their being all of equal diameters, they take tolerably well into the teeth of the thick wheel E, because they are made of soft wood.[194] But if they were made of metal, the wheel E in Fig. 1, ought to be made of the shape of E (seen edgewise) in Fig. 2, with very deep teeth: and the wheels F, G, and H in Fig. 1, of diameters proportioned to their respective numbers of teeth, as F, G, and H in Fig. 2; And then, the teeth of these three wheels would be of equal sizes with those of the wheel E wherein they work; and the motions would be free and easy, without any pinching or shake in the teeth." (Vide "The description and use of a new machine called The

[194] This is undoubtedly the proper method of constructing such a machine; but as the *thick wheel* would require to be made up of *three distinct wheels*, it could no longer be called a machine of "*five wheels*," as it would then have SEVEN WHEELS in it. To have wheels made of soft wood, so that they might act better, is a method that cannot be recommended. When orreries of any description whatever, are to be made, brass will be found the best and most suitable material for the wheels, &c. The only recommendation in favour of wooden wheels is, that as the teeth in them are generally cut into a triangular shape, they the more readily accommodate themselves to irregularities of *the pitch* line of action.

As formerly noted, we have in our possession one of these Mechanical Paradox Orreries, made by Ferguson. It belonged to his son, the late John Ferguson, Esq., after whose death, it came into the possession of James L. Rutherfurd, Esq., Edinburgh, from whom we received it in February 1864, which we here acknowledge with thanks.

The large immoveable circle in this machine is made of hard mahogany, and is 15 inches in diameter, by 1 inch thick. The frame containing the wheel-work is also of mahogany, 10 inches long, 4⅞ inches broad, and ¼-inch thick, held together by four short pillars of the same material. (The five wheels are all of brass, and in high numbers).

Round the edge of the large immoveable circle there is a paper edging 1 inch broad, having on it, in pen and ink printing, the Names of the Mouths, with their Divisions, &c. The Sun, in the centre, is represented by a gilded ball 2 inches in diameter. The Earth is an ivory ball, having engraven on it 24 meridian lines, polar, tropical, and equinoctial circles. Surrounding the Earth is a mahogany horizon 3¼ inches in exterior diameter, the surface of which is covered with paper, and has on it, printed by pen, the Names of the Signs of the Ecliptic, and its Divisions; exterior to this is another circle of mahogany, lying at the angle of 5⅓ degrees; this represents the Moon's orbit—its surface is papered, and has printed on it, by pen, figures of the Sun and Moon, and the usual degrees, &c., the exterior diameter of which is 5¾ inches. Directly under the Sun, there is a paper disc, with a variety of circles described on it, and within them, neatly printed by pen, the years, in figures, from 1760 to 1800 inclusive. Within them is a lunar circle, having the following:—"*The Time of the Mean New Moon in January, and of the Conjunction of the Sun with the Moon's Ascending Node;*" and in another space is, "*In Leap Years, on Jan. and Feb., add 1 day to the time found by this instrument;*" and near to this is "*J. Ferguson, invt. et fecit.*" All the pen and ink printing is in the autograph of Ferguson, very neatly done; there is no date, but as its Tables begin with the year 1760, it may be inferred, that it would be made about that time. The machine is still in excellent condition and working order, and is a curiosity.

Mechanical Paradox," pp. 9—16; also, his "Select Mechanical Exercises," pp. 57—71.)

In the year 1755, Ferguson was in Cambridge—delivering his course of lectures on Astronomy, Mechanics, &c. As to his success, we have no accounts; while there, he frequently visited the Rev. Professor Ludlam and Mr. Edward Waring, the Mathematician; from the former he received a sketch plan of his ingenious "Perpetual Day of the Month Shifting Apparatus;" and from the latter, "An Astronomical Problem, with the Rule for its Solution." As the problem is a curious one, we here give it a place, along with another way of solving it, by Dr. John Ford of Bristol.

Mr. Waring's Problem on the Conjunction of the Planets; and Formula for its Solution.[195]

"*The periodical times of the six primary planets being given, and supposing them to have been all at once in a line of conjunction with the Sun, to find how much time would elapse before they were all in a line of conjunction with the Sun again.*"

"Let a, b, c, d, e, f, be respectively equal to the periodical times or revolutions of the six planets about the Sun; a being the longest, or Saturn's period; b the next longest, or Jupiter's; c the next, or Mars's; d the next, or the Earth's; e the next, or Venus's; and f the shortest period of all, which is Mercury's; and let p, $q\,p$, $r\,q\,p$, $s\,r\,q\,p$, &c., be equal to the difference or time between the succeeding conjunctions of any two, three, four, &c., of them. 'Tis evident that q, r, s (the multipliers) must be whole numbers, because the numbers of conjunctions are so.

"The time between the conjunctions of the first two is $\frac{a\,b}{a-b}=p$; that of the first three is $\frac{n\times a\,c}{a-c}=q\,p$ (where n is any number assumed, to make q, a whole number) or, which is the same, $\frac{n\,a\,c}{\overline{a-c}\times p}=q$; $\frac{a\,c}{\overline{a-c}\times p}$ being reduced to its lowest denominator, q will be equal to the numerator of that fraction. In the same

195 Mr. Edward Waring, an eminent Mathematician, who afterwards became Professor of the Mathematics at Cambridge, and was the author of the following works:—"*Problems Concerning Interpolations;*" "*On the General Resolution of Algebraical Equations;*" "*On the Resolution of Attractive Powers;*" "*Problems in Equations;*" "*On some Properties of the Sum of the Division of Numbers;*" "*On Finding the Values of Algebraical Quantities by Converging Series,*" &c. He died in the year 1798, aged 64; and consequently, in the year when he met Ferguson in Cambridge, he would be in the 21st year of his age.

manner, r will be equal to the numerator of the fraction $\frac{a\,d}{a-d \times q\,p}$ reduced to its lowest denominator, s will be equal to the numerator of the fraction $\frac{a\,e}{a-e \times r\,q\,p}$, reduced to its lowest denominator; and so on, from the slowest to the quickest revolving bodies in the system; by which means, the times of all their conjunctions may be found." (Ferguson's "Tables and Tracts." Lond., 1767, pp. 141, 142).

Dr. John Ford's method for Solving the same Problem.—(Sent by him to Ferguson—probably about the same period—1755.)

"This problem may be solved by a different method as follows, for which I am obliged to my generous friend Mr. John Ford, Surgeon in Bristol.[196]

"Let A, B, C, D, E, F, stand for the six planets, beginning with Saturn, and ending with Mercury; and, a, b, c, d, e, f, be the times of their periodical revolutions respectively; Then, by a known rule,[197] the synodical period, or conjunction, of A and B, will be the time $\frac{a\,b}{a-b}$; and that of B and C will be $\frac{b\,c}{b-c}$; that of C and D will be $\frac{c\,d}{c-d}$; that of D and E will be $\frac{d\,e}{d-e}$; and that of E and F will be $\frac{e\,f}{e-f}$.

"Now it is obvious, that A and B can never be in conjunction but in the time $\frac{a\,b}{a-b}$, or some multiple of it; neither can B and C be in conjunction but in the time $\frac{b\,c}{b-c}$, or some multiple of that time. A, B, and C, will therefore be in conjunction when $\frac{m\,a\,b}{a-b}$ is equal to $\frac{n\,b\,c}{b-c}$, where m and n represent two integer numbers, prime to each other; which, being respectively multiplied into $\frac{a\,b}{a-b}$ and $\frac{b\,c}{b-c}$ shall make the two products equal. And these two numbers are easily discovered; for, by supposition, $\frac{m\,a\,b}{a-b}$ is equal to $\frac{n\,b\,c}{b-c}$; therefore, $m\,n :: \frac{b\,c}{b-c} : \frac{a\,b}{a-b}$. Reduced therefore, $\frac{b\,c}{b-c}$ and $\frac{a\,b}{a-b}$ into integers of the least dimensions (as minutes, or seconds of time), which shall have the same proportion to each other as these numbers have; and you will have the multipliers m and n, and consequently, the synodical period or conjunction of A, B, and C; which we shall call R. In the same

[196] Mr. John Ford was an eminent Surgeon in Bristol in the middle of last century, "was an excellent Mathematician, and had a general knowledge of the several branches of Natural and Experimental Philosophy, and assisted Mr. Ferguson in disposing of tickets for his lectures in Bristol, and was altogether a most worthy man." (Ext. letter from Bristol in 1832).

[197] The rule is,—After the planetary periods are made *prime*, multiply them together, and then divide by their difference, and the quotient will be the time of conjunction. See also Mr. Ford's Explanations in his paper here quoted.

manner may the synodical period of C, D, and E, be investigated, which we call S; then find two prime numbers r and s in their lowest dimensions, which shall have the same proportion to each other as the times R and S; then will r S, or its equal S r, give the synodical period, or conjunction, of the five planets A, B, C, D, and E, which characterize by T. Find lastly, the synodical period of E and F, by the rule $\frac{ef}{e-f}$, which denote by X; and the least integer number t, x, in the same proportion to each other as T and X being found, t X or x T will be the synodical period, or conjunction of the six primary planets A, B, C, D, E, F; or the time that must elapse between any conjunction of them all, and the next succeeding conjunction. Which time, being divided by the time of the periodical revolution of each planet, will show how many revolutions each planet has then made.

"There are several ways of finding the above-mentioned prime integer numbers or multipliers; but the following is very convenient and easy.

"Let $\frac{a}{b}$ and $\frac{c}{d}$ be two of the fractions. Multiply the denominator of the first into the numerator of the second, and *vice versâ;* then strike out both the denominators, by which process the above fractions become $a\,d$ and $b\,c$; which numbers are in the same proportion as the fractions; and, if they are prime to each other, are the numbers required. But if they are not prime, divide them by their greatest common divisor, in order to reduce them to their lowest denomination.

"The reason why these numbers must be prime integers is plain: for, if they were not so, we should not have the synodic period required, but some multiple of it; and if they were not integers, we should not have exact multiples of the lower synodic periods from which we deduce the higher.

"To facilitate calculations which may be made on these principles, I shall subjoin the following Table, which shows the annual periods of the primary planets, reduced to hours; and their synodic periods, taken two by two progressively. But although the synodical periods of the planets, taken two by two, is so short, it must not be imagined that the synodical periods of three planets must be proportionably so too. The synodic period of the Earth and Venus (by the Table) is 1 year 218 days 17 hours; and that of Venus and Mercury is 144 days, 12 hours;

but the synodical period of these three planets is upwards of 5,500 years.[198]

"If the periods of the three planets be so incommensurate, how much more so must be the periods of the six revolving primaries of our system? Indeed we here cannot but see and admire the wisdom and providence of the SUPREME BEING! For, had the times of the annual revolutions of the several planets been more commensurate, the present arrangement of our system would doubtless have been greatly disturbed by the conspiring attractions of the six bodies, when they happened to be in conjunction; an arrangement which, from the goodness of the Almighty, we must conclude to be, in its present state, the best adapted to answer the purposes for which the system was created."

Names of the Planets.	Their Periodical Revolutions reduced to hours.
Saturn, ...	258223 h. $=a$
Jupiter, ...	103980
Mars, ...	16487
Earth, ...	8766
Venus, ...	5393
Mercury, ...	2111

Their synodic periods or conjunctions with each other (as under),

Names of the planets.	Y.	D.	H.	Hours.
Saturn and Jupiter,	19	313	10	$=174076=\frac{ab}{a-b}$
Jupiter and Mars,	2	85	21	$=19593=\frac{bc}{b-c}$
Mars and the Earth,	2	49	10	$=18718=\frac{cd}{c-d}$
Earth and Venus,	1	218	17	$=14015=\frac{de}{d-e}$
Venus and Mercury,	0	144	12	$=3468=\frac{ef}{e-f}$

198 The following is an arithmetical illustration of the process:—

For the conjunction of the Earth and Venus,

365 days 6 hours = 8766 hours in the Earth's year.

224 do. 17 do. = 5393 do. Venus's do.

Multiply these hours together, and then divide by their difference.

$\frac{8766-5393 = 3373}{8766\times5393 = 47275038}$, and 47275038÷3373 = 14015·7 hours for the conjunction of the Earth and Venus.

For the conjunction of Venus and Mercury.

224 days 17 hours = 5393 hours in Venus's year.

87 do. 23 do. = 2111 do. Mercury's do.

As before, multiply the hours together, and divide by their difference.

$\frac{5393-2111 = 3282}{5393\times2111 = 11384623}$, and 11384623÷3283 = 3468·8 hours for the conjunction of Venus and Mercury.

Then, as 14015·7 and 3468·8 are prime to each other, multiply them together, and reduce to years, &c.

To illustrate the use of this Table, let it be required to find the synodical period or conjunction of the Earth, Venus, and Mercury.

That of the Earth and Venus 14015 hours$=\frac{de}{d-e}$; and that of Venus and Mercury 3468 hours$=\frac{ef}{e-f}$.

Therefore, from what has been already laid down, the synodical period of the three planets will be when $m \times 14015$ is equal to $n \times 3468$; or when $m : n :: 3468 : 14015$; but these numbers being integers, and in their lowest terms already, they require no reduction. Therefore, 3468×14015 gives the synodical period of the three planets$=48604020$ hours$=5544$ years 221 days 12 hours. The reader may proceed to find out the synodic periods or conjunctions of the rest, according to the foregoing rules." [199] (Vide "Tables and Tracts." Lond., 1767, pp. 141—149).

$14015{\cdot}7 \times 3468{\cdot}8 = 5546{\cdot}13964$ years $= 5546$ y. 50 d. 53 m. for the conjunction of the Earth, Venus, and Mercury. Or if we take the rough figures of Mr. Ford, as here given, $14015 \times 3468 = 48604020 \div 8766 = 5544{\cdot}60643$ years $=$ 5544 y. 221 d. 11 h. 58 m., or in round numbers, 5544 y. 221 d. 12 h. as above.

[199] This question, in order to be properly treated, ought to have its yearly periods of revolutions reduced to seconds. About 40 years ago, we reduced the sidereal revolutions of all the then known planets to seconds, and made them *prime* to each other (as by the foregoing rule), and found out, to great minuteness, the conjunctive period of all the planets. At this distance of time, we still recollect the tedious operation. But, were the following put, and an answer demanded, we would not envy the party who proposed to himself the task of solving it,—viz. Suppose Mercury, Venus, the Earth and Moon, Mars, Jupiter and satellites, Saturn and satellites, Uranus and satellites, and Neptune and satellites, *all* to be in a line of conjunction with the Sun. How much time would elapse before they were all in a conjunctive line again?

The subject of planetary revolutions may be familiarly illustrated thus:—Suppose two men propose to walk round a circle, and that one of the men can walk round it in 3 minutes, the other in 5 minutes. After starting to walk round the circle—with the relative velocities given, how much time will elapse before they meet or come into conjunction?—As the 3 and the 5 are prime to each other; Rule.—Multiply the times of velocity into each other, and divide by their difference, thus,—$3 \times 5 = 15$, and $3—5 = 2$ for a divisor; hence, we have $\frac{2}{15}$, or $15 \div 2 = 7\frac{1}{2}$ minutes for the time that elapses before they meet.

Again, Suppose, as in the *first case*, the times of revolution are 3 and 5 minutes, and that both men start from a given point to walk round the circle in *contrary directions*—how much time would elapse before they both met? Rule—the numbers 3 and 5 being prime to each other.—Multiply the times of revolution together, and *add* the times for a divisor,—thus, $3 \times 5 = 15$, and $3+5 = 8$ for a divisor; therefore, $\frac{8}{15}$ is the fractional expression—reduced, $15 \div 8 = 1\frac{7}{8}$ minutes, or in 1 minute $52\frac{1}{2}$ seconds, for the time of a conjunctive meeting when going round the said circle in contrary directions. One example more.—The minute-hand of a watch revolves round the dial-plate in an hour; the hour-hand in 12 hours. Suppose they are exactly in conjunction at XII. o'clock; when, after leaving the XII., will these two hands again meet? Here we have 1 hour and 12 hours, and $1—12=11$; hence, we have $\frac{11}{12}$, that is, the two hands come into conjunction 11 times in 12 hours. Therefore, if 11 : 60 m. : : 12?—$60 \times 12 = 720 \div 11 = 65\frac{5}{11}$ minutes, the time that must elapse before the two hands meet, that is, they meet at $5\frac{5}{11}$ minutes past 1 o'clock; and by multiplying the $5\frac{5}{11}$ minutes by 2, 3, 4, 5, &c., it will give the number of minutes past 2, 3, 4,

One of our notes informs us that "Ferguson was of a serious turn of mind;—besides writing on his favourite sciences, his pen was not unfrequently employed on theological subjects, explaining hard questions, or in refutation of infidel opinions." He wrote at least papers on the following topics:—1st, "*On the Existence of Light before the Creation of the Sun.*" 2d, "*On the going back of the Sun on the Sun-Dial of Ahaz.*" 3d, "*On the Name and Number of the Apocalyptic Beast;*" and "*On the Birth and Crucifixion of Christ.*" Of the two first mentioned papers, we can find no trace; but we were successful in obtaining copies of the two last papers, from the original in the possession of our late friend, Mr. Upcott, Islington, London, who, when we were on a visit to him in 1831, kindly allowed us to take transcripts of them. Paper No. 3 will be found under date 1759. The following is a copy of No. 4 *paper:*

"ON THE BIRTH AND CRUCIFIXION OF CHRIST.—The Years of our Saviour's Birth and Crucifixion ascertained, and the Darkness at the time of the Crucifixion proved to be Supernatural.

The Vulgar Era of Christ's birth was never settled till the year 527, when Dionysius Exiguus, a Roman Abbot, fixed it to the end of the 4713th year of the Julian Period, which was four years too late. For our Saviour was born before the death of Herod, who sought to kill him as soon as he heard of his birth. And according to Josephus (B. XVII., ch. 8), there was an eclipse of the Moon, in the time of Herod's last illness; which Eclipse appears, by our Astronomical Tables, to have been in the year of the Julian Period 4710, March 13th, at 3 hours past midnight at Jerusalem. Now, as our Saviour must have been born some months before Herod's death, since in the interval he was carried into Egypt, the latest time in which we can fix the true Era of his birth is about the end of the 4709th year of the Julian Period.

and 5 o'clock, when they meet after these hours, in the 2d, 3d, 4th, and 5th revolutions, &c., or as follows:—There are 60 minutes round the dial-plate of a watch; therefore, $\frac{11}{12} \times \frac{60}{60} = \frac{660}{720}$, and $720 \div 660 = 1\frac{60}{660}$ minutes $= 1\frac{5}{11}$ hours; that is, at $5\frac{5}{11}$ minutes after 1 o'clock; also, 660 : 60 : : 720 ?—$720 \times 60 = 43200 \div 660 = 65\frac{5}{11}$ minutes past XII., which is 1 hour $5\frac{5}{11}$ minutes for the time of first meeting of the minute and hour-hand after they leave XII. These examples and rules are given to familiarise the subject of planetary motion, and it is to be hoped that, to some, they may prove useful on certain occasions. (See note 59).

Concerning the time of our Saviour's entering upon his public ministry (which may be called the time of his appearance, because, till then, he was not publicly known, so as to be talked of), and also concerning the time of his death, there is a remarkable prophecy in the IXth Chapter of the Book of Daniel, from the 24th verse to the end; which is in our English translation as follows:—

Ver. 24. *Seventy weeks are determined upon thy people, and upon thy holy city, to finish the transgression, and to make an end of sins, and to make reconciliation for iniquity, and to bring in everlasting righteousness, and to seal up the vision and prophecy, and to anoint the most Holy.*

Ver. 25. *Know therefore and understand, that from the going forth of the commandment to restore and to build Jerusalem, unto the Messiah the Prince shall be seven weeks; and threescore and two weeks the street shall be built again, and the wall, even in troublous times.*

Ver. 26. *And after threescore and two weeks shall Messiah be cut off, but not for himself: and the people of the prince that shall come, shall destroy the city and the sanctuary, and the end thereof shall be with a flood, and unto the end of the war desolations are determined.*

Ver. 27. *And he shall confirm the covenant with many for one week: and in the midst of the week he shall cause the sacrifices and the oblation to cease, and for the overspreading of abominations he shall make it desolate, even until the consummation, and that determined shall be poured upon the desolate.*

In the Hebrew, there are no stops nor pointings to any words or sentences; and in the above translation, one part of the 25th verse is most injudiciously pointed with a semicolon at *seven weeks;* which ought to run thus,—*seven weeks and threescore and two weeks.*

In the 24th verse, what we have rendered *prophecy,* is *prophet* in the original; and in some translations, which I have procured from those who understand the Hebrew very well, instead of *vision and prophecy,* it is rendered *visions and prophets.*

In ver. 27th, where we have it *the midst of the week,* all the translations I have procured render it the *half part of the week;* which may be taken either for the first or last part of it.

In the same verse, where we have it, *And he shall confirm the covenant with many for one week;* some translations render it, *And in one week a covenant shall be confirmed with many.* Now let the whole be put together, agreeable to this translation, without dividing it into verses (which is only a modern invention), but pointing it here and there for the sake of reading, and it will run thus,—

Seventy weeks are determined upon thy people and thy holy city, to finish the transgressions and to make an end of sins; and to make reconciliation for iniquity, and to bring in everlasting righteousness, and to seal up the visions and prophets, and to anoint the most holy. Know therefore, and understand, that from the going forth of the commandment to restore and build Jerusalem, unto the Messiah the prince, shall be seven weeks and threescore and two weeks: the street shall be built again, and the wall, even in troublous times.† And after threescore and two weeks shall Messiah be cut off, but not for himself. (And the people of the prince that shall come shall destroy the city and sanctuary, and the end thereof shall be with a flood; and unto the end of the war desolations are determined). And in one week a covenant shall be confirmed with many, and in half part of the week* HE ‡ *shall abolish the sacrifices and offerings. And for the overspreading§ of abominations he shall make desolate even unto consuming; and that which is determined shall be poured upon the desolate.*

'Tis evident that the first part of this prophecy relates to the coming of Christ; to his being put to death, *not for himself,* but for the sins of mankind, by which great sacrifice he was to put an end to all other sacrifices and offerings; to his introducing the righteousness of ages, and sealing up, or putting an end to prophecies. And that the latter part mentions the destruction of Jerusalem, in a very emphatic and striking manner.

* Some translate this *the holy of holies,* and Mr. Purver, *the very holy one.*
† By most translators, *in the straightness of times.*
‡ The Messiah.
§ *Wing,* in the Hebrew language. J. F.

In the seventh chapter of *Ezra*, we have an account of a very ample and full commission (or commandment) which was given by King *Artaxerxes* (*Artaxerxes Longimanus* ‖) to Ezra, to go up to Jerusalem, in order to repair that city, and restore the state of the Jews; and that Ezra took his journey on the first day of the first month, viz. the month *Nisan*, which began about the vernal equinox. And on the 14th day of that month (reckoned from the New Moon, at which the month began) the passover was always kept; for Josephus expressly says, 'the passover was kept on the 14th day of the month *Nisan*, according to the Moon, when the Sun was in Aries.' ¶ And the Sun always enters the sign Aries at the time of the vernal equinox.

This commandment was given in the 7th year of *Artaxerxes's* reign, and that year—according to *Ptolemy's* canon, the rectitude of which was scarce ever called in question—was the 4256th year of the Julian period; and from the vernal equinox in that year, we are to count the above-mentioned seventy weeks to the death of Christ. For, as the accomplishment of the prophecy must end with the expiation of sin, we cannot suppose these weeks to end at any other time.

But, if we count many revolutions of 70 common weeks, from the time of the Jewish passover in the year of the Julian period 4256, we shall find that no Messiah or Saviour did appear on Earth within that space of time: nor will these reckonings lead us from one passover to another. And it is certain, from the four Gospels, that Christ was crucified at the time of the passover; and St. John, chap. xviii., ver. 28, is so particular, as to inform us that our Saviour was crucified on the very day that the Passover was to be eaten by the Jews, who would not defile themselves by mixing with the multitude early in the morning, at the time of his trial. From these circumstances it is plain that these prophetic weeks mean something very different from the weeks by which we commonly reckon.

In the Old Testament, we read of weeks of years, as well of weeks of days. For, as every seventh day was to be a Sabbath for man, on which he was to rest from his labour, so every seventh year was to be a Sabbath for the land, in which it was

‖ In the Book of *Esther* he is called *Ahasuerus*.
¶ Lib. i., cap. 10. J. F.

to rest from tillage. Let us therefore take these 70 weeks to be weeks of years, making 490 years in all; and the reckoning will lead us from the Passover in the year of the Julian period 4256, to the Passover in the year 4746, which was the 33d year of our Saviour's age, accounted from the vulgar era of his birth.

It is expressly foretold in this prophecy, that from the time of the commandment's being given out to restore and build Jerusalem, to the Messiah the prince, (or to the time of his appearing in his public character) there would be seven weeks and threescore and two weeks, or 69 weeks in all: the first seven of which, being the straitest or shortest of the times, consisting of 49 years, we may very well allot to the repairing of Jerusalem; after which, there should be threescore and two weeks, or 434 years, to the public appearance of the Messiah: and then there remained only one week, or seven years, for the public ministry; which, I apprehend, is meant by *confirming the covenant with many.*

But as some of the translations which I have procured, say, concerning that week, *And in one week a covenant shall be confirmed with many;* and all of them have it, *and in half part of the week* (which might be either the first or last half of it) HE *shall abolish the sacrifices and offerings;* it does not appear that the Messiah is brought in for the whole of the seventieth week, but only for one half of it, *in confirming* (or establishing) *the* new and everlasting *covenant* of the Gospel; by which, *the righteousness of ages,* mentioned in the first verse of the prophecy, seems to be plainly meant.

And when we consider that CHRIST'S messenger, *John* the Baptist, preached so long before Christ took the public ministry upon himself, as that he acquired great fame in many countries around, which could not be done in a short time, we may believe that the last verse of the prophecy allots the first half of the seventieth week, or three years and a half, to the time of *John's* preaching; at the end of which time he baptized Christ, who was then entering into the thirtieth year of his age—according to St. Luke—and then Christ took his public ministry upon himself for the remaining half of the seventieth week; at the end of which he was cut off by the wicked and self-hardened Jews, and so put a *virtual* end to all their sacrifices and offerings; which *finally* ended with the destruction of their city and temple about 37 years after.

So that, in the first place, taking the whole of the prophecy together, as in *ver.* 25, and then dividing it into four different periods or parts as above-mentioned, it will very naturally run thus,—

	Weeks.		Years.
From the time of Ezra's receiving the Commandment to repair Jerusalem, until the expiation of sin by Christ,	70	or	490
For the time of these repairs, ...	7	or	49
From the finishing of these repairs to the coming of Christ by his messenger, *John* the Baptist,	62	or	434
From that time to the end of *John's* ministry, and the baptism of Christ,	½	or	3½
From thence to the end of Christ's ministry, by his death on the cross,	½	or	3½
In all,	70	or	490

For a very full illustration of this matter, I refer the reader to Dr. *Prideaux's* Connection of the Histories of the Old and New Testament.

The beginning of these seventy weeks of years being found to be the year of the Julian period 4256, at the time of the Jewish Passover, their ending must have been at the Passover in the year of the Julian period 4746, in the 33d year after the year of Christ's birth; and consequently, according to this prophecy, our Saviour was crucified in the 4746th year of the Julian period.

'Tis plain from all the four Gospels, that the Crucifixion was on a Friday; because it was on the day next before the Jewish Sabbath; and, as above-mentioned, on the day the Passover was to be eaten (at least) by many of the Jews.

The Jewish year consisted of twelve months, as measured by the Moon, which contains 354 days; to which they either added 11 days every year, in order to make their years keep pace with the Sun's course of 365 days; or 30 days in three years. So that, although their months were lunar, their years were solar. And they always celebrated the Passover on the fourteenth day of the first lunar month, reckoning from the first time of their seeing the New Moon; which, especially at that time of the year, might be when she was about 24 hours old;

and consequently, their fourteenth day of the month fell upon the day of full Moon; and, according to *Josephus*, they always kept the Passover at the time of the full Moon next after the vernal equinox. But the full-Moon day on which our Saviour was crucified fell on Friday. And as 12 lunar months want 11 days of 12 solar months, the Passover full Moons (as well as all others) fall 11 days back every year; which, being more than a week by four days, makes it, that in a few neighbouring years, there cannot be two Passover full Moons on the same day of the week. And when this anticipation would have made the Passover full Moon fall before the equinoctial day, they set it a whole month forward, to have it at the first full Moon after the vernal equinox; which puts it off the same day of the week again.

The dispute among Chronologers, about the year of our Saviour's crucifixion, is limited within four or five years at most. And it certainly was in the year in which the Passover full Moon fell on a Friday. And I find, by calculation, that the only Passover full Moon which fell on a Friday, from the 20th year after our Saviour's birth to the 40th, was in the 4746th year of the Julian period, which was the 33d year of his age, reckoning from the beginning of the year next after that of his birth, according to the vulgar era; but the 37th, reckoning from the true era thereof; and the said Passover full Moon was on the third day of April.

And when we reflect on what the *Jews* told him, some time before his death, (John viii. 57) '*thou art not yet fifty years old*,' we must confess that it should seem much likelier to have been said to a person near forty, than to one but just turned of thirty. And we may easily suppose that St. Luke expressed himself only in round numbers, when he said that *Christ was baptized about the* 30*th year of his age;* when he began his public ministry; as our Saviour himself did, when he said he should lie *three days and three nights in the grave.*

And thus we have an Astronomical demonstration of the truth of this ancient prophecy, seeing that the prophetic year of the Messiah's being cut off was the very same with the Astronomical. Besides, we have the testimony of a heathen author, which agrees with the same year. For *Phlegon* informs us, that in the fourth year of the 202d *Olympiad* (which was the

4746th year of the Julian period, and the 33d year after the year of Christ's birth) there was the greatest eclipse of the Sun that ever was known, for the darkness lasted *three hours* in the *middle of the day*, which could be no other than the darkness on the crucifixion-day; as the Sun never was totally hid *above four minutes of time*, from any part of the Earth, by the interposition of the Moon. If *Phlegon* had been an Astronomer, he would have known that the said darkness could not have been occasioned by any regular eclipse of the Sun, as the Moon was then in the opposite side of the heavens, on account of her being full. And as there is no other body than the Moon that ever comes between the Sun and the Earth, it is evident that the darkness at the crucifixion was miraculous, being quite out of the ordinary course of nature.

There have been great difficulties about our Saviour's eating the Paschal lamb on the evening of the day before it was eaten by the Jews. But I apprehend this difficulty may be easily removed, when we consider that the Jews began their day in the evening, and ended it in the next following evening. So that, although it was on a different day, according to our way of reckoning, it was still *the same day* according to theirs. And we do not find that they brought in his eating the lamb on the Thursday evening, as any accusation againt him, which they would have been glad to do, if they could have made a handle of it for that purpose. J. Ferguson, 5th August, 1755." (See Ferguson's Astronomy, Sect. 395; "Tables and Tracts," pp. 180—194, 1st Edit.; his "Syllabus of Lectures," 1763; and MS. "*Common Place Book*," pp. 154, 155.

1756.

We have no memoranda of Ferguson for the first months of the year 1756, but it is likely that he would be busy with his Astronomical and other lectures, and also attending to his old profession of limner. He would likewise have much of his attention devoted to correcting the proof-sheets of his great work, "Astronomy Explained upon Sir Isaac Newton's Principles," which was then in the press,—as also to the examination of the many engravings which illustrate that work. In the midst of these multifarious engagements, he would hear, with great regret, of the death of one of his earliest friends and patrons when he first came to London, viz. that of Sir Dudley Ryder, who died on 25th May, 1756. It will be remembered that Sir Dudley, shortly after Ferguson's arrival in London, purchased from him his small orrery made with ivorywheels. (See note 70).

FERGUSON'S ASTRONOMY.—This, "*the great work of Ferguson,*" appears to have been published in June 1756; there are no advertisements to be found regarding it in any of the newspapers or magazines of 1756, *before* July, although there are many to be found in July, and for many months afterwards. The earliest notice of it we have seen is in Martin's "*Miscellaneous Correspondence,*" for July, 1756; it is briefly announced thus:

> "Astronomy Explained upon Sir Isaac Newton's Principles, in 4to, 16s., sewed. By James Ferguson."
> Martin's Miscellaneous Correspondence for July 1756.

We have not been able to get a sight of the first edition, but have a copy of the third edition in 4to. Lond., 1764. It is a volume of 354 pages, exclusive of 10 pages of index-matter; and a few pages at the commencement, embracing Dedication and Contents, and 18 folding engravings; sewed, 16s., bound in calf, 18s. It was published by A. Millar, in the Strand, London, who afterwards printed his other works, and became one of Ferguson's most intimate friends.[200]

The following is the TITLE-PAGE of this work:—

> "Astronomy Explained upon Sir Isaac Newton's Principles, and made easy to those who have not studied Mathematics; to which are added, a Plain Method of finding the Distances of all the Planets, from the Sun, by the Transit of Venus over the Sun's Disc, in the year 1761; An Account of Mr. Horrox's Observation of the Transit of Venus in the year 1639; and of the Distances of all the Planets from the Sun, as deduced from Observations of the Transit in the year 1761. By James Ferguson.
>
> Heb. xi. 3. The Worlds were framed by the word of God.
> Job xxvi. 7. He hangeth the Earth upon nothing.
> ————13. By his Spirit he garnished the Heavens.
>
> (Illustrated with Engravings.) London: Printed for A. Millar, in the Strand. MDCCLVI."

Many writers justly esteem this treatise on Astronomy as his

200 Mr. Andrew Millar, a celebrated publisher in London, "was of Scotch parentage, and literally the artificer of his own fortune. By persevering industry, and a happy train of successive patronage and connection, he became one of the most eminent booksellers and publishers of the eighteenth century." He died at Kew Green, near London, June 8th, 1768, aged 61 years, and was buried in the cemetery at Chelsea.

"*happiest effort*," and his "*great work*." It at once took a high position, and, for a great many years, superseded all other treatises on Astronomy. It still continues to be held in high esteem; and the edition by Dr. Brewster sustains its reputation.

Ferguson's Astronomy has been translated into several languages,—and in our own country, it has passed through a great many editions. The following are the dates of the publication of the various editions, in quarto and octavo, as far as we have been able to ascertain, viz. 1756—1757—1764—1770—1772—1773—1785—1790—1794—1799—1809—1811—1821. The work is dedicated to the Earl of Macclesfield, who was then President of the Royal Society of London. The annexed is a copy of the DEDICATION:

> "To the Right Honourable GEORGE, EARL OF MACCLESFIELD, Viscount Parker of Ewelme, in Oxfordshire; and Baron of Macclesfield, in Cheshire, President of the Royal Society of London, Member of the Academy of Sciences at Paris, of the Imperial Academy of Sciences at St. Petersburg, and one of the Trustees of the British Museum, Distinguished by his Generous Zeal for Promoting every Branch of Useful Knowledge, this Treatise of Astronomy is Inscribed, with the most profound respect, by his Lordship's most obliged, and most humble servant, JAMES FERGUSON."

Our notes inform us, that the publication of his Astronomy "at once placed Ferguson in a high position in the temple of fame; that his "name, as an Astronomical writer, was now established, and held in universal respect;" and that "his friendship was now eagerly sought after, and was highly appreciated by men of learning and science." Even artists appear to have found him out, as one or two likenesses of him were done this year (1756).

One of these likenesses, in oil, is still extant, and in the possession of Mrs. Thomson, Nethercluny, Mortlach, Banffshire. The canvas on which this likeness is painted is 23 inches by 19 inches. Mrs. Thomson has very kindly sent us, through the Rev. Mr. Annand of Keith, an excellent photograph, taken from this portrait. On the back of the likeness is written, "Painted by John Beatson, 1756." This appears to be one of the earliest of Ferguson's likenesses; but it differs from any we have seen; both as to physiognomy, and its extreme plainness of attire.

According to an entry in an old "Day Book," which belonged

to Mr. Nairne, Optician, London,[201] we find, that in December, 1756, Ferguson purchased from him an Air-Pump, with an extensive apparatus belonging to it, for the sum of £32, in order to add the subject of Pneumatics to his Course of Lectures. An engraving of this Air-Pump and Apparatus is to be found in Ferguson's "Lectures on Select Subjects in Mechanics, Hydrostatics, Pneumatics, and Optics."

1757.

Ferguson's Residence.—It appears that Ferguson, early this year, removed from Broad Street, Golden Square, to Duke Street, Bloomsbury, London. In Kent's London Directory for 1757, we find

"James Ferguson, Astronomer, Duke Street, Bloomsbury."

Accident to Ferguson.—Sometime early in the month of February, 1757, he accidentally got his right leg severely bruised; but *how*, *when*, or *where*, is not known. The only notice we have of this accident is given in a letter which he sent to the Rev. Mr. Birch, which we subjoin. The original letter is among the "Birch Letters," British Museum, Vol. 9, p. 145, and Museum, No. 4308.

"Rev. Sir,

I have lately had the misfortune (by accident) of bruising my right leg, and regardless of any ill consequence, I took little notice of it till lately. I have since applied to a Surgeon, who assures me there is no real danger. However, as his skill alone (under God) must effect a cure, I must submit to his orders—some days confinement under the circumstances. I was in hopes I should have been able to wait upon you to-morrow, but find now, I must be obliged to your goodness to take the fatigue upon your own hands.

Rev. Sir,
Your most obliged, obedient servant,
James Ferguson."

Feb. 20th, 1757.
To the Reverend Mr. Birch,
at his lodgings,
in Norfolk Street, by ye Strand.

201 Mr. Edward Nairne, a celebrated Optician, and Philosophical Instrument Maker, in Cornhill, London. In 1771, he published "*An Account of the Equatorial Telescope;*" in 1777, "*An Account of Experiments made with an Air-Pump;*" and, in 1780, "*A Treatise on Electricity.*" Mr. Nairne was one of Ferguson's most intimate friends. From Mr. Nairne he purchased the greater part of his Philosophical Apparatus. Ferguson, in his Treatise on Electricity, p. 9,

WHIRLING TABLE.—Early in 1757 he made a very efficient *Whirling Table*, with an extensive apparatus for demonstrating "The Propensity of Matter to keep the state it is in.—That Bodies moving in orbits, have a tendency to fly out of these orbits.—That Bodies move faster in small orbits than in large ones.—The Centrifugal Forces.—That a Double Velocity in the same Circle is a balance to a quadruple power of gravity.—Kepler's Problem.—The absurdity of the Cartesian Vortexes.—That if one body move round another, both of them must move round their common centre of gravity.—That the squares of the periods of the planets are as the cubes of their distances.—That the Central Forces are inversely as the squares of the periods;—also, a demonstration of the cause of the Tides, and the motion of the Earth."

The following description and figure of the Whirling Table are by Ferguson.

"The Whirling Table is a machine contrived for showing experiments of Central Forces, &c. A A is a strong frame of wood, B a winch or handle fixed on the axis C of the wheel D, round which is the cat-gut string F, which also goes round the small

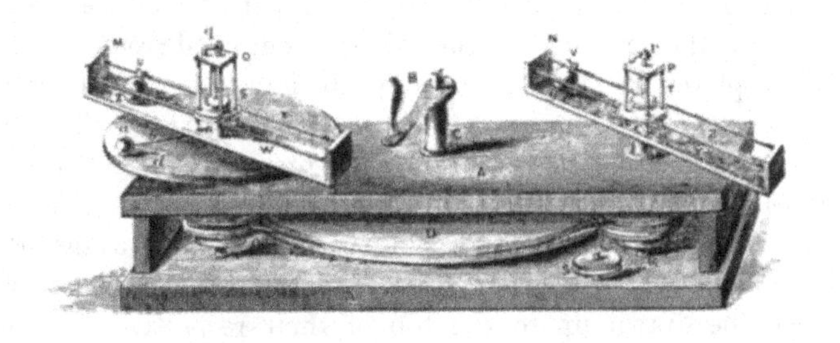

Ferguson's Whirling Table.

wheels G and K, crossing between them and the great wheel D. On the upper end of the axis of the wheel G, above the frame,

strongly recommends Electrical Machines—he says, "*The Electrical Machine mostly in use is made in the greatest perfection by Mr. Edward Nairne, Optician, in Cornhill, London.*" As formerly noted, Ferguson apprenticed his eldest son, James, to Mr. Nairne, in 1763, "to learn, practically, the business of Optician and Mathematical Instrument Maker;" and at Ferguson's funeral, 23d November, 1776, we find Mr. Nairne classed as one of the six "pall bearers." See account of Ferguson's funeral under date November 23, 1776.

is fixed the round board *d*, to which the bearer M S X may be fastened occasionally, and removed when it is not wanted. On the axis of the wheel H is fixed the bearer N T Z; and it is easy to see that when the winch B is turned, the wheels and bearers are put into a whirling motion.

"Each bearer has two wires W, X, and Y, Z, fixed and screwed tight into them at the ends by nuts on the outside. And when these nuts are unscrewed, the wires may be drawn out in order to change the balls U and V, which slide upon the wires by means of brass loops, fixed into the balls, which keep the balls up from touching the wood below them. A strong silk line goes through each ball, and is fixed to it at any length from the centre of the bearer to its end, as occasion requires, by a nut-screw at the top of the ball; the shank of the screw goes into the centre of the ball, and presses the line against the under side of the hole that it goes through. The line goes from the ball, and under a small pulley fixed in the middle of the bearer; then up through a socket in the round plate (see S and T) in the middle of each bearer; then through a slit in the middle of the square top (O and P) of each tower, and going over a small pulley on the top, comes down again the same way, and is at last fastened to the upper end of the socket fixed in the middle of the above-mentioned round plate. These plates, S and T, have each four round holes near their edges for letting them slide up and down upon the wires which make the corners of each tower. The balls and plates being thus connected, each by its particular line, it is plain that if the balls be drawn outward, or towards the ends M and N of their respective bearers, the round plates S and T will be drawn up to the top of their respective towers O and P.

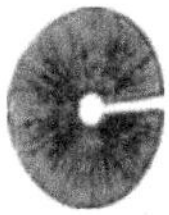

"There are several brass weights, some of two ounces, some of three, and some of four, to be occasionally put with the towers O and P, upon the round plates S and T; each weight having a round hole in the middle of it, for going upon the sockets or axes of the plates, and is slit from the edge to the hole, (as in the annexed figure) for allowing it to be slipped over the foresaid line which comes from each ball to its respective plate."—For a full account of the experiments exhibited by this Whirling Table. we refer the reader to his "Lectures on Select Subjects in Mechanics, Hydrostatics, Pneumatics, and Optics."

ASTRONOMICAL CLOCK.—In the year 1757, Ferguson contrived and made an Astronomical Clock for "Showing the Equation of Time—The apparent Daily Motions of the Sun, Moon, and Stars; with the Times of their Rising, Southing, and Setting."

The following is his description of this Clock:—

"About ten years ago,[202] I made a wooden model of a Clock for showing the apparent diurnal motions of the Sun and Stars, with the times of their rising and setting for every day of the year; and the days of the months all the year round, without any need of shifting by hand in the short months, as is always done in common clocks. I copied the dial-plate of this model from a clock that Mr. Ellicott had made for the King of Spain;[203] but although Mr. Ellicott showed me the whole inside of the clock, I did not ask him what the numbers of teeth in the wheels of it were, although, I am convinced, he would have told me, if I had; nor do I, in the least, remember how many wheels there were in the uncommon or Astronomical part of it; and so I set about contriving wheels and numbers for performing the like motions.

The Dial-plate contains twice twelve hours, and within the circle of hours there is a large opening in the plate, a little elliptical; the edge of this opening serves for a horizon.

Below the Dial-plate, and seen through the large opening in it, is a flat plate on which the equator, ecliptic, and tropics, are drawn; and all the stars of the first, second, and third magnitudes are laid down, that are visible in the horizon (of Madrid in Mr. Ellicott's, and of London in mine), according to their right ascensions and declinations; the centre of the plate being the north pole. The ecliptic is cut out into a narrow groove in the plate; and a small Sun slides in the groove by a pin, and is carried round by a wire fixed in the axis, which comes a little way out through the centre of the plate. The edge of this plate

202 The description of this Clock is taken from Ferguson's "Tables and Tracts," first published in London in 1767. He, thus writing the description in 1767, says it was contrived and made by him "*about ten years ago*," that is, in 1757.

203 Regarding Mr. John Ellicott, one of Ferguson's earliest friends in London, see note 78.

is divided into the months and days of the year, and the Sun's wire shows the days of the months in these divisions. This star-plate goes round in a sidereal day, making 366 revolutions in a year; in which time, the Sun makes 365, and consequently shifts a division or day of the month every 24 hours.

A small wire is stretched from over the centre of the sidereal plate to the upper XII on the fixed Dial-plate. This wire is for the meridian.

When the Sun, or any Star, comes to the Eastern edge of the horizon, the hour-index is at the time of rising of the Sun or Star, for the day of the year, pointed to by the wire, that carries the Sun; and when the Sun or Star comes to the western edge of the horizon, the hour-index is at the time of its setting. The Sun always comes to the meridian at the instant of the solar noon; but every star comes sooner to the meridian every day, than it did on the day before, by 3 minutes 55 seconds 54 thirds of mean solar time, as it revolves from the meridian to the meridian again in 23 hours 56 minutes 4 seconds 6 thirds.[204]

When any star is on the meridian in the clock, the star which it represents is on the meridian in the heavens; the time whereof is seen by the hour-index on the Dial-plate. And, as the stars have their revolutions on the plate, one may look at the clock at any time, and see what stars are then above the horizon, what stars are then on the meridian, and what stars are then rising or setting.

My contrivance for showing these motions and phenomena, in the model, consists of no more than two wheels and two pinions, as follows:—

The wheels are of equal diameters, and so are the pinions; the numbers of teeth are 61 in one wheel, and 73 in the other. The pinions are both fixed on one axis, the one having 20 leaves and the other 24.[205] The wheel of 73 teeth is fixed to the back of the sidereal plate, and the axis of the wheel of 61 comes through the wheel of 73, and through the sidereal plate, and carries the wire round on which the Sun slides in the ecliptic groove, and also the hour-hand on the Dial-plate.

204 The true period of a sidereal revolution is 23 h. 56 m. 4 s. 5 thirds 26 fourths, or in 23·93446967 hours.

205 Ferguson says that his contrivance for showing the motions and phenomena in his model, consisted of "*no more than two wheels and two pinions;* properly speaking, what he calls pinions of 20 and 24, are in reality wheels; therefore, the

The wheel of 61 teeth turns the pinion of 20, and the pinion of 24, (fixed on the same axis with that of 20) turns the wheel of 73;—(as shown in the annexed figures which represent the arrangement of the wheel-work).

Now, if the wheel of 61 teeth be turned round in 24 hours, and carry the Sun and hour hand, the wheel of 73 teeth will be turned round in 23 hours 56 minutes 4 seconds

Plan of the Wheel-work of Ferguson's Solar and Sidereal Clock.

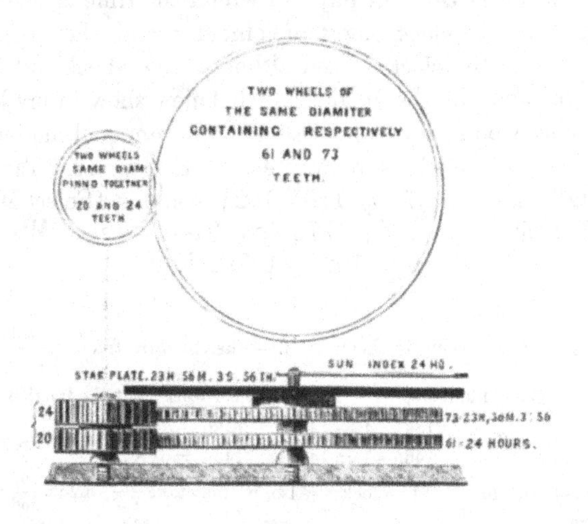

Section of the Wheel-work of Ferguson's Solar and Sidereal Clock.

6 thirds.[206] And so, the sidereal plate will make just 366 revolutions, in the time that the Sun makes 365.

Mr. Ellicott had the prettiest, and most simple contrivance

motions are exhibited by an addition of *four wheels* to the common clock movement.

[206] These wheels are not derived from a natural root, as they are $\frac{73 \times 20 = 1460}{61 \times 24 = 1464}$ and if $1464 : 24\text{ h.} :: 1460 ? = 1460 \times 24 = 35040 \div 1464 = 23{\cdot}93442623$ h., or 23 h. 56 m. 3 s. 56 thds. 3 fourths 56 fifths, and so on. Ferguson calls the period arising from these wheels, 23 h. 56 m. 4 s. 6 thirds, which is 10 thirds more than the wheels really produce.

The true apparent diurnal revolution of the stars is accomplished in 23 h. 56 m. 4·0906 s., or in 23·93446967 h., or $\frac{23{\cdot}93446967}{24{\cdot}00000000}$, which, on being reduced by continuous fractions, as shown in several preceding notes, we obtain the fractions $\frac{1}{1}$, $\frac{365}{366}$, $\frac{1461}{1465}$, $\frac{12053}{12086}$, $\frac{49673}{52809}$, $\frac{111332}{117704}$, &c. These fractions are *prime* and cannot be reduced, which renders them useless for wheel-work; but we may

I ever saw, in his clock, for showing the difference between equal and solar time (generally termed the equation of time) on all the different days of the year. He generously allowed me to copy that part into my model, and I have quite concealed it within one of my wheels, not to show how it is done unless he publishes an account of it. The Sun, by that simple contrivance, even in my model, comes as much sooner or later to the meridian, than when it is noon by a well regulated clock, as the Sun in the heavens does, at all the different times of the year, excepting the four days on which the time of noon shown by the Sun and clock ought to coincide; and then there is no difference in the clock. And although the wheel-work is quite open to sight in the model which I now show in my lectures, no person who sees it can guess how the unequal motion of the Sun, in the model, is performed." (Ferguson's "Tables and Tracts." Lond., 1767, pp. 116—122; also, his "Select Mechanical Exercises." Lond., 1773, pp. 31—33; also, MS. "Common Place Book," page 102, Col. Lib. Edin.)

produce approximations to them, thus,—take the *third* fraction $\frac{1461 \div 3 = 487}{1465 \div 5 = 293}$, which are prime numbers,—too large for wheel-work; but the fraction $\frac{1461}{1465}$ may be treated as follows:—$\frac{1461-1 = 1460}{1465-1 = 1464}$ and $\frac{1460 = 73 \times 20}{1464 = 61 \times 24}$, which are Ferguson's numbers in this model of his clock;—Also, we may take $\frac{1461}{1465} \times 2 = \frac{365}{366} = \frac{3287}{3296} = \frac{19 \times 173}{42 \times 103} =$ 23 h. 56 m. 4 s. 077 dec., $\frac{365}{366} \times 3 + \frac{1}{1} = \frac{1096}{1099}$, and $\frac{1461}{1465} \times 2 + \frac{1096}{1099} = \frac{3018}{4029} = \frac{49 \times 82}{51 \times 79}$. Let a wheel of 79 turn round in 24 hours and drive a wheel of 51 teeth, to which is made fast a wheel of 49 teeth (turning round with it in the same time), which drives a wheel of 82;—This wheel of 82 teeth will turn once round in 23 h. 56 m. 4 s. 1102 dec.; Again, let us take $\frac{1461}{1465} \times 7 + \frac{365}{366} = \frac{10592}{10621} = \frac{32 \times 331}{43 \times 247}$. If a wheel of 247 teeth go round in 24 hours and drive a wheel of 43, and this wheel to have one of 32 made fast to it, and made to turn a wheel of 331, this wheel of 331 will turn round in 23·93446945 h. or 23 h. 56 m. 4 s. 5 thds. 24 fourths, an extremely near approximation discovered by us many years ago, and which was at the time communicated to the members of the Royal Society by G. B. Airy, Esq., Astronomer-Royal. Since then, we have discovered a still closer approximation to the true period by taking $\frac{12053}{12086} \times 3 + \frac{1461 = 37620}{1465 = 37723}$, reducible to $\frac{180 \times 209}{119 \times 317}$ which gives a period of 23 h. 9344696869 decimal = 23 h. 56 m. 4 s. 5 thirds, 27 fourths,—the true period being 23 h. 56 m. 4 s. 5 thds. $26\frac{8}{10}$ fourths. About ten years ago we constructed, in brass, a model with the latter set of wheel-work, and have it still in our possession. (Vide Philosophical Transactions for February, 1850, and Denison's Treatise on Clock and Watch Making, pp. 38, 39).

FERGUSON'S ASTRONOMY—Second Edition.—The *first* Edition, published in 1756, having been nearly all disposed of by the Spring of the year 1757—a *second edition*, with corrections, was published in June, 1757. About the end of this year he sold "the remaining part of the copy of his work to his publisher,—Mr. Andrew Millar, Bookseller and Publisher, Strand, London, for £300." (See Ferguson's letter, date, January, 1758).

NEW ASTRONOMICAL INSTRUMENT.—Ferguson, on 9th August 1757, published a new *Rotula*, under the designation of "A NEW ASTRONOMICAL INSTRUMENT;" like his previously published Rotulas, it consisted of a series of three large moveable cards, and "showed the Day of the Month, Change and Age of the Moon, Places of the Sun and Moon in the Ecliptic; the Rising, Southing, and Setting of the Sun, Moon, and Stars, in the Latitude of London. Price Five Shillings and Sixpence." This New Astronomical Instrument has been long out of print. Among the drawings, &c., which belonged to the late King George III., now deposited in the British Museum, there is a copy of it in excellent preservation.

"*On* THE YEARS *of the* BIRTH *and* DEATH *of* CHRIST."—DISPUTES, &c.—"In the months of March, April, and May, 1757, several *masked* writers sent queries to the newspapers and magazines, pointing out what they considered to be inaccuracies in his Astronomy, and requested Ferguson's answer." The following note from Ferguson to Mr. R. Baldwin, the then printer and publisher, London, is in reference to this subject. The paper which accompanied the note is too long for insertion here—it is somewhat similar to that already given in pp. 206—213, with a few parenthetical explanations.

"SIR,—I have seen a serious letter in your Magazine for last April, concerning the method I have taken in my Astronomy to settle the years of the birth and death of Christ, which is now reprinted, with some alterations in the second edition of this work. If you think proper to insert the following to satisfy the author of said letter, you are entirely at liberty to do so.

SIR,

Your humble servant,

LONDON, August, 1757. JAMES FERGUSON."

The original is in pencil—was written in August, 1757, and is now to be found among the MSS. letters in the British Museum.

ASTRONOMICAL TABLES AND PRECEPTS—in MS.—In the year 1757, Ferguson calculated and arranged a series of Tables for finding the times of new and full Moon, and Eclipses. They made a thin quarto of 28 pages, with 2 folding pen and ink sketches of Eclipses. These Tables are still in existence, beautifully written, in excellent preservation, and half bound (size 10¾ by 8½ inches). They were never published in a separate volume,—so far as we are aware—but are to be found, in detached portions, in several of his works. The manuscript is entitled,

> "Astronomical Tables, with Precepts, for Calculating the True Times of New and Full Moon and Eclipses. By James Ferguson. M.D.CCLVII."

These Tables, along with two other manuscript works, were discovered in Edinburgh towards the end of the year 1865, and were deposited in the College Library there in December of the same year. For particulars regarding the other manuscript works, see dates 1758 and 1776, as also the Appendix.

HAND-MILL for GRINDING CORN.—Sometime in the summer of 1757, "The Society for the Encouragement of Arts, Manufactures, and Commerce, offered a Premium to the person who should contrive and make the most effectual Hand-mill, to grind corn into meal, for making bread for the poor; fourteen different mills were produced, to the said Society, at the end of November, 1757; and a Committee appointed to examine the same, having seen them severally grind corn, and called to their assistance some of the most able judges to determine on the meal so ground, they reported in favour of Mr. Gordon's Mill." (Hinton's Universal Magazine, Vol. xxii. p. 357.) Ferguson resolved to compete for the premium. He contrived and made an efficient hand-mill in the autumn of 1757, and sent it to the Society. The award was made in the middle of December, 1757, in favour of Gordon, a "*friend*" of Ferguson; through whose duplicity Ferguson's mill was rejected. See the following letter of Ferguson's to the Rev. Alexander Irvine of Elgin.

1758.

THE earliest notice we have of Ferguson, in 1758, is through the medium of a very long and interesting letter which he wrote to his old friend, the Rev. Alexander Irvine, Elgin, which letter was presented some years ago to the Elgin Museum, by Mr. Alexander Duffus, Cabinet-maker, Elgin. It will be seen that it refers to himself and family; to his thoughts of leaving London; to his apparatus; to his newly made Whirling Table; to Mr. Harrison's Clock and Corn Mill; to his being "*egregiously bit by one Gordon*" in the matter of the Corn Mill model; to a balance of £300 from his publisher; to a newly published Rotula, and to a *foolosopher*, and the then expected Comet. We give the letter in full.

Copy of FERGUSON'S LETTER to the REV. ALEXANDER IRVINE, of ELGIN.

LONDON, *Jan.* 17*th*, 1758.

"DEAR REVEREND FRIEND,—After so long a silence, I at last write to you again, and inform you that the children and I are in very good health; but poor Bell has been extremely low in her spirits since the end of last June, and seems only now to be getting a little better.[207]

"I am still going on in the old way; only my eyes are rather too much failed to draw pictures; and indeed I cannot say that I have drawn six these last twelvemonths. And as to Astronomy, there are at present more than double the number which might serve the place—people's taste lying but very little in that way; so that unless something unforeseen happens, I believe my wisest course will be to leave London soon—everything being so excessively dear, and the taxes so oppressive, that there is no living. I assure you I am not quite in jest when I now request of you to inform me for how much rent one might have a tolerable house in or about Elgin, and what a man with a wife and three children might soberly live upon by the year;[208] also,

207 The "*poor Bell*" refers to Mrs. Ferguson. As previously mentioned, her maiden name was Isabel Wilson. The "*children*" were AGNES, aged 12 years and 5 months; JAMES, aged 9 years and 3 months; and MURDOCH, aged 5 years and 2 months—three children. Ferguson, at this critical period of his life, was in his 48th year; his wife in her 38th year; and they had (in January 1758) been resident in London for 14 years and 8 months.

208 Elgin, an old Episcopal city in the north of Scotland, about 540 miles NNW of London, and 17 miles west of Keith—the locality of his youthful days. At the period Ferguson writes, 1758, Elgin had a population of about 2,200.

whether it would be prudent to dispose of my Astronomical machinery here, for a third part of what it cost me, or bring them north, with any view of having now and then some lectures in your part of the country? I own, I had much rather choose the latter—for I should be like a fish out of water without my apparatus.[209] But all this *entre nous;* and when you have considered it, rather let me know as it were by way of invitation and information from yourself, than as answering a question.[209]

I believe I told you already that I have added a Whirling Table to my apparatus, for showing the central forces, or laws by which the planets move, and are retained in their orbits; and it even goes so far as to demonstrate that the squares of the periods are as the cubes of the distances; and the central forces inversely are as the square of the periods. I have lately procured an Air-pump, with a most complete apparatus.[210]

Mr. Harrison's clock has lately been examined and much approved of at the Admiralty, by the Board for the longitude. He has thereupon received a pretty large sum of money, and will soon be ordered out for a trial.[211]

I have been most egregiously bit just now by one Gordon,

209 The following is an inventory of the apparatus Ferguson had at this period (1758), viz. 1. A "*Trajectorium Lunare.*" 2. Season Hoops for illustrating the Change of the Seasons. 3. A Large Wooden Orrery. 4. A small Three-Wheeled Orrery. 5. "*The Four-Wheeled Orrery.*" 6. "*An Improved Celestial Globe.*" 7. "*The Planetary Globes.*" 8. Wooden Model of a curious Astronomical Clock. 9. A Crank Orrery. 10. "*The Calculator.*" 11. "*The Tide-Dial.*" 12. The Centrifugal Machine. 13. Mechanical Paradox. 14. "*The Cometarium.*" 15. Simple Lunar Machine. 16. A "*Satellite Machine.*" 17. "*The Eclipsareon.*" 18. A Seven-Wheeled, or rather, *Eight-Wheeled* Wooden Orrery. 19. Mechanical Paradox Orrery. 20. The Whirling Table, with an extensive apparatus. 21. An Air-Pump and complete set of apparatus. 22. Astronomical Clock Model for showing the motions of the Sun and Stars, the Equation of Time, &c., besides models of levers, wedges, wheel and axles, screws, wheels, waggons, pumps, mills, &c., &c., as also a great many astronomical and other diagrams and maps.

210 The Whirling Table, with a figure, has already been described; and the Air-Pump and apparatus have been already noticed.

211 Mr. John Harrison, a self-taught genius (like Ferguson), was born at Pontefract, in Yorkshire, in 1693, and was bred a carpenter. At an early age, he displayed considerable ingenuity in constructing several curious pieces of mechanism; before he had attained the age of 21, he had made two wooden clocks without having received any instruction in horology. He had a particular fondness for clockwork, and afterwards became celebrated as the inventor of the Marine Chronometer and "Gridiron Pendulum." At various times he received considerable sums from Government for the excellent performance of his chronometers. In 1767, when he was paid the balance of his reward, he had altogether received from Government the sum of £20,000. He died in London, in 1776, aged 83. (See also note 30).

a countryman of ours, concerning which, I shall here give you a short abstract of the history.[212]

The poor people of this country have been long abused by the millers, who would not grind their corn, but forced them to give theirs in exchange for meal so adulterated, as has been proved to be slow poison. In compassion to them, a set of gentlemen entered into an agreement lately, to have hand-mills made and given in presents, that they might grind their own corn at home; and proposed a reward of £50 sterling to any one who would bring them a hand-mill at this last Christmas.[213] As I know mill-work tolerably well, I made a model of one, and showed it to the gentlemen; they liked it, and desired me to bring a working mill after it, thinking I would stand a chance for the premium.

Soon after I had shown this model, the above said Gordon called upon me, and told me that a gentleman in the country had heard of it, and wrote to him to have a mill made after it, if I would allow it to be copied. As I had then never heard anything against this man's character, I lent him the model—knowing thereby I should have an opportunity of seeing how such a mill would perform without being at the charge of making one. It was made, and performed so well, that a man of a very ordinary degree of strength could grind three-quarters of a bushel of wheat within an hour, and make as good flour as could be desired. I then had a mind to get such another mill made, and let it take its chance with the rest.

But Gordon told me that he could greatly improve my scheme, and lay it before the Society as his own; and if it would gain the premium, he would give me one half. What could I do? He had got it in his hand, and I could not hinder him. What he called an improvement was an additional part for boulting the flour as the mill ground it. This the Society told me

[212] One of our correspondents, in referring to Gordon, says, "My father knew the man well; he was a man of great cunning, very plausible, little or no conscientiousness, and had no inventive powers. He prowled about in a cunning way, picking up information on anything that was new, and by adding a *pin*, a *nail*, or a *screw*, to any new contrivance, he did not scruple to call the whole his own. He appears to have died in straitened circumstances about the year 1785."

[213] The "set of gentlemen" here mentioned were members of "*the Society of Arts*," established in London in 1754, now so celebrated for its utility and high standing.

they did not like, because it would make such mills too expensive for them to give away, and so desired me to bring mine without it. I told Gordon this, and asked him whether he would make me a mill, just after the model, by the end of November; he assured me he would, and so undertook to do it. I believed it was in hand, and was always told so at his house, till within a fortnight of the time, when himself told me it was not begun. And at that very time I observed the mill, which he said was made for a gentleman in the country, but had never been sent: and, indeed, I much question whether he ever had such a commission. I then desired him to bring in that mill (knowing it was then too late for me to employ another person); and though it had the boulting-work, he could easily inform the gentlemen how much cheaper it could be made without that part. But he refused to bring it in, till the Duke of Argyle huffed him into it; and he, being the Duke's cabinet-maker, could not refuse. But he had taken previous care to make another mill which I knew nothing of, and had raised the structure in so foolish a manner, that his mill required at least double the power to work it that mine did.

There were about twenty different sorts of mills brought into a great room, taken by the Society for their reception, and Mr. Harrison had one among the rest. Many trials were made, and it was long thought by most of the committee appointed for examining these mills, that the dispute would have been confined to Mr. Harrison's and mine. Gordon was always there; his mill was set just by mine, and he kept a man for working them both. Mr. Harrison's mill, by working it too hard, had broke, and was therefore rejected. Upon Monday, the 24th December, I was given to understand that my mill had been rejected the Saturday before, on account of a trial then made, when it was judged to have failed all at once, because it hardly produced what deserved the name of flour; and I not knowing of any trial to have been on that day, was unluckily absent.

The news did not much surprise me, because I immediately suspected roguery, and then upon dropping my hand upon the cog-wheel I could easily perceive that the upper mill-stone had been raised so high that the flour then ground by it must have been very bad indeed; upon which, I told the gentlemen of the committee that the mill had been rejected upon a very unfair trial, and mentioned the circumstances, telling the gentlemen that they might soon be convinced of the truth of this by grinding a little wheat in the mill as the stones were then set, and upon allowing me to set them properly close, it would still be found to make as good flour as ever. But they declared, as the mill was once rejected it must for ever stand so, and blamed me for having been absent, telling me also that if they should indulge me with another trial, all the rest whose mills were rejected would insist upon the same, and so there would be no end of it.

I shall trouble you no further with this disagreeable subject, than to inform you that Gordon's mill, in which I had neither hand nor share, has gained the prize. But I imagine the committee begin to find they have given it for very little; the mill, as I am told, begins already to fail, and some parts of it are more than half worn out. And no great wonder; for in it the heavy stones of two feet diameter, and seven inches thick, are turned by a wheel with inclined teeth working in an endless screw, just as the fly of a common jack is turned. I leave you to judge what sort of a mechanical conjuror he is. Mine is turned by a cog-wheel and trundle.[214]

I have just sold the remaining part of the copy of my book

[214] In looking into "*The Universal Magazine,*" Vol. xxii. pp. 357, 358, dated June, 1758, we find the following regarding Gordon's mill:—

Your committee, having examined all the hand-mills produced in consequence of your premium, came unanimously to a resolution,—That the stone mill made by Mr. Gordon (the diameter of the stones, called Cologne stones, being 23 inches, and the price £6 10s. with a fly, and £5 15s. without a fly), and the steel mill made by Mr. Peter Lyon (the diameter of which is five inches and three-quarters; the price, with a fly, £2 15s., and without a fly, £2), are the two best mills; that either of them will grind corn in a proper manner for the poor. But, all circumstances considered, as your committee are doubtful to which to give the preference, and the two above candidates consenting to divide the premium of £50 between them, your committee recommend it to be so divided, and also, recommend the giving a premium, for hand-mills, another year. The Society agreed with their committee, and the premium of £50 was accordingly so divided. And there being still great room for invention and improvement in the making of hand-mills, £50 is again offered to the person who shall make for the Society, on or before the first Wednesday in November, 1758, an hand-mill, which will most effectually and expeditiously grind wheat and other grain into meal, in a cheap manner, for making bread for the poor." Then follows, "A Description of Mr. Gordon's Stone Mill, with a Copperplate print." In "The Universal Magazine," for July, 1758, Vol. xxiii. page 33, there is "A Description of Mr. Lyon's Steel Mill, with a Copperplate print." In April, 1758, the Society of Arts has, among other advertisements, the following in the Universal Magazine, Vol. xxii. p. 203:—"Hand-mills.—There being still great room for invention and improvement in the making of hand-mills, there will be given to the person who shall make for the Society, on or before the first Wednesday in November, 1758, an hand-mill, which will most effectually and exto a bookseller for £300; for as I design to leave London soon, I should have been but embarrassed by it.[215]

I wish I could find an opportunity of sending you a plate which I have just published, something in the nature of the late Rotula. It shows the day of the month, age, and change of the Moon; the places of the Sun and Moon in the ecliptic, with the times of the rising, southing, and setting of the Sun, Moon, and Stars of first, second, and third magnitude, from A.D. 1756 to A.D. 1805.

Be so good as to convey the enclosed letter to Keith by post, and write me as soon as you can with convenience.—I am, with united compliments, &c.,

Your most obliged, humble servant,

JAMES FERGUSON.

From the Club, opposite Cecil Street,
in the Strand.

We have no news as yet of the expected comet. A certain *foolosopher* has almost frightened many people out of their senses about it. I tell them I wish they would look into their Bibles, and they would be satisfied that too much is to be done before the 12th of next May, for the world to come then to an end."

The expected Comet here alluded to was Halley's Comet. It made its appearance on 25th December, 1758—fully eleven months after this letter was written.

It will have been observed that Ferguson, near the commencement of his letter, just quoted, expresses to the Rev. Mr. Irvine a wish to leave London, and remove to Elgin, and contemplates doing so, "*unless something unforeseen happens.*" Ferguson

peditiously grind wheat and other grain into meal, in a cheap manner, for making bread for the use of the poor, £50." Whether Ferguson responded to this second call of the Society, by making another mill for competition for the £50 premium, is not now known; but most likely he did not; and thus prevented a repetition of conduct such as Gordon's.

215 From what we have heard of the great liberality of Mr. Millar to authors, we are inclined to believe that he would make offer to Ferguson to publish his Astronomy at his own risk—a risk, involving, perhaps, an outlay of at least £500; he would also, probably, promise to buy the copyright of the work, provided the 1st edition went quickly off his hands, and cleared all expenses, then to pay part of the price agreed on for it, and the remainder of the money after the 2d edition was issued. How much money Ferguson received as a first payment from Mr. Millar, is uncertain, but the second or remaining portion was £300, as here noted.

did not leave London; the inference therefore is, that "*something unforeseen did happen*" to keep him to London. Entertaining this idea, we made search through a great many of the London newspapers and magazines of this period, in order to find if there were any notice or paragraph in them regarding Ferguson and his movements, so as to account for his continuing to reside in London. In our search, "*The Morning Post and General Advertiser*," of May 2d, 1758, came into our hands, in which we found the following notice :—

THE PRINCE OF WALES SENDS FOR FERGUSON.—" H. R. H. The Prince of Wales (Prince George) yesterday sent for the celebrated Mr. James Ferguson, who waited on his Royal Highness, at Leicester House, with his new Astronomical apparatus, with which the Prince appeared to be much interested." [216]

We think there can be no doubt that this message from the Prince of Wales to Ferguson, and the result of his interview with him, became the "*unforeseen something*," which at once made him abandon all thoughts of going to Elgin, and resolve on taking up his permanent residence in London. The 1st May, 1758, was a great "*Red-Letter Day*" for Ferguson; it was certainly the grand turning point in his career, as it opened the way to his future advancement.

This newspaper paragraph alludes to Ferguson's "New Astronomical Apparatus," which he showed the Prince at Leicester House. This was, probably, either his *Whirling Table* and *apparatus*, then recently made, or his New Rotula, published on 9th August in the preceding year, and entitled, "A New Astronomical Instrument," &c. (See date 9th August, 1757).

FERGUSON'S CLOCK.—The Clock usually called "*Ferguson's Clock*," appears to have been invented and made by Ferguson in 1758, and is often alluded to in works on Horology, &c. It was contrived by him as an improvement on a singular Clock which had then been recently invented by the celebrated Dr. Franklin;—now known as "*The Franklin Clock*." [217]

[216] George, Prince of Wales, who, in October 1760, became King George III.

[217] Benjamin Franklin, LL.D., the celebrated Philosopher and Statesman, was born in Boston, North America, in the year 1706; besides being the author

Annexed is a view of the dial, a plan of the wheel-work, and description of Ferguson's Clock, from his "*Select Mechanical Exercises.*"

A "Clock that shows the Hours, Minutes, and Seconds, by means of only Three Wheels and Two Pinions in the whole Movement.

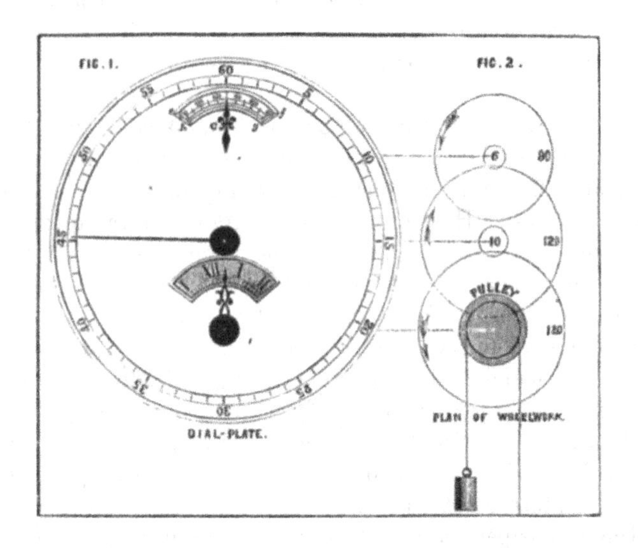

Ferguson's Simple Clock, 1758.

As Dr. Franklin, whom I rejoice to call my friend, is perhaps the last person in the world, who would take anything amiss that looks like an amendment or improvement of any scheme

of several works, he discovered "*The Identity of Lightning and Electricity, On the Smoothing of the Surface of a Ruffled Sea by throwing Oil on the Waters;* he was also the *Inventor of the Harmonicon*, a musical instrument, and *The Franklin Clock.* He died in 1790, aged 84."

Franklin came to England in 1724, and for many years worked at his trade as a printer in London. He left London some years after this date, but we find him again in London in 1755—1757—1766—1774, &c. Franklin and Ferguson appear to have been introduced to each other in 1757. Ferguson's death, in 1776, put an end to the intimate friendship which had so long subsisted between them. A curious old Circular Horologe, or Clock, once in our possession, (now in the Museum at Banff) has engraven on it, John T. Desaguliers, LL.D., 1729, Lect. on Nat. et Exp. Phil., London. *Benjamin Franklin,* LL.D., 1757.—"*James Ferguson,* 1766,"—"*Kenneth M'Culloch,* 1774;" and the initials G. W.—See note 167.—Ferguson, in his "Select Mechanical Exercises," pages 1—4, describes the "Franklin Clock," and gives engravings of its Dial-face and wheel-works.

he proposes, I have ventured to offer my thoughts concerning his clock, and how one might be made as simple as his, with some advantages. But I must confess, that my alteration is attended with some inconveniences, of which his are entirely free. I shall mention both, to the best of my knowledge, that they who choose to have such simple and cheap clocks may have them made in either way they please.[218]

The Doctor's clock cannot well be made to go a week without drawing up the weight; and if a person wakes in the night, and looks at the clock, he may possibly be mistaken four hours in reckoning the time by it,[219] as the hand cannot be upon any hour, or pass by any hour, without being upon or passing by four hours at the same time. To avoid these inconveniences, I have thought of the following method :—

In Fig. 1 the dial-plate of such a clock is represented, in which there is an opening *a b c d* below the centre. Through this opening, part of a flat plate appears, on which the twelve hours are engraved, and divided into quarters. This plate is contiguous to the back of the dial-plate, and turns round in 12 hours; so that the true hour, or part thereof, appears in the middle of the opening, at the point of an index A, which is engraved on the face of the dial-plate. B is the minute-hand, as in a common clock, going round through all the 60 minutes on the dial in an hour; and, in that time, the plate seen through the opening *a b c d* shifts one hour under the fixed engraven index A. By these, you always know the hour and minute, at whatever time you view the dial-plate. In this plate is another opening *e f g h*, through which the seconds are seen on a flat moveable ring, almost contiguous to the back of the dial-plate; and, as the ring turns round, the seconds upon it are shown by the top point of a *fleur-de-lis* C, engraved on the face of the dial-plate.

218 "*The Ferguson Clock*" was invented and made by Ferguson in 1758. We scarcely think a second clock on such a construction would be made—such a clock would never keep accurate time. The Clock is now only known as "*one of the curiosities of Horology.*"

219 "*The Franklin Clock.*"—This is also a clock now reckoned amongst "*The Curiosities of Horology.*" Owing to the arrangement of the hours on its dial-plate, it is rendered almost unfit for general use. We have in our possession a small Franklin Horologe; the dial-plate is of thick brass, 3 inches in diameter, with Roman, or Chapter hours engraven on it *a la* Franklin. The works are those of an extraordinary strong watch attached to the back of the dial. (Benjn. Franklin, LL.D., 1757, is engraven on the back of the dial outside of the watch works).

Fig. 2 represents the wheels and pinions in this clock. A is the first or great wheel; it contains 120 teeth, and turns round in 12 hours. On its axis is the plate on which the 12 hours above-mentioned are engraved.

This plate is not fixed on the axis, but only put tight upon a round part thereof, so that any hour, or part of an hour, may be set to the top of the fixed index A, without affecting the motion of the wheel. For this purpose, twelve small holes are drilled through the plate, one at each hour, among the quarter divisions; and, by putting a pin into any hole in view, the plate may be set, without affecting any part of the wheel-work. This great wheel A, of 120 teeth, turns a pinion B of 10 leaves round in an hour; and the minute-hand B (Fig. 1) is on the axis of this pinion, the end of the axis not being square, but round; that the minute-hand may be turned occasionally upon it, without affecting any part of the movement. On the axis of the pinion B is a wheel C of 120 teeth, turning round in an hour, and turning a pinion D of 6 leaves in 3 minutes; for 3 minutes is a 20th part of an hour, and 6 is the 20th part of 120. On the axis of this pinion is a wheel E of 90 teeth, going round in 3 minutes, and keeping a pendulum in motion that vibrates seconds, by pallets, as in a common clock, where the pendulum wheel has only 30 teeth, and goes round in a minute. But, as this wheel goes round only in 3 minutes, if we want it to show the seconds, a thin plate must be divided into 3 times 60, or 180 equal parts, and numbered 10, 20, 30, 40, 50, 60; 10, 20, 30, 40, 50, 60; 10, 20, 30, 40, 50, 60; and fixed upon the same axis with the wheel of 90 teeth, so near the back of the dial-plate, as only to turn round without touching it; and these divisions will show the seconds, through the opening *e f g h* in the dial-plate, as they slide gradually round below the point of the fixed *fleur-de-lis* C.

As the great wheel A, and pulley on its axis over which the cord goes (as in a common 30 hour clock) turns round only once in 12 hours. this clock will go a week with a cord of common length, and always have the true hour, or part of that hour, in sight at the upper end of the fixed index A on the dial-plate. These are two advantages it has beyond *Dr. Franklin's* clock; but it has two disadvantages of which his clock is free. For, in this, although the 12-hour wheel turns the minute-index B, yet, if *that* index be turned by hand, to set it to the proper minute for any time, it will not move the 12 hour plate to set the corresponding part of the hour even with the top of the index A; and therefore, after having set the minute-index B right by hand, the hour-plate must be set right by means of a pin put into the small hole in the plate just below the hour. 'Tis true, there is no great matter in this; but I have some suspicion that the pendulum wheel E having 90 teeth instead of the common number 30, may be some disadvantage to the 'scapement, on account of the smallness of the teeth; and 'tis certain, that it will cause the pendulum-ball to describe but small arcs in its vibrations. Indeed some men of science think small arcs are best; but if they really are, I must confess myself ignorant of the reason. For, whether the ball describes a large or a small arc, if the arc be nearly *cycloidal*, the vibrations will be performed in equal times; the time then depending entirely on the length of the pendulum-rod, not on the length of the arc the ball describes. The larger the arc is, the greater is the *momentum* of the ball; and the greater the *momentum* is, the less will the times of the vibrations be affected by any unequal impulse of the pendulum wheel upon the pallets.

But the worst thing about this clock (and what every one will allow to be a disadvantage) is, that the weight of the flat ring on which the seconds are engraved, will load the pivots of the axis of the pendulum-wheel with a great deal of friction, which ought by all possible means to be avoided; and yet I have seen one of these clocks (lately made) that goes very well, notwithstanding the weight of this ring. For my own part, I think it might be quite left out; for they are of very little use in common clocks, not made for Astronomical observations, and table-clocks never have them." (Vide Ferguson's "Select Mechanical Exercises." Lond., 1773, 1st Edit., pp. 4—11).

In his recently discovered "Common Place Book," Ferguson says, "*I showed this Clock to Dr. Franklin, who approved very much of the alteration.*" "Common Place Book," p. 98 (Col. Lib. Edin.)

MANUSCRIPT Copy of FERGUSON'S ASTRONOMY.—Under date 1757, page 224, allusion is made to three manuscript books, in Ferguson's autograph, which were discovered near the end of the year 1865. The second one of the three, in the order of date,—is a quarto copy of his Astronomy, carefully and beautifully written —an exact copy of his quarto Astronomy then in print. It is illustrated by engravings same as in the printed copies, with four additional ones, done by him in pen and ink,—viz. A view of the Dial-face of an Astronomical Clock;—Plan of the wheel-work of his large Orrery;—and two projections of the Solar Eclipses in 1761 and 1764. The title-page of this volume is the same as in the printed copy (see page 214), and has the date MDCCLVIII lettered with the pen at the foot of the page. It has been supposed that this MS. is the original from which the *first edition* of his Astronomy was printed (in 1756)—if we admit this, then Ferguson must have added the title-page to it in 1758. We are of opinion that the MS. *was written* in 1758, from the *second* and *corrected edition* of 1757. The MS. is *very carefully* and *neatly written*, free from erasures and interlineations, such as are usually to be found in *copy* for the press —*too clean* to have passed through the printing office. We can form no opinion as to Ferguson's motive for writing this MS.

The manuscript volume is a thick quarto of 328 pages, followed by 24 pages of printed matter, and Tables by Tobias Mayer. It is full bound in calf, with gilded edges—in excellent preservation—size, $10\frac{7}{8}$ by $8\frac{1}{2}$ inches, and $1\frac{3}{4}$ inch thick—now in College Library, Edinburgh, with the other two volumes. For an account of the third MS. Vol., see date 1776.

The year 1758 opens with a long and interesting letter from Ferguson to his much esteemed friend, the Rev. Mr. Irvine of Elgin. The year closed on Ferguson with the sad intelligence of his friend's death. The Rev. Alexander Irvine died at Elgin on the 22d December, 1758, in the 59th year of his age, and 34th of his ministry. (See also note 69).

1759.

BIRTH: JOHN FERGUSON born.—Our first memorandum of Ferguson, for the year 1759, refers to an "interesting domestic occurrence," viz. the birth of a son, baptized *John*. The following extract of the birth is in the autograph of Ferguson, taken from a small pocket Bible once belonging to the Ferguson family, and to which we have been indebted for a record of all the births in *his* family. As previously noted, this Bible is now the property of Dr. George, Surgeon, Keith.

"JOHN; (born), Tuesday, 27th Feby. 1759, N.S."

It is very probable that Ferguson's son, John, was named after either his grandfather, whose name was John, or in compliment to his uncle *John*, who, at the time, was residing in the parish of Keith, Banffshire. Ferguson had now a family of *one daughter* and *three sons*, viz. *Agnes*, *James*, *Murdoch*, and *John*; there was no after addition to his family.[222]

PYROMETER.—In 1759, Ferguson invented and made a Pyrometer, for showing the expansion of metals by heat. The following description, and annexed figure of this instrument are from Ferguson's "*Lectures on Select Subjects.*"

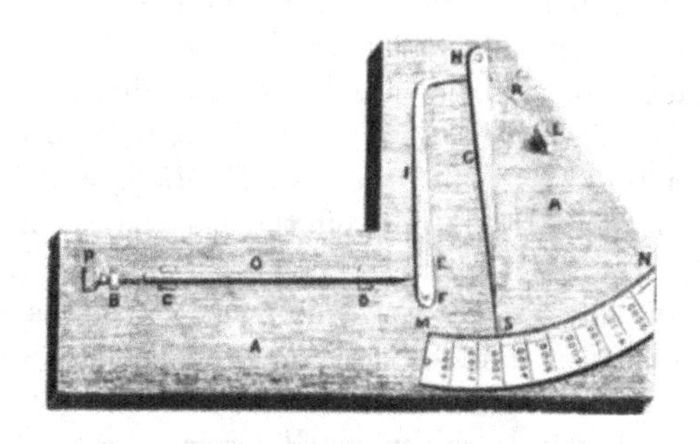

Ferguson's Pyrometer.

"A A is a flat piece of mahogany, in which are fixed four brass studs B, C, D, L; and two pins, one at F, and the other at H. On the pin F turns the crooked index E I, and upon the pin H, the straight index G K, against which, a piece of watch-spring R bears gently, and so presses it toward the beginning of the scale M N, over which the point of that index moves. This scale is divided into inches and tenth parts of an inch; the first inch is marked 1000, the second 2000, and so on. A bar of metal O, is laid into notches in the top of the studs C and D; one end of the bar bearing against the adjusting screw P, and the other end against the crooked index E I, at a 20th

[222] For notices of John Ferguson, see date 1773; and the *Appendix.*

part of its length from its centre of motion F. Now it is plain, that however much the bar O lengthens, it will move that part of the index E I, against which it bears, just as far; but the crooked end of the same index, near H, being 20 times as far from the centre of motion F, as the point is against which the bar bears, it will move 20 times as far as the bar lengthens. And as this crooked end bears against the index G K at only a 20th part of the whole length G S from its centre of motion H, the point S will move through 20 times the space that the point of bearing near H does. Hence, as 20 multiplied by 20 produces 400, it is evident that if the bar lengthens but a 400th part of an inch, the point S will move a whole inch on the scale; and as every inch is divided into 10 equal parts, if the bar lengthen but the 10th part of the 400th part of inch, which is only the 4000th part of an inch, the point S will move the tenth part of an inch, which is very perceptible.

To find how much the bar lengthens by heat, first lay it cold into the notches of the studs, and turn the adjusting screw P until the spring R brings the point S of the index G K to the beginning of the divisions of the scale M; then, without altering the screw any farther, take off the bar and rub it with a dry woollen cloth till it feel warm; and then, laying it on where it was, observe how far it pushes the point S upon the scale by means of the crooked index E I; and the point S will show exactly how much the bar has lengthened by the heat of rubbing. As the bar cools, the spring R bearing against the index K G, will cause its point S to move gradually back towards M, in the scale; and when the bar is quite cold, the index will rest at M, where it was before the bar was made warm by rubbing. The indexes have small rollers under them at I and K; which, by turning round on the smooth wood, as the indexes move, make their motion the easier, by taking off a great part of the friction, which would otherwise be on the pins F and H, and of the points of the indexes themselves on the wood."[223] (Vide Ferguson's "*Lectures on Select Subjects.*" Lond. 1760, p. 246).

[223] This Pyrometer is "defective both in principle and execution;" it was found so shortly after its introduction. Mr. Charles F. Parkington, who edited an edition of Ferguson's Mechanics, in 1825, has a foot-note regarding this Pyrometer, in which he says, "The Pyrometer here described has been justly considered as defective both in principle and execution. If it be merely intended to illustrate the expansion of metals by heat, the fact may be shown by fitting a piece of large (thick) wire into a hole, and it will be found that after it has been

In 1759, Ferguson published "*An Astronomical Rotula*," with printed description of its several parts and how to use it. This Rotula, like all its predecessors, has been long out of print —the only account we have seen of this one is to be found in "The King's Library," British Museum.

The late Adam Walker, Esq., Lecturer on Natural and Experimental Philosophy, &c., London—a friend of Ferguson's—had many of his MS. papers, which, in 1831, were shown to us by his son, Deane F. Walker, Esq., Lecturer on Astronomy and the Eidouranion; at the time, we took notes of all we saw with the dates affixed to each.[224]

One of these manuscript papers refers to date 1759, and illustrates by calculation the conjunctive period of the hour, minute, and seconds-hand of a clock, which we give as follows; as also, The Conjunctions of the Hour and Minute Hands of a Watch,—illustrating the Conjunctive Periods of the Sun and Moon.

"Suppose the Hour, Minute, and Seconds-hand of a Clock to be in Conjunction at the Hour of XII. How much time will elapse ere they come into Conjunction again?"

"Here we have the periodical revolution of the Hour-hand $=720$ minutes $=a$; the periodical revolution of the Minute-hand $=60$ minutes $=b$; and *that* of the Seconds-hand $=1$ minute $=c$; from whence we collect $\frac{ab}{a-b}$, $=\frac{43200}{660}$ min. $=\frac{7200}{110}=\frac{720}{11}$ min. for the synodical period or conjunction of the Hour and Minute-hands; and $\frac{bc}{b-c}=\frac{60}{59}$ for the synodical period of the Minute and Seconds-hands. Then, to find the synodical period

brought to a red heat, that its bulk will have so far increased as to prevent its passing through the hole. When it is found necessary, however, to ascertain the comparative amount of expansion in various metals under the same degree of temperature, a variety of more accurate instruments may be resorted to." Ferguson himself appears to have been convinced of the defects of this Pyrometer, as he, in 1764, invented another of a superior order. See Pyrometer under date 1764.

224 Mr. Adam Walker, an eminent lecturer on Natural and Experimental Philosophy, and author of "Familiar Philosophy," and other works; was the inventor of the Eidouranion, or Transparent Orrery. In early life, he was a friend of Ferguson's, from whom he received many MSS. and pieces of apparatus. He died in 1821, at the great age of 90. These MSS. &c., at his death, came into the possession of his son (our late friend) Deane Walker, Esq., lecturer on Astronomy and the Eidouranion, London,—to whose kindness on many occasion we were much indebted. In 1831 we received from him several of his "Ferguson MSS. and pen and ink sketches," and allowed to copy those he retained. Mr. Deane Walker died at Tooting, near London, on 10th May, 1865, aged 87. For an account of his celebrated Eidouranion, and lecture on it, see the Glasgow Mechanics' Magazine, vol. 1st, pp. 20—23, 1824.

of all the three hands, we must suppose $\frac{m\times720}{11} = \frac{n\times60}{59}$, from whence we have $m:n::\frac{60}{59}:\frac{720}{11}$. Now, the least integer numbers, represented by m and n, in the proportion of $\frac{60}{59}$ to $\frac{720}{11}$ are 11 and 708. Therefore, $11:708::\frac{60}{59}:\frac{720}{11}$; and consequently, $\frac{11\times720}{11}$ $(=708\times\frac{60}{59})$ = the synodical period of the three hands of the clock = 720 minutes, or just 12 hours.[225]

The periodical revolutions of the Sun and Moon, round the ecliptic, and their synodical period of conjunction with each other, may be familiarly represented by the motions of the hour and minute hands of a watch, round its Dial-plate. For, the Dial-plate is divided into 12 hours, as the ecliptic is divided into 12 signs; the hour-hand goes round in 12 hours, as the Sun does in 12 months, and the minute-hand goes round in 1 hour, as the Moon does in (somewhat less than) a month. And, as the Moon never is in conjunction with the Sun in that point of the ecliptic where she was at the last conjunction before, so the minute-hand never is in conjunction with the hour-hand at that point of the Dial-plate where it was at the last preceding conjunction. So that, the 12 hours on the Dial-plate may represent the 12 signs of the ecliptic; the hour-hand the Sun, and the minute-hand the Moon. Only, the motion of the minute-hand is too slow for the Moon, in proportion to *that* of the hour-hand compared with the motion of the Sun. For, in the time of the Sun's going round the ecliptic, which is 12 calendar months, there are 12·36 conjunctions of the Sun and Moon;[226] but in the time the hour-hand goes round the Dial-plate, the minute-hand is only 11 times conjoined with it.

These hands are always in conjunction at XII o'clock. The first column of the following Table shows the number of their conjunctions in 12 hours, and the collateral lines show in how many hours, minutes, &c., after XII, they come to their succeeding conjunctions marked in the first column; the time between any conjunction and the next being 1 hour $5\frac{5}{11}$ minutes."

2) 2) 3) 5)

225 The above may also be done as follows:—$\frac{720}{60} = \frac{360}{30} = \frac{180}{15} = \frac{60}{5} = \frac{12}{1}$, and $12\times1\times1 = 12$ hours, the period of a continual conjunction of these three hands.

226 There are 12·368269 conjunctions of the Moon with the Sun in 365 d. 5 h. 48 m. 51·6 s. (See notes 59 and 199).

Conjunctions.	Hour.	Min.	Sec.	Thds.	Fourths.	Fifths.	Sixths.	Sevenths.	Eighths.	Ninths.	Tenths.
1	I.	5,	27,	16,	21,	49,	5,	27,	16,	21,	$49\frac{1}{11}$.
2	II.	10,	54,	32,	43,	38,	10,	54,	32,	43,	$38\frac{2}{11}$
3	III.	16,	21,	49,	5,	27,	16,	21,	49,	5,	$27\frac{3}{11}$.
4	IIII.	21,	49,	5,	27,	16,	21,	49,	5,	27,	$16\frac{4}{11}$
5	V.	27,	16,	21,	49,	5,	27,	16,	21,	49,	$5\frac{5}{11}$
6	VI.	32,	43,	38,	10,	54,	32,	43,	38,	10,	$54\frac{6}{11}$
7	VII.	38,	10,	54,	32,	43,	38,	10,	54,	32,	$43\frac{7}{11}$
8	VIII.	43,	38,	10,	54,	32,	43,	38,	10,	54,	$32\frac{8}{11}$
9	IX.	49,	5,	27,	16,	21,	49,	5,	27,	16,	$21\frac{9}{11}$
10	X.	54,	32,	43,	38,	10,	54,	32,	43,	38,	$10\frac{10}{11}$
11	XI.	0,	0,	0,	0,	0,	0,	0,	0,	0,	0.

If the above process was carried to infinity, in the horizontal lines, the numbers would circulate at every fifth column.

JAMES FERGUSON.

LONDON, *June* 18*th*, 1759."

(Vide also Ferguson's "*Tables and Tracts.*" Lond., 1767, pp. 149—152).

TROY AND AVOIRDUPOISE WEIGHT TABLES.—In the month of June, 1759, Ferguson calculated and published, in large type, on a single sheet,

> "Tables for Changing Troy Weight into Avoirdupoise Weight, and Avoirdupoise Weight into Troy Weight."

This sheet has been out of print for upwards of a century—it is now not to be had—the price it sold at, and its success in the literary market, is unknown. In his MS. "Common Place Book," pp. 238, 239, he gives copies of these Tables. At the end of them, *we are told that they* were "*Calculated by James Ferguson, June* A.D. 1759." We find the MS. Tables in the "Common Place Book," similar to those he afterwards gave in his "Tables and Tracts," pp. 232—235, to which the reader is referred.

Under date 1755 (page 206), it is remarked that he sometimes turned his attention to Theological subjects, and there is given his paper, "*On the Birth and Crucifixion of Christ.*" The present paper discusses the question of "the NAME AND NUMBER OF THE BEAST," which results in a very remarkable coincidence of sums total. We copied it in 1831 from a MS. of Ferguson's in the possession of the late William Upcott, Esq., Islington, London.

2 H

"On the NAME and NUMBER of the APOCALYPTIC BEAST."—"Whoever reads the description of the '*Scarlet* Woman,' and of the Beast, in the Book of Revelations, will allow that both these do agree so exactly with the Romish Constitution or Establishment, that they can hardly mean anything else. The name written upon her forehead is MYSTERY; (see Rev. chap. xvii., v. 5, and chap. xiii., v. 18) the number of the Beast, which is there also said to be the number of a Man, is 666.

In the Hebrew language the word סתור (*Sethor*) signifies MYSTERY; and the Talmudists use the word רומיית (*Romith*) to signify the Romish Constitution or Establishment. All the letters of these words are numerals, and their sum in each is 666.

Among the Greeks, the Church of Rome was called ΛΑΤΕΙΝΟΣ (*Lateinos*) or the Latin Church. All these letters are numeral ones too, and their sum is 666.

The Papists call the Pope VICARIVS, FILII, DEI (*The Vicar of the Son of God*). And, if we take the sum of all the numeral letters in these three words, we shall find *it* also to be 666—(See

Hebrew.

SETHOR in HEBREW, and Numerical Value of the Letters.	
ס	60
ת	400
ו	6
ר	200
Sum; 666	

Hebrew.

ROMITH in HEBREW, and Numerical Value of the Letters.	
ר	200
ו	6
מ	40
י	10
י	10
ת	400
Sum; 666	

Greek.

LATEINOS in GREEK, and Numerical Value of the Letters.	
Λ	30
Α	1
Τ	300
Ε	5
Ι	10
Ν	50
Ο	70
Σ	200
Sum; 666	

Latin.

VICAR of the SON of GOD in Latin, and Numerical Value of the Letters.	
V	5
I	1
C	100
A	.
R	.
I	1
V	5
S	.
F	.
I	1
L	50
I	1
I	1
D	500
E	.
I	1
Sum; 666	

Thus, the Sum Total of the Numerical value of the Letters in each of the Tablets, is exactly 666,—A very fatal coincidence against Popery.

JAMES FERGUSON.
LONDON, *7th November*, 1759."

(See also Ferguson's recently discovered MS. "*Common Place Book*," Col. Lib. Edin., pp. 7 and 244).

We have in our possession a scrap of paper on which is written, in Ferguson's autograph, "An additional coincidence of the number 666 in connection with the name of a man." In this "additional coincidence," Ferguson introduces the *motto* on the Palace of the Pope at Rome, viz. "*Vicarivs Dei Generalis in Terris*," (the Vicar General of God on Earth),—

"VICARIVS DEI GENERALIS IN TERRIS.
5,1,100, 1,5, 500, 1,———50,1. 1,—,———1,—.
Thus, 5+1+100+1+5+500+1+50+1+1+1=666.

The Sum total of the letters in the *motto* '*Vicarivs Dei Generalis In Terris*' is 666.

JAMES FERGUSON, *5th Dec.* 1759."

MECHANICAL PARADOX ORRERY.—Sometime in the year 1759, Ferguson made a new Mechanical Paradox Orrery, similar to the one which he made in 1755, the only difference between the two being, that in this new one the wheels are all brass, with double the number of teeth in each of them; and also, in having a circular dial on the top surface of the frame containing the wheels, directly under the Sun, the axis of the Sun being in the centre of the dial, which has laid down on it, in a series of concentric circles, a set of Lunar Tables from 1760 to 1800. This paradox orrery is now in our possession—for a short description of it, see conclusion of note 194.

SUN-DIALS—MODELS, &c.—Several of our notes show that Ferguson, during the year 1759, delineated a great many projections of Sun-dials, and made models of them to exhibit in his lectures on Dialling, he having just then added that subject to his course. Among the models he then made, we find—A Horizontal Dial,—Vertical Dials,—A South Dial,—Inclining and Reclining Dials,—Pardie's Universal Dial, &c.

Dial on a Card.—Somewhere about the end of the year 1759, Ferguson delineated a Dial on a Card, which, when rectified, showed The Hour of the Day,—The Time of the Sun's Rising and Setting,—The Sun's Declination, and the Days on which the Sun entered the Signs of the Zodiac. The original Dial is in our possession, having been kindly presented to us by James L. Rutherfurd, Esq., Edinburgh, in October, 1863. It is done on mill-board ⅛-inch thick, 13¼ inches by 9, covered with foolscap paper (on which the Dial is delineated). *James Ferguson, fecit*, is lettered in the lower left-hand corner of the dial. On the back of the dial, neatly written by him, are the following directions:—

"How to rectify and use the Dial.

Set the cross line on the brass slider to the day of the month, and stretch the thread from thence over the angular point XII, where the curve lines meet; then shift the bead on the thread to that point, and the Dial will be rectified for the following purposes.

1. To find the Hour of the Day when the Sun Shines.—Raise the Gnomon (no matter how much or how little), and hold the edge of the Dial next the Gnomon toward the Sun, so as the uppermost edge of the shadow may just cover the *Shadow Line;* and the bead then playing freely on the face of the Dial (by the weight of a plummet hung occasionally to the thread), will show the time of the day among the hour-lines as it is before or after noon.

2. To find the Time of Sun-rising and Sun-setting.—Move the thread among the hour-lines, till it either covers some one of them or lies parallel betwixt any two; and then it will cut the time of Sun-rising among the forenoon hour-lines, and of Sun-setting among the afternoon hour-lines, for the day of the year indicated by the cross line on the slider.

3. To find the Sun's Declination.—Stretch the thread over the angular point at XII, and it will cut the Sun's Declination for the day of the year at which the cross line on the slider stands.

When the bead, rectified as above, moves along any of the curve lines on which the Signs of the Zodiac are marked, the

Sun enters those Signs on the days pointed to by the cross line on the slider.

This Dial answers only for those Places which have the same Latitude with London. To construct such Dials for other Latitudes, see my Mechanical Lectures.—JAMES FERGUSON."

Referring to his "Mechanical Lectures," we find a description of the dial, with two figures, showing how it is constructed. "The Rules for Use" are nearly the same as those already given. What follows, not in the original, but in his Lectures, being important, we give as addenda,—

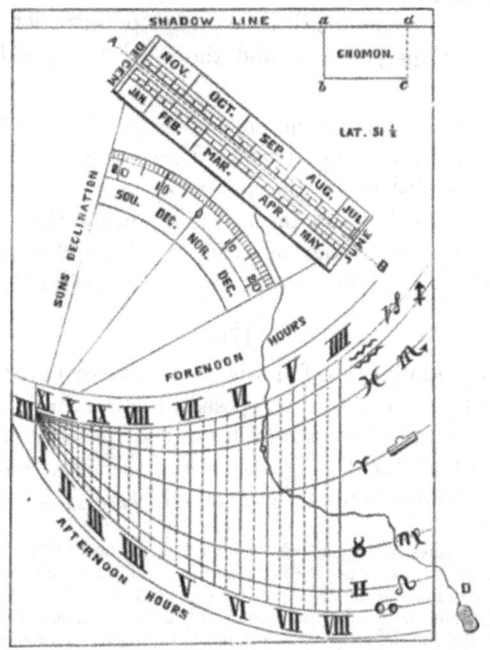

We are told this is "a portable dial, which may be easily drawn on a card, and carried in a pocket-book. The lines *a d*, *a b*, and *b c*, of the gnomon must be cut quite through the card; and as the end *a b* of the gnomon is raised occasionally above the plane of the dial, it turns upon the uncut line *c d* as on a hinge. The line dotted A B must be slit quite through the card, and the thread must be put through the slit, and have a knot tied behind to keep it from being easily drawn out. On the other end of this thread is a small plummet, and on the middle of it a small bead for showing the hour of the day." ("Lectures on Select Subjects in Mechanics," &c.)

In the same year (1759), Ferguson published this Dial on a card 4¼ inch by 3 inches, having in the left hand lower corner, *J. Ferguson, delin.* and under it PEREUNT ET IMPUTANTUR. This ingenious Card-Dial has been long out of print, and is very rarely to be had. We, however, understand that it has frequently been copied and sold since Ferguson's time. We are indebted to the kindness of Mr. Robert Sim, Keith, for our copy—it came into his hands from the last relatives of Ferguson in Keith, and is still in excellent preservation; the green silk cord for the plummet and the bead being still attached to it.

We have no other memoranda of Ferguson for 1759, but may add, that as his celebrated work "Lectures on Select Subjects," &c., was published in the following year, it is to be presumed that much of his leisure time, during the later part of 1759 and the beginning of 1760, was devoted to writing that work, and making the drawings necessary for its illustration.

1760.

IN MEMORIAM.—The following memorial-tribute to the worth of a departed friend, was composed by Ferguson in 1760. The original, in his own autograph, written on the inside of the first board of the "Ferguson Bible," so often referred to, is now in the possession of Dr. George of Keith:—

> "Mr. James Mackenzie of Oxford Road, in the Parish of St. Mary le bonne, Grocer, died the 11th February 1760, aged 25 years 5 days, and was interr'd in the same Parish churchyard. He was of solid Judgment and good sense, remarkably sober, virtuous and constant; a dutiful son, a loving Bro^{r}. and most sincere friend; and may justly be said to have been a young man of real worth. His soul took its flight of a Monday a quarter past three o'clock in the afternoon from this Earth of trouble to a Heaven of eternal Happiness."

Note.—One of the pall-bearers at Ferguson's funeral in 1776 is named Mackenzie, probably a relative of the subject of the above record.

"LECTURES ON SELECT SUBJECTS," &c.—In the year 1760, Ferguson published his second-best and celebrated work, *Lectures on Select Subjects in Mechanics,* &c. Like his Astronomy, his LECTURES are written in clear and plain language, so much so, that the merest tyro in the sciences of which he treats cannot possibly fail to understand the author's meaning and demonstrations. Until lately, "*The Lectures*" were the generally acknowledged *text-book* on Mechanics in Britain, but the now advanced state of the sciences has called into existence other works on the subjects of which it treats. This work has gone through numerous editions. In our memoranda, we have the following editions and dates,—viz. 1st edition, published in 1760,—the 2d in 1764,—the 3d in 1767,—the 4th in 1772,—the 5th in 1776,—the 6th in 1784,—the 7th in 1793; also, republications in 1806,—1823,—1825, and 1827. The edition of 1806 is edited by "David Brewster, LL.D." (now Sir David Brewster) in two volumes. "*The Lectures*" have also been frequently published in America, and also translated and published in several foreign countries.

he annexed is the title-page of *Lectures on Select Subjects.*

Lectures on Select Subjects in Mechanics, Hydrostatics, Hydraulics, Pneumatics, and Optics, with the Use of the Globes, the Art of Dialling, and the Calculation of the Mean Times of New and Full Moons and Eclipses. Illustrated with Copperplates. By James Ferguson, London. MDCCLX.

It is dedicated to "PRINCE EDWARD," in the style so peculiar to last century,—viz.

"To His Royal Highness PRINCE EDWARD, Sir,—As Heaven has inspired your ROYAL HIGHNESS with such a love of ingenious and useful arts, that you not only study their theory, but have often condescended to honour the professors of mechanical and experimental philosophy with your presence and particular favour, I am thereby encouraged to lay myself and the following work at your ROYAL HIGHNESS's feet;[227] and at the same time beg leave to express that veneration with which I am, Sir, your ROYAL HIGHNESS'S most obliged, and most obedient humble servant, James Ferguson."

[227] PRINCE EDWARD was the second son of Frederick, Prince of Wales (eldest son of George II.), was born at Norfolk House on 14th March, 1739. He was seized with a malignant fever at Genoa, in Upper Italy, and died there on September 17th, 1767, aged 28. He was a great patron of the arts and sciences, and had great delight in conversing with Ferguson on Astronomy and kindred subjects; he also occasionally attended Ferguson's lectures.

THE ARMILLARY TRIGONOMETER.—Sometime in the year 1760, Ferguson made a model, in wood, of an Astronomical instrument called *The Armillary Trigonometer*, the invention of Mr. Mungo Murray, Shipwright, at Deptford.[228] It solved the following problems:—" 1st, The Time of the Day, Forenoon and Afternoon, and the Sun's true Azimuth from the South at that time. 2d, The variation of the Compass. 3d, The time of the day being given, to find the Sun's altitude and azimuth at that time. 4th, The time of the Sun's rising and setting, on any day of the year, in any given North latitude less than 66½ degrees. 5th, To find when the Morning Twilight begins, and when the Evening ends. 6th, A place being in the north frigid zone (that is, in more than 66½ degrees of North latitude), to find on what day of the year the Sun begins to shine constantly on that place without setting; and how long he continues to do so. 7th, How long Twilight continues at the Poles of the Earth. 8th, The Sun's depression below the horizon, at any time of the night, in any given latitude less than 66½ deg. 9th, To find in what North latitude the longest day is of any given length less than 24 hours. 10th, The Sun's amplitude at rising and setting, in any given latitude less than 66½ deg. 11th, The length of the longest day being given, at any place whose latitude is North; to find the latitude of that place. 12th, In the Summer months, to find an East and West line; and consequently, a Meridian line, for a place of any given latitude. 13th, The distances of all the Forenoon and Afternoon hours from XII, on a horizontal dial, for any given latitude. 14th, The distances of the Forenoon and Afternoon hours from XII, on a vertical South dial, for any given latitude. 15th, The distances of the Forenoon and Afternoon hours from XII, on a vertical dial, declining from the South toward the East or West, by any given number of degrees."

In Tables and Tracts we are informed, " Mr. Mungo Murray, Shipwright, contrived a very useful instrument which he calls '*The Armillary Trigonometer;*' and I had it some months by me in the year 1757. *Since* that time, he showed me a paste-

[228] Mr. Mungo Murray, an ingenious Shipwright, at Deptford, near London. He is the author of an excellent work entitled, "*A Treatise on Shipbuilding and Navigation, with numerous Plates and neatly-printed Tables.*" London, 1754. Price 10s. 6d. *Supplement to the Treatise on Shipbuilding*, translated from M. Bouguer's "Traité du Naivre." Plates, 3s. London, 1765.

board model of an instrument, much of the same sort, but of a much smaller size; which, I believe, he has not yet made, either of wood or metal. And, as it is a thing that deserves well to be known, on account of its great utility, I have made it of wood. The only addition that I have made to Mr. Murray's scheme is a circular scale of the Sun's declination, for the different days of the year, to save the trouble of referring to Tables of the Sun's declination in printed books." (Vide "Tables and Tracts," pp. 80—104, with copperplate of the instrument).

The following, by Ferguson, ON COLOURS and the RAINBOW, were extracted by us, in 1835, from the original in the possession of the late Dawson Turner, Esq., of Yarmouth—both papers are dated 1760.

"ON COLOURS.—Colours produced by the mixture of colourless fluids :—

1. Spirit of wine mixed with spirit of vitriol make a . . . *red.*
2. Solution of mercury mixed with oil of tartar, . . . *orange.*
3. Solution of sublimate and lime-water. . . . . . *yellow.*
4. Tincture of roses and oil of tartar, ... *green.*
5. Solution of copper and spirit of sal-ammoniac, . . . *purple.*
6. Tincture of roses and spirit of wine, ... *blue.*
7. Solution of sublimate and spirit of sal-ammoniac, . . . *white.*
8. Solution of sugar of lead and solution of vitriol, . . . *black.*

Colours produced by the mixture of coloured fluids :—

1. Tincture of saffron, which is yellow, mixed with tincture of red roses, make a . . . . *green.*
2. Tincture of violets, which is blue, and spirit of sulphur, which is brown, make a . . . . *crimson.*
3. Tincture of red roses, which is red, and spirit of hartshorn, which is brownish, make a . . . . *blue.*
4. Tincture of violets, which is blue, and solution of Hungarian vitriol, which is blue, make a . . . . . . *purple.*
5. Tincture of cyanus (blue-bottle flower), which is *blue*, and spirit of sal-ammoniac coloured blue, make a . . . *green.*
6. Solution of Hungarian vitriol, which is blue, and lixivium, which is brown, make a . . . . *yellow.*
7. Tincture of cyanus, which is blue, and solution of copper, which is green, make a . . . . *red.*

Colours Changed and Restored :—

1. Solution of copper, which is *green*, by spirit of nitre is made colourless; nd is again restored by oil of tartar.
2. Limpid infusion of galls is made *black* by a solution of vitriol, and transparent again by oil of vitriol; and then *black* again by oil of tartar.

3. Tincture of *red* roses is made black by a solution of vitriol, and becomes *red* again by oil of tartar.
4. A slight tincture of red roses by spirit of vitriol becomes a fine *red;* then, by spirit of sal-ammoniac turns green; and then by oil of vitriol becomes *red* again.
5. Solution of verdigris, which is *green,* by spirit of vitriol becomes colourless; then by spirit of sal-ammoniac becomes *purple;* and then by oil of vitriol becomes colourless again.

ON THE RAINBOW.—The proportional breadth of each colour in the Rainbow, supposing the whole breadth thereof to be divided into 360 equal parts.

The red is found to have	45	parts.
The orange	,,	27
The yellow	,,	48
The green	,,	60
The blue	,,	60
The indigo	,,	40
The violet	,,	80
	Total,	360

If the flat upper surface of a top be divided into 360 equal parts, all round its edge, and be divided by 7 lines into so many portions or sectors of circles, in the above proportions, and the respective colours be lively painted in these spaces, but so as the edge of each colour may be made nearly like the colour next adjoining, that the separation may not be well distinguished by the eye; and the top be made to spin, all these colours together will so blend as to appear white. And if a large round black spot be painted in the middle, so as there may be only a broad flat ring of colours around it, the experiment will succeed the better.

Red is the least refrangible of all the colours, orange next, and *violet* the most of all.

Mr. Edward Delaval, F.R.S., has found, by experiments on melting different metals with pure glass, that they colour the glass according to their different densities or specific gravities; the most dense giving a red colour to the glass, and the least, a blue or violet.

Thus, gold melted with glass, gives it an orange colour; silver a yellow; copper a green, and iron a blue colour.

JAMES FERGUSON.

LONDON, *June,* 1760."

Vide also "Tables and Tracts," pp. 296—298.

FERGUSON ABANDONS the PROFESSION of LIMNER.—In his Memoir, he mentions that he followed his profession of limner for 26 years. He adopted this profession when in Edinburgh in 1734; consequently, he abandoned it in 1760 —now confining his whole attention to that of Lecturer on Experimental Philosophy and Astronomy. In the art of limning, Ferguson says, "*he never strove to excel,*" as his mind was ever "*pursuing things more agreeable;*" besides, his sight was beginning to fail; and in 1758, in his letter which we have quoted under that date, he had then little or no business as a limner. In this letter of 1758 he says that he had not drawn "*six pictures* these *last twelvemonths.*" These facts before him were brightened by the success of his Astronomy—he had sold the remainder of his right in that work to his publisher for £300, and it is understood that he, in 1760, received about £350 for the copyright of his Lectures on Select Subjects in Mechanics, &c., (then just published)—shows that, when he abandoned his old profession of limner in 1760, he had about £650 at command, and as a Lecturer on Experimental Philosophy and Astronomy he was meeting with great success.

The only other memoranda which we have of Ferguson for 1760 is in the shape of an advertisement of what he taught in 1760, and his fees for the same,—viz.

> "Mr. Ferguson teaches the Use of the Globes in a Month, *attending one* Hour every Day (Sunday *excepted*) for *Two Guineas at home,* or *Four Guineas abroad, if not more than a mile from his house. Those who are taught abroad provide Globes for themselves.* He *also teaches* Practical Geometry by the Scale and Compasses, the Construction of Maps, the Projection of the Sphere, the Principles of Dialing, and the Calculation of New and Full Moons and Eclipses; *on the same Terms.*"

The above is extracted from a mutilated printed hand-bill of date 1760. See also the advertisement on the fly-leaf, in his "Analysis of a Course of Lectures." Printed at Bristol in 1763.

1761.

FERGUSON'S RESIDENCE, 1761.—In Kent's London Directory for the year 1761, is inserted the following entry of his residence,—viz.

(1761).—"James Ferguson, Astronomer, Great Russell Street, Bloomsbury."

In 1761 it would appear, from an advertisement at the foot of one of his publications of this year, that he either had a residence in Red-Lion Court, Fleet Street, London, or had a room in that *Court* as a *Class-room.* A quarto work published by Ferguson early in 1761, entitled, "*A Plain Method of Determining the Parallax of Venus,*" has the following at foot of title-page :—" London : Printed for, and sold by the AUTHOR, in *Red-Lion Court, Opposite Serjeant's Inn, Fleet Street.*"

PAMPHLET on the PARALLAX of VENUS, published in quarto. —This pamphlet, published by Ferguson early in 1761, is a thin quarto of 54 pages, with 4 illustrative folding-plates, shows how to determine the Parallax of Venus by the then forthcoming transit of that planet over the Sun's disc, calculated to take place on June 6th, 1761; and consequently, the Distances of all the Planets from the Sun. This work is now very scarce; the title is as follows :—

> "A Plain Method of determining the Parallax of Venus by her Transit over the Sun, and from thence, by Analogy, the Parallax and Distance of the Sun, and of all the rest of the Planets. By James Ferguson. London: Printed for and sold by the Author, in Red-Lion Court, Opposite Serjeant's Inn, Fleet Street; and also by Mr. Millar, Bookseller in the Strand; Mr. Nairn, Optician, near the Royal Exchange; and Mr. Watkins, Optician, near Charing Cross. MDCCLXI."

It appears to have been published in February, 1761, as it is then, for the first time, advertised in the Gentleman's Magazine, vol. 31, p. 95, Feb. 1761.

The following curious calculations on the Cycles of the Sun, Moon, and Roman Indiction, are taken from a copy we made in 1834 from the original in the possession of William Upcott, Esq., Islington, London. They were made by Ferguson in 1761. We observe a similar calculation in his "Common Place Book" without a date.

"A PROBLEM.—The Cycle of the Sun is 28 Years,—the Cycle of the Moon 19 Years,—and the Roman Indiction 15 Years. Now, supposing all these Cycles to begin together, or to be 1 each;—Query, In what time afterwards will they all begin together again, and how many Revolutions will each Cycle have then made?

Years.

Let $a=28$ $\frac{ab}{a-b}=\frac{532}{9}$, and $\frac{bc}{b-c}=\frac{285}{4}$, by the method of solving such Questions.

$b=19$ $=2128$ and 2565 by cross multiplication of the denominators into the numerators.

$c=15$ $=112$ and 135, in lowest terms, by their greatest common divisor, which is 19.

Then $ab=532$ And $\frac{532\times135}{9}=\frac{285\times112}{4}=7980$ years, the time when the Cycles will begin together again.

a-$b=$ 9 $\frac{7980}{28}=285$ years, the rev. of a $\frac{7980}{19}=420$ years; the rev. of b; and $\frac{7980}{15}=532$ y. the rev. of c.

b-$c=285$ So that the Cycle of the Sun will have gone through 285 periods.

b-$c=$ 4 revolutions,—the Cycle of the Moon 420,—And the Roman Indiction through 532. And $a\times b\times c=7980$ years, but $\frac{7980}{7980}=1$. So that there is but one Conjunction of all these three Cycles in the multiplication of them all into one another, which, by the foregoing calculations will appear very singular. This 7980 years make the Great Julian Period, in which there are not any two years wherein all the three Cycles will be equal.

JAMES FERGUSON, 11 *May*, 1761."

(See also his MS. "*Common Place Book*," Col. Lib. Edin., p. 172).

We may here note that as the years 28, 19, and 15 are *prime* to each other, nothing more is necessary than to multiply them into each other,—$28\times19\times15=7980$, the years in Julian Period—then $7980\div28=285$ rev. of Solar Cycle. $7980\div19=420=$ rev. of Lunar Cycle, and $7980\div15=532$ rev. of Indiction—respectively in the great period of 7980 years.

On the back of the half sheet of paper on which the foregoing is written, we found the following somewhat similar sort of calculation on the conjunctive period of three hands on the dial-plate of a clock, which at same was time copied—it is entitled,

"The Question on the Conjunctive Period of the Three Hands *Answered*."

"Supposing a Clock to have three hands A, B, C, all going the same way round the Dial-plate. That A goes round in 365 days 5 hours 48 minutes 54 seconds (or 31556934 seconds), the time in which the Sun goes round the

Ecliptic, or length of the Solar Tropical year. That B goes round in 27 days 7 hours 43 minutes 5 seconds (or 2360585 seconds), the time in which the Moon goes round the Ecliptic; and that C goes round in 7 days (or 604800 seconds. If all these hands set out from a conjunction at any one point of the Dial-plate; Query, In what time afterward will they all come to a conjunction again, at the very same point of the plate, and how many revolutions will each have then made?

Multiply the Solar Year A=	31556934	Seconds.
By the Lunar Revolution B=	2360585	
The Product will be	74492825046390	
Mult. this Product by the Week-hand C=	604800	
And the Product will be=	45053260588056672000	

Which is the number of seconds, in the Grand Period, or time in which all the three hands will be in a conjunction again at the same point of the Dial-plate they set out from at first.

Divide this period by A=31556934, and the quotient will be 1427681808000 for the number of Solar Tropical Years contained in it; equal to the number of revolutions of the hand A in that time.

Divide the same period by B=2360585, and the quotient will be 19085633683200, for the number of revolutions of the Moon, in that period of so many Solar Tropical Years.

Divide the same period by C=604800, the number of seconds in a week, and the quotient will be 74492825046390, which is the number of revolutions of the hand C, or number of weeks in the Grand Period.

Hence, if the Sun and Moon were in conjunction at any assigned instant of a given Day of the Year and Week, they would not be in conjunction again, at the like instant of the same Day of the Year and Week, in less than 1427681808000 Solar Tropical Years. This period contains 1427651677822 Julian years 2 days 10 hours 40 min. 48 sec.

JAMES FERGUSON, 14 *May*, 1761."

(See also Ferguson's MS. "Common Place Book," Col. Lib. Edin., from page pasted on the inside of the first board, in which the paper is simply entitled, "*A Question Answered.*")

FERGUSON LECTURES on ASTRONOMY, &c., in LONDON and the

PROVINCES.—During the first months of the year 1761, Ferguson delivered a great many Lectures on Astronomy, &c., in London and the provinces, particularly explaining and illustrating, by diagrams, the then forthcoming Transit of Venus over the Sun's disc, to take place on 6th June, 1761. In March, 1761, we find him lecturing in Chelmsford, from which place he addresses Mr. Urban, of the Gentleman's Magazine, on the subject of Twilight. The following is a copy of his note accompanying this communication :—

"Mr. Urban,—Sir,—If the following on the Twilight, &c., be found sufficiently curious and entertaining to merit a space in your learned and ingenious Collection, they are at your service.—Yours, &c., J. Ferguson.—Chelmsford, March 8th, 1761."
(Gent. Mag. Vol. 31, p. 124).

This paper on Twilight occupies two columns of the Gentleman's Magazine, at page noted.

TRANSIT OF VENUS OVER THE SUN'S DISC, JUNE 6th, 1761.—For a considerable length of time before June 6th, 1761, the subject of the expected Transit of Venus had excited very general interest and speculation. When "the Transit-Day" arrived, many parties all over the country were early on the alert on the morning of Saturday, June 6th, in order to get a glimpse of the rare sight. Ferguson, on this occasion had his station, with other observers, on the top of the British Museum, as the following extracts from the manuscript papers of the Royal Society will show.

"Observations on the Transit of Venus over the Disc of the Sun on Saturday morning, the 6th June, 1761:

Interior Contact 17½ minutes past 8 o'clock A.M.
Exterior Contact 34 minutes past 8 o'clock A.M.

"The Watch, corrected by a meridian in the Museum and by a line drawn circumscribing the shadow of the Door on the Leads.

Present

"Dr. Woulfe; from Poland.
Mr. Moore; of Jermyn Street.
Mr. Ferguson; Russell Street, Bloomsbury.
Dr. Morton; of the Museum.
Wortley Montague; Esq.

Mr. James Empson; of the Museum.
Stephen Fuller, Esqr. of Hart Street, Bloomsbury.
Revd. Mr. Forster; Rector of Chisleborough, Somerset."

(Vide manuscript papers of the Royal Society, Dr. Birch's Collection, British Museum, being additional MS. No. 4440, 604, entitled, "Observations of the Transit of Venus over the Sun's Disk, June 6th, 1761, at the British Museum.")

Then follows,—

"Mr. James Ferguson's own remarks on the foregoing observations,—viz.

"The idea I had formed of the internal contact was that the planet would touch the edge of the Sun in an instant, like two drops of quicksilver meeting on a plane, and that in an instant the black contact would appear; but in this I was deceived, the particulars of the phenomenon being as follows:—

"June 6th heavy with clouds, till six o'clock, when the clouds began to dissipate, but not enough to afford a plain sight of Venus on the Sun, till more than ½ past seven, and the planet got nearer the limb of the Sun, than I had described to see it on the disk.

"With a Six-feet Reflector, and its magnifying power of 110 and also 220 times, I carefully examined the Sun's disk to discover a satellite of Venus, but saw none; for I had a very clear dark glass next my eye, and the Sun's limb appeared most perfectly defined, and at the distance of about a sixth part of Venus's diameter from its edge, was the darkest part of Venus's phasis, from which, to the centre, an imperfect light increased and illuminated about the centre.

"At 8ho. 16m. per clock from the top of the Museum, I was prepared to observe the internal contact; and as Venus drew nearer to the limb of the Sun, the penumbra near the limb of Venus became darker, and threatened to obscure the point of contact at the instant it would happen; the circumstances of which, for each of the moments of time, are imperfectly delineated, on account of the nearness of the lines; but more truly described as follows; (a right line representing that part of the Sun's limb near where the contact happened, and an arc, the approaching limb of Venus,—for each three seconds of time, from the loss of the thread of light) in words for each second of time by the clock, thus:—

ho. m. s.	
At 8 16 41	no diminution of light between the limb of Venus and that of the Sun.
8 16 42	slight penumbra, diminution of light, where the contact was to be.
8 16 43	penumbra of a grey colour near the same place.
8 16 44	penumbra almost brown, and thread of light very narrow, and almost lost.
8 16 45	penumbra brown, and the thread of light in the contact point indistinct or lost.
8 16 46	penumbra more brown, and the touch the smallest possible.
8 16 47	penumbra almost black, and the touch a little broader.
8 16 48	slight black in the point of contact, and the edges a little broader.
8 16 49	true black in the point of contact, and the edges a little broader.
8 16 50 more so	Here I concluded with myself that observers would differ in their
8 16 51 more so	judgments about the moment of contact, some seconds of time,
8 16 52 more so	or that some would estimate the contact sooner than others.

From these observations, I concluded that the thread of light in the point of contact was so obscured as to be indiscernible at $8^{h.}$ $16^{m.}$ $46^{s.}$, and that true black did not succeed in the same point till $3^{sec.}$ after,—namely, $8^{h.}$ $16^{m.}$ $49^{sec.}$; and from both of these properties, I concluded that the real internal contact was at $8^{h.}$ $16^{m.}$ $47^{sec.}$ by the clock; which makes $8^{h.}$ $16^{m.}$ $11^{sec.}$ equal time, and $8^{h.}$ $18^{m.}$ $2^{sec.}$ apparent time at Greenwich.

"At $8^{ho.}$ $35^{min.}$ per clock, the external contact was near and not encumbered with such a penumbra, or partial light, as the internal contact had been. At $8^{ho.}$ $35^{m.}$ $4^{s.}$ the least dent possible quite black, appeared in the Sun's limb, and at $8^{h.}$ $35^{m.}$ $6^{s.}$ the limb was restored to its perfect form, there having been a small trembling light between the narrow watery border of Venus, and the vanishing point of contact in the Sun's limb for these two seconds of time, from which the external contact at Chelsea was $8^{h.}$ $34^{m.}$ $30^{s.}$ apparent time; which makes $8^{h.}$ $37^{m.}$ $2^{s.}$ apparent time at Greenwich. From the aforegoing circumstances, it appeared to me that the external contact was more easily to be determined than the internal one, which was contrary to what I had before expected; and because the point of contact must have appeared through such a telescope as I observed with, in its proper colour, dark or black, sooner than through a small magnifying power of equal light, I concluded that, through my telescope, the internal contact was visible sooner than through a two-foot reflector, ten or twelve seconds of time.

"As these observations reconcile a seeming contradiction in Mr. Short's numbers of the internal contact, and, whilst I am very certain with respect to the particulars of the external contact, cannot determine why they differ from that ingenious observer's numbers, or any others.

JAMES FERGUSON."

PROJECTION of the TRANSIT of VENUS.—The annexed cut exhibits the Sun and the path of Venus over his disc, as seen at London on Saturday morning, June 6th, 1761, at the times mentioned by Ferguson in the foregoing account of the transit.

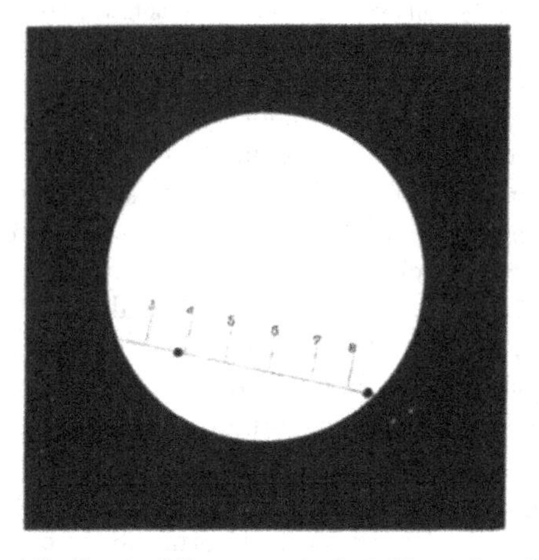

Projection of the Transit of Venus over the Sun's Disc, 6 June, 1761, seen from about 12 min. of 4 morning until 8 hours 44 min. do.

Note.—The white disc represents the Sun—the dark spot under 4 the position of Venus on the Sun at Sunrise—the black small spot to the right of 8 o'clock, Venus at the termination of the transit

Capel Lofft, in his "*Poem on the Universe*," referring to this transit, says,—

"England and rival France, admiring, watch'd
The important prodigy; when the *sixth* day
Had dawn'd propitious on the Junian month.
Let me explain the cause and the design
Which pointed Observation's eagle eye
To this phenomenon; and thou direct
O Ferguson! than whose no beam more clear
Pierces the gloom, where science is conceal'd."

Eudosia, or a Poem on the Universe, Book 2d, p. 41
(by Capel Lofft, Esq., 1781).

ROYAL PENSION.—According to several memoranda, it would

appear, that on or about 20th June, 1761, Ferguson received official intimation that his Majesty (Geo. III.) had graciously bestowed on him an annuity of £50 a-year. The first half-year's payment of £25 was tendered to Ferguson on January 22d, 1762, to which date we refer.

MULTUM IN PARVO.—We have in our possession one of Ferguson's manuscripts with this title—(much in little space). It contains eight different articles, which occupy about a page and a half of foolscap paper, and bears the date of Dec. 1761. The following is a copy:

1. "ON THE QUANTITY of LAND and WATER on the EARTH'S SURFACE.—The Seas and unknown parts of the Earth (by a measurement of the best maps) contain 160,522,026 square miles. The inhabited parts, 38,990,569. Europe contains 4,456,065; Asia, 10,768,823; Africa, 9,564,807; and America, 14,110,874 square miles. In all, 199,512,595; which is the number of square miles on the whole of the Earth's surface."

We may here observe that a globe whose surface contains 199,512,595 square miles, must have a diameter of 7,964¼ miles nearly—the most accurate measure of the Earth's mean diameter is 7,912 miles, or 57¼ miles less than the above.

2. "ON the VELOCITY of LIGHT.—It has been proved, by the Eclipses of Jupiter's satellites, that light takes 8 minutes of time to come from the Sun to the Earth. And as the Earth's distance from the Sun is 95,000,000 miles, in round numbers, 'tis plain that the velocity of light is 11,875,000 miles in a minute; and consequently, 197,916 miles in a second, which is 1,486,458 times as swift as the motion of a cannon ball, and 10,440 times as swift as the Earth moves in its annual orbit."

3. "ON the VELOCITY of a CANNON BALL COMPARED with the EQUATORIAL *parts* of the *Earth*.—The Earth's circumference, in the nearest round numbers, is 25,000 English miles; and the mean velocity of a cannon ball is 8 miles per minute. Divide 25,000 by 8, and the quotient will be 3,125 for the number of minutes in which a cannon ball would go round the Earth. But as the Earth turns round its axis in 1,440 minutes

(or 24 hours), every place on the Equator goes through 17·361 miles per minute. Therefore, the velocity of the Equatorial parts of the Earth is to the velocity of a cannon ball as 17·361 is to 8, or as 2·17 is to 1. So that any place on the Equator moves with somewhat more than double the velocity of a cannon ball."

4. "ON the VELOCITY of SOUND.—According to Dr. Halley, Mr. Flamsteed, and Dr. Derham, sound moves 1,142 feet in a second of time, 68,520 feet in a minute, and 778,636 miles in an hour. Hence we know how far a thunder cloud is from us, if we have a watch that shows seconds. Thus, suppose there were four seconds from the moment we see the flash of *lightning* to the moment we hear the clap of *thunder*, 'tis plain that the cloud which produced the thunder is four times 1,142 feet, or about 4,568 feet from us, which is about four-fifths of a mile.

As just observed, sound moves with a velocity of 1,142 feet in a second of time, and consequently, 5,280 feet—or 1 mile—in 4·631579 seconds. If a pinion of 19 leaves turns round in a second of time, and takes into a wheel of 88 teeth, it will turn that wheel round in 4·631579 seconds."

5. "On the Diameter and Circumference of the visible part of a cloudy Sky.—The greatest distance of the cloud from the horizon at sea is 94 miles from the diameter all around; and consequently, the whole extent or diameter of the horizon, reaching to the clouds, is 188 miles; and the circumference thereof is 590·97 miles."

This ought to be 590·63 miles.—EDIT.

6. "ON the WEIGHT of the whole ATMOSPHERE.—On a square inch, the weight is 15 pounds; on a square foot, 2,160; on a square yard, 19,440; on a square mile, 60,217,344,000; and on the whole surface of the Earth and Sea together 12,014,118,565,447,680,000 pounds.

The surface of the body of a middle-siz'd man is about 14 square feet; and as the weight or pressure of the air is equal to 2,160 pounds on every square foot on, or near, the Earth's surface; and as the pressure of the air is equal in all directions, its pressure on the whole body of a middle-siz'd man is equal to

30,240 pounds, or 13½ tons. But, because the spring of the internal air is of equal force with the pressure of the external, the pressure is not felt."

Referring to No. 1, it will be seen that the Earth's surface contains 199,512,595 square miles; therefore, 199,512,595 × 60,217,344,000 = 12,014,118,565,447,680,000 pounds.—EDIT.

7. "ON SQUARING THE CIRCLE.—Although there has not yet been any method found for doing this to mathematical exactness, yet, by means of the following numbers, it may be brought so very near the truth as to be within a grain of sand in a square mile, supposing 100 grains of sand—placed on a straight line, and touching one another—to be equal to the length of an inch; and consequently, 40,144,896,000,000 to cover a square mile.

If the diameter of a circle be given, and the length of the side of a square so nearly equal to the circle as to be true to 14 places of figures be required; say, As 1 is to the diameter of the given circle, so is 0·88622692545276 to the side of the square required, in such measures as the diameter of the circle was taken.

If the length of the side of a square be given, and the diameter of a circle equal—as nearly as above-mentioned—to the square be required; say, As 1 is to the side of the given square, taken in any measure, as feet, inches, &c., so is 1·12837916709551 to the diameter of a circle—taken in the same kind of measures—whose area is equal to the area of the given square.

In practice, it is sufficient to take out the decimal parts to four places of figures; for, even by so small a number, we come so near the truth as to be within a ten thousandth part of the whole area of being perfectly true. And this is nearer than any one can pretend to delineate on paper.

Thus, supposing the diameter of a circle to be 12 inches, and that it is required to find the length of the side of a square—or make a square—whose area shall be equal to the area of the circle; say, As 1 is to 12, so is ·8862 to 10·6344 inches, the length of the side of the square required. Or, supposing the side of a square to be 12 inches, and that it is required to find the diameter of a circle whose area shall be equal to the area of a square; say, As 1 is to 12, so is 1·1284 (instead of 1·128379) to 13·5408 inches, the diameter of the circle required.

Hence, as a square vessel, just one foot wide and one foot deep, would hold a cubic foot of water, a cylindrical vessel 13·54 inches in diameter and one foot deep would hold a cubic foot of water too; at least so near the truth, that no difference could be perceived.

The diameter of any circle is in proportion to its circumference, as 1 is to 3·1415926535897932384626434, &c., or as 1 is to 3·1416, near enough for practice.

Any circle is equal to a parallelogram, whose length is equal to half the circumference of the circle, and breadth equal to half the diameter. Therefore, multiply half the circumference by half the diameter, and the product shall be equal to the area of the circle, in square measure. The square root of this area is the side of a square equal to the circle. Also, multiply the square of the diameter by 3·1416 and the product will be the area."

8. "PHILOSOPHICAL PROPOSITIONS.—

Rules.

1st, No more causes of natural appearances are to be admitted than what are real, and sufficient to account for these appearances.

2d, Natural effects of the same sort are to be accounted for by the same causes.

3d, Such qualities as are found to belong to all bodies that we can make experiments upon, are to be looked upon as qualities belonging to all bodies whatsoever.

JAMES FERGUSON.
LONDON, 9 *Dec.* 1761."

On looking into "Tables and Tracts," we find Nos. 1, 2, 4, 5, 6, discussed therein inter pp. 300—303. No. 7 will be found in his "Select Mechanical Exercises," pp. 126—129, and Nos. 3 and 8 in his MS. "Common Place Book," (Col. Lib. Edin.) pp. 47 and 195.

ELECTRICAL MACHINE, &c.—Referring to our memoranda, we find that, in the year 1761, being desirous of extending his course of lectures, so as to embrace the subject of Electricity, he purchased from Mr. Nairne, Philosophical Instrument Maker, Cornhill, London, an Electrical Machine and apparatus for experiments. It is very probable that this is the Electrical Machine and apparatus we find in Plate 1st of his "Introduction to Electricity," published in 1770.

1762.

ROYAL PENSION to FERGUSON.—In searching among the State Papers in the State Paper Office, for 1762, we found an entry of date January 22d, 1762, as that on which he was paid his first half-year's pension from the King's Privy Purse. The entry is as follows:—

"Privy Purse Accounts, 1762.

Jany. 22d. Pensions.
Mr. James Ferguson, £25."

again on

"July 24th, Mr. James Ferguson, £25."

The last payment from the Privy Purse was made to him about four months before his death, viz. "July 26th, 1776, Mr. James Ferguson, £25."

This pension placed Ferguson on firm ground, and at same time gave him a position in the world honourable to himself, and of advantage in prosecuting his profession of lecturer on Astronomy and other branches of science. His pecuniary difficulties were now at an end, and from this date, says an old friend of his, "*he went ever on afterwards prospering and to prosper.*"

The kindness of "the good old king" placed Ferguson in easy circumstances, and when at any time in his after life the Royal pension was alluded to, "he expressed himself as being under the deepest obligations of gratitude to his Majesty, as it had made him easy and comfortable." (See letter in Appendix).

The total sum received by him—from "*the Privy Purse*"—from first payment on 22d January, 1762, to 26 July 1776, inclusive, was £750; having drawn his pension for 15 years at the rate of £50 per annum.

PLANETARIUM.—In the year 1762, Ferguson was busily engaged upon a Planetarium for "*showing the periodical revolutions of the Earth and all the other planets round the Sun, in a Clock, so as to agree nearly with the periodical revolutions of the planets about the Sun in the Heavens.*" The following is the description he gives of this machine, but as there is no plate of the wheel-work, we have made a sectional view to enable the reader the more readily to understand its arrangement while he reads the description.

Ferguson says, "Let six hollow sockets, or arbors, be made to fit and turn within one another, and all of them to turn upon a fixed spindle, or axis; on the top of which let there be a ball to represent the Sun. Let the widest arbor be the shortest, and have an arm on its uppermost end to carry a ball representing Saturn, and a wheel of 206 teeth on its lowermost end.

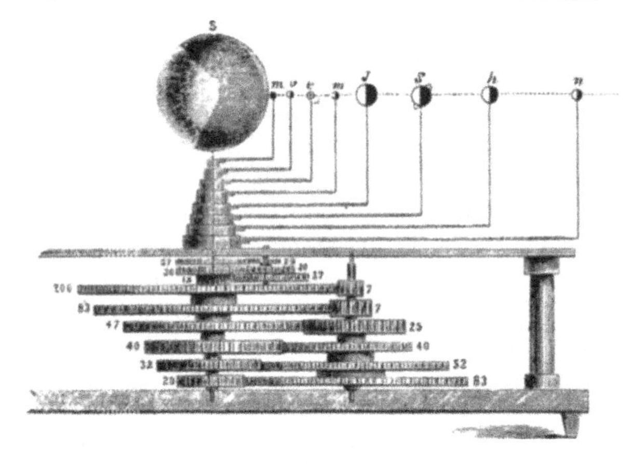

Section of the Wheel-work of Ferguson's Planetarium.

Let the next sized arbor be so much longer than the above one, as to have a wheel (of 83 teeth) put upon it, below the wheel of 206, and an arm on the other end (above Saturn's) for carrying a ball to represent Jupiter.

Let the third socket be so much longer than the second, as to have a wheel on it (of 47 teeth) below the wheel of 83, and an arm on its other end, above Jupiter's for carrying a ball to represent Mars.

Let the fourth arbor be so much longer than the third, as to have a wheel (of 40 teeth) on its lower end, and an arm on its upper end, above Mars's, for carrying a ball to represent the Earth.

Let the fifth arbor be so much longer than the fourth, as to have a wheel (of 32 teeth) on its lower end, below the wheel of 40, and an arm on its upper end, above the Earth's, for carrying a ball to represent the planet Venus.

Let the sixth (which is the smallest) arbor, be so much longer than the fifth, as to have a wheel (of 20 teeth) on its lower end below the wheel of 32, and an arm on its upper end, above Venus's, for carrying a ball to represent the planet Mercury.

Saturn's arm must be the longest of all, because that planet is the farthest of all from the Sun;[229] Jupiter's the next longest, Mars's the next, the Earth's the next, Venus's the next, and Mercury's the next or shortest of all, because Mercury is the nearest of all the planets to the Sun.[230]

The wheels must be fixed on their respective arbors, and diminish in their sizes from the highest numbers to the lowest; so that, when they are all put together, they may form somewhat of the appearance of a cone.

And, to give these wheels and planets their proper motions, they must be turned by six wheels (or rather four wheels and two pinions), all fixed on one solid axis, in a conical manner, inverted with respect to the other six wheels, so as the wheels and pinions on the solid axis may take into those on the arbors, and turn them.

The solid axis, with all its wheels and pinions, will turn round in the same time together, because the wheels and pinions are all *fixed* on the axis, and must be turned round once in a year by clock-work.

Then, if the uppermost and smallest pinion on the axis has 7 leaves, taking into Saturn's wheel of 206 teeth, Saturn will be carried round the Sun in 10,748 days 18 hours 43 minutes; for as 7 is to 206, so is 365·25 to 10748·78.

If the next pinion on the axis (which must be bigger size than the pinion of 7 above it) has 7 leaves also, and takes into Jupiter's wheel of 83 teeth; Jupiter will be carried round the Sun in 4,330 days 19 hours 40 minutes; for as 7 is to 83, so is 365·25 to 4,330·82.

If the wheel below this pinion on the axis has 25 teeth, and takes into Mars's wheel of 47, Mars will be carried round the Sun in 686 days 16 hours 5 minutes; for, as 25 is to 47, so is 365·25 to 686·67.

229 In Ferguson's time, Saturn was the most distant known planet from the Sun. Since his day, Uranus, Neptune, and nearly 100 asteroids have been discovered, regarding which, see recent works on Astronomy.

230 It is suspected that there is a planet between Mercury and the Sun; it has even been stated that such a planet has been seen and that its period round the Sun is performed in 19 d. 17 h.

If the next bigger wheel on the axis, which turns round in 365·25 days, has 40 teeth, and takes into the Earth's wheel of 40 teeth, the Earth will be carried round the Sun in 365·25 days (365 days 6 hours).

If the next bigger wheel has 52 teeth, and takes into Venus's wheel of 32 teeth, Venus will be carried round the Sun in 224 days 18 hours 29 minutes; for, as 52 is to 32, so is 365·25 to 224·77.

And lastly, if the largest wheel on the axis has 83 teeth, and takes into Mercury's wheel of 20, Mercury will be carried round the Sun in 88 days 0 hours 14 minutes; for, as 83 is to 20, so is 365·25 to 88·01.[231]

I have seen a calculation of this sort in a printed book; but the numbers there are so faulty for Mars and Saturn, that I was obliged to alter them, Saturn's period being wrong by 51 days. How near these are to the truth will appear by comparing them with the annual periods in the following Table." (Ferguson's "*Tables and Tracts.*" Lond., 1767, pp. 129—134).

In this Planetarium, the revolution of the planets round the Sun extends to that of Saturn only. To bring it up to the present day, we have calculated wheel-work for the periods of Uranus and Neptune, as is shown in the engraving, above wheel 206.

Thus, on Saturn's wheel of 206 teeth is made fast a small wheel of 13 teeth, which turns round with wheel 206 in Saturn's period of 10748·78 days, and drives a wheel of 37 teeth, to which is riveted a wheel of 30, which drives another wheel of 30 round in 30592·68 days, or 30592 days 16 hours 19 minutes. On the first mentioned wheel of 30 there is riveted a wheel of 29, which turns a wheel of 57 once round in 60130·44 days, or

231 The diurnal rotation of Mercury is accomplished in 24 h. 5 m. 28 s. Hourly motion of Mercury's equator, 407 miles. Saturn's rotation is performed in 10 h. 16 m. 2 s. Hourly motion of Equator, 23,730 miles. It may be here remarked, that wheel-work of higher numbers than those given here by Ferguson would produce more accurate periods—thus, for example, instead of Mercury's wheel of 20 being driven by a wheel of 83, let a wheel of 33 be used instead of 20, and a wheel of 137 for that of 83, and the result will be 87 d. 23 h. 28 m. for Mercury's period. Some years ago we sent a sectional plan of the wheel-work of a planetarium, for showing the motions of Mercury, Venus, the Earth, Mars, Jupiter, Saturn, Uranus, and Neptune, to an Orrery maker in London, and we believe he now makes them with these higher numbers and with the additional planets. The wheel-work is enclosed within a circular box supported on a tripod stand. The Sun is a large gilded ball, the planets, of ivory, set in motion by a spring clock, and the price he charged for one he made was £26. In Ferguson's MS. "Common Place Book," at pp. 82 and 189, we find rough sketches of a Planetarium, with wheels $\frac{137}{33}$ for Mercury's period.

A TABLE, showing the times contained in the annual and diurnal revolutions of the planets, with their relative and true distances from the Sun, the circumference of their orbits, the number of miles they move every hour therein, and their daily mean motions in degrees and parts of a degree. Their distances in miles from the Sun are here set down, as they were found to be by the Transit of Venus over the Sun, June 6th, 1761.

The Six Primary Planets.	Their Annual Periods round the Sun.	Their Diurnal Rotations.	Hourly motion of their Equators.	Their relative Mean Distances from the Sun.	Their real distances from the Sun in Eng. M.	The circumference of their orbits in English Miles.	The number of miles they move in each hour in their orbits.	Their daily mean motions in their orbits.
			Miles.	Parts.			English Miles.	° ′ ″
Mercury,	87 d. 23 h.	unknown	unknown	3871	36,841,468	231,574,940	109699·16	4 5 32
Venus,	224 17	24 d. 8 h.	43	7233	68,891,486	433,032,198	80295·24	1 36 8
The Earth,	365 6	1 0	1042	10000	95,173,000	598,230,286	68243·24	0 59 8
Mars,	686 23	24 h. 40 m.	556	15237	145,014,148	911,517,502	55287·00	0 31 27
Jupiter,	4332 12	9 h. 56 m.	25920	52009	494,990,976	3,111,371,849	29083·60	0 4 59
Saturn,	10759 7	unknown	unknown	95400	907,956,130	5,707,152,817	22101·64	0 2 1

60130 days 10 hours 33 minutes, being about 3 days too slow in 84 years (Uranus's period), and about 5 days too fast in

Neptune's period of 165 years, which is near enough for motions produced by simple pairs of wheels. For a Planetarium, with very accurate wheel-work and periods, we refer the reader to Dr. Dick's "*Practical Astronomer*," (pp. 528—537) where will be found an engraving and full description of the Planetary Machine, from original calculations made in London, in 1842, by the present writer.

ASTRONOMICAL LECTURES in the Midland Counties, and "THE GREAT SILK MILL" at Derby.—One of our notes informs us that Ferguson, in the year 1762, made a tour through the Midland Counties of England, and delivered in several towns there, his usual Course of Lectures on Astronomy, Mechanics, Hydraulics, &c. In his progress he visited Derby, and delivered his course in that town; during his sojourn, he frequently visited the ingenious Mr. Whitehurst, Clockmaker, from whom he received several curious papers and drawings of *Escapements, Pendulums,* and *Hydraulic Engines.* He also visited "*The Great Silk Mill at Derby,*" (erected by Lombe, in 1719,) and appears to have been much gratified with his inspection of the far-famed mill. The following account of it, by Ferguson, was copied in 1831, from the original belonging to Deane Walker, Esq., Lecturer on Astronomy, &c., London.

"NOTES, and VERSES, on the GREAT SILK MILL at DERBY.

In this Silk Mill there are 26,586 wheels, and 97,746 movements, continually at work, except on Sundays. The machinery is actuated by one great water-wheel, which goes round *three times* in a minute. In each time of its going round, 73,728 yards of silk are twisted, so that, in 24 hours, 318,504,960 yards are twisted. The Water-wheel is kept constantly going, except on Sundays, when it is disengaged from all the rest of the works. Any part of these movements may be stopt without the least prejudice or interruption to the rest. This grand machine is disposed of in four stories of large rooms above one another.

J. FERGUSON, 1762."

Wondrous Machine! Thy curious Fabric shows
How far the power of human wisdom goes!
Where many thousand movements all attend
Upon a WHEEL, and on THAT cause depend.
Sceptic, advance! propose thy scheme of wit,
That faith to reason always must submit.

Whence learn'd these movements to obey command?
Who taught them how to roll, and when to stand?
Was it by chance this curious fabric came?
Or did some thought precede, and rule the Frame?
Worthy the Mortal, on whose soul, confest,
His GREAT CREATOR'S Image stands imprest!
Now turn from Earth to Heaven thy doubting eyes,
And read th' amazing Glories of the Skies!
Worlds without number roll in different Spheres,
Keep to their Seasons, and complete their years.
Five thousand circuits, made with equal force,
The Earth has finish'd by its annual Course.
The Sun dispenses beams of genial Light,
And lends his rays to cheer the gloomy night.
STUPENDOUS POWER and THOUGHT! Enquire no more:
Own the FIRST MOVER; and, convinc'd, ADORE.

Ferguson subjoins these lines to his account of the mill, copied probably from some manuscript or print shown to him by his Derby friends. (See also, Ferguson's "Tables and Tracts," pp. 172, 173).

We may here note that Hutton, the late learned Bookseller of Birmingham, who served his apprenticeship in this mill, says that there were only 13,384 wheels in it.

The following article is taken from the original, in Ferguson's autograph, in our possession:—

"ON THE PLATONIC BODIES.—The diameter of a sphere being given, to find the side of any of the Platonic bodies that may be either inscribed in the sphere or circumscribed about it, or that which shall be equal to it.

As 1 is to the number given in the following Table, respecting the thing sought, so is the diameter of the given sphere to the side of the Platonic body sought:

The diameter of a Sphere being unity, the side of a	That may be inscribed in the Sphere, is	That may be circumscribed about the Sphere, is	That is to be *equal* to the Sphere, is
Tetraëdron,	0·816497	2·44948	1·64417
Octaëdron, .	0·707107	1·22474	1·03576
Hexaëdron,	0·577350	1·00000	0·80610
Icosaëdron,	0·525731	0·66158	0·62153
Dodecaëdron,	0·356822	0·44903	0·40883

The side of any of the five Platonic bodies being given, to find the diameter of a sphere that may be inscribed in that body, or circumscribed about it, or that is equal to it. As the respective number in the Table, under the title *Inscribed, Circumscribed, Equal,* is to 1, so is the side of the given Platonic body to the diameter of its inscribed, circumscribed, or equal sphere, in solidity.

The side of any of the five Platonic bodies being given, to find the side of either of the Platonic bodies which are equal in solidity to that of the given body.—As the number under the title *equal,* against the given Platonic body, is to the number under the same title against the body whose side is sought, so is the side of the given Platonic body to the side of the Platonic body sought.

To find the solid contents of any of the five Platonic bodies. —As 1 is to the cube of the side of any of these bodies, so is 0·117851 to the solid contents of the Tetraëdron, 0·417404 to that of the Octaëdron, 1·00000 to that of the Hexaëdron, 2·181695 to that of the Icosaëdron, and 7·663199 to the solid content of the Dodecaëdron.

JAS. FERGUSON.

LONDON, 16 *December*, 1762."

Vide also Ferguson's "Tables and Tracts," pp. 220, 221, 224.

FERGUSON receives a LETTER from LADY ISABELLA FINCH.— Our memoranda for 1762 concludes with a short note from the Lady Isabella Finch to Ferguson, dated from "*Berkeley Square* (London), *Nov. 27th,* 1762," from which it is seen, that Ferguson had sent her a copy of one of his works to present to the Princess Amelia. The following is a copy of the note, (the original is in the "*Birch Collection of Letters,*" British Museum, Vol. viii., No. 58, and Museum No. 4,307):—

"Lady Isabella Finch returns Mr. Ferguson many thanks for his valuable book. Which y^t^ he may have an opportunity of presenting in forme to Her R. Hss the Pss Amelia, she will give him timely notice of ye first Drawing Room she has at Cavendish Square.[232]

BERKELEY SQUARE,
Novr. 27th, 1762."

[232] Her Royal Highness the Princess Amelia, a daughter of King George II., born 30th May, 1711. She constantly resided in England. Died, unmarried, on 31st October, 1786, aged 75. Lady Isabella Finch was one of the PRINCESS'S *ladies in waiting.*

The note is thus addressed, "*Mr. James Ferguson, at his House in Russell Street, Bloomsbury,*" (London). Shortly afterwards, Ferguson removed to Mortimer Street, Cavendish Square.

1763.

In glancing over our memoranda for 1763, we find it one of the most remarkable years of Ferguson's life—a year of great success, and a year of great sorrow. Of success, as shown by the pecuniary proceeds of his lectures at Bristol and Bath; and in being elected a Fellow of the Royal Society;—and of sorrow, consequent on the sudden and mysterious disappearance of his accomplished and only daughter, whom, after the middle of this year, he never saw more! This, and other interesting matters, will be noticed, and satisfactorily cleared up, as we proceed with our details.

FERGUSON APPLIES for the VACANT SITUATION of CLERK to the ROYAL SOCIETY, and is REJECTED.—The first entry in our memoranda informs us that, early in January, 1763, Ferguson was one of the candidates for the then vacant situation of a "Clerk to the Royal Society,"—Salary, £47 7s. per annum. At a meeting of the Royal Society on February 3d, the members proceeded to take into consideration their vacant "Clerkship" situation. It was found that there were six candidates for the situation, but only four went to ballot; and as "*the petition*" of Ferguson for said situation was the first on the roll, it was first read before the members—it began thus:—

"1st, The Petition of Mr. James Ferguson of Mortimer Street, Cavendish Square, Lecturer on Geography, Astronomy, &c., and author of several works in this latter Science, and of divers communications to the Society, was read (84 members being present),—

For Ferguson, . . . Yea, 33.
Nay, 51."

Ferguson was consequently rejected, and the ballot went on with the rest. It resulted in favour of Emanuel Mendes Da Costa; his ballot being, *Yea*, 62—*Nay*, 18; Da Costa was most likely elected because of his extensive knowledge of languages. He, however, proved a bad bargain to the Society, seeing that he embezzled their funds to the amount of more than £1,000. He was dismissed in December, 1767.

For these particulars, we are indebted to the kindness of Dr. Sharpey and Professór Stokes, present Secretaries of the Royal Society.

FERGUSON'S paper on the TRANSIT of VENUS—READ before the ROYAL SOCIETY.—On referring to "the Philosophical Transactions" of the Royal Society, we find that Ferguson, on February 10th, 1763, sent a paper to be read before the members, entitled,

> "A Description and Delineation of the Transit of Venus Expected in 1769." Read before the Royal Society, on 10th Feb. 1763. (See Phil. Trans., Vol. 53, page 30).

Ferguson, early in March, went to Bath and Bristol, and remained in these cities until the beginning of May, delivering several courses of Day and Evening Lectures on Astronomy, Mechanics, Hydrostatics, Hydraulics, Pneumatics, and Optics, which courses were illustrated by a very efficient and extensive apparatus. His lecture-room in Bath was the large room of the "*Lamb Inn*," Stall Street; and at Bristol, in the large room of the "*Bell Inn*," Broad Street. He appears to have been at this period about six weeks in Bath and Bristol, and the receipts from his lectures were, "*Total*, £138 15s." From this may be deducted, say £38 15s. for travelling expenses, &c., leaving, as *nett proceeds*, £100,—on the whole a pretty successful lecturing tour.

FERGUSON APPLIES to be ADMITTED a FELLOW of the ROYAL SOCIETY.—It has generally been supposed that Ferguson was elected a Fellow of the Royal Society without solicitation, but such was not the case. On 26th March, 1763, he, while in Bristol, wrote a letter to his friend the Rev. Dr. Birch (who had then recently been elected Secretary of the Society), expressing to him a desire to become a member, and offering to give "*a bond*" to ensure payment of the usual entry fees, &c., and annual payments, &c. The following is a copy of this letter:—

"BRISTOL, *March 26th*, 1763.

"REVEREND SIR,—As with the highest respect and gratitude I remember my friends of the Royal Society who voted for my succeeding the late Mr. Hauksbee,[233] so I do assure you that my mind justly approves of the determination made by a great majority for Mr. Da Costa, on account of his much superior talents for that important office.

"Do you think I might now presume to offer myself as a member of that respected and illustrious Society? It is an honour I very much desire to have; and besides which, the Improvements I should have by attending their meetings, and by access to the great Library, could not fail to be very great for me, in those Sciences wherein I find myself defective.

"I beg you will communicate to the Right Honble. the Earl of Macclesfield, and to others whom you think proper, this my earnest desire of becoming a member of the Royal Society; and if that learned and venerable Body should think me any way deserving of that honour, and in consequence thereof will favour me with their election, I shall, on returning to London, give my Bond for what is to be paid on that account.—I am, with the greatest respect,

Reverend Sir,
Your much obliged,
and very humble servant,
JAMES FERGUSON."

The original letter is in "*the Birch Collection of Letters*," British Museum, Vol. viii., No. 227 and 228, Museum, No. 4307.

We have never seen Dr. Birch's reply to this letter, but from the following "*Certificate*," it would appear that the Rev. Doctor lost no time in the furtherance of Ferguson's wishes. It would seem that he wrote out the Certificate to the Royal Society, and got it signed by members to whom Ferguson was personally known, as the rules of the Society required. The following is a copy of Ferguson's Certificate for membership entered in the books of the Royal Society, for which we have again kindly to thank Dr. Sharpey and Professor Dr. Stokes, present Secretaries of the Royal Society:—

"Mr. James Ferguson,
of Mortimer Street,

Well known to the Public by his writings and Lectures on the subject of Astronomy, being desirious of Election into the Royal Society, is recommended by us on our personal knowledge, as well as on account of his various

233 Mr. Francis Hauksbee was the author of "Physico-Mechanical Experiments on Light, Electricity, &c., with plates, Lond., 1709;" joint author of "Hauksbee and Whiston's Course of Mechanical, Optical, Hydrostatical, and Pneumatical Experiments;" inventor and maker of a curious orrery; and, in the latter end of his life, was much celebrated for his Chemical Experiments. He was long Clerk and Librarian to the Royal Society. He died on January 11th, 1763, aged 75.

communications to the Society, as highly deserving the honour. London, 14th April, 1763.

D. Wray.
Gowin Knight.
P. Newcome.
(*Signed*) Ja. Short.
Cha. Morton.
John Belchier.
Nic^s^. Munckley." [234]

The foregoing Certificate having been laid before the members of the Society at the close of their Session 1762-1763, it could not be entertained until the re-opening of the Session at the end of the year. Accordingly, at one of their earliest meetings at the end of the year 1763, the Certificate was again read, and before a full meeting of members, when on

"Novr 24, 1763, It was resolved by ballot unanimously, that Mr. James Ferguson of Mortimer Street is this day elected a Fellow of the Society on account of his singular merit, and of his circumstances, be excused the usual admission fee, and also the annual contributions."

The next and only other note on this election is from an entry in the books of the Royal Society, notifying that, on

"December 8th, 1763, Mr. James Ferguson admitted a member."

That is, this was the date of his first admission in person before the members, on which occasion, some one or other of his friends would, agreeably to form, present him to the President and Fellows.

FERGUSON versus KENNEDY'S "ASTRONOMICAL CHRONOLOGY." —Early in May, 1763, Ferguson finished his Courses of Lectures at Bath and Bristol, and returned to London after an absence of about two months. Shortly after his return, he appears to have had his attention called to that very absurd production, Kennedy's "COMPLETE SYSTEM OF ASTRONOMICAL CHRONOLOGY, UNFOLDING THE SCRIPTURES," then recently published.[235] The

[234] Five out of the above seven names appended to the certificate are unknown to us; but "*Ja. Short*" was the well-known Optician, and maker of "Short's Reflecting Telescopes," London. "*Cha. Morton*" is understood to be Dr. Morton, a Fellow of the Royal Society, and of the British Museum.

[235] The Reverend John Kennedy was Rector of Bradley, in Derbyshire. Shortly after Ferguson published his Astronomy, in 1756, he was introduced to Mr. Kennedy. Ferguson, in one of his printed letters to Kennedy, says that Kennedy,

Solar and Lunar periods on which, for a great part, the calculations of this work rests, are assumed, if not manufactured periods, and the *axioms* which guide its logic are absurd and ludicrous. For instance, Kennedy boldly asserts the following as axioms:—That no Astronomer but himself could tell the true length of 24 hours as measured either by the revolution of the Sun or of a Star,—That Dr. Keill has given sophistical directions concerning it,—That the precise mathematical difference between the sidereal and solar day is *four minutes*,[236]—That all solar days of 24 hours each are *precisely of equal length*,[237]—That all equations of time are *unastronomical*, and *ought to be rejected!* [238]—That all solar tropical years have been of the same *precise* mathematical length of 365 days 5 hours 49 minutes, ever since the Creation, and will continue to be to the end of time,[239]—That the synodical revolution of the Moon round the Earth consisted *precisely* of 29 days 12 hours 44 minutes 1 second 45 thirds,[240]—That the *first meridian* should cut 156

before he published his "Astronomical Chronology," "*was many times at my house, and sometimes disputed, or rather cavilled with me for hours together* (*for you never would be quiet*), *and often thought you had conquered me when you only confounded my head with flashes of noisy words, and would never hear what I had to say.*" (See Ferguson's 2d letter to Kennedy, page 29).

236 The true length of a sidereal day, which is always precisely of the same measure, consists of 23 h. 56 m. 4 s. 5 thirds 27 fourths, or decimally, 23 h. 56 m. 4·0908 s.

237 The *mean* length of a solar day (that is, taking the mean for all the year round), it will be found 24 hours. Mr. Kennedy appears to have been ignorant of the astronomical fact that the Sun, at particular seasons of the year, passes *twice* over the meridian in the 24 mean hours; at other times it does not pass over the meridian within the 24 hours at all. Sometimes the day exceeds 24 hours in length, sometimes it falls short of it.

238 When we find that the Earth's motion in its orbit is unequal, and the obliquity of the ecliptic so considerable, *equations* become absolutely essential to add to, or subtract from, the time shown by a well-regulated clock and a Sun-dial. We feel surprised at Mr. Kennedy having dogmatically made such an unastronomical assertion—it shows how little he knew of Physical and Practical Astronomy.

239 Mr. Kennedy here adopts as a truth, that the precise mathematical length of a solar or tropical year, from the Creation to the end of time, has been, and must be, 365 days 5 hours 49 minutes. Here again we have to refer to the unequal motion of the Earth in its orbit, as also to the Earth being acted on by several of the planets when they approach a conjunctive line with the Sun and Earth—these cause *perturbations*, and consequently, prevent the Earth from moving round the Sun so as to accomplish its circuit in *any exact period of time*. But taking the mean average of 250 revolutions of the Earth round the Sun, the mean length of one of these revolutions is 365 d. 5 h. 48 m. 51·6 s. Mr. Kennedy's assumed period of 365 d. 5 h. 49 m. is therefore very absurd, and used in a calculation for even limited epochs, would lead to serious errors.

240 Mr. Kennedy assumes the synodical revolution of the Moon to consist of 29 d. 12 h. 44 m. 1 s. 45 thds., but he does not inform us from whence he found this measure. There can be no doubt that it is some *mean revolution* he had

degrees west of Greenwich, in the Great South Sea; and that the Moon was created on the *third* day of the Creation week at farthest, &c.

On these, or several of these points, Ferguson resolved to address Kennedy through the medium of the *Critical Review.* Accordingly, in May and June, 1763, Ferguson's letter of remarks on Kennedy's "Astronomical Chronology," appeared in the "CRITICAL REVIEW;" being a long letter, part of it was inserted in May, the concluding portion of it in June. Instead of waiting until the concluding portion of Ferguson's letter was published (which the editor assured his readers would be done in next Review for June), Kennedy rushed into print (regardless of what the concluding part of Ferguson's letter would touch on), and published a quarto pamphlet of 20 pp., entitled, "AN EXAMINATION OF MR. FERGUSON'S REMARKS (*inserted in the Critical Review for May,* 1763) UPON MR. KENNEDY'S SYSTEM of ASTRONOMICAL CHRONOLOGY. By JOHN KENNEDY." This work of Kennedy's is no "Examination" at all; after a few introductory remarks, he asks Ferguson not simply one, two, or three queries, but puts no less than "XXIII Queries" to Ferguson, waiting an answer to them; and concludes with a few arithmetical questions which he proposes and answers as favourable to his assumed or manufactured periods and their results.

We may here notice that Ferguson, in his second letter to Kennedy (pp. 6, 7), says to Kennedy, in reference to the letter he had sent to the Critical Review,

"You know that I laid the whole of my remarks before you in manuscript before I gave them to be printed, and submitted them to your perusal, promising at same time to alter any part thereof in which you might think yourself misrepresented; so that you might think me at least a fair opponent. But you returned them immediately, telling me that you would not read them till they were in print. To show still as great a degree of

adopted, or a manufactured period, to answer some result. The true mean synodical revolution of the Moon, by Meyer's Lunar Tables, consists of 29 d. 12 h. 44 m. 2 s. 53 thds. 25 fourths 23 fifths 25 sixths. Ferguson, in all his lunar calculations, adopted the period given by Dr. Pound,—viz. 29 d. 12 h. 44 m. 3 s. 2 thds. 58 fourths, which is a period about 10 thirds of time *too slow,* while that by Kennedy is about 1¼ s. too fast. The period, as given by Meyer, is still recognised as the correct period of a lunation, excepting in a small correction recently made, and the mean synodic revolution of the Moon is now acknowledged as consisting of 29 d. 12 h. 44 m. 2 s. 52 thds.

fairness as possible, I ordered my name to be put to those remarks in the Critical Review, a thing seldom, if ever, done by those who send a paper to be printed in these monthly productions.

"Had you not," continues Ferguson, "even without the least degree of modesty in your voluminous system, fallen so unmercifully upon all those who have written on Astronomy before you,—Had you not (to use your own words, for I cannot find better) 'rashly impeached the veracity of all our solar and lunar tables,' and imposed your own calculations upon the world with such a high degree of assurance and magisterial sufficiency,—And had you not, with the greatest degree of boldness, asserted things to be true, which the merest novice in Astronomy knows to be false, your *system* might have slept in silence and oblivion without any disturbance from me."

Thus, we see from the above that Ferguson sent his manuscript remarks to Kennedy before they went to the Critical Review, and promised to alter anything in which Kennedy might fancy himself misrepresented. But instead of receiving them in a generous spirit, Kennedy indignantly returns the manuscript to Ferguson, refusing to read them until they were in print. We also see what caused Ferguson to send his "*Remarks*" on Kennedy's "*Astronomical Chronology*" to the Critical Review—a very proper motive and excuse.

As a matter of course, Ferguson declined Kennedy's very modest request,—viz. to answer all the 23 queries he had put to him. But in November of the same year (1763), Ferguson sent another letter to the Critical Review, "*in order*," he tells us, "*to avoid all further disputes between us about Astronomical matters.*" Kennedy took no notice of this; and nothing more of this dispute, or of Kennedy, is heard of until the end of year 1775, when he re-appears as the opponent of Ferguson, to which date the reader is referred.

"ASTRONOMICAL TABLES AND PRECEPTS.—About mid-summer of 1763, Ferguson published his "ASTRONOMICAL TABLES AND PRECEPTS," an octavo pamphlet of 64 pages, embellished with an illustrative frontispiece folding-plate, *On the Geometrical Construction of Solar and Lunar Eclipses.* Like many of Ferguson's pamphlets, this one is now rarely met with. We kept inquiring for it in many of the old book shops in London for several years before we succeeded in getting a copy. The annexed is a copy of Title-page :—

"Astronomical Tables and Precepts for Calculating the true Times of New and Full Moons, and showing the method of projecting Eclipses, from the Creation of the World to A.D. 7800, to which is prefixed a Short Theory of the Solar and Lunar Motions. By James Ferguson. London: Printed for the author. MDCCLXIII."

REMARKABLE FISH CAUGHT near BRISTOL—Ferguson inspects it.—During Ferguson's sojourn in Bristol, a very singular fish was caught in King-Road—(the Long Angler of Pennant, or Sophius Conubicus of Shaw)—the fish was brought to Hot-Wells, near Bristol, and exhibited. Ferguson went to see the fish,—when he made a drawing of it, which, along with a short description, he sent to his friend, the Rev. Dr. Birch, Secretary of the Royal Society, who read the paper to the members on June 2d. On referring to the Philosophical Transactions for May, 1763, we find an engraving and description of this fish;—the paper is entitled,

"An account of a Remarkable Fish taken in the King's Road near Bristol." Read before the Royal Society, 2d June, 1763. (See Phil. Trans. Vol. 53, page 170);

Ferguson's letter to Dr. Birch is dated "Bristol, May 5th, 1763."

FERGUSON APPRENTICES his ELDEST SON to a LONDON OPTICIAN.—In the month of June, 1763, Ferguson apprenticed his eldest son, James (now in his 15th year), to Mr. Edward Nairne, Optician and Philosophical and Mathematical Instrument Maker, Cornhill, London, "*to learn the practical departments of said business in its different branches.*"

"ANALYSIS of a COURSE of LECTURES."—During his temporary lecturing sojourn at Bristol, Ferguson found time to write out and publish, at a Bristol Press, an Analysis of the Course of Lectures he was then engaged upon, under the following title :—

"Analysis of a Course of Lectures on Mechanics, Pneumatics, Hydrostatics, Spherics, and Astronomy, read by James Ferguson. Bristol: Printed by S. Farley, in the Castle Green, 1763. Price Sixpence."

It is now an exceedingly rare tract, a copy of which we had much difficulty in procuring. It is an octavo of 44 pages, containing full notes of his Lectures on Mechanics, Pneumatics, Hydrostatics, Spherics, and Astronomy. Like his other works, it is written in plain language, and is very full in all its details.

DOMESTIC CALAMITY—MYSTERIOUS DISAPPEARANCE of FERGUSON'S DAUGHTER.—In the midst of these his multifarious pursuits, in the height of his prosperity and renown, the hand of adversity fell heavily and suddenly upon Ferguson. Unlike those afflictions which time smooths or eradicates—*his* was destined to be life-enduring—a never-ending grief to himself, his wife, and family, by the sudden and mysterious disappearance of his only daughter, AGNES FERGUSON, an elegant and accomplished lady, in her eighteenth year.

It is now upwards of forty years since we first read in "Nichol's life of Bowyer," that Ferguson's "*only daughter was lost in a very mysterious and singular manner*," but without saying *how lost, when, or where.* Shortly afterwards, we saw Partington's Ferguson's Mechanics (Lond. 1825), and read his short addenda to Ferguson's Life, but it threw no new light on the matter, being simply a condensed extract from Bowyer. Partington merely says, "*His daughter, an elegant and accomplished young lady, suddenly disappeared.*" Bowyer was one of Ferguson's most intimate friends; a frequent visitor at his house, and often formed part of the family circle: he would thus often see his daughter, converse with her on philosophical subjects; and therefore, from personal observation, be enabled to give a true sketch, and to say that "*she was remarkable for the elegance of her person, the agreeableness and vivacity of her conversation, and in philosophical knowledge worthy of such a father.*"

Being on intimate terms of friendship with the family, Bowyer would learn from Ferguson himself all that he knew of the disappearance of his daughter;—all that Ferguson and his family knew of the sad calamity would be by them told to Bowyer; and all that Bowyer in his turn can tell is, that "*she was lost in a very mysterious and singular manner.*" It must be observed that this notice is obscure. Bowyer says, "*she was lost*," &c.; from this, many have been led to conjecture that "*she was lost somewhere abroad;*" while others again had heard that "*she had been lost on the streets of London.*" Surely Bowyer must have known *when* and *where* and *under what circumstances* she was last seen. Had Bowyer communicated these particulars, which he easily could have done, it would have prevented the sad calamity from being afterwards covered with so much mystery, conjecture, and surmise.

In order, if possible, to clear up this mysterious affair, we lately resolved on a thorough investigation of the matter. In commencing with researches we found our way obstructed with many difficulties; for instance, to narrow our period of search, it was necessary to ascertain the year, or as near to the year of the *disappearance* as was possible; or what would have answered the same purpose, to obtain a copy of *date of the daughter's birth.* After correspondence on the subject, and also much tedious research, we found that neither the year of *the disappearance,* nor *the year of her birth,* could be then ascertained.

We had in consequence to fall back on what was known, and therefrom deduce the limits of our period of search,—viz. Ferguson, in his Memoir, mentions that he was married in May, 1739; therefore, by adding 19 years to this date, we get at 1758 for the earliest probable date of *the disappearance;* and as the occurrence took place in Ferguson's lifetime, we can readily ascertain the limit for our research. Ferguson died in November, 1776; consequently, between 1758 and 1776, *the disappearance* occurred, and between these years we directed our search. Shortly afterwards it occurred to us that even these limits might be much narrowed. We recollected that Ferguson, in one of his interviews with Mr. Poyntz regarding a mathematical school appointment, mentioned to him that he did not know "how to maintain his wife during the time he must be under the master's tuition." This was in the summer of 1743, and if Ferguson had then any family he would have mentioned the circumstance to Mr. Poyntz; as he did not, the reasonable inference appeared to be that Ferguson *then,* viz. in 1743, had *no children;* and as all accounts of *the disappearance* of the daughter concurred in placing it "*some years before Ferguson's death,*" in 1776, we deducted six years from 1776, which gave 1770 for the new limit of search; and by adding the daughter's age, 18, to 1743, we obtained 1761 as the proper date to commence with; we had therefore to direct our attention to a thorough search between the years 1761 and 1770, a period of ten years; and subsequent discovery showed the correctness of our views of these limits.

We had but got on a short way in our inquiries when we received from our indefatigable friend, Mr. Robert Sim of Keith, a letter and extract of a remarkable occurrence which seemed to *solve the mystery*, and as it was within the limits of 1761-1770, confirmed us in this opinion. It had sometimes been said, traditionally, that Ferguson's daughter had been kidnapped and taken to some foreign country; and this theory was somewhat strengthened by the outrageous abduction-case extract sent us by Mr. Sim, as follows:—

"ABDUCTION.

This offence was by no means uncommon in England some years ago. In the London Chronicle for 1762 there is an extract from a letter dated *Sunday, Highgate, June 6th*, from which it appears that on that morning, between twelve and one, a post-chaise, in which was a lady, was driven through that place very furiously by two postilions, and attended by three persons who had the appearance of gentlemen, from which she cried out, *Murder! Save me! Oh save me!* Her voice subsided from weakness into faint efforts of the same cries of distress; but as there was at that time no possibility of relief, they hastily drove towards Finchley Common. From another quarter, says the London Chronicle, 'we have undoubted intelligence of the same carriage being seen, and the same outcries heard as it passed through Islington, with the additional circumstance of the two postilions being in their shirts. Is this outrage to be suffered in England?'"

Extracted from "Hone's Every-Day Book, &c." Vol. i., pp. 767, 768.
London, 1762.

We had now no doubt that Ferguson's daughter had been subjected to an atrociously forcible abduction, and that "*the mysterious disappearance*" was now solved, and had therefore written down the date of the event as having occurred on June 6th or 7th, 1762; and consequently suspended farther inquiry.

Some short time afterwards, however, we received another letter from our friend, Mr. Sim, mentioning, that in a pocket

Bible, which had formerly belonged to Ferguson,[241] he had found, in Ferguson's own handwriting, on one of the fly-leaves, an entry of all the births of his children, and there it is recorded that AGNES, the lost daughter, was *born on Thursday, 29th August,* 1745. By adding 18 years to which date, we get 1763 as the year of her disappearance; and therefore, the abduction case just quoted had no reference to Ferguson's lost daughter. Bowyer mentions that she was in her 18th year when lost, and we have the testimony of the lost one herself that she was not quite eighteen years of age when she *disappeared.* See the following account of her by Dr. Blake.

FERGUSON'S DAUGHTER'S LIFE, AFTER HER DISAPPEARANCE; and HER MISERABLE DEATH.—Having discovered that 1763 was the date of Agnes Ferguson's disappearance, we wrote to Mr. Augustus Burt, London, requesting that a search be made in the London newspapers and magazines for 1763, in the British Museum, regarding the occurrence. After a very tedious examination, he could find nothing relating to the event. Some considerable length of time after this, Mr. Burt discovered in the British Museum, No. $978\frac{c}{11}$ a volume of tracts of the "London Corresponding Society," in which is included "The Female Jockey Club," which belonged to Dr. Blake, a Surgeon in St. Martin's Lane, London,—towards the end of last century. Dr. Blake, who attended Miss Ferguson in her last illness, has written on a blank leaf at the end of "The Female Jockey Club" tract, a short account of the fate and sad end of this unfortunate lady. From which, it will be seen that she had been forcibly abducted—was afterwards taken to the continent—subsequently returned to England—was sometime in the Fleet Prison for debt, and at last, on January 27th, 1792, she ended her sad career in a garret, amid squalid poverty, in Old Round Court,—a Court off the Strand, near Charing Cross—(destroyed in 1823-1826 to make way for new improvements then in

[241] By the discovery of this Bible we have been enabled to record the dates of birth of the whole of Ferguson's children, which the reader will find in their proper places.

progress). The following is a copy of Dr. Blake's memorandum entry:—

"Some time since I was call^{d} to attend a dying female in Old Round Court, whom I found in the last stage of consumption; prior some days to her dissolution, she told me her sad tale. Her name was Ferguson; years before, she had been inveigled from her father's house by gentlemen whom she had often seen at the Lectures of her father, who proved to be none other than James Ferguson the eminent astronomer and mathematician. She thus was launched out into this vast world before she attained the age of eighteen,—leaving home, relations and every tie, for this nobleman, who at first, devoted all his attention to her, and took her to Italy, where she became a star of attraction, but her lover being prone to pleasure himself,—pecuniary difficulties cast a gloom over her apparent prosperity, until the resources were completely stopped,—they returned to England, and she to maintain her pretensions wrote sonnets, odes, elegies, and other small works indiscriminately.

"The noble Lord left her, then came her time of toil,—she was too proud to return to those she had disgraced, she could not appear to retain the least remembrance of her earlier days, —though not totally debauched by the flatteries and vanities of a deceitful world, yet disabled and no longer qualified to appear in her former sphere, the spectacles of London were frequented by her. Nature had been liberal towards her, and she was conscious of her attractions,—there she gave full sway to her powers, till the cruel unrelenting arm of the law stopt her career and drove her to the Fleet Prison;—Here she continued in an abject state of poverty until one who finds the wretched out liberated her from her confinement,—she then forgot the dreadful scenes she had passed, and again ventured on the theatre of boundless dissipation, and thus again became involved in debt; yet her personal claims were indisputable, and conscious of which she applied and obtained permission of Mr. Garrick to appear at the Theatre.—Various characters were attempted, but all her efforts were faint and languid,—but if she failed to please, she never disgusted; the attendance and fatigue inseparable from a theatrical profession were ill suited to a system of the wildest dissipation, and so was dissolved her engagement,—from thence she declined, one step below the other, until when I was called in to the miserable room in Round Court, and found the deplorable remains or ruins of that beauteous symmetry which must have charmed all beholders,—to serve only as a sad memorial of the fatal effects of inordinate excesses;—let us hope, one salutary caution to her sex.

"This poor creature, after months of agony, died 27th of Jany. 1792." (See vol. Tracts, British Museum, No. 978$\frac{c}{11}$).

Thus, the mysterious disappearance of Ferguson's accomplished and only daughter, is now, after a lapse of nearly a hundred years, satisfactorily cleared up. Although a most sad and melancholy tale, it is yet satisfactory to know the whole history and mystery of the affair.

Dr. Blake shows that Ferguson's daughter had led a disreputable life in London, &c., from the close of the year 1763 till near the end of the year 1791, a period of nearly 28 years. As far as is now known, she lived almost the whole of this long period in London, and would probably now and then see her father, her mother, and brothers on the crowded streets, when following out her sad career, and known by another name. During this long period of 28 years, her father and mother and her eldest brother had died. An esteemed correspondent writes—"This may suggest the following but now unanswerable questions,—viz. Did she ever, or did she often see, at a distance, her parents or her brothers, whilst walking through the streets of London? did she, when her parents and brother died, at the time hear of their deaths? did she, afar off, follow their remains to the tomb? did she ever, in after days, enter the churchyard of St. Marybone where they lay, and drop the tears of repentant sorrow over their graves?"

We shall conclude our notice of Agnes Ferguson with a few additional notes sent to us by Mrs. Casborne, to whom we are much indebted for several memorabilia which she has from time to time sent us. We may here mention that Mrs. Casborne is the daughter of Ferguson's "Eudosia." Ferguson taught Mrs. Casborne's mother, when Miss Emblin, and early in her teens, the science of Astronomy; her lessons in that science were afterwards in a volume, entitled, The Young Gentlemen and Ladies Astronomy, wherein the elements of Astronomy are discussed by dialogues—Ferguson taking the cognomen "*Neander*," and giving Miss Emblin that of "*Eudosia.*" Mrs. Casborne, the daughter of "Eudosia," informs us that her mother often spoke of Ferguson, but was too young, when her mother died, to remember particulars. Mrs. Casborne's aunt, the sister of Eudosia, who died at an advanced age in 1839, used frequently to allude to the disappearance of Ferguson's daughter. Mrs. Casborne's letter, of date July, 1861, in reference to Ferguson's daughter, says,—"*My mother's sister, who has been only dead 22 years, has often mentioned to me the sudden disappearance of Ferguson's daughter, with whom he was walking, I believe, in the Strand,—Ferguson being occupied in some calculations did not perceive when his daughter withdrew her arm; and when he did, he thought she had gone home to arrange some household affairs, and returned home in that idea, and expected her for days and weeks, but she never returned—he never saw her more.*

"*The date of this occurrence I do not know,—some circumstances attending the death of a young lady, a few years afterwards, and dying in London upon her arrival from the West Indies, led Ferguson to believe that she might be his daughter; but nothing definite was ever known.*"

It appears that Ferguson, somewhere about 1770, had heard some tidings of his lost daughter being in London. He, and his two sons, James and Murdoch, made some search or inquiries after her about that period, but without success. We find that Mrs. Casborne makes mention of this search in one of her letters to us, and think it too probable that the daughter lived under an assumed name.

It is a remarkable circumstance, that the daughter parted from her father for the last time while in the Strand, and that in about 28 years thereafter she died in an obscure garret leading off the same street.

At the time when the daughter disappeared, Ferguson and family had their residence in Mortimer Street, Cavendish Square, London.

PAPER on the GREAT SOLAR ECLIPSE of APRIL 1st, 1764. —The only other note which we have for 1763 refers to a paper he drew up regarding a great Solar Eclipse which was to take place on 1st April, 1764. This paper, with a delineation of the Eclipse accompanying it, is dated "Mortimer Street, Novr. 16th, 1763," and was read before the members of the Royal Society on November 17th. The following is the title of this paper and projection, taken from the Books of the Royal Society:—

> "Account of an Eclipse of the Sun, 1st April, 1764." Read before the Royal Society, 17th November, 1763. (Phil. Trans., Vol. 53, page 240).

1764.

A New Crane invented by Ferguson.—Our first note for 1764 refers to a "New and Safe Crane" having "Four different Powers," a description of which, accompanied by a large drawing of it, was sent to the Royal Society early in January, 1764. The following is the title of the paper in Philosophical Transactions:—

> "Description of a New and Safe Crane, which has Four Different Powers." Read before the Royal Society, 19th January, 1764. (See Phil. Trans., Vol. 54, page 24; also, see "The Universal Magazine" for Feb. 1766, which has a full description of it, and an engraving).

The following description of the New and Safe Crane is taken from Supplement to his Lectures on Select Subjects, and the annexed cut is reduced from the drawing that accompanied the paper:—

"The Description of a New and Safe Crane, which has four different Powers, adapted to different Weights.

"The common Crane consists only of a large wheel and axle; and the rope, by which goods are drawn from ships, or let down from the quay to them, winds or coils round by the axle, as the axle is turned by men walking in the wheel. But, as these engines have nothing to stop the weight from running down, if any of the men happen to trip or fall into the wheel, the weight descends, and turns the wheel rapidly backward, and tosses the men violently about within it, which has produced melancholy instances, not only of limbs broke, but even of lives lost, by this ill-judged construction of cranes. And besides, they have but one power for all sorts of weights; so that they generally spend as much time in raising a small weight as in raising a great one.

"These imperfections and dangers induced me to think of a methòd of remedying them. And for that purpose, I contrived a Crane with a proper stop to prevent the danger, and with different powers suited to different weights; so that there might

Ferguson's Crane.

be as little loss of time as possible; and also, that when heavy goods are let down into ships, the descent may be regular and deliberate.

"This Crane has four different powers: and, I believe, it might be built in a room eight feet in width: the gib being on the outside of the room.

"Three trundles, with different numbers of staves, are applied to the cogs of a horizontal wheel with an upright axle, and the rope, that draws up the weight, coils round the axle. The wheel has 96 cogs, the largest trundle 24 staves, the next largest 12, and the smallest has 6. So that the largest trundle makes 4 revolutions for one revolution of the wheel; the next 8, and the smallest 16. A winch is occasionally put upon the axis of either of these trundles for turning it; the trundle being then used that gives a power best suited to the weight; and the handle of the winch describes a circle in every revolution equal to twice the circumference of the axle of the wheel. So that the length of the winch doubles the power gained by each trundle.

As the power gained by any machine, or engine whatever, is in direct proportion as the velocity of the power is to the velocity of the weight; the powers of this Crane are easily estimated, and they are as follows:

If the winch be put upon the axle of the largest trundle, and turned four times round, the wheel and axle will be turned once round; and the circle described by the power that turns the winch, being in each revolution double the circumference of the axle, when the thickness of the rope is added thereto, the power goes through eight times as much space as the weight rises through; and therefore, (making allowance for friction) a man will raise eight times as much weight by the crane as he would by his natural strength without it; the power, in this case, being as eight to one.

If the winch be put upon the axis of the next trundle, the power will be as sixteen to one, because it moves 16 times as fast as the weight moves.

If the winch be put upon the axis of the smallest trundle, and turned round, the power will be as 32 to one.

But, if the weight should be too great, even for this power to raise, the power may be doubled by drawing up the weight by one of the parts of a double rope, going under a pulley in the moveable block which is hooked to the weight below the arm of the gib; and then the power will be as 64 to one. That is, a man could then raise 64 times as much weight by the crane as he could raise by his natural strength without it; because, for every inch that the weight rises, the working power will move through 64 inches.

By hanging a block with two pulleys to the arm of the gib, and having two pulleys in the moveable block that rises with the weight, the rope being doubled over and under these pulleys, the power of the crane will be as 128 to one. And so, by increasing the number of pulleys, the power may be increased as much as you please: always remembering, that the larger the pulleys are, the less is their friction.

Whilst the weight is drawing up, the ratchet teeth of a wheel slip round below a catch or click that falls successively into them, and so hinders the crane from turning backward, and detains the weight in any part of its ascent, if the man who works at the winch should accidentally happen to quit his hold, or choose to rest himself before the weight be quite drawn up.

In order to let down a weight, a man pulls down one end of a lever of the second kind, which lifts the catch out of the ratchet-wheel, and gives the weight liberty to descend. But, if the descent be too quick, he pulls the lever a little farther down, so as to make it rub against the outer edge of a round wheel; by which means, he lets down the weight as slowly as he pleases: and, by pulling a little harder, he may stop the weight if needful, in any part of its descent. If he accidentally quits hold of the lever, the catch immediately falls, and stops both the weight and the whole machine.

This Crane is represented in the annexed cut, where A is the great wheel, and B its axle on which the rope C winds. This rope goes over a pulley D in the end of the arm of the gib E, and draws up the weight F, as the winch G is turned round. H is the largest trundle, I the next, and K is the axis of the smallest trundle, which is supposed to be hid from view by the upright supporter L. A trundle M is turned by the great wheel, and on the axis of this trundle is fixed the ratchet-wheel N, into the teeth of which the catch O falls. P is the lever, from which goes a rope Q Q, over a pulley R to the catch, one end of the rope being fixed to the lever, and the other end to the catch. S is an elastic bar of wood, one end of which is screwed to the floor, and, from the other end goes a rope (out of sight in the figure) to the further end of the lever, beyond the pin or axis on which it turns in the upright supporter T. The use of this bar is to keep up the lever from rubbing against the edge of the wheel U, and to let the catch keep in the teeth of the ratchet-wheel. But, a weight hung on the farther end of the lever would do full as well as the elastic bar and rope.

When the lever is pulled down, it lifts the catch out of the ratchet-wheel, by means of the rope Q Q, and gives the weight F liberty to descend. But if the lever P be pulled a little farther down than what is sufficient to lift the catch O out of the ratchet-wheel N, it will rub against the edge of the wheel U, and thereby hinder the too quick descent of the weight; and will quite stop the weight if pulled hard. And if the man, who pulls the lever, should happen inadvertently to let it go, the elastic bar will suddenly pull it up, and the catch will fall down and stop the machine.

W W are two upright rollers, above the axis or upper gudgeon of the gib E: their use is to let the rope C bend upon them, as the gib is turned to either side, in order to bring the weight over the place where it is intended to be let down.

N.B.—The rollers ought to be so placed, that if the rope C be stretched close by their outmost sides, the half thickness of the rope may be perpendicularly over the centre of the upper gudgeon of the gib. For then, and in no other position of the rollers, the length of the rope between the pulley in the gib and the axle of the great wheel will be always the same in all positions of the gib; and the gib will remain in any position to which it is turned.

When either of the trundles is not turned by the winch in working the crane, it may be drawn off from the wheel, after the pin near the axis of the trundle is drawn out, and the thick piece of wood is raised a little behind the outward supporter of the axis of the trundle. But this is not material: for, as the trundle has no friction on its axis but what is occasioned by its weight, it will be turned by the wheel without any sensible resistance in working the crane." ("*A Supplement to Mr. Ferguson's Book of Lectures on Mechanics, Hydrostatics, Pneumatics, and Optics.*" London, 1767, pp. 3—11).

Paper on the Great Solar Eclipse of 1764.—At one of their meetings, early in 1764, the Royal Society agreed to print Ferguson's paper on the Great Solar Eclipse (read on 17th November of the preceding year), for which purpose, it was remitted to Mr. Bowyer, printer to the Society. Dr. Birch, the Society's Secretary, and friend of Ferguson, appears to have either seen or written to Ferguson regarding the matter, which brings from him the following letter to Dr. Birch:—

"Reverend Sir,—I called at Mr. Bowyer's, according to your directions, last Tuesday morning, but he was not at home; however, I saw his Compositor, who told me that he had only received the enclosed paper, which gives a general account of the periodical returns of the Sun's Eclipse, April 1st, 1764; but had not got the Table therewith which I gave in to the Royal Society.

"On looking over the paper again, I think the general account given therein might do without the Table. But if the Royal

Society has thought that the Table should likewise be printed, I imagined it might be proper to divide the Table into four quarto pages, which perhaps the printer might be at some loss to do,—and therefore, as it was given all on one great page, and the paper refers only to the particular columns thereof, I have drawn lines with a black-lead pencil under those words in the paper, which are to be altered according to the words standing against them in an additional paper which I have fastened to the margin. So that if the Royal Society pleases to print the Table on four pages, as I herewith send it you divided, you'll be so good as to dash out those words in the original paper under which the black-lead lines are drawn; and then the printer will understand, that instead of them, he is to take in the words on the margin; and, although I hope the transmitted copy of the Table on four pages is correct, yet I could wish, that the Compositor had the original one before him, to go by the numbers therein.

"But if the Royal Society think proper to print only the general account in the paper, without the Table, then, I beg you will tear off the additional paper on the margin, and rub out the above-mentioned black-lead lines against it in the paper, in the second page whereof I find that some gentleman has put in a very proper correction,—viz. A.D. 2665, which I had omitted, and for which I am really obliged to him.—I am, with great respect,

Reverend Sir,

Your most obliged humble servant,

MORTIMER STREET, *Feb.* 16*th*, 1764. JAMES FERGUSON."

"The Compositor told me that he could spare the paper for a fortnight."

The original of this letter is among the "*Birch Letters*," British Museum, No. 4308.

FERGUSON at LIVERPOOL—LECTURES there, and OBSERVES ECLIPSES in MARCH and APRIL.—Early in the month of March, 1764, Ferguson left London for Liverpool to deliver his course of Lectures on Astronomy, Mechanics, Hydrostatics, Pneumatics, Optics, &c. His lectures were delivered in the large room of the Golden Lion Inn, Dale Street, commonly called "*Buck's*

Rooms;" and during his month's sojourn, he resided with his friend Captain Hutchinson, Dock-master.[242]

The Great Solar Eclipse of April 1st took place during Ferguson's stay in Liverpool. We shall give his own account of it, as observed by him in Liverpool, in the following long and interesting letter to his friend, the Reverend Dr. Birch, Secretary, Royal Society:—

"LIVERPOOL, *April 2d*, 1764.

"REVEREND SIR,—Having been in this Town since the beginning of March, and hoping that the sky would prove favourable (as to my great joy it did) for observing both the Lunar Eclipse of March the 17th and the Solar Eclipse of yesterday, I proposed to Captain Hutchinson, at whose house I stay, to have a meridian line drawn on the leads, on the top of his house, in order to adjust his clock for observing the times of these Eclipses by,—and we got Mr. Holden, who is master of a mathematical school here, to do it for us, by several observations of the altitude and azimuth of the Sun by day, and of the Stars by night, and there were such exact agreements found by many repeated observations, that no doubt could remain of the meridian being very well ascertained. The same gentleman, who is justly esteemed to be a very accurate observer, and an able calculator, finds the latitude of Liverpool to be 53° 22′, and its longitude is generally thought to be three degrees west of Greenwich,—but he believes to be somewhat less.[243]

242 Captain William Hutchinson was a native of Newcastle-on-Tyne. In 1750 he had the command of the Government Frigate Leostaffe. In 1758 he invented mirrors for Lighthouses. In 1760 he was appointed Dock-master of Liverpool; and in same year, fitted up Bidstone Lighthouse with Reflectors—the first ever used—(Bidstone, in Cheshire, is about 3½ miles NW of Liverpool). From the 1st January, 1768, to 18th August, 1793, he kept a Daily Register of the Tides, Barometer, Weather, and Winds, at Liverpool. The Tidal Register was used by Mr. Richard Holden, teacher in Liverpool, and "aided him in making out the 3000 observations mentioned in the preface to his Tide-Table. In 1794 he published "*A Treatise on Naval Architecture.*" He died at an advanced age on 11th February, 1801, after having been 41 years Dock and Harbour Master at Liverpool, and was buried there, in St. Thomas's Churchyard. Mr. John Perris, present librarian Lyceum Library, Liverpool, informs us, that Captain Hutchinson's Tidal Tables and Calculations are deposited in the Lyceum Library. Mr. Perris also mentions that the Captain retired from the situation of Dock and Harbour Master in 1793, and ceased to make observations in 1796. In Gore's Liverpool Directory, for 1776, we find the Captain's residence,—viz. "William Hutchinson, near the Dock Gates, at No. 1 Old Dock," (near the foot of South Castle Street, *then* called *Pool Lane*).

243 Mr. Hartnup, Astronomer, Observatory, Liverpool, from the most accurately-tested observations, &c., makes his Observatory to be in 53° 24′ 48″ N, and in 3° 0′ 1″ W of Greenwich. If, therefore, Captain Hutchinson's house was

The clock being duly adjusted by our meridian line, at noon, and the time being found by observations of several stars in the evening of March 17th, the apparent time of the beginning of the Moon's eclipse was observed to be at 10ho. 27′ P.M., and the end at 13ho. 11′.

On the next day I calculated the time of the ecliptic conjunction of the Sun and Moon for April 1st, by Meyer's Tables, as we have them published by Mr. Maskelyne, and then made a projection of the Sun's eclipse for that time by them, for this place, according to its latitude, as determined by Mr. Holden, and supposing its longitude to be 3 degrees west from Greenwich, and put up this projection in the council-room, that it might be seen, in order to find how it might agree with observation.

Being provided with a good reflecting telescope at Captain Hutchinson's, I cut a round hole in a paste-board which would go tight on the tube, and took the Sun's image on a paper behind it, as large as I could have the image of the Sun, sharp and well defined around the edge, which was included in a circle of 4 inches in diameter. I divided the diameter into 12 equal parts, for digits, and each digit into 4 parts, the half of every fourth part being left to be estimated by the eye.

Mr. Holden, and two other gentlemen, who are esteemed good observers, and were provided with refracting telescopes, and Hadley's quadrants, were with me on Sunday morning, and I desired a third gentleman to note down the times, and to be careful not to mistake the minutes of time, as one might be more apt perhaps to mistake the minutes than the seconds. The clouds threatened us disappointment till about ten minutes before the calculated time of the beginning of the Eclipse, and then the Sun shone out very clear, and during the time of observation we were but seldom interrupted by thin flying clouds.

The first and last contacts of the Moon and Sun were so sharp and instantaneous, that it seemed possible to determine them within one second of time,—several altitudes of the Sun were taken during the Eclipse by reflecting the Sun's image from a

at No. 1 Old Dock, Liverpool, the latitude given by Mr. Holden has an error of at least 1′ 21″ when the Liverpool Observatory is reduced to the site No. 1 Old Dock, the true latitude of which is 52° 23′ 21″. If the Captain's private residence was in the southern extremity of the suburb of Toxteth Park, then the latitude given by Holden might be about correct.

basin of treacle, and the quantities eclipsed were plainly visible on the fore-mentioned image of the Sun on the paper, even to the eighth part of a digit. But the altitudes want yet to be corrected by their respective refractions.

Several people came into the room to see the Eclipse, some of whom were subscribers to my lectures, and I could not well refuse them admittance. But I told them before-hand that they must neither speak nor move till the Eclipse was found to be begun,—this they strictly complied with, and gave no manner of disturbance, and, after it was begun, I desired them all to come and view it by the telescope, which hindered me from observing the number of digits eclipsed for the first hour.

I kept by the reflecting telescope, and watched the Sun carefully for about five minutes before the calculated time of the beginning of the Eclipse. Our watches were adjusted to the mean or equal time, and two of them kept exactly alike during the whole time of the Eclipse,—the Observations were as follows:—

h.	′	″	Digits.	Sun's Altitude.		
8	59	0	Eclipse Begun.	28°	37′	00″
10	2	0	8½			
10	5	0	9	35	49	30
10	11	0	9¾			
10	13	0	10	36	29	0
10	18	0	10⅜			
10	21	0	10½	37	0	0 uncertain.
10	25	0	10½	37	20	0
10	30	0	10			
10	38	0	9			
10	40	30	Cusps perpendicular by a plumb-line's shadow on the Sun's image. Sun's altitude then,	38	48	0
10	43	30	8½	38	57	30
10	47	0	8	39	8	30
10	54	30	7			
11	0	0	6½			
11	3	45	6			
11	12	0	5			
11	15	45	4½			
11	19	15	4			
11	28	0	3 uncertain, on account of a thin flying cloud.			
11	35	2	2			
11	45	15	0½ uncertain, by another cloud.			
11	50	45	Eclipse ended, Sky quite clear—Sun's Altitude	47°	27′	0″,

all wrote down by Mr. Baxtonden, who kept a copy thereof. At night, Mr. Holden returned and examined the clock by the Stars, and found the time shown by the clock to be true.

Between the beginning and the middle of the Eclipse, we could plainly perceive inequalities in the Moon's eastern limb on the Sun, by means of the reflecting telescope, and I often observed little tremulous bright specks of the Sun's lowermost cusp, but they vanished in an instant, except one, which was considerably larger than any of the rest, and was visible for about two or three seconds of time by estimation;—but I was so intent upon observing it, and looking for others, that I forgot to have the time of its appearance marked down. This undoubtedly was owing to a dent or valley in that part of the limb of the Moon which no hill beyond it took off from the sight.

But as the Eclipse was drawing toward the end, we could perceive no inequalities of the Moon's western limb on the Sun, nor any such specks in the Sun's edge about either of the cusps.[244]

As the Moon's latitude was north ascending, and the cusps not perpendicular till after the middle of the Eclipse, I apprehend that when they were so, the present altitudes of the centres of the Sun and Moon were equal. But whether they were then so or not I leave to better judgments to determine.

I shall now set down the times of the beginning, middle, and

244 Probably these "*little tremulous specks*" were no *other* than what is now known as "*Baily's Beads;*" if so, and this the first record of them, "*Ferguson's Beads*" might be substituted for that of Baily's. The following is an account of a similar phenomenon, partly produced by artificial means, observed by Martin, who says,

"I cannot here omit to mention a very *unusual Phœnomenon* that I observed about ten years ago (1737) in my darkened room. The window looked toward the west, and the spire of Chichester Cathedral was directly before me at the distance of about 50 or 60 yards. I used very often to divert myself in observing the pleasant manner in which the Sun passed behind the spire, and was eclipsed by it for some time; for the image of the spire and Sun were very large, being made by a lens of 12 feet focal distance. And once as I observed the occultation of the Sun behind the spire, just as the disk disappeared, I saw several small bright round bodies or balls running towards the Sun from the dark part of the room, even at the distance of 20 inches; I observed their motion was a little irregular, but rectilinear, and seemed accelerated as they approached the Sun. These luminous globules appeared also on the other side of the spire, and preceded the Sun, running out into the dark room, sometimes more, sometimes less together, in the same manner as they followed the Sun at its occultation. They appeared to be in general about $\frac{1}{20}$ of an inch in diameter, and must therefore be very large luminous globes in some part of the heavens, whose light was extinguished by that of the Sun, so that they appeared not in open day light; but whether of the meteor kind, or what sort of bodies they might be, I could not conjecture." Martin's "Philosophia Britannica," London, 1747, vol. 2d, pp. 298, 299.

ending of the Eclipse, as predetermined by the above-mentioned projection thereof for Liverpool from Meyer's Tables, which were the apparent times; and shall reduce the observed equal times to the apparent, by subtracting 3 minutes 48 seconds (which we suppose here was the equation of time) from the equal times as observed by the clock and two watches, which kept equally going together.

	Apparent Times.					
	By Projection.			By Observation.		
Beginning, . .	8 ho.	56′	0″	8 ho.	55′	12″
Middle, . .	10	21	45	not certain.		
End, . . .	11	48	0	11	46	57
Duration, . .	2	52	0	2	51	45
Digits Eclipsed, .		$10\frac{25}{60}$		$10\frac{1}{2}$ exactly.		

We wish for the accounts of the observed times at the Royal Observatory and at London; because, by comparing the difference, and making allowance for the velocity of the penumbra between Liverpool and London, the longitude of Liverpool might be known.

As the observed quantity was somewhat greater than the projected, as so the digits and the projection which I gave in some time ago to the Royal Society, from Meyer's Elements, made the lower edges of the Sun and Moon to be nearly in contact at the greatest obscuration at Greenwich, I am apt to think that the appearance at Greenwich was annular,—and am with the greatest esteem,

Reverend Sir,

Your most obliged humble servant,

JAMES FERGUSON."

"To the Rev. THOMAS BIRCH, D.D.,
Secretary, F.R.S."

The original letter is among the "*Birch Letters*," British Museum, No. 2020.

The annexed woodcut is a representation of the Great Solar Eclipse, as seen at Liverpool on April 1st, 1764, at the time of middle of the Eclipse,—viz. 10 hours 21 minutes 45 seconds, when $10\frac{1}{2}$ digits were eclipsed. This representation of the Eclipse is copied from one of that year.

The foregoing letter was read by Dr. Birch to the members of the Royal Society on April 5th. In the Society's books the paper is entitled,

> "Observations on the Eclipse of the Sun, April 1st, 1764." Read before the Royal Society on 5th April, 1764. (See Phil. Trans. vol. 54, pp. 107, 108).

Solar Eclipse, as seen at Liverpool, 1st April, 1764.

The beginning of this letter on the Solar Eclipse shows, that Ferguson was about a month in Liverpool at this period, and that during his sojourn he resided with Captain Hutchinson.

ASTRONOMICAL CLOCK.—In Ferguson's "*Tables and Tracts,*" we read—

"In the year 1764, when I happened to be in Liverpool, I contrived a clock for Captain Hutchinson, who is Dock-master of the place, for showing the age and phases of the Moon, and the time of High and Low water at Liverpool, every day of the year, with the state of the tides at any time of the day; by looking at the Clock."

"At the right and left hand lower corners of the Dial-plate, under the common circles for the hours and minutes, there are two small circular plates. On the plate at the left hand are two circular spaces, the outermost of which is divided into twice twelve hours, with their halves and quarters; within which, the second circular space is divided into 29½ equal parts for the days of the Moon's age; each day standing under the time of the Moon's coming to the meridian on that day, in the circle of 24 hours. An axis comes through the centre of this plate, and carries two indexes round it in 29 days 12 hours 45 min., or from change to change of the Moon: and these indexes are set as far asunder as the time of High water at Liverpool differs from the time of the Moon's coming to the meridian. So that, by looking on this plate in the morning, one may see at what time the Moon will be on the meridian, and at what time it will be High water at the place.

Dial-plate of Astronomical Clock.

On the right hand plate, around its edge, all the different states of the tides are marked, from High to Low, and from Low to High; and within appellations is a shaded ellipsis, the highest points of which represent High water, and the lowest parts Low water. An index goes round this plate in the time of the Moon's revolving from the meridian to the meridian again; and at all different times, points out the state of the tide, as it may be then High or Low, rising or falling.

In the arch of the Dial-plate above the hour of XII, a blue plate rises and falls as the tides do at Liverpool; and over this plate, in a painted sky, a globular ball, half black, half white, shows the phases of the Moon for every day of her age throughout the year.

The wheel-work for showing these appearances is as follows:—

A wheel of 30 teeth is fixed on the axis of the twelve-hour hand, and turns round with it. This wheel turns a wheel of 60 teeth round in 24 hours, and on its axis is a wheel of 57

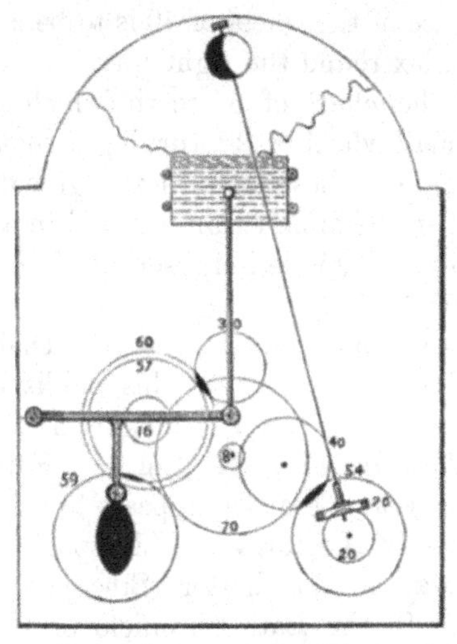

Plan of the Astronomical Wheel-work.

teeth, which turns round in the same time, and turns a wheel of 59 teeth round in 24 hours 50½ minutes, on whose axis is the index on the right hand corner-plate, going round the plate in the time of the Moon's revolving from the meridian to the meridian again; and showing the state of the tide at any time, when the clock is looked at. On the axis of the same wheel of 59 teeth is fixed an elliptical plate, which raises and lets down the Tide-plate in the arch twice in 24 hours 50½ minutes, in which time there are two ebbings and two flowings of the Tides.

The above-mentioned wheel of 57 teeth has a pinion of 16 leaves on its axis, turning round a wheel of 70 teeth, on whose axis is a pinion of 8 leaves, turning a wheel of 40 teeth, which turns a wheel of 54 teeth round in 29 days 12 hours 45 minutes; and on the axis of the wheel of 54 teeth are the two indexes on the left hand corner plate, for showing the Moon's age on that plate, with the time of her southing and of High water.

The wheel of 40 teeth here mentioned might have been of any other number, and might have been left out, if the pinion of 8 leaves had taken into the wheel of 54 teeth; but then the index would have gone the wrong way round the dial-plate. So that the only use of the wheel of 40 is to be a leading wheel for turning the index round the right way.

On the axis of the wheel of 54 teeth (which turns round in a lunation) is a small wheel of 20, turning a contrate wheel of the same number, on whose axis is the globular Moon (half black, half white) in the arch, turning round in a lunation, and showing all her phases." (Vide Ferguson's "*Tables and Tracts.*" Lond., 1767, pp. 122—126).

As Ferguson gives no engravings of the Dial-face and the wheel-work of this Clock in any of his published works, we here give them—on a reduced scale—from his unpublished MS. "Common Place Book," p. 94, (Col. Lib. Edin.)

In April of this year, Ferguson appears to have finished his Course or Courses of Lectures in Liverpool, when he left for London, by way of Manchester, Stockport, Birmingham, Bristol, Bath, &c. In the Bath Chronicle of 10th May, 1764, Ferguson's name appears among the "*arrivals*" at Bath on that day.

It is not known if he lectured at any of the above-mentioned places, but from an old memorandum, it is obvious that he sojourned several days in Stockport, examining the ingenious Mr. Earnshaw's Mechanical curiosities, and Astronomical Clocks,[245] and in conversing with him on such matters. During one of these interviews, Mr. Earnshaw gave Ferguson a calculation he had made for producing a motion to show the Sun's place in the ecliptic every day in the year (or the apparent annual revolution of the Sun). Ferguson appears to have estimated very highly the simple set of wheel-work deduced from Mr. Earnshaw's calculation; so much so, that he gives it a

[245] Mr. Edward Earnshaw (not Arnshaw) was an ingenious Clock and Watchmaker at Stockport, near Manchester. He was the inventor and maker of a great many astronomical clocks, orreries, and philosophical apparatus. A writer says that "his shop was a regular Noah's Ark of Mechanical nick-nacks." So in Mr. Earnshaw's company Ferguson would be quite at home.

place in his "Tables and Tracts," published three years afterwards, in an article entitled,

"An easy method for showing the Sun's place in the Ecliptic every day of the year, in a Clock; and his motion round the Ecliptic in a Solar year."

The following is his short description of this annual motion, and as he gives no drawing of it, we here annex a section of the wheel-work, drawn from the description.—Ferguson says,

"Let a pinion of 12 leaves be turned round once every ten hours, and, this pinion take into a wheel of 67 teeth, on whose axis let there be a single-threaded screw taking into a wheel of 157 teeth. This last wheel will turn round in 365 days 5 hours

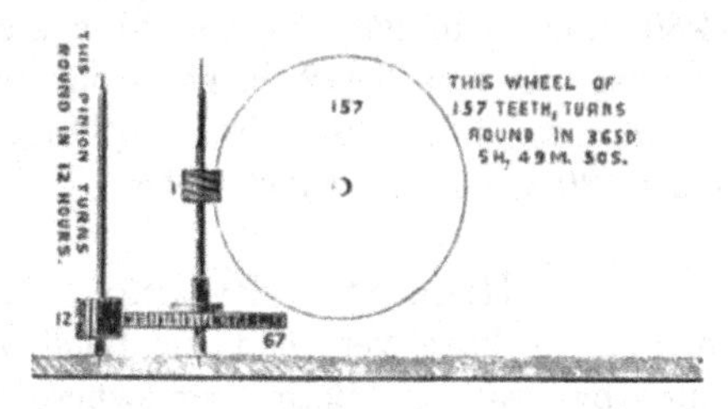

Mr. Earnshaw's Annual Train.

49 min. 50 sec. And an index on its axis will carry a Sun through the whole 360 degrees of an ecliptic, engraven on the dial-plate, in the same time; and may show the days of the months on another circle within the ecliptic.[246] This was the contrivance of Mr. Arnshaw, near Manchester, who communicated it to me." ("*Tables and Tracts.*" Lond., 1767, pp. 128, 129).

[246] This is a very awkward train of wheels—one which few would adopt; it is here indispensable that the first mover *turn* round in 10 hours, and to get this done would, we presume, be troublesome; also, endless screws ought to be avoided. The train is stated by Ferguson to produce 365 d. 5 h. 49 m. 50 s.; on trial, we find it 365 d. 5 h. 49 m. 59 s., for $\frac{12}{67} \times \frac{1}{157} = \frac{12}{10519}$, and $10519 \div 12 =$ 8765·83333 h., or 8765 h. 49 m. 59 s. = 365 d. 5 h. 49 m. 59 s. being a period too slow by 1 m. 8 s. Ferguson's own old orrery period, besides being more convenient, comes much nearer to the truth, and could be got up much easier than Earnshaw's,—thus, Ferguson, in his orreries, has $\frac{8}{25} \times \frac{7}{69} \times \frac{7}{83} = \frac{392}{143175}$, and 143175 $\div 392 =$ 365 d. 5 h. 48 m. 58·77 s., being only about 6 s. too slow, while Earnshaw's is 1 m. 8 s. (= 68 seconds); and by adopting Ferguson's ratio $\frac{392}{143175}$, to a common clock with a 12 hour mover, the wheels would require to be $\frac{4}{25} \times \frac{7}{69} \times \frac{7}{83}$, or $\frac{8}{50} \times \frac{7}{69} \times \frac{7}{83}$, this being wheel-work to be applied to 12 hours as a first mover, and $\frac{8}{25} \times \frac{7}{69} \times \frac{7}{83}$ to 24 hours as a first mover, which is the first mover in the orrery.

TABLE of PENDULUM LENGTHS.—In 1764, Ferguson calculated and published "*A Table for showing what length a Pendulum must be to make any given number of Vibrations in a minute, from* 1 *to* 300, *in the latitude of London; Calculated to the hundredth part of an Inch, by James Ferguson,* A.D. 1764." MS. "Common Place Book," (Col. Lib. Edin.) p. 56.

The half of this Table was inserted in his "Tables and Tracts," pp. 108, 109 (published in 1767), under the following title:—"*A Table, showing of what length a Pendulum must be to make any given number of Vibrations in a minute, from* 1 *to* 180, *in Lat.* 51° 30′."

The following are extracted from this Table:—

A pendulum to make 1 vibration in a minute must be 11738 feet 4·800 inches in length,—one to make 30 vibrations in a minute, 13 feet 0·512 inches,—one to make 60 vibrations in a minute, 3 feet 3·128 inches (39·128 inches),—and one to make 180 vibrations, 4·347 inches.

DESCRIPTION of the MECHANICAL PARADOX—PUBLISHED.—Ferguson appears to have returned to London from his Liverpool tour, by way of Bath, somewhere about the middle of May, 1764, when we find him publishing a pamphlet of 16 octavo pages, with large frontispiece plate of the Paradox, describing its use, &c. He, at the conclusion of his long letter of date April 10th, 1776 (inserted in this work under date 1750), mentions, that after having kept the Paradox beside him for several years, *without finding any one who could explain the principles on which it acted,* converted it into a simple Orrery, and then published this pamphlet, describing its several parts, and its use as an orrery, "in order to save himself the trouble of explaining it any longer." This pamphlet has been for the last eighty years or so, out of print; consequently, it is rarely to be found, and when it is, a very high price is asked for it. Our search extended over a period of ten or twelve years; and when at last a copy was found, ten shillings was paid for it, the original cost being one shilling only. The following is a copy of the title-page of this pamphlet:—

"The Description and Use of a New Machine called the Mechanical Paradox, invented by James Ferguson, F.R.S. London: Printed for the author, and sold by A. Millar, Bookseller, in the Strand. MDCCLXIV. Price One Shilling."

NEW PYROMETER.—In 1764 Ferguson invented and made a new Pyrometer, as he, and others, had found his old one of 1759 defective. This Pyrometer is of a different construction, and indicates with greater accuracy, and more strikingly, the expansion of metals by heat. The following are his figure and description of this instrument:—

"A Pyrometer, that makes the Expansion of Metals by Heat Visible to the five-and-forty-thousandth Part of an Inch.

The upper surface of this machine is represented in Fig. 1 in the annexed cut. Its frame A B C D is made of mahogany

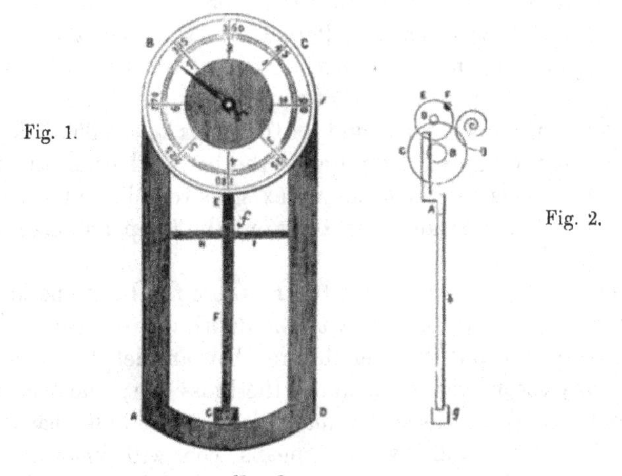

New Pyrometer.

wood, on which is a circle divided into 360 equal parts; and within that circle is another, divided into 8 equal parts. If the short bar E be pushed one inch forward (or toward the centre of the circle), the index *e* will be turned 125 times round the circle of 360 parts or degrees. As 125 times 360 is 45,000, 'tis evident, that if the bar E be moved only the 45,000th part of an inch, the index will move one degree of the circle. But as in my Pyrometer, the circle is 9 inches in diameter, the motion of the index is visible to half a degree, which answers to the ninety-thousandth part of an inch in the motion, or pushing of the short bar E.

One end of a long bar of metal F is laid into a hollow place in a piece of iron G, which is fixed to the frame of the machine; and the other end of this bar is laid against the end of the short bar E, over the supporting cross-bar H I; and, as the end *f* of the long bar is placed close against the end of the short bar, 'tis plain, that if F expands, it will push E forward and turn the index *e*.

The machine stands on four short pillars, high enough from a table, to let a spirit-lamp be put on the table, under the bar F; and, when that is done, the heat of the flame of the lamp expands the bar, and turns the index.

There are bars of different metals, as silver, brass, and iron; all of the same length as the bar F, for trying experiments on the different expansion of different metals, by equal degrees of heat applied to them for equal lengths of time; which may be measured by a pendulum that swings seconds,—Thus,

Put on the brass bar F, and set the index to the 360th degree; then put the lighted lamp under the bar, and count the number of seconds in which the index goes round the plate, from 360 to 360 again; and then blow out the lamp, and take away the bar.

This done, put on an iron bar F where the brass one was before, and then set the index to the 360th degree again. Light the lamp, and put it under the iron-bar, and let it remain just as many seconds as it did under the brass one; and then blow it out, and you will see how many degrees the index has moved in the circle; and by that means you will know in what proportion the expansion of iron is to the expansion of brass; which I find to be as 210 is to 360, or as 7 is to 12. By this method the relative expansion of different metals may be found.

The bars ought to be exactly of equal size; and to have them so, they should be drawn, like wire, through a hole.

When the lamp is blown out, you will see the index turn backward; which shows that the metal contracts as it cools.

The inside of this Pyrometer is constructed as follows:

In Fig. 2 A *a* is the short bar, which moves between the rollers; and, on the side *a* it has 15 teeth in an inch, which takes into the leaves of a pinion B (12 in number) on whose axis is the wheel C of 100 teeth, which take into the 10 leaves of the pinion F, on the top of whose axis is the index above-mentioned.

Now, as the wheels C and E have 100 teeth each; and the pinions D and F have ten leaves each; 'tis plain, that if the wheel *c* turns once round, the pinion F and the index on its axis will turn 100 times round.[247] But, as the first pinion B has only 12 leaves, and the bar A *a* that turns it has 15 teeth in an inch, which is 12 and a fourth part more; one-inch motion of the bar will cause the last pinion F to turn an hundred times round, and a fourth part of an hundred over and above, which is 25. So that, if A *a* be pushed one inch, F will be turned 125 times round.[248]

A silk thread *b* is tied to the axis of the pinion D, and wound several times round it; and the other end of the thread is tied to a piece of slender watch-spring G which is fixed into the stud H. So that, as the bar *f* expands, and pushes the bar A *a* forward, the thread winds round the axle, and draws out the spring; and as the bar contracts, the spring pulls back the thread, and turns the work the contrary way, which pushes back the short bar A *a* against the long bar *f*. This spring always keeps the teeth of the wheels in contact with the leaves of the pinions, and so prevents any shake in the teeth.

In Fig. 1, the eight divisions of the inner circle are so many thousandth parts of an inch in the expansion or contraction of the bars; which is just one thousandth part of an inch for each division moved over by the index." ("*Supplement to Ferguson's Book of Lectures on Mechanics*," &c. Lond., 1767, pp. 11—15; also, Supplement to his "*Lectures on Select Subjects in Mechanics*," &c.)

HYDROSTATIC MACHINE for showing the UPWARD PRESSURE of FLUIDS.—According to an old drawing and description, in

247 These wheels and pinions may be expressed, and velocities ascertained, as follows:—$\frac{10}{100} \times \frac{10}{100} = \frac{100}{10,000}$, and $10,000 \div 100 = 100$ times the pinion F turns round for one turn of the first mover.

248 This may also be similarly illustrated,—viz. $\frac{10}{100} \times \frac{10}{100} \times \frac{12}{15} = \frac{1200}{150,000}$, and $150,000 \div 1200 = 125$ times as above.

manuscript, of this Hydrostatic Machine, it appears that it was invented and made by Ferguson in July, 1764, as a substitute for the common Hydrostatic-bellows. Annexed is a woodcut of this machine and its details, with his description, taken from the "Supplement" to his "Lectures on Select Subjects in Mechanics," &c.

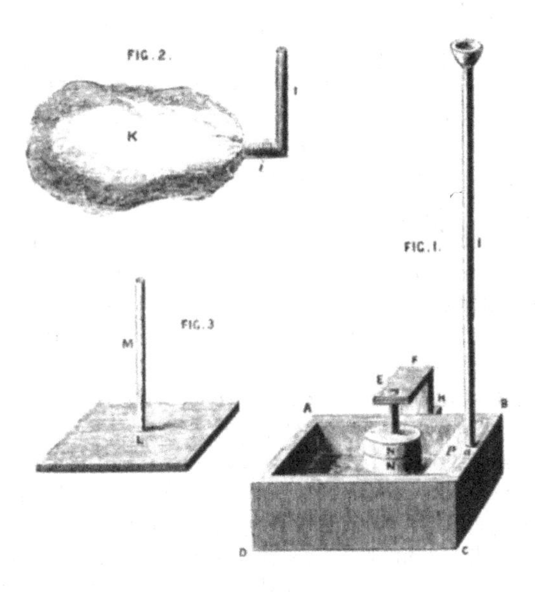

"In Fig. 1, A B C D is an oblong square box, in one end of which is a round groove, as at *a*, from top to bottom, for receiving the upright glass tube I, which is bent to a right angle at the lower end (as at *i* in Fig. 2), and to that part is tied the end of a large bladder K (Fig. 2), which lies in the bottom of the box. Over this bladder is laid the moveable board L (Figs. 1 and 3), in which is fixed an upright wire M; and leaden weights N N, to the amount of 16 lbs., with holes in their middle, are put upon the wire, over the board, and press upon it with all their force.

The cross-bar *p* is then put on, to secure the tube from falling, and to keep it in an upright position. And then the piece E F G is to be put on, the part G sliding tight into the dovetail'd groove H, to keep the weights N N horizontal, and the wire M upright; there being a round hole *e* in the part E F for receiving the wire.

There are four upright pins in the four corners of the box within, each almost an inch long, for the board L to rest upon; to keep it from pressing the sides of the bladder below it close together at first.

The whole machine being thus put together, pour water into the tube at top; and the water will run down the tube into the bladder below the board; and after the bladder has been filled up to the board, continue pouring water into the tube, and the upward pressure which it will excite in the bladder, will raise the board with all the weight upon it, even though the bore of the tube should be so small, that less than an ounce of water would fill it.

The upward pressure against every part of the board (which the bladder touches) is equal in area to the area of the bore of the tube, and will be pressed upward with a force equal to the weight of the water in the tube; and the sum of all these pressures, against so many areas of the board, will be sufficient to raise it, with all the weights upon it.

In my opinion, nothing can exceed this simple machine, in making the upward pressure of fluids evident to sight." ("Supplement to Ferguson's Book of Lectures on Mechanics," &c. Lond., 1767, pp. 25—27; also, "Supplement to Lectures on Select Subjects in Mechanics," &c.)

The following we copied in 1831, from the original in Ferguson's autograph, then in the possession of Deane Walker, Esq., Lecturer on Astronomy, London. They appear to have been all written on the same day in 1764, and are entitled *Curious Memoranda.* The *first* and the *last* are the same as those found in his Select Mechanical Exercises—No. 1 at p. 168, and No. 3 at p. 169; No. 2 is to be found at p. 235 of his "Tables and Tracts."

"CURIOUS MEMORANDA.

No. 1. ON the COMBINATION of THE LETTERS of the ALPHABET.

"The Number of different Ways in which all the Letters of the Alphabet might be combined, or put together, from 1 Letter to 25. Or, the Number of Changes which might be rung on any Number of Bells not exceeding the Number of Letters in the Alphabet.

1	A	1
2	B	2
3	C	6
4	D	24
5	E	120
6	F	720
7	G	5040
8	H	40320
9	I	362880
10	K	3628800
11	L	39916800
12	M	479001600
13	N	6227020800
14	O	87178291200
15	P	1307674368000
16	Q	20922789888000
17	R	355687428096000
18	S	6402373705728000
19	T	121645100408832000
20	U	2432902008176640000
21	V	51090942171709440000
22	W	1124000727777607680000
23	X	25852016738884976640000
24	Y	620448401733239439360000
25	Z	15511210043330985984000000

Thus, 2 letters may be put 2 different ways together; 3 letters, 6 different ways; 4 letters, 24 ways; 5 letters, 120 ways; 6 letters, 720 ways; and so on to the 25th letter, which may be put 15511210043330985984000000 ways together.

No. 2 "ON THE QUESTION of FOUR WEIGHTS to WEIGH ANY NUMBER of POUNDS from 1 to 40.

How 40 pounds weight, or any number of pounds from 1 to 40 may be weighed by four weights, viz. of 1 lb., 3 lbs., 9 lbs., and 27 lbs.

Pounds.	Scale A	Scale B	Pounds.	Scale A	Scale B
1	1	0	21	27, 3	9
2	3	1	22	27, 3, 1	9
3	3	0	23	27	3, 1
4	1, 3	0	24	27	3
5	9	3, 1	25	27, 1	3
6	9	3	26	27	1
7	9, 1	3	27	27	0
8	9	1	28	27, 1	0
9	9	0	29	27, 3	1
10	9, 1	0	30	27, 3	0
11	9, 3	1	31	27, 3, 1	0
12	9, 3	0	32	27, 9	3, 1
13	9, 3, 1	0	33	27, 9	3
14	27	1, 3, 9	34	27, 9, 1	3
15	27	3, 9	35	27, 9	1
16	27, 1	3, 9	36	27, 9	0
17	27	9, 1	37	27, 9, 1	0
18	27	9	38	27, 9, 3	1
19	27, 1	9	39	27, 9, 3	0
20	27, 3	9, 1	40	27, 9, 3, 1	0

EXPLANATION.

The two columns under *Pounds* express the number of pounds to be weighed: to the right hand of which, the column under A shows what weights are to be put into one scale of the balance, and the column under B shows what weights are to be put into the other: by which means, the scale B will be so much lighter than the scale A, as to require a weight to be put into it, equal to the given number of pounds to be weighed, as stated at the left hand, under *Pounds*, against the weights in the scales; and then the balance will be even.

If, to the above four mentioned weights one of 81 pounds be added, you may weigh 121 pounds; or any number from 1 to 121 pounds—If to weights 1, 3, 9, 27, 81.

No. 3. THE CISTERN QUESTION.

Suppose a square Cistern to be a Mile wide and a Mile deep, or to contain a Cubic Mile of Water; and that a Cubic Yard of Water should run Off from it every Minute until it was quite emptied.—*Query,* How much Time would all the Water take to run out of the Cistern?

Answer.—5451776000 minutes; .because there are so many cubic yards in a cubic mile—or 10365 Julian Years, 139 days, 7 hours, 20 minutes.

SOLUTION:—1760×1760×1760=5451776000=cubic yards in a cubic mile,—and $365\frac{1}{4}$ days=a Julian Year,—and $365\frac{1}{4}$×24×60=525960 minutes in a Julian Year—therefore 5451776000÷525960=10365·381398 or 10365 Julian Years 139 Days, 7 Hours, 20 Min.

JAMES FERGUSON.
LONDON, 22 *August*, 1764."

HYDRAULIC MACHINE for SHOWING the CAUSE of EBBING and FLOWING WELLS.—In a visit to Deane Walker, Esq., London, in 1831, we were shown a large and beautiful drawing of this machine in Indian ink by Ferguson, having the sectional view of a hill above it, and so as to show to the eye, the cause of the ebb and flow of Wells or Reciprocating Springs. The drawing had no description, being only designated as above, and having "*James Ferguson, delin., London,* 1764," neatly written in the lower right hand corner. Ferguson does not allude to this machine in his "*Lectures on Select Subjects,*" &c., published in 1760; an engraving and description of it, however, is to be found in his "*Supplement*" to these Lectures, published in 1767. From this it appears that this curious and now well known apparatus was invented and made between the years 1760 and 1767, and therefore, there is little or no doubt that the machine was constructed when the drawing of it was made—sometime in 1764.

Ferguson, during his lecturing tour in Derbyshire, in 1762, would no doubt visit the celebrated Ebbing and Flowing Well at Tideswell, near Chester-le-Street; and, as a matter of course, would soon afterwards begin to think of the possibility of making a machine to imitate its ebb and flow, for exhibiting to his audience in his Hydraulic lectures. As shown above, the machine was matured and constructed in 1764, and ever after, continued to be one of the most interesting machines amongst his hydraulic apparatus.

The drawing here alluded to appearing to be the original, from which the engraving was taken in the "*Supplement*" to his Lectures, no copy was taken. The following engraving and description of the machine is taken from the Supplement:

"THE CAUSE of RECIPROCATING SPRINGS, and of EBBING and FLOWING WELLS, EXPLAINED.—In Fig. 1, let *a b c d* be a hill, within which is a cavern A A near the top, filled or fed by rains or melted snow on the top *a*, making their way through chinks and crannies into the said cavern, from which proceeds a small stream *c c* within the body of the hill, and issues out in a spring at G on the side of the hill, which will run constantly whilst the cavern is fed with water.

From the same cavern A A let there be a small channel D, to carry water into the cavern B; and from that cavern let there be a bended channel E *e* F, larger than D, joining with the former channel *c c*, as at *f*, before it comes to the side of the hill; and let the joining at *f* be below the level of the bottom of both these caverns.

As the water rises in the cavern B, it will rise as high in the channel E *e* F; and when it rises to the top of that channel at *e*, it will run down the part *e* F G, and make a swell in the spring G, which will continue till all the water is drawn off from the cavern B, by the natural syphon E *e* F (which carries off the water faster from B than the channel D brings water to it), and then the swell will stop, and only the small channel *c c* will carry water to the spring G, till the cavern B is filled to B again by the rill D; and then the water being at the top *e* of the channel E *e* F, that channel will act again as a syphon, and carry off all the water from B to the spring G, and so make a swelling flow of water at G as before.

To illustrate this by a machine (Fig. 2), let A be a large wooden box, filled with water; and let a small pipe *c c* (the upper end of which is fixed into the bottom of the box) carry water from the box to G, where it will run off constantly, like a small spring. Let another small pipe D carry water from the same box to the box or well B, from which let a syphon E *e* F proceed, and join with the pipe *c c* at *f*: the bore of the syphon being larger than the bore of the feeding pipe D. As the water from this pipe rises in the well B, it will also rise as high in the syphon E *e* F; and when the syphon is full to the top *e*, the water will run over the bend *e*, down the part *e* F, and go off at the mouth G; which will make a great stream at G; and that stream will continue, till the syphon has carried off all the water from the well B; the syphon carrying off the water faster from B than the pipe D brings water to it: and then the swell at G will cease, and only the water from the small pipe *c c* will run off at G, till the pipe D fills the well B again; and then the syphon will run, and make a swell at G as before.

And thus, we have an artificial representation of an Ebbing and Flowing Well, and of a Reciprocating Spring, in a very natural and simple manner." ("A Supplement to Ferguson's Book of Lectures on Mechanics," &c. Lond., 1767, pp. 27—30; also, "Supplement to his Lectures on Select Subjects in Mechanics," &c.)

HYGROMETER.—Towards the end of the year 1764, Ferguson invented and made a New Hygrometer or instrument for measuring the degrees of moisture, an account and drawing of which were sent to the Royal Society, and read before that body on November 8th, 1764. In the Philosophical Transactions, Ferguson's paper is entitled,

> "Description of a New Hygrometer." Read before the Royal Society, 8th Nov., 1764. (See Phil. Trans., vol. 54, page 259).

In the year 1751, Ferguson invented and made a Hygroscope —an instrument for *showing* changes in the air from dry to moist, and from moist to dry. (See pp. 158, 159). The present instrument is called a Hygrometer—an instrument for *measuring* the degree of humidity in the atmosphere.

1765.

Our first memorandum for 1765 refers to a simple method for finding the quantity and weight of water in pipes, and was read to the members of the Royal Society. In the Philosophical Transactions, vol. 55, p. 61, the paper is entitled, a

"Short and Easy Method for Finding the Quantity and Weight of Water in a full pipe of any given height and diameter of bore. By James Ferguson." Read before the Royal Society, 7th February, 1765. See Phil. Trans., vol. 55, page 61; also, Universal Magazine, vol. 14.

We have no copy of, or extract from, this paper on Pipes; but on referring to "Tables and Tracts," we find a similar paper, which, if not a copy of the original in all its details, certainly contains all that is necessary for solving what he proposed in the title of his paper. The article referred to is as follows:—

"A Table, by which the quantity and weight of water in a cylindrical pipe of any given diameter of bore, and perpendicular height, may be found; and consequently, the power may be known that will be sufficient to raise the water to the top of the pipe, in any pump, or other hydraulic machine.

Feet High.	Diameter of the cylindric bore 1 inch.		
	Quantity of Water in cubic inches.	Weight of Water in Troy ounces.	In Avoirdupois ounces.
1	9·4247781	4·9712340	5·4541539
2	18·8495562	9·9424680	10·9083078
3	28·2743343	14·9137020	16·3624617
4	37·6991124	19·8849360	21·8166156
5	47·1238905	24·8561700	27·2707695
6	56·5486686	29·8274040	32·7249234
7	65·9734467	34·7986380	38·1790773
8	75·3982248	39·7698720	43·6332312
9	85·8230029	44·7411060	49·0873851

For tens of feet high, remove the decimal points one place forward; for hundreds of feet, two places; for thousands, three places; and so on.

Then multiply the sums by the square of the diameter of the given bore, and the product will be the answer.

Example.—Query, The quantity and weight of water in a cylindrical pipe 85 feet high, and 10 inches diameter?

Feet High.	Cubic inches.	Troy Ounces.	Avoird. Ounces.
80 5	753·982248 47·123890	397·698720 24·856170	436·332312 27·270769
85	801·106138 mult. by 100	422·554890 100	463·603081 100
Ansr.	80110·613800	42255·489000	46360·308100

The square of 10 is 100, which number (80110·6) of cubic inches being divided by 231, the number of cubic inches in a wine gallon, gives 342·6 for the number of gallons; and the respective weights (42255·489 and 46360·3) being divided by 12, and the latter 16, give 3521·29 for the number of Troy pounds, and 2897·5 for the number of Avoirdupois pounds, that the water in the pipe weighs. So much power would be required to balance or support the water in the pipe, and as much more to work the engine as the friction thereof amounts to." (pp. 228—230) "*Tables and Tracts.*" Lond., 1767.

Table of Standard Weights of Money, &c.—In 1765, Ferguson computed and compiled

"A Table, Showing the Standard Weight, Value, and comparative View of English Silver Money, from King William the First, A.D. 1066, to A.D. 1765,—According to the Mint Indentures." (Vide MS. "Common Place Book," p. 83, Col. Lib. Edin. "Select Mechanical Exercises," pp. 163, 164).

In the same year Ferguson discovered a method for showing how much time is contained in any given number of mean lunations, which he made the subject of a paper sent to the Royal Society. The paper, in the Philosophical Transactions, is entitled,

> "Short and Easy Method for Finding the Quantity of Time contained in any given number of Lunations, and number of Mean Lunations in any given quantity of time. By James Ferguson."
>
> *March*, 1765.

Ferguson, in "Tables and Tracts," gives an abstract from this paper in the form of a Table, with explanatory examples of its use, of which the following is a copy.

"A Table, showing how much time is contained in any given number of mean Lunations, the Lunation being 29 days 12 hours 44 minutes 3 seconds $2'''$ 58^{iv}, or 29·53059085108 days.[249]

Lunations.	Days : Decimals of a day.
1	29·53059085108
2	59·06118170216
3	88·59177255324
4	118·12236340432
5	147·65295425540
6	177·18354510648
7	206·71413595756
8	236·24472680864
9	265·77531765972

"Although the Table seems to go no further than nine mean Lunations, yet it will do for any number from 1 Lunation to 900,000,000,000, by removing the decimal point one place forward for ten of Lunations, two places forward for hundreds of Lunations, three places for thousands: and so on, as in the following Examples. For, if we wanted to know how much time is contained in 10 Lunations, then suppose a cypher put to 1 in the first column, to make it 10, and the decimal point in the first line to be put one place forward, it will be 295·3059085108, for the number of days and decimal parts of a day in 10 Lunations. The decimal parts may be reduced to the known parts of an integral day, by the common method of reducing decimals.

249 It is now long since it was discovered that this lunation period of Ferguson's time is slightly "*a plus period*,"—viz. a period of about 10 thirds of time in excess of the true mean period in each lunation, or about 2 seconds a-year.

The synodic revolution of the Moon, or the time elapsing between change and change, is at a mean rate 29 d. 12 h. 44 m. 2·8765 s. The Chaldeans have rceord-

Example I.—In 10 Lunations, Query, How much time?

Lunations.	Days. Decimals of a day.
10	295·3059085108
	mult. by 24 h.
	12236340432
	6118170216
	Hours 7·3418042592
	mult. by 60 m.
	Min. 20·5082555520
	mult. by 60 s.
	Sec. 30·4953331200
	60 thds.
	Thds. 29·7199872000

Answer—295 days 7 hours 20 min. 30 sec. 29·719 thds.

Example II.—In 74212 mean Lunations, Query, How many days, hours, minutes, &c.?

Lunations.	Days. Decimals of a day.
70000	2067141·3595756
4000	118122·36340432
200	5906·118170216
10	295·3059085108
2	59·06118170216
74212	2191524·20824034896
Lun.	Days. mult. by 24 h.
	83296139584
	41648069792
	Hours. 4·99776837504
	mult. by 60 m.
	Min. 59·86610250240
	60 s.
	Sec. 51·96615014400
	60 thds.
	Thds. 57·96900864000

By reduction, 2191524 days÷365·25 days=6000 Julian Years and 24 days.

	Years.	da.	ho.	min.	sec.	thds.
Answer;	6000.	24.	4.	59.	51.	57.

ed in their Historical Annals an account of an eclipse of the Moon which happened on 19th March, B.C. 721.—An eclipse of the Moon occurred on the 28th April, 1771, having the position of the Moon, the nodes, and apogee, similar to their position in the Chaldean Eclipse. From 19th March, B.C. 720, and 28th April, A.D. 1771, there are 910044·152581 days, or 30817 synodic revolutions nearly. Therefore, 910044·152581÷30817 = 29·530588 days = 29 d. 12 h. 44 m. 2·88 s., which is the modern standard period for a mean synodic revolution.

Example III.—In 100000000000 mean Lunations, Query, How much time?

Lunations.	Days.
100000000000	2953059085108, Ansr.

In Example III. the number of cyphers annexed to 1 are equal to the number of decimal parts in the first line of the Table; and therefore the whole of that line becomes a whole number of integral days, without any fraction. So that, in 100,000,000,000 mean Lunations, there are just 2953059085108 days.

It is somewhat remarkable, that every 49th mean new Moon falls but 1 min. 30 sec. 34 thirds short of the same time of the day as before.

Example IV.

Lunations.	Days. Decimals of a day.
40	1181·2236340432
9	265·77531765972
49	1446·99895170292
Lun.	Days. mult. by 24 h.
	399580681168
	199790340584
	Hours 23·97484087008
	mult. by 60 m.
	Min. 58·49045220480
	60 s.
	Sec. 29·42713228800
	60 thds.
	Thds. 25·62793728000

which wants only 1 minute 30 seconds 34·4 thds. of 1447 days.

A Table showing how many mean Lunations are contained in any given quantity of time.

Years.	Lun. Decimals of a Lunation.	Hours.	Lun. Decimals of a Lunation.	Sec.	Lun. Decimals of a Lunation.
1	12·368530038627	1	0·0014109662	1	0·0000003919
2	24·737060077255	2	0·0028219345	2	0·0000007838
3	37·105590115882	3	0·0042328987	3	0·0000011758
4	49·474120154510	4	0·0056438649	4	0·0000015677
5	61·842650193137	5	0·0070548312	5	0·0000019597
6	74·211180231765	6	0·0084657974	6	0·0000023516
7	86·579710270392	7	0·0098767637	7	0·0000027435
8	98·948240309020	8	0·0112877299	8	0·0000031355
9	111·316770347647	9	0·0126986962	9	0·0000035274
Days.	**Lun. Decimals of a Lunation.**	**Min.**	**Decimals of a Lunation.**	**Thds.**	**Decimals of a Lunation.**
1	00·033863189760	1	0·0000235161	1	0·0000000065
2	00·067726379520	2	0·0000470322	2	0·0000000131
3	00·101589659280	3	0·0000705483	3	0·0000000196
4	00·135452759040	4	0·0000940644	4	0·0000000262
5	00·169315948800	5	0·0001175805	5	0·0000000327
6	00·203179138560	6	0·0001410966	6	0·0000000392
7	00·237042328320	7	0·0001646127	7	0·0000000457
8	00·270905518080	8	0·0001881288	8	0·0000000522
9	00·304768707840	9	0·0002116449	9	0·0000000587

For tens of Julian years, days, hours, &c., remove the decimal points one place forward; for hundreds, two places; for thousands, three places; for tens of thousands, four places; and so on, as in the following Example. It appears by the first line of the above Table that in 10000 Julian years (which contains 365·25000 days) there are 1236853 mean Lunations, and ·0038627, or $\frac{38627}{1000000}$ parts of a Lunation, which small fraction may be neglected.

(In common working, 'tis sufficient to take in only four or five of the decimal figures).

Example V.—In 6000 Julian years, 24 days 4 hours 59 min. 52 sec., Query, How many mean Lunations?

		Lunations. Decimals.
Years	6000	74211·180231765
Days	20	0·677263795
	4	0·135452759
Hours	4	0·005643864
Minutes	50	0·001175805
	9	0·000211645
Seconds	50	0·000019597
	2	0·000000784

Answer, 74212·000000014.

More Examples would be superfluous."

Besides being read before the Royal Society, the paper was printed in quarto, entitled, "An Easy Method for Finding the quantity of time contained in any number of Mean Lunations." Lond., 1765.

The following particulars of the effects produced by Villette's large concave burning mirror, are taken from Ferguson's autograph;

"VILLETTE'S BURNING MIRROR.—A short account of M. VILLETTE'S *Concave Burning Mirror.* This Mirror is 3 feet 11· inches in diameter, and its focal distance is 3 feet 2 inches. It is made of copper, tin, and bismuth.

The effect of the Sunbeams on different bodies held in its focus were as follows:

1. A piece of Roman tile began to melt in 3 seconds, and was ready to drop in 100 seconds.
2. Chalk fled away in 33 seconds.
3. A Fossil shell calcined in 7 seconds.
4. Copper ore vitrified in 8 seconds.
5. Iron ore melted in 24 seconds.
6. A great tooth of a fish melted in 33 seconds.
7. Welch asbestos was a little calcined in 28 seconds.
8. A King George's halfpenny melted in 16 seconds.
9. Tin melted in 3 seconds, and had a hole in it in 6.
10. A bone calcined in 4 seconds, and was vitrified in 33.
11. A diamond weighing 4 grains lost $\frac{7}{8}$ parts of its weight.

Thus, the effects on Chalk, on the Tooth, and on the Bone, were accomplished in the same time.

The solar beams are condensed 1700 times in the focus of this mirror (the condensation in the focus being as the area of the mirror is to the area of its focus), and their heat, in the focus, is 433 times as great as the heat of common fire.

JAMES FERGUSON.

LONDON, 26 *July*, 1765."

See also Ferguson's "Tables and Tracts," pp. 295, 296.

1766.

NEW ORRERY.—Our memoranda for 1766 commence with a note copied from one of Ferguson's papers (in autograph), which mentions that he, early in that year, commenced to make a new Orrery for the illustration of his Astronomical lectures, to supersede the one he had hitherto used. This orrery was made with brass wheels, and was inclosed in a box of twelve sides, with a brass ecliptic supported on twelve short brass pillars, and in all its parts, interior and exterior, resembled the orrery usually represented in the large frontispiece to his Astronomy. It also showed the same Astronomical motions,—viz. "The annual revolutions of Mercury, Venus, and the Earth, round the Sun, in their proper periodical times; the Moon's motion round the Earth, showing her periodical, sidereal, synodical, and diurnal motions, with her phases, age, and southings; The motions of the Sun, Venus, and the Earth, round their respective axes; the Vicissitudes of Seasons; the retrograde motion of the nodes of the Moon's orbit; with the times of all the new and full Moons, and of all the Solar and Lunar Eclipses." This orrery was surmounted by a Sphere, and the whole was placed under a glass cover of twelve equal sides—corresponding with the twelve sides of the box;—the paper concludes with a quotation from *Claudin's* epigrammatic description of Archimedes's Sphere in Latin, followed by two translations of it in English, as follows:—

"Jupiter in parvo cum ceneret athera vitro,
Risit et ad Superos talia dicta dedit:
Huccine mortalis progressa potentia curæ?
Jam meus in fragili luditur orbe labor.
Jura pole, rerumque fidem, legesq; Deorum
Ecce Syracusius transtulit arte Senex.
Inclusus variis famulatur Spiritus astris,
Et vivum certis motibus urget opus.
Percurrit proprium mentitus Signifer annum.
Et simulata novo Cynthia mense redit.
Jamo, suum volvens audax industria mundum
Gaudet, et humana Sidera mente regit.

Quid falso insontem tonitru Salmonea miror?
Æmula Naturæ parva reperta manus."

TRANSLATIONS.

"When Jove espy'd in Glass his Heavens made,
He smiled, and to the other Gods thus said:
'Tis strange that human art so far proceeds,
To ape in brittle Orbs my greatest deeds.
The heavenly motions, Nature's constant course,
Lo! here old Archimede to art transfers.
Th' inclosed Spirit here each Star doth drive;
And to the living work sure motions give.
The Sun in counterfeit his year doth run,
And Cynthia too her monthly circle turn.
Since now bold man hath Worlds of's own descry'd
He joys, and th' Stars by human art can guide.
Why should we so admire proud Salmon's cheats,
When one poor hand Nature's chief work repeats!"

(Derham's Artificial Clock Maker, p. 88, 3d Edit.)

"When in a glass's narrow sphere confined,
Jove saw the fabric of the Almighty mind,
He smiled, and said, 'Can mortal's art alone
Our heavenly labour mimic with their own?
The Syracusan's brittle work contains
The eternal law, that through all nature reigns.
Framed by his art, see stars unnumbered burn,
And, in their courses, rolling orbs return:
His Sun, through various signs describes the year,
And, every month, his mimic Moons appear.
Our rival's laws his little planets bind,
And rule their motions by a human mind.
Salmoneus could our thunder imitate,
But Archimedes can a world create.'"

JAMES FERGUSON, *Feb. 5th*, 1766.

One of Ferguson's Orreries—with brass wheel-work—was gifted to University College, London, in 1851, by Mr. George Walker of Port Louis, probably the orrery just referred to. The following particulars were supplied by Mr. Atkinson, Secretary to the College.

Height of brass feet, ...	3½	inches.
„ Box, containing the wheel-work, . .	8½	„
„ From surface of revolving plates to the summit,	17	„
Total,	29	inches.

Diameter, ... 30 inches.

ECLIPTIC CIRCLE; all of Brass and Engraved, and is supported by 12 Brass Pillars.

BOX, containing the wheel-work, is of Stained Wood—has 12 sides, ornamented with brass plates—each plate having a name of one of the Signs of the Zodiac.

THE PLANETS are of Ivory; the Sun is of Brass.

THE INSCRIPTION.—The inscription is on an oblong plate of White Metal, placed above the Sign *Virgo* on the panel of the box, and reads as follows:—

"This Orrery was made by the self-taught astronomer, James Ferguson; and was purchased after his death by the Rev. George Walker, F.R.S., and President of the Literary and Philosophical Society of Manchester, who repaired it with his own hands. It was presented to University College by his son, George Walker, of Port Louis, France, June 27th, 1851."

FERGUSON AND HIS ASSISTANT, KENNETH M'CULLOCH.—We do not know when this orrery was finished; probably it would be during 1766. Ferguson not only worked at it with his own hands, but was ably assisted by an ingenious workman named Kenneth M'Culloch, who was "*well skilled in all the intricate motions of wheel-work*," and who appears to have worked for Ferguson occasionally, from this period, till Ferguson's death in 1776.[250]

ON the DIVISION of a CIRCLE INTO ANY GIVEN NUMBER of EQUAL PARTS.—The following paper for dividing a Circle into any given number of equal parts, and which may be found useful on many occasions, is dated 1766, and belonged to the late Adam Walker, Esq., Lecturer on Natural Philosophy, London. This method of division probably suggested itself to Ferguson while employed upon his orreries, as the numbers of wheel's teeth selected for illustration, are, in part, the numbers he adopted in the annual train of his orreries.

[250] Kenneth M'Culloch was a native of Scotland, and long in the employ of Ferguson, assisting him in the construction of his apparatus, clocks, orreries, &c. and died at about the age of 80 in the year 1808. Ferguson was exceedingly fond of Kenneth (or *Kenny*, as he used to call him), and from time to time bestowed on him marks of favour, presenting him now and then with mechanical nick-nacks, such as Card Rotulas, &c., and shortly before his death, Ferguson gave him a curious table-clock, which showed the motion of the Earth by the rotation of a small globe.

This note is from Mr. Andrew Reid, who also sometimes worked for Ferguson about 1772, and who has been often quoted. In Brewster's Edinburgh Encyclopædia, article "*Planetary Machines*," p. 636, Kenneth M'Culloch is mentioned as having been employed to make a Planetary Machine for the Royal Institution, London;—the reference to him notes, that "*About the year* 1801, *one of the original proprietors of the Royal Institution suggested a plan for exhibiting the* EQUATED *motions of all the planets at that time discovered. This suggestion, being made soon after the two planets Ceres and Pallas had been discovered, was adopted; and Kenneth M'Culloch, an aged workman, brought up under James Ferguson, was employed in the construction of the machine in the workshops of the Institution.*"

"To divide the circumference of a circle into any given number of equal parts, whether even or odd.

As there are very uncommon and odd numbers of teeth in some of the wheels of astronomical clocks, and which consequently could not be cut by any common engine used by clockmakers for cutting the numbers of teeth in their clock-wheels, I thought proper to show how to divide the circumference into any given odd or even number of equal parts, so as that number may be laid down upon the dividing-plate of a cutting engine.[251]

There is no odd number but from which, if a certain number be subtracted, there will remain an even number, easily to be subdivided. Thus, supposing the given number of equal parts divisions of a circle on the dividing-plate to be 69; subtract 9, and there will remain 60.

Every circle is supposed to contain 360 degrees; Therefore, say, as the given number of parts in the circle, which is 69, so is 9 parts to the corresponding arc of the circle that will contain them; which arc, by the Rule of Three, will be found to be $46\frac{95}{100}$.[252] Therefore, by the line of chords on a common scale, or rather on a sector, set off $46\frac{95}{100}$ (or $46\frac{9}{10}$) degrees with your compasses, in the periphery of the circle, and divide that arc or portion of the circle into 9 equal parts, and the rest of the circle into 60; and the whole will be divided into 69 equal parts, as was required.

Again, suppose it were required to divide the circumference of a circle into 83 equal parts; subtract 3, and 80 will remain. Then, as 83 parts are to 360, so (by the Rule of Proportion) are 3 parts to 13 degrees and one hundredth part of a degree; which small fraction may be neglected. Therefore, by the line of chords, and compasses, set off 13 degrees in the periphery of the circle, and divide that portion or arc into 3 equal parts, and the rest of the circle into 80; and the thing will be done.[253]

Once more, suppose it were required to divide a circle into 365 equal parts; subtract 5, and 360 will remain. Then, as

[251] We had, for upwards of 30 years, the engine with which Ferguson cut his wheels; it has all the numbers on it he here illustrates. But the circles with 69 and 83 divisions, and some others, appear to have been divided by hand—most probably by Ferguson—as no clockmaker has the slightest use for such numbers, in wheels, as 69 and 83. This engine is now in the Museum of Banff.

[252] Thus, 69 : 360 : : 9 ?—$360 \times 9 = 3240 \div 69 = 46^\circ.9565$.

[253] 69 : 360 : : 9 ?—$360 \times 9 = 3240 \div 69 = 46^\circ.9565$.

365 parts are to 360 degrees, so are 5 parts to $4\frac{93}{100}$ degrees. Therefore, set off $4\frac{93}{100}$ (or $4\frac{9}{10}$) degrees in the circle; divide that space into 5 equal parts, and the rest of the circle into 360; and the whole will be divided into 365 equal parts as was required.[254]

I have often found this rule or method very useful in dividing circles into an odd number of equal parts, or wheels into odd numbers of equal siz'd teeth with equal spaces between them; and now I find it just as easy to divide any given circle into any odd number of equal parts, as to divide it into an even number. And, for this purpose, I prefer the line of chords on a sector to that on a plain scale; because the sector may be opened so as to make the radius of the line of chords upon it equal to the radius of the given circle, unless the radius of the circle exceeds the whole length of the sector when it is opened, so as to resemble a straight line, ruler, or scale; and this is what very seldom happens.

Any person who is used to handle the compasses, and the scale or sector, may very easily, by a little practice, take off degrees, and fractional parts of a degree, by the accuracy of his eye, from a line of chords, near enough the truth for the above-mentioned purpose." (Vide also "Select Mechanical Exercises," London, 1773, pp. 38—42).

Ferguson has useful directions appended to the foregoing, showing how to determine the working distance between the centres of two given wheels, and how to make *the teeth* and *spaces* of the same gauge.

He puts it as a question thus:—

"Supposing the distance between the centres of two wheels, one of which is to turn the other, be given; that the Number of Teeth in one of these Wheels is different from the Number of Teeth in the other, and it is required to make the Diameters of these Wheels in such Proportion to one another as their Numbers of teeth are, so that the Teeth in both Wheels may be of equal Size and the Spaces between them equal, that either of them may turn the other easily and freely; it is required to find their Diameters.

Here it is plain that the distance between the centres of the

[254] 365 : 5 : : 360 ?—$360 \times 5 = 1800 \div 365 = 4^{\circ}.9315$.

wheels is equal to the sum of both their radii in the working parts of the teeth. Therefore, as the number of teeth in both wheels, taken together, is to the distance between their centres, taken in any kind of measure, as feet, inches, or parts of an inch; so is the number of teeth in either of the wheels to the radius or semi-diameter of that wheel, taken in the like measure, from its centre to the working part of any one of its teeth.

Thus, suppose the two wheels must be of such sizes as to have the distance between their centres 5 inches; that one wheel is to have 75 teeth, and the other to have 33, and that the sizes of teeth in both the wheels is equal, so that either of them may turn the other. The sum of the teeth in both wheels is 108; therefore, say as 108 teeth is to 5 inches, so is 75 teeth to $3\frac{47}{100}$ inches; and as 108 is to 5, so is 33 to $1\frac{53}{100}$ inches. So that, from the centre of the wheel of 75 teeth to the working part of any tooth in it, is 3 inches and 47 hundred parts of an inch; and from the centre of the wheel of 33 teeth to the working part of either of its teeth, is 1 inch and 53 hundred parts of an inch.

JAMES FERGUSON.

July 24th, 1766."

(Also, see "Select Mechanical Exercises." Lond., pp. 42—44).

Our next note for 1766 records the death of the Reverend Dr. Birch, Secretary of the Royal Society, *Ferguson's intimate and influential friend*, as he has been designated, whose death was greatly regretted by Ferguson.[255]

UNIVERSAL DIALLING CYLINDER.—Towards the end of the year 1766, Ferguson appears to have invented and made his celebrated Universal Dialling Cylinder. In the Memoir, by himself, it will be recollected that, after mentioning that the *Eclipsareon* was the best machine he ever contrived, he says, "my next best contrivance is *the Universal Dialling Cylinder* —the instrument now under consideration. The following description and figure of this Cylinder are taken from his "Supplement" to "Lectures on Select Subjects."[256]

255 The Reverend Thomas Birch, D.D., died in 1766, aged 61 years. He was the author of "The History of the Royal Society of London," in which the most considerable of those papers communicated to the Society, which have not been published, are inserted, 4 vols. Lond., 1756, &c. Ferguson appears to have become acquainted with Dr. Birch about the year 1750.

256 The *Dialling Sphere* was sold in July, 1806, at the public sale of the As-

"In the annexed figure, A B C D represents a cylindrical glass tube, closed at both ends with brass plates, and having a wire or axis E F G fixed in the centres of the brass at top and bottom. This tube is fixed to a horizontal board H, and its axis makes an angle with the board equal to the angle of the Earth's axis, with the horizon of any given place, for which the cylinder is to serve as a dial. And it must be set with its axis parallel to the axis of the world in that place, the end E pointing to the elevated pole; or, it may be made to move upon a joint; and then it may be elevated for any particular latitude.

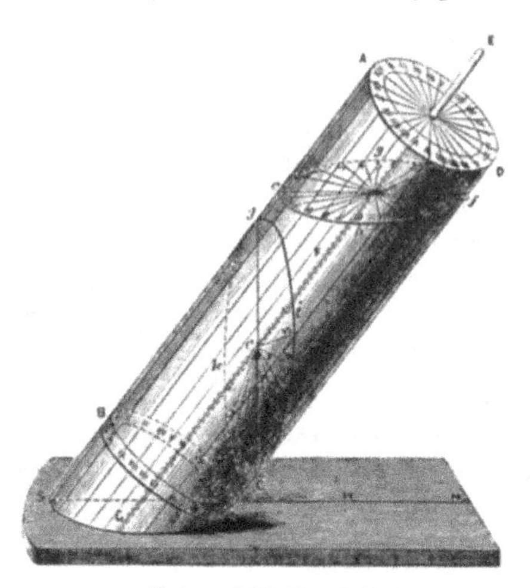

Universal Dialling Cylinder.

"There are 24 straight lines, drawn with a diamond, on the outside of the glass, equidistant from each other, and all of them parallel to the axis. These are the hour-lines; and the hours are set to them as in the figure: the XII next B stands for midnight, and the opposite XII, next the board H, stands for mid-day or noon.

"The axis being elevated to the latitude of the place, and the foot-board set truly level, with the black line along its middle

tronomical and Mathematical collection of the late Alexander Aubert, Esq., Highbury House, Islington. It was sold on the third day of the sale (Tuesday, 22d July, 1806), and is thus entered in a priced Catalogue of Messrs. Leigh & Sotheby:—

"*Third day's Sale*—Catalogue, p. 14—Lot 195.—Ferguson's Dialling Sphere and a piece of Brass Machinery in a wooden frame—this lot sold for £1 2s."

in the plane of the meridian, and the end N toward the north; the axis E F G will serve as a stile or gnomon, and cast a shadow on the hour of the day, among the parallel hour lines, when the Sun shines on the machine. For, as the Sun's apparent diurnal motion is equable in the heavens, the shadow of the axis will move equably in the tube; and will always fall upon *that* hour-line which is opposite to the Sun at any given time.

"The brass plate A D, at the top, is parallel to the equator, and the axis E F G is perpendicular to it. If right lines be drawn from the centre of this plate, to the upper ends of the equidistant parallel lines on the outside of the tube; these right lines will be the hour-lines on the equinoctial dial A D, at 15 degrees distant from each other; and the hour-letters may be set to them as in the figure. Then, as the shadow of the axis within the tube comes on the hour-lines of the tube, it will cover the like hour-lines on the equinoctial plate A D.

"If a thin horizontal plate *e f* be put within the tube so as its edge may touch the tube all round; and right lines be drawn from the centre of that plate to those points of its edge which are cut by the parallel hour-lines on the tube; these right lines will be the hour-lines of a horizontal dial, for the latitude to which the tube is elevated. For, as the shadow of the axis comes successively to the hour-lines of the tube, and covers them, it will then cover the like hour-lines on the horizontal plate *e f*, to which the hours may be set, as in the figure.

"If a thin vertical plate, *g c*, be put within the tube, so as to front the meridian, or 12 o'clock line thereof, and the edge of this plate touch the tube all round; and then, if right lines be drawn from the centre of the plate to those points of its edge which are cut by the parallel hour-lines on the tube; these right lines will be the hour-lines of a vertical south dial; and the shadow of the axis will cover them at the same times when it covers those of the tube.

"If a thin plate be put within the tube, so as to decline, or incline, or recline, by any given number of degrees, and right lines be drawn from its centre to the hour-lines on the tube; these right lines will be the hour-lines of a declining, inclining, or reclining dial, answering to the like number of degrees, for the latitude to which the tube is elevated.

"And thus, by this simple machine, all the principles of dialling are made very plain, and evident to the sight. And the axis of the tube (which is parallel to the axis of the world in every latitude to which it is elevated) is the stile or gnomon for all the different kinds of Sun-dials.

"And lastly, if the axis of the tube be drawn out, with the plates A D, *e f*, and *g c* upon it, and set up in sunshine, in the same position as they were in the tube, you will have an equinoctial dial A D, a horizontal dial *e f*, and a vertical south dial *g c;* on all which, the time of the day will be shown by the shadow of the axis or gnomon E F G.

"Let us now suppose that, instead of a glass tube, A B C D is a cylinder of wood; on which the 24 parallel hour-lines are drawn all around, at equal distances from each other; and that from the points at top, where these lines end, right lines are drawn toward the centre, on the flat surface A D. These right lines will be the hour-lines on an equinoctial dial, for the latitude of the place to which the cylinder is elevated above the horizontal foot or pedestal H; and they are equidistant from each other, as in Fig. 2, which is a full view of the flat surface or top A D of the cylinder, seen obliquely in Fig. 1. And the axis of the cylinder (which is a straight wire E F G all down its middle) is the stile or gnomon, which is perpendicular to the plane of the equator.

"To make a horizontal dial, by the cylinder, for any latitude to which its axis is elevated; draw out the axis and cut the cylinder quite through, as at *e h f g*, parallel to the horizontal board H, and take off the top part *e* A D *f e;* and the section *e h f g e* will be of an elliptical form, as in Fig. 3. Then, from

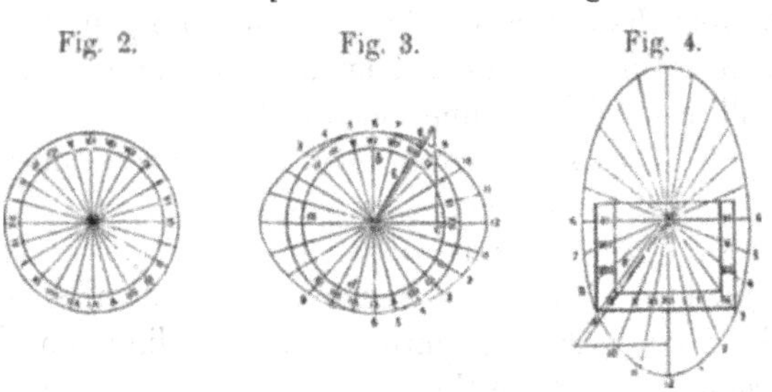
Fig. 2. Fig. 3. Fig. 4.

the points of this section (on the remaining part *e* B *c f*) where the parallel lines on the outside of the cylinder meet it, draw right lines to the centre of the section; and they will be the true hour-lines for a horizontal dial, as *a b c d a* Fig. 3, in which may be included in a circle drawn on that section. Then put the wire into its place again, and it will be a stile for casting a shadow on the time of the day, on that dial. So, E (Fig. 3) is the stile of the horizontal dial, parallel to the axis of the cylinder.

"To make a vertical south dial by the cylinder, draw out the axis, and cut the cylinder perpendicularly to the horizontal board H, as at *g i* C *k g*, beginning at the hour-line (B *g e* A) of XII, and making the section at right angles to the line S H N on the horizontal board. Then, take off the upper part *g* A D C, and the face of the section thereon will be elliptical, as shown in Fig. 4. From the points in the edge of this section, where the parallel hour-lines on the round surface of the cylinder meet it, draw right lines to the centre of the section; and they will be the true hour-lines on a vertical direct south dial, for the latitude to which the cylinder was elevated: and will appear as in Fig. 4, on which the vertical dial may be made of a circular shape, or of a square shape, as represented in the figure. And F will be its stile parallel to the axis of the cylinder.

"And thus, by cutting the cylinder any way, so as its section may either incline, or decline, or recline, by any given number of degrees; and from those points in the edge of the section where the outside parallel hour-lines meet it, draw right lines to the centre of the section; and they will be the true hour-lines, for the like declining, reclining, or inclining dial. And the axis of the cylinder will always be the gnomon or stile of the dial. For, whichever way the plane of the dial lies, its stile (or the edge thereof that casts the shadow on the hours of the day) must be parallel to the Earth's axis, and point toward the elevated pole of the heavens." (Vide "*A Supplement to Mr. Ferguson's* Book of Lectures on Mechanics," &c. Lond., 1767, pp. 42—48).

This descriptive account of the Universal Dialling Cylinder, accompanied with drawings, was sent to the Royal Society, and read before that body on 2d July, 1767. In the records of the Society the paper is entitled,

"A New Method of Constructing Sun-Dials for any given Latitude, with the assistance of Dialling Scales, or Logarithmic Calculation. Read before the Royal Society on 2d July, 1767." See Phil. Trans., vol. 57, p. 389; also, Gentleman's Magazine for March, 1769, where an engraving and description are given.

The Editor of the "Gentleman's Magazine," for March, 1769, after announcing the publication of Ferguson's new method of constructing Sun-dials, says, "This is (like all Mr. Ferguson's performances) very simple, very ingenious, and explained with such perspicuity, that to read English is all that is necessary to understand it; and consequently, all who read this article will be able to make a Dial. It is, however, necessary to insert the cut to which this article refers, for illustration, which perhaps in a future number we may do, and insert this curious and useful article at length." (Gentleman's Magazine for March, 1769, vol. 39, p. 101; also, vol. 39, pp. 143, 144, where it is inserted with copperplate illustrations).

TABLES OF SPECIFIC GRAVITY, AND FORCE OF THE WIND.—The following Tables appear to have been computed and compiled by Ferguson sometime in the year 1766.

A TABLE OF THE SPECIFIC GRAVITIES OF BODIES.

A cubic inch of	Troy Weight.			Avoird.		Comparative Weight.
	oz.	pw.	gr.	oz.	drams.	
Very fine Gold,	10	7	4·45	11	5·85	19·639
Standard Gold,	9	19	6·06	10	14·88	18·887
Guinea Gold,	9	7	17·18	10	4·76	17·793
Quicksilver,	7	3	4·39	7	13·16	13·565
Lead,	5	19	16·32	6	8·86	11·325
Pure Silver,	5	17	0·00	6	6·69	11·090
Standard Silver,	5	11	3·25	6	1·55	10·534
Copper,	4	14	22·62	5	3·33	9·000
Plate Brass,	4	8	2·05	4	13·31	8·344
Cast Brass,	4	5	10·76	4	10·08	8·001
Steel,	4	2	20·21	4	8·71	7·835
Block Tin,	3	17	5·52	4	3·99	7·320
Diamond,	1	15	21·07	1	13·35	3·400
Fine Marble,	1	8	14·11	1	9·30	2·710
Common Glass,	1	7	5·20	1	7·88	2·579
Alabaster,	0	19	18·43	1	2·03	1·873
Dry Ivory,	0	19	5·83	1	0·89	1·823
Dry Boxwood,	0	10	20·77	0	9·54	1·201
Sea Water,	0	10	20·51	0	9·51	1·035
Common Water,	0	10	13·18	0	9·23	1·000
Red Wine,	0	10	11·42	0	9·20	·993
Proof Spirits,	0	9	19·73	0	8·62	·931
Pure Spirits,	0	9	3·27	0	8·02	·866
Æther,	0	7	14·00	0	7·46	·720
Cork,	0	2	12·77	0	2·21	·240
Air,	0	0	0·25	0	0·009	·001

Take away the decimal points from the numbers in the right hand column, and reckon them to be whole numbers; and they will show how many Avoirdupois ounces are contained in a cubic foot of each of the above bodies in the Table." (Tables and Tracts, p. 237; MS. "Common Place Book," p. 236, Col. Lib. Edin.)

A TABLE OF THE DIFFERENT VELOCITIES AND FORCES OF THE WINDS.

Velocity of the Wind.		Perpendicular force on one foot Area, in Pounds Avoirdupois.	Common appellations of the forces of Winds.
Miles in one Hour.	= Feet in one Second.		
1	1·47	·005	Not perceptible.
2	2·93	·020	Just perceptible.
3	4·40	·044	
4	5·87	·079	Gentle pleasant Wind.
5	7·33	·112	
10	14·67	·492	Pleasant brisk Gale.
15	22·00	1·107	
20	29·33	1·968	Very brisk.
25	36·67	3·075	
30	44·00	4·428	High Winds.
35	51·33	6·027	
40	58·67	7·872	Very high.
45	66·00	9·963	
50	73·33	12·300	A storm or tempest.
60	88·00	17·712	A great storm.
80	117·33	31·488	A hurricane.
100	146·70	49·200	A hurricane that tears up trees, and carries buildings, &c., before it.

The force of the Wind is as the square of its velocity.

That the Force of the Wind is as the square of its velocity, I have often proved by experiments made on my *Whirling Table*.

JAMES FERGUSON.

LONDON, 1766."

(Tables and Tracts, p. 238; MS. "Common Place Book," Col. Lib. Edin., p. 74).

"On PUMPS, and DIRECTIONS for PUMP MAKERS.—The following paper on Pumps, &c., appears to have been partly writ ten in 1765, but not completed until near the end of 1766.

"In all pumps," says Ferguson, "the pressure of the column of water, or its weight felt by the working power, when raised to any given height above the surface of the well, is in proportion to the height of the column, considered throughout, as if it were equal in diameter to that part of the bore in which the piston or bucket works.

The advantage or power gained by the handle of the pump is the same as in the common lever; that is, as great as the length from the axis of the handle to its end where the power is applied, exceeds the length of the other part of the handle, from the axis on which it turns to the pump-rod wherein it is fixed, for lifting the piston and water.

In the making of pumps, the diameter of the bore where the bucket works should be proportioned to the height which the pump raises the water above the surface of the well, so that a man of ordinary strength might work all pumps equally easy, let their heights be what they may. The annexed Table shows

A TABLE FOR PUMP-MAKERS.

Height of the pump above the surface of the Well.	Diameter of the bore where the Piston works.	Water discharged in a minute in Gallons and Pints.	
Feet.	Inches.	Gal.	Pints.
10	6·93	81.	6.
15	5·65	54.	4.
20	4·90	40.	7.
25	4·38	32.	6.
30	4·00	27.	2.
35	3·70	23.	3.
40	3·47	20.	4.
45	3·26	18.	1.
50	3·10	16.	3.
55	2·95	14.	7.
60	2·83	13.	5.
65	2·71	12.	4.
70	2·62	11.	5.
75	2·53	10.	7.
85	2·44	10.	2.

JAMES FERGUSON.

LONDON, *Nov.* 1766

how this may be done, and what quantities of water may be raised in a minute by one man, supposing the handle of the pump to be a lever increasing the power five times.

N.B.—In my book of Lectures, pag. 75, last paragraph, and line 3 of column 1, in pag. 76 for bucket, read surface of the water in the well.

1st, Find the given height of pump, in the first column of the Table; and against it in the second column, you have the diameter which the bore must be of in inches and hundredth parts of an inch.

2d, In the third column, you have the quantity of water in English gallons and pints, that a man of common strength can raise to that height in a minute.

With respect to the power required to work the pump, or the quantity of water discharged thereby, it matters not what the diameter of the bore be in any other part than that wherein the piston or bucket works." ("Tables and Tracts." Lond., 1767, pp. 230—232).

MODEL OF BLAKEY'S ENGINE.—According to one of our notes, we find that in the year 1766, Mr. Blakey, the inventor of an "Engine for Raising Water," waited on Ferguson with a working model of his engine, and set it in motion before him—the engine was not *perfected*; but such as it was, Ferguson seems to have been pleased with its action, and obtained leave from its inventor to make a model of it to be shown in his lectures —a model appears to have been accordingly made in the same year. In his MS. "Common Place Book," Ferguson gives a lengthened description of this engine, accompanied with a large drawing in Indian ink. Two years after making his model he appears to have got tired of waiting for Blakey *perfecting* his engine, as we find at the conclusion of his description, dated 1768, a note as follows :—

"Mr. Blakey has been upwards of two years about this Simple Machine, and has never yet finished one to work. If I live till he does, I shall give an account of it on the following page. JAMES FERGUSON.
Nov. 16*th*, 1768."

See MS. "Common Place Book," pp. 80, 81; and "Select Mechanical Exercises, pp. 107—109.

Ferguson died in 1776—just eight years after the date of

this note—and as there is no second "*account of it on the following page*" of his "Common Place Book," it is to be inferred that Blakey never *perfected* his engine, at least in Ferguson's time.

FERGUSON REMOVES FROM MORTIMER STREET.—Sometime during the year 1766, Ferguson removed from Mortimer Street, Cavendish Square, to No. 4 Bolt Court, Fleet Street, which proved his *Ultima Domus*. (See date 1776).

ON THE TIDES.—We have two papers, in the autograph of Ferguson, on the Tides. The following is a copy of the most interesting one:—

"On the Cause of the Ebbing and Flowing of the Sea at the same time on Opposite sides of the Earth.

The reason why the tides rise on the side of the Earth which is at any time turned towards the Moon, is plain to every one; because her attraction must occasion a swelling of the waters toward her on that side; but the cause of so great a swell, at the same time, on the opposite side of the Earth, which is then turned away from the Moon, has been very hard to account for; because the rising of the tide there is in a direction quite contrary to the attraction of the Moon. But this difficulty is immediately removed, when we consider, that all bodies moving in circles have a centrifugal force, or constant tendency to fly off from the centres of the circles they describe; and this centrifugal force is always in proportion to the distance of the body from the centre of its orbit, and the velocity with which it moves therein; when the body is large, the side of it which is farthest from the centre of its orbit will have a greater degree of centrifugal force than the centre of the body has; and the side of it which is nearest the centre of its orbit will have a less degree of centrifugal force than its centre has.

As the Moon goes round the Earth every month in her orbit, the Earth also goes round an orbit every month, which is as much less than the quantity of matter in the Earth, which is 40 times. For, by the laws of nature, when a small body moves round a great one, in free and open space, both these bodies must move round the common centre of gravity between them.

The Moon's mean distance from the Earth's centre is 240,000

English miles: divide therefore this distance by 40, the difference between the quantity of matter in the Earth and Moon, and the quotient will be 6000 miles, which is the distance of the common centre of gravity (between the Earth and Moon) from the centre of the Earth.

Now, as the Earth and Moon move round the common centre of gravity between them, once every month, 'tis plain, that whilst the Moon moves round her orbit, at 240,000 miles from the Earth's centre, the centre of the Earth describes a circle of 6000 miles radius, round the centre of gravity between the Earth and the Moon, the Moon's attraction balancing the centrifugal force of the Earth's at its centre.

The diameter of the Earth is 8000 miles, in round numbers, and consequently its semi-diameter is 4000; so that the side of the Earth, which is at any time turned toward the Moon, is 4000 miles nearer the common centre of gravity between the Earth and Moon than the Earth's centre is; and the side of the Earth which is then farthest from the Moon, is 4000 miles farther from the centre of gravity between the Earth and Moon than the Earth's centre is at that time.

Therefore, the radius of the circle described by the parts of the Earth which come about toward the Moon, by the Earth's diurnal motion, is 2000 miles; the radius of the circle described by the Earth's centre is 6000; and the radius of the circle described by those parts of the Earth which, in revolving on its axis, are furthest from the Moon, is 10,000 miles.

The centrifugal forces of the different parts of the Earth being directly as their distances from the above-mentioned common centre of gravity, round which both the Earth and Moon move, these forces may be expressed by 2000 for the side of the Earth nearest the Moon, by 6000 for the Earth's centre, and by 10,000 for the side of the Earth which is farthest from the Moon. But the Earth's attraction is greatest on the side of the Earth next her, where the centrifugal force or tendency to fly off from the common centre of gravity (and consequently, from the Moon) is least; and therefore, the tides must rise on the side of the Earth which is nearest the Moon, by the excess of the Moon's attraction.

As her attraction balances the centrifugal force at the Earth's centre, 'tis plain that the centrifugal force of the side of the

Earth which is farthest from the Moon, is greater than her attraction; and therefore, the tides will rise as high upon that side from the Moon, by the excess of the centrifugal force, as they rise on the side next her by the excess of her attraction.

And as the Earth is in constant motion on its axis, so as that any given meridian revolves from the Moon to the Moon again in 24 hours, $50\frac{1}{2}$ minutes, each place will come to the two eminences of water, under and opposite to the Moon, in 24 hours, $50\frac{1}{2}$ minutes, or have two tides of flood and two of ebb in that time. For, as much as the waters rise above the common level of the surface of the sea, under and opposite to the Moon, so much they must fall below that level half way between the highest places; or at 90 degrees from them.

On these principles, it is equally easy to account for the rising of the tides, at the same time, on both sides of the Earth; and this rising is made evident to sight in my Lecture on the central forces; and the principles on which it depends are made obvious to the understandings of all observers."

JAMES FERGUSON.

LONDON, *Sep. 4th,* 1766.

See also Tables and Tracts, pp. 303—309.

We may here observe, that Ferguson, in this paper, gives 40 as the number of times that the solid contents of the Earth exceed that of the Moon. This is an error; for the mass of the Earth is 49 times the mass of the Moon.

All globular bodies are to each other as the cubes of their diameters. The Earth is 7,912 miles in diameter, the Moon 2,160 miles; therefore, according to the rule, $\frac{7912 \times 7912 \times 7912 = 495289174528}{2160 \times 2160 \times 2160 = 10077696000}$ and $495289174528 \div 10077696000 = 49\frac{1}{7}$ times nearly. Also, he makes the distance of the common centre of gravity between the Earth and Moon 6000 miles from the Earth's centre, whereas the distance is only 4800 miles; but assuming the 40 and the 240,000 to be correct, 6000 miles would not be the result. Let $a = 240,000$, $x =$ distance of the centre of gravity from the Earth's centre. Now, from the principles of the lever, an equilibrium can only take place when mass of the Earth : Moon's mass :: $a - x : x$, that is, when $1 : \frac{1}{40} :: a - x : x$, from which we find $x = \frac{a}{41} = 5853\frac{27}{41}$ miles $(240000 \div 41 = 5853\frac{27}{41})$. If we apply the same rule to the true mass 49, then the 49 becomes $\frac{a}{50}$, that is, $240,000 \div 50 = 4800$ miles, the distance of the common centre of gravity from the Earth's centre, as given above. This centre of gravity is therefore 844 miles distant from the surface of the Earth—in a straight line always, with the centre of the Moon.

BAROMETER TABLE.—The following Table by Ferguson, is copied from the original, in our collection. It appears to have been computed and written about the end of the year 1766:—

"A TABLE,

Showing the height to which a Barometer must be raised above the plane surface of the Earth, in order that the Mercury may stand at any given height in the Tube,—this is shown in the first part of the Table. The second part shows at what height the Mercury will stand in the Tube, when the Barometer is raised to any given height above the Earth's plane surface.

PART I.		PART II.	
Height of the Mercury in inches.	Height of the Barometer in feet above the Earth's plane surface.	Height of the Barometer above the Earth.	Height of the Mercury in inches.
30·000	FEET 0	FEET 0	30·00
29·000	915	1000	28·91
28·000	1862	2000	27·86
27·000	2844	3000	26·85
26·000	3863	4000	25·87
25·000	4922	5000	24·93
20·000	10947	MILES 1	24·67
15·000	18715	2	20·29
10·000	29662	3	16·68
5·000	48378	4	13·72
1·000	91831	5	11·28
0·5	110547	10	4·24
0·25	129262	20	1·60
0·1	153120	25	0·95
0·001	216480	30	0·23
0·000	279840	40	0·08

By the first part of this Table, and a common Barometer or Weather Glass, the perpendicular height of a hill above the plane surface of the Earth may be nearly found. Thus, suppose the Mercury was observed to stand at 30 inches in the Tube when at the foot of the hill, and at 27 inches when carried up to the top: against this sinking of three inches, you have 2844 feet (or 948 yards) for the perpendicular height of the hill. The second part is too plain to need any description or example.

JAMES FERGUSON.

LONDON, 1766."

(See also MS. "Common Place Book," Col. Lib. Edin., p. 82; "Tables and Tracts," pp. 294, 295; and "Select Mechanical Exercises," p. 121, for a more extended and complete Table.

1767.

BATH AND BRISTOL LECTURES.—Ferguson, in the beginning of January, 1767, left London for Bath and Bristol to deliver several Courses of Philosophical Lectures, in these cities, illustrated by his now very extensive apparatus. In the Bath Chronicle (deposited in the British Museum), for January, February, and March, 1767, are the following advertisements:—

"BATH, *Jany. 22d*, 1767.

"MR. FERGUSON gives notice that his SECOND COURSE of LECTURES on EXPERIMENTAL PHILOSOPHY and ASTRONOMY, will begin this day, precisely at TWELVE O'CLOCK at the LAMB INN, in STALL STREET, and be continued every day afterward (Sundays excepted), at the same hour, till the whole be finished. Subscribers pay a Guinea each for the whole course, which consists of TWELVE LECTURES. Non-subscribers pay Half-a-Crown each for every Lecture they attend.

If those who sometime ago proposed to subscribe for an EVENING COURSE, still intend to do it, they are desired to fill up the subscription at the LAMB INN by SATURDAY next at TWELVE O'CLOCK; for if they delay it any longer, Mr. Ferguson must then begin to pack up his apparatus gradually, as he has done with its different parts, in order to send it to London when the present Course is over.

N.B.—No Gold will be changed at the Lecture Room."

Pope's "Bath Chronicle, Jan. 22d, 1767."

It is doubtful whether the parties, to whom allusion is made in the foregoing advertisement, subscribed for an Evening Course; but to the same paper he sent the following advertisement:

"BATH, *February 9th*, 1767.

"For the LAST TIME DURING THE PRESENT SEASON, MR. FERGUSON will begin a LECTURE on the ORRERY at the LAMB INN, in STALL STREET, at SIX O'CLOCK *this evening;* in which all VICISSITUDES OF SEASONS and the TIMES, CAUSES, and RETURN of all the ECLIPSES of the SUN and MOON will be explained and demonstrated on the principles of *Nature*, together with the PHENOMENA of SATURN'S RING. The Year of our SAVIOUR'S CRUCIFIXION will be astronomically ascertained, and the DARKNESS at the TIME of the CRUCIFIXION proved to have been *out of the Common Course of Nature.*

☞ Each Non Subscriber who attends this Lecture is to pay Half-a-Crown; and no Gold will be changed at the Lecture Room."

Pope's Bath Chronicle, February 12th, 1767.

The foregoing advertisement, of date February 9th, was not inserted into the Bath Chronicle until the 12th—this he notices in his next advertisement in the same newspaper, as follows:

"BATH, *February* 19*th*, 1767.

"Mr. Ferguson gives notice that he will return to this City, and read a COURSE of LECTURES on EXPERIMENTAL PHILOSOPHY and ASTRONOMY, any TIME before the Beginning of April next, if FIFTY PERSONS will subscribe a Guinea each for the same at Mr. Leak's, Mr. Frederick's, or Mr. Taylor's, Booksellers in Bath.

☞ The advertisement which appeared in last Thursday's Paper, concerning his reading a Lecture *that* Evening, was put in by mistake; for all his Machinery was packed up before that Time, and he was then in Bristol."

In order to keep his name prominently before the public, before returning to Bath to deliver another Course of Lectures, he advertised two of his works in the Bath newspaper, as follows:

"Just Published, by Mr. Ferguson.

1st, LECTURES on SELECT SUBJECTS in MECHANICS, HYDROSTATICS, PNEUMATICS, and OPTICS, with the USE OF THE GLOBES; The ART OF DIALLING, and the CALCULATIONS of the MEAN TIMES of New and Full Moons and ECLIPSES, illustrated with 23 Copperplates of the author's MACHINERY and DIAGRAMS. Price 12s. 6d. in Boards, Quarto.

2d, ASTRONOMY explained upon SIR ISAAC NEWTON'S PRINCIPLES, and made easy to those who have not studied MATHEMATICS, to which are added —The method of finding the DISTANCES of all the PLANETS from the SUN, by the TRANSIT of VENUS, over the SUN'S DISK in the year 1761. These Distances deduced from Observations of that Transit, and an account of Mr. Horrox's Observations of the Transit of the year 1639, Illustrated with 18 large Copper Plates; Price in Boards 16s. 6d. Quarto."

Ferguson was in Bristol from about 12th February until about the beginning of April, 1767, delivering Courses of Lectures there to considerable audiences; but the Bristol newspapers for this period we have been unable to find. The following is his advertisement in the Bath Chronicle:

"BATH, *March 26th*, 1767.

"MR. FERGUSON, F.R.S., will return to this City in the beginning of April this year, and read a COURSE of TWELVE LECTURES on EXPERIMENTAL PHILOSOPHY and ASTRONOMY, if fifty Persons will subscribe a Guinea each for the same.

The principal subjects will be as follows:—To show how human strength may be assisted by art, and how to construct Machines and Engines for that purpose.—To compute the Degree of Powers gained by any Machine or Engine already constructed; To show the best methods of making Cranes, Wheel Carriages, and Mills for Grinding Corn and Sawing Timber; all which will be exemplified by a great variety of Models. To demonstrate the surprising properties of water, by HYDROSTATIC MACHINES, and of AIR, by the AIR-PUMP: To construct ENGINES for raising WATER to GENTLEMEN'S SEATS above the Level of the SPRINGS or RIVERS. To explain the Laws by which the DEITY REGULATES and GOVERNS all the MOTIONS of the PLANETARY SYSTEM: To represent the MOTIONS of the PLANETS and the COMETS by MACHINERY: To show the CAUSES of the different SEASONS, the MOTIONS and PHASES of the MOON, The HARVEST MOON, The TIDES, and all the ECLIPSES of the SUN and MOON, by means of an Orrery.

N.B.—If the Subscription should not be compleat by next Tuesday, the 31st March, Mr. Ferguson will be under the necessity of having his apparatus sent to London, as he has been invited by many gentlemen to give a Course of Lectures there, and will be obliged to go to Liverpool the latter end of April for the same purpose."

It is probable that this advertisement did not succeed, for in about a fortnight afterwards we find him again advertising in the Bath Chronicle, proposing to deliver a Course of Lectures in Bath, if *only* Twenty-five subscribe a Guinea each. The advertisement is as follows:

"BATH, *April 16th*, 1767.

"MR. FERGUSON gives notice that if TWENTY-FIVE PERSONS will subscribe a GUINEA each, at MR. LEAK'S, or MR. FREDERICK'S, Booksellers in this City, by TUESDAY next, the 21st Inst., for a COURSE of LECTURES on EXPERIMENTAL PHILOSOPHY and ASTRONOMY; he will begin the first LECTURE of that COURSE on WEDNESDAY the 29th Inst. at TWELVE O'CLOCK at the LAMB INN, in STALL STREET, and continue to read every following day (Sunday excepted) at the SAME HOUR till the COURSE be finished, unless the Company appoint a more convenient Hour for their attendance. But if the subscription be not full by the 21st Inst., he must then begin to pack up his apparatus in order to send it to London."

It is likely that the Twenty-five subscribers were obtained, and that he commenced his mid-day lectures on April 21st.

Resolving to have evening lectures, he sent the following advertisement to the Bath Chronicle:

"BATH, *April 23d*, 1767.

"MR. FERGUSON will begin a LECTURE on the ORRERY to morrow at SEVEN O'CLOCK, in the EVENING, at the LAMB INN, in STALL STREET; and will continue to read at the same HOUR, on SATURDAY, MONDAY, and TUESDAY next on that GRAND MACHINE.[257] These four LECTURES will be on the SOLAR SYSTEM, the SEASONS,—the MOTIONS and PHASES of the MOON, and the Doctrine of Eclipses.

Subscribers are to pay eight Shillings each, for attending these Lectures; non-subscribers, Half-a-Crown for each Lecture they attend.

(N.B.—No Gold will be changed at the Lecture-Room)."

It will be recollected that Ferguson, at the conclusion of his advertisement of March 26th, mentions that he had to be in London and Liverpool before the end of April; since these four lectures would keep him in Bath until at least April 28th, it is likely that he commenced his London Course in *May*, as he did not get to Liverpool until near the end of that month.

He left Bath for London on or about May 1st, 1767. Shortly after his arrival, he began to deliver a Course of Lectures, and at intervals revised the proof sheets of his Supplement to "Lectures on Select Subjects," an octavo volume of 68 pages, containing 13 large copperplate engravings—only one edition of the Supplement was published, as it was afterwards appended to Lectures on Select Subjects. We have a copy of the original edition, now difficult to be had. The following is copy of the title-page:—

"A Supplement to Mr. Ferguson's Book of Lectures on Mechanics, Hydrostatics, Pneumatics, and Optics, with the Use of the Globes, and the Art of Dialling; containing Thirteen Copperplates, with descriptions of the Machinery, which he has added to his Apparatus, since that Book was printed. By James Ferguson, F.R.S. Lon-

[257] This *Grand Machine* probably refers to the new Orrery which he and Kenneth M'Culloch finished making in the previous year (1766). At Ferguson's sale in March, 1777, this orrery appears to have been bought by Thomas Hawys, Esq., who died in 1807. The orrery was again put up for sale in 1807. In the "Catalogue of the Collection of Mathematical and Philosophical Instruments, the property of Thomas Hawys, Esq., late of Charter House Square, deceased, to be sold by public auction, by Messrs. King and Lochee, at their Great Room, No. 38 King Street, Covent Garden, on Thursday, October 13th, 1807, at 12 o'clock," we find at page 5, "Lot 68, *A large Orrery with glass shade, made by Mr. Ferguson.*" Under date March 27th, 1769, the reader will find copy of a letter from Ferguson to Mr. Hawys.

don: Printed for A. Millar, and sold by T. Cadell, opposite Catherine Street, in the Strand. MDCCLXVII."

FERGUSON IN LIVERPOOL AND IN SCOTLAND.—He appears to have been in Liverpool delivering a Course of Lectures, from about May 20th until the middle of June, again enjoying the company and hospitality of his philosophical friend, Captain Hutchinson, Dock-master. After finishing a Course of Lectures on Experimental Philosophy and Astronomy, he went to Scotland either on business, or for recreation. It is not known what districts he visited—probably his native district, where still remained some relatives and the friends of his early years. His parents had been long removed from the scene; but his elder brother, John, and four of his sisters, were still alive; also, the Rev. Mr. Cooper, of Glass, and Mr. James Glashan, his old and kind-hearted master and enthusiastic admirer. Probably Alexander Cantley might still be found, busy with his Geometry and Sun-Dials.

It is certain that Ferguson was in Edinburgh in 1767, and called on the late Mr. Thomas Reid, Watch and Clock Maker in that city, who, in his "Treatise on Horology," alludes to the visit, as follows:—

"The late James Ferguson, who was eminent as a writer and a lecturer on Natural Philosophy, being in Edinburgh about the year 1767, was so obliging as to communicate to us a description of the wheels, pinions, and endless screws, in a train of wheel-work to produce a mean synodical revolution of the Moon, mentioning that they were (as he supposed), computed and invented by Mudge, but at that time, no account of them had been published, nor was till long afterwards. The effect of this train makes the revolution equal to 29 days 12 hours 44 minutes and 3 seconds." Mr. Reid adds, "It is perhaps as ingenious a mechanical contrivance as can well be imagined, and was executed and put to a clock by Mudge himself, for his sincere noble friend and patron, his Excellency the late Count Bruhl," &c.[258] ("Treatise on Clock and Watch Making, by

[258] Had a public reward been offered for the most complicated and round-about method for producing, by means of wheel-work, a mean synodic revolution of the Moon, this plan by Mudge would undoubtedly have gained the prize, for a more needlessly complicated train of wheels, for such a purpose, we never saw—not to mention the difficulty of its execution. This train-work by Mudge consists of 6 wheels, 4 pinions, and 2 endless screws,—viz. four wheels of 45 teeth each, one wheel of 60, and one of 42 teeth, with pinions of 8, 8, 19, and 3, and

Thomas Reid," 4th Edit., 1849, p. 70; also, MS. "Common Place Book," pp. 86 and 264, Col. Lib. Edin.)

Whilst in Edinburgh, Ferguson was much in the company of Dr. Buchan (afterwards celebrated for his work, "Domestic Medicine"), and Dr. Lind, an eminent Physician and ingenious experimental Electrician.

"TABLES AND TRACTS"—PUBLISHED.—During Ferguson's absence from London, his "Tables and Tracts" were published. This is an octavo volume of 320 pages, illustrated with three copperplate engravings.

Ferguson had from his earliest years taken notes of everything interesting, curious, and useful, and this work entitled "Tables and Tracts," is the result of part of his researches and notanda. The work consists of 75 papers, on a great variety of subjects; and was perhaps the most easily got-up of all his works. There have been several editions of it published,—viz. 1st Edit., 1767.—2d Edit., 1771.—3d Edit., 1777, &c. The annexed is copy of the title-page:—

> "Tables and Tracts relative to several Arts and Sciences. By James Ferguson, F.R.S. London: Printed for A. Millar and T. Cadell, in the Strand. M.DCC.LXVII. Price 5s."

2 endless screws. In his MS. "Common Place Book," where Mudge's Lunar wheel-work is described and illustrated by a pen and ink plan of the wheels, he says,

"I have ventured to make a little alteration in this (Mudge's) train, by making the fixed pinion have only 11 leaves (instead of 19), and the wheel H to have only 26 teeth (instead of 45), and by keeping to all the rest of Mr. Mudge's numbers, I find it makes the mean lunation to consist of 29 d. 12 h. 44 m. 2 s. 53‴, which differs only 7 thirds of a second of time from Mudge's mean lunation numbers; and consequently will differ but one day therefrom in 740571 lunations, or 59875 Julian years."

Ferguson's MS. "Common Place Book" (Col. Lib. Edin.), p. 86.

No train can be found so simple and accurate as that derived from the fraction $\frac{1749}{51649} = \frac{3}{13} \times \frac{11}{29} \times \frac{53}{137}$ or $\frac{11}{29} \times \frac{15}{65} \times \frac{53}{137}$ or $\frac{15}{58} \times \frac{22}{65} \times \frac{53}{137} = 29$ d. 12 h. 44 m. 2·88096 s., a train we used above 40 years ago in one of our astronomical machines. Taking this last train,—viz. $\frac{15}{58} \times \frac{22}{65} \times \frac{53}{137}$,—15 turns round in a day of 24 hours, and drives wheel 58, which has made fast to it wheel 22, which, in its turn drives wheel 65, to which is made fast wheel 53, which drives wheel 137 once round in 29·5305889 days, or 29 d. 12 h. 44 m. 2·88 s. (error in 100 years, about 27 seconds too much). The annexed cut is a sectional view of the

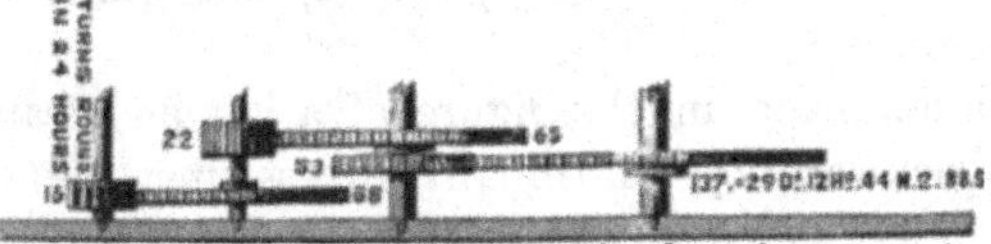

arrangement of the wheel-work of this very simple and accurate lunation train. See also note 142.

(The greater portion of this work was republished in 1823, in Brewster's Ferguson's Essays).

FERGUSON'S CATALOGUE of APPARATUS, and TERMS for READING LECTURES.

Ferguson appends to "Tables and Tracts," (1767) a catalogue of his apparatus for illustrating and demonstrating his Course of Lectures, and his Terms for reading them in London, and within an hundred miles thereof, of which the following is a copy :—

" A List of the Apparatus on which Mr. Ferguson reads his Course of Twelve Lectures on Mechanics, Hydrostatics, Hydraulics, Pneumatics, Dialling, and Astronomy.

The numbers relate to the Lectures read on the machinery to which they are prefixed.

I.

Simple machines for demonstrating the powers of the lever, the wheel and axle, the pullies, the inclined plane, the wedge and the screw.

A compound engine in which all these simple machines work together.

A working model of the great crane at *Bristol*, which is reckoned to be the best crane in Europe.

A working model of a crane that has four different powers, to be adapted to the different weights intended to be raised; invented by Mr. Ferguson.

A Pyrometer that makes the expansion of metals by heat visible to the 90-thousandth part of an inch; so as to be seen by the bare eye at two feet distance from the machine.

II.

Simple machines for showing the centre of gravity of bodies, and how far a tower may incline without danger of falling.

A double cone that seemingly rolls up-hill, whilst it is actually descending.

A machine made in the figure of a human creature, that tumbles backward by continually oversetting the centre of gravity.

Models of wheel carriages; some with broad wheels, others

with narrow; some with large wheels, and others with small; for proving experimentally which sort is best.

A machine for showing what degree of power is sufficient to draw a loaded cart or waggon up-hill; when the quantity of weight to be drawn up, and the angle of the hill's height are known.

A machine for diminishing friction; and showing that the friction is not in proportion to the quantity of the surface that either rubs or rolls; but in proportion to the weight with which the machine is loaded.

A model of a most curious Silk-reel, invented by *Mr. Verrier*, near *Wrington*, in *Somersetshire*.

A large working model of a Water-mill for sawing timber.

A model of a Hand-mill for grinding corn.

A model of a Water-mill for winnowing and grinding corn, drawing up the sacks, and boulting the flour.

A model of Dr. Barker's Water-mill (for grinding corn), in which mill there is neither wheel nor trundle.

A machine for demonstrating that the power of the wind, on windmill sails, is as the square of the velocity of the wind.

A model of the engine by which the piles were driven for a foundation to the piers of *Westminster* bridge.

III.

A machine for showing that fluids weigh as much in their own elements as they do in air.

A machine for showing that, on equal bottoms, the pressure of fluids is in proportion to their perpendicular heights; let their quantities be ever so great or ever so small.

Machines for showing that fluids press equally and in all manner of directions.

A machine for showing how an ounce of water in a tube may be made to raise and support sixteen pound weight of lead.

A machine for showing that, at equal heights, the smallest quantity of water whatever will balance the greatest quantity whatever, if the columns join at bottom.

A machine for showing how solid lead may be made to swim in water, and the lightest wood to sink in water.

Machines for showing and demonstrating the hydrostatical paradox.

A machine for demonstrating that the weight of the quantity of water displaced by a ship is equal to the whole weight of the ship and cargo.

Machines for showing the working of syphons, and the *Tantalus'* cup.

A large machine for showing the cause and explaining the phenomena of ebbing and flowing wells, and of intermitting and reciprocating springs.

IV.

Machines for showing that when solid bodies are immersed and suspended in fluids, the solid loses as much of its weight as its bulk of the fluid weighs; and that the weight lost by the solid is imparted to the fluid.

A hydrostatic balance, for showing the specific gravities of bodies, and detecting counterfeit gold or silver.

A working model of *Archimedes'* spiral pump.

Glass models for showing the structure and operations of sucking, forcing, and lifting pumps.

A working model of an engine for extinguishing fire.

A working model of a quadruple Pump-mill for raising water by means of water turning a wheel.

A working model of the *Persian* wheel for raising water.

A model of the great hydraulic engine under London bridge; that goes by the tides, and raises water by forcing pumps.

V and VI.

An Air-pump, with a great apparatus to it for experiments, showing the weight and spring of the air.

A Wind-gun.

VII.

A large armillary sphere, for showing the apparent motions of the Sun and Moon, with the times of their rising and setting, in all latitudes, and on all the days of the year.

A wooden model of an astronomical clock, showing the apparent motions and times of rising and setting of the Sun, Moon, and Stars, with the age and phases of the Moon, at all times.

Another model of a clock, for showing the apparent motions of the Sun and Stars, with the times of their rising and setting, and the equation of natural days.

A simple machine, by which all the principles of dialling are made evident to sight.

Pardie's universal dial, for finding a meridian line, and showing the true solar time of the day.

Three dials of different kinds joined together; on all of which, the time of the day is shown by the shadow of one stile.

A collection of nine dials, all in one portable instrument, showing the time of the day in all latitudes; and all the places of the Earth where it is day, and where it is night, at any time when the Sun shines on the dial; and all the places of the Earth to which the Sun is rising, and to which it is setting at that time.

An universal dial in the form of a plain cross.

An instrument for finding the true distances of all the Forenoon and Afternoon hours from XII, on horizontal and vertical dials, for all latitudes. And also for finding the hour of the day, and the variation of the compass, at any place; together with the Sun's declination, azimuth, amplitude, and time of rising and setting, in any given latitude.

VIII.

A whirling table, for explaining and demonstrating the laws by which the planets move, and are retained in their orbits: that the Sun and all the planets move round their common centre of gravity: that the Earth and Moon move round their common centre of gravity once every month: that the Earth moves round the Sun, in common with the rest of the planets, and turns round its own axis: that the power of gravity diminishes in proportion as the square of the distance from the attracting body increases: that a double velocity in any orbit would require a quadruple power of gravity to retain the body in that orbit: that the squares of the periodical times in which the planets move round the Sun are in proportion to the cubes of their distances from the Sun. A plain experimental demonstration of the doctrine of the tides; and the cause of their rising equally high at the same time, on opposite sides of the Earth.

IX, X, XI, and XII.

A machine for showing the motions of comets.

An Orrery, showing the real motions of the planets round

the Sun; the apparent stations, direct and retrograde motions of Mercury and Venus, as seen from the Earth: the different lengths of days and nights, and all the vicissitudes of seasons, arising from the diurnal and annual motions of the Earth: the motions and various phases of the Moon: the Harvest-moon: the tides: the causes, times, and returns of all the eclipses of the Sun and Moon: the eclipses of Jupiter's satellites, and the phenomena of Saturn's ring.

In London, any number of persons, not less than twenty, who will subscribe one Guinea each, may have a course of twelve Lectures read on the above-mentioned Apparatus, provided they agree to have at least three Lectures a-week; in which they may appoint the days and hours that are most convenient for themselves.

Within ten miles of London, any number, not less than thirty, may have a course; each subscriber paying one Guinea. And

Within an hundred miles of London, any number of subscribers, not less than sixty, may have a course; each paying as above." ("Tables and Tracts." Lond., 1767, pp. 318—328).

It will thus be seen that Ferguson had upwards of *fifty pieces* of apparatus to illustrate his course of Twelve Lectures. This is the earliest Catalogue of his apparatus we have been able to find.

ASTRONOMICAL CLOCK.—Sometime during this year, Ferguson invented and made the Dial-works of a new Astronomical Clock.

In his MS. "Common Place Book," he gives a lengthened description of the wheel-work of this Clock, accompanied by a fine large sectional drawing in Indian ink, showing the positions of its wheels and acting parts, as also a miniature plan of the Dial-face; annexed to which, written cross-ways on the page, is the following:—

"The Dial-work of a Clock, for showing the Phases of the Moon, the motion of the Earth, the Vicissitudes of Seasons, the Places of the Earth which are enlightened by the Sun at any time of Inspection, with the lengths of Days and Nights at all times of the year, and at all Places of the Earth.—JAMES FERGUSON, 1767."

MS. "Common Place Book," p. 93 (Col. Lib. Edin.)

Of the wheel-work, we may note that the astronomical part

of the Clock has five wheels and one very long pinion,—that a wheel turning round in 24 hours shifts forward a tooth daily, of a large wheel of 365 ratchet teeth, thereby causing this wheel to turn round on its axis in 365 days. A pin in the rim of this wheel works a large terrestrial globe, giving it the proper positions to a stationary Sun opposite its centre, in front of the Dial. This globe projects half way through a large circular opening in the under part of the Dial-face; and the pin in the large wheel just mentioned operates on the machinery of the globe, and causes it to show the Vicissitudes of the Seasons, and the different lengths of Day and Night at all places of the Earth.

1768.

THE YOUNG GENTLEMAN AND LADYS ASTRONOMY PUBLISHED.—Our first note for 1768 refers to the March number of the "Gentleman's Magazine" for that year, from which it would appear that this very simple and celebrated work on Astronomy was published before March, 1768. Of this work, many editions have been published, both in this and other countries. It is written in the form of a Dialogue between "*Neander*" and "*Eudosia*,"—Ferguson being *Neander*, the tutor; his youthful pupil Miss Emblin, *Eudosia*.[259] It is written in language so plain and simple, that "*he who runs may*

[259] As here noticed, this work is in the form of a Dialogue; Neander, the tutor; Eudosia (the youthful Miss Anne Emblin), the pupil, represented as sister of Neander. It may be here noted that Neander was a celebrated physician of Germany in the 16th century (died in 1558). Eudosia was the daughter of Leontius the Sophister, and wife of Theodosius the younger. She wrote the work "Centones Homerici," applying his verses to the History of Christ, (died A.D. 460, aged 60).

Being informed that one of the daughters of "Eudosia" (Mrs. Anne Casborne) resided at New House, Packenham, Bury St. Edmunds, we took leave to address a note to her, respectfully inquiring if she could furnish any interesting particulars respecting her mother, the Eudosia of Ferguson's Dialogues (afterwards the wife of Capel Lofft, Esq.) In her courteous reply, of date July 1861, is the following. * * * "*You are quite right in your conjecture that Anne Emblin, my mother, the pupil of Ferguson, was* (*the first*) *wife of Capel Lofft; and I have often heard from her sister, Martha Hatch, the wife of the Rev. George Avery Hatch, Rector of St. Matthew and St. Peter, Cheapside, London, that my mother was the pupil mentioned, under a Greek name, in Ferguson's Dialogues on Astronomy, and that it suggested to my father, Capel Lofft, the name 'Eudosia' for his poem,* 'THE UNIVERSE.' *I do not know the year my mother became the pupil of Ferguson; but, from my aunt's account, she was very young—had only entered her teens about* 1767—*and Ferguson then resided in the City, in Bolt Court, Fleet Street. She was a little younger than my father, who was born in Nov.* 1752. *From the same authority, I heard some years afterwards that my mother delighted to converse with Ferguson, and often went to stay with him, at which time, she was in all her youth and beauty, and when King George the Third used to call her*

read," and at once acquire a knowledge of the elements of Astronomy. Madame Genlis remarks, "*this book is written with so much clearness, that a child of ten years old may understand it perfectly from one end to the other.*" Editions of this work were published in England, as follows:—1st Edit., 1768. —2d Edit., 1769.—3d Edit., 1772.—4th Edit., 1779.—5th Edit., 1790, &c. Our Edition is the seventh (1804). It is an octavo volume of 182 pages, and embellished with 7 illustrative copperplate engravings. The following is copy of the title-page:—

> "An Easy Introduction to Astronomy for Young Gentlemen and Ladies—Describing The Figure, Motions, and Dimensions of the Earth; The different Seasons; Gravity and Light; The Solar System; The Transit of Venus, and its Use in Astronomy; The Moon's Motion and Phases; The Eclipses of the Sun and Moon; The Cause of the Ebbing and Flowing of the Sea, &c. By James Ferguson, F.R.S. London: Printed for A. Millar and T. Cadell, in the Strand. MDCCLXVIII. Price 5s."

We never saw the first edition of this work; but a friend who had mentions, that the title-page of the first edition was simply as follows:—

> "The Young Gentleman and Lady's Astronomy in Ten Dialogues. By James Ferguson, F.R.S. London: Printed for A. Millar and T. Cadell, in the Strand. MDCCLXVIII. Price 5s."

The third edition we have seen, and its title-page and that of all subsequent editions is the same. At the end of the second Dialogue in this Astronomy, *Eudosia* asks *Neander*—"*What shall I do? I fear I cannot remember much of what you have told me this morning, so as to write it down,*" to which *Neander* replies, "*Never mind* that, *Eudosia, for I believe I shall publish these our conversations, for the sake of other young ladies; many of whom are, no doubt, willing to learn Astronomy, but have nobody to teach them. And then you can have the whole together in print.*" To this *Eudosia* replies, "*If you do, Sir, I must insist upon your not mentioning my name.*" *Neander* assures her that her "*desire shall be complied with.*" Ferguson, in a foot-note, adds, "*Several years ago, I had the pleasure of instructing a young lady, who, for goodness of heart, acute-*

'*the flower of Windsor.' She died on the* 8*th September*, 1801, *aged about* 48 *years.*" See also note 14, p. 13.

ness of judgment, and strong inclinations to learn Astronomy, answered exactly to the account here given of Eudosia." (See conclusion of the second Dialogue of this Astronomy).[260] As both our Neander and Eudosia have been dead for upwards of half a century, Eudosia's name is now under no restriction, and we have therefore given her name in this notice, and also alluded to her in another part of this work.

ELECTRICAL ORRERY.—During this year, Ferguson appears to have been employed in making Electrical Apparatus, amongst which were a curious electrical clock and orrery, figures of which were afterwards given in his Introduction to Electricity. A new kind of electrical orrery was about this period contrived by Mr. Edward King of Lincoln's-Inn, with whom Ferguson was intimate. This is a very simple machine—duly appreciated and highly recommended by Ferguson in his future lectures.

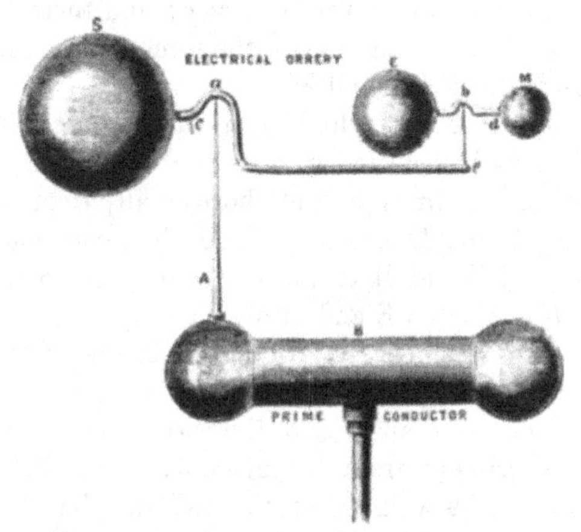

From some old notes, we find that he made several of these electrical orreries, and gave them away as presents to some of his friends who were electricians.

The above woodcut is a representation of "King's Orrery," and is really the original of the electrical orreries sold by Opticians, &c., of the present day. The following is Ferguson's description of it:—

260 In the foregoing extract from Mrs. Casborne's interesting letter, it is noted that "Eudosia" died in 1801, aged about 48. Therefore, it would be sometime in the year 1767, when in her 15th year, that she became Ferguson's pupil.

"A new Experiment in Electricity, showing the Motions of the Sun, Earth, and Moon; by Edward King, Esq., of Lincoln's-Inn.

The Sun and Earth go round the common centre of gravity between them, in a solar year, and the Earth and Moon go round the common centre of gravity between *them* in a lunar month. These motions are represented by an electrical experiment as follows:

The ball S represents the Sun, E the Earth, and M the Moon, connected by bended wires *a c* and *b d*; *a* is the centre of gravity between the Sun and Earth, and *b* is the centre of gravity between the Earth and Moon. These three balls, and their connecting wires, are hung and supported on the sharp point of a wire A, which is stuck upright in the prime conductor B of the electrical machine; the Earth and Moon hanging upon the sharp point of the wire *c a e*, in which is a pointed short pin, sticking out horizontally at *c*; and there is just such another pin at *d*, sticking out in the same manner, in the wire that connects the Earth and Moon.

When the globe (or cylinder) of the electrical machine is turned, the above-mentioned balls and wires are electrified; and the electrical fire flying off horizontally from the points *c* and *d*, cause S and E to move round their common centre of gravity *a*; and E and M to move round their common centre of gravity *b*. And, as E and M are light when compared with S and E, there is much less friction on the point *b* than upon the point *a*; so that E and M will make many more revolutions about the point *b* than S and E make about the point *a*. I had this experiment from my ingenious friend Mr. King; and have adjusted the weights of the balls so that E and M go twelve times round *b* in the time that S and E go only once round *a*.

It makes a good amusing experiment in electricity; but is so far from proving that the motions of the planets in the heavens are owing to a like cause, that it plainly proves they are not. For the real Sun and Planets are not connected by wires or bars of metal; and, consequently, there can be no such metallic points as *a* and *b* between them. And without such points, the electric fluid would never cause them to move: for, take away these points in the above-mentioned experiment, and the balls will continue at rest, let them be ever so strongly electrified." (Vide "Select Mechanical Exercises." Lond., 1773, pp. 132—134).

The wheel-works of Ferguson's electrical orreries were principally constructed with card-board,—we have a card-wheel of one of these electrical orreries, but in a mutilated state. About same time, he made a "*Thunder-House*," of which he in his MS. "Common Place Book," p. 106 (Col. Lib. Edin.), gives a pen and ink drawing, and speaks of it as "*A Capital Experiment.*"

DEATH OF FERGUSON'S PUBLISHER.—Our next note refers to the death of Mr. Andrew Millar, Ferguson's Publisher and Bookseller. He died at Kew on June 8th, 1768, aged 61. The business was afterwards carried on under the firm of "Strachan and Cadell," (successors to Mr. Millar) in the Strand. (Vide note 200).

FERGUSON'S second son, MURDOCH, sent to EDINBURGH UNIVERSITY to STUDY MEDICINE, &c.—Ferguson's second son, Murdoch Ferguson, went to Edinburgh at the end of this year to attend the Medical Classes of its University. Alexander Smith, Esq., Secretary to the University, informs us, by letter, that Murdoch Ferguson is entered in the books of the University of Edinburgh as a Medical Student, and as being in attendance in the classes of Anatomy and Surgery during the session 1768–1769. Murdoch attended one session only. He appears to have been a rather unstable youth, and was the cause of grief to his parents, as shall be again adverted to in the proper place and under the proper date. Ferguson, it will be recollected, was in Edinburgh sometime in the summer of the previous year (1767), and was much in the company of Doctors Buchan and Lind, and it may be assumed he would mention to them his son's predilection for the medical profession, and that they would advise Ferguson to send him to Edinburgh.

VISITS TO THE KING AT KEW, &c.—Ferguson this year, 1768, began, by command of the king, to visit his Majesty at Kew and St. James's. Indeed, from this time, till his death, Ferguson had the honour of frequent invitations to converse with his Majesty on Philosophical and

Mechanical topics, and on the turning of wood, ivory, &c. The king, who took great delight in turning, had at Kew, a turning lathe, with a full complement of tools; and it was admitted by the wood, ivory, and brass turners of that period, that the king was a most expert and ingenious turner in wood and metal.[261]

THE SOLAR SPOTS.—The following engraving of the Sun's disc and spots, as observed by Ferguson on December 3d, 1768,

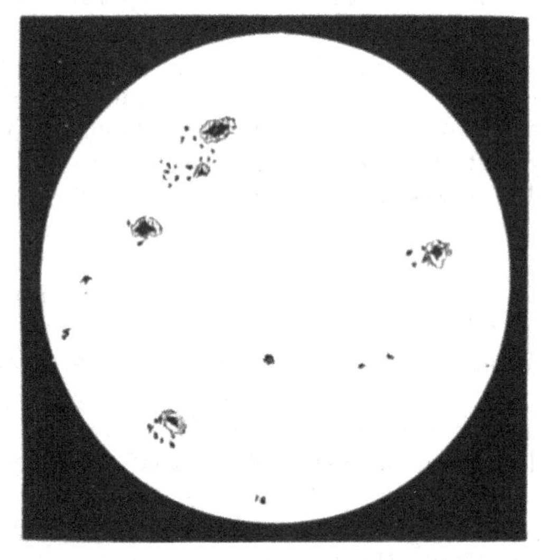

is taken from a short leaf inserted between pp. 256, 257 of his MS. "Common Place Book," (Col. Lib. Edin.) Above the MS. sketch are the words,

"*Spots on the Sun seen on Saturday, Decr. 3d, at Ten o'clock,* A.D. 1768."

And appended is the following note:—

[261] Mr. Henry Mayhew, in one of his works, states that an old turner, gossiping over the reminiscences of his trade, said, "I have given gentlemen lessons in turning. Many gentlemen, and some peers, are very good turners. I gave lessons to a gentleman who had the lathe and all the turning tools and apparatus that old George III. used to work with. They sold for £500 at a sale. I have seen some of the old King's turning, and it was very fair. With industry, he might have made 40s. or 50s. a-week as a hard-wood and ivory turner." (Vide "Novelties, Inventions, and Curiosities in Arts and Manufactures," 3d Edit. Lond., 1853, p. 254).

Regarding Ferguson's visits to the King at Kew, vide Dr. Houlston's letter in the Appendix; also, conclusion of Memoir in Partington's edition of Ferguson's "Lectures on Select Subjects."

"*I observed the Sun at Mr. Nairne's, Cornhill, and made the above drawing, which was compared with the Sun by Mr. Nairne. I never saw so many spots upon the Sun at any time before.—J. F.*"

1769.

PUBLICATIONS.—Early in the year 1769, Ferguson published the sixth edition of his "Analysis of a Course of Lectures," &c., see page 278; also, a Syllabus of his Course, with the following title:—

> "Syllabus of a Course of Lectures on the most interesting parts of Mechanics, Hydrostatics, Hydraulics, Pneumatics, Electricity, and Astronomy. By James Ferguson, F.R.S. (*Philosophia mater omnium bonarium artium est,*—Cicero, 1 Tus). London: Printed in the year MDCCLXIX."

TRANSIT OF VENUS.—Sometime in March, 1769, Ferguson published, on a large sheet, a projection and description of the then forthcoming Transit of Venus, being a copy of the paper which he had some time before sent to the Royal Society. It has been long out of print; we understand that there is a copy of it in the British Museum.

LETTER FROM FERGUSON TO MR. HAWYS.—Mr. Hawys, in March, 1769, sent a letter to Ferguson respecting some errors, &c., in "Tables and Tracts." The following is his reply, copied from the original in our collection. It is addressed on the cover, "To Mr. Charles Hawys, Charter House Square, London."

"SIR,

I was in the Country when my Tables and Tracts were printed. Hence some errors arose that probably would not if I had been in Town, and had had proper time to correct the Press.

In page 110, the lengthening of the Pendulum in Lat. 90 deg. should be 0·2065 inches. The integer 2 should be 0.

In the Table of remarkable Æras and Events (at the end, pag. 179), Please to correct as in the margin, where the numbers under Julian Period 4618 and 4640 are wrong.

Julian Period
For 4618 read 6418
— 4640 —— 6440

But (just now) on looking at the Errata, I find these last numbers are mentioned.

There is no absolute determinate proportion between the weight of a pendulum Ball and the wire by which it hangs—only, the less the wire weighs in proportion to the Ball, the nearer will the length of the wire (or rather the distance between the point of suspension and point of oscillation) be to what is calculated in the Table you mention, with respect to the standard Pendulum for vibrating seconds.

You have not told me whether your Clock in the Country swings seconds or not. But this you may depend upon; that, whether it does or not, the alteration of the Barrel will not alter the going of the Clock, as to its keeping time. Only, if you reduce the size of the Barrel so as to make it spend less time, you must add as much to the present weight that keeps the Clock going as you diminish the barrel below its present size. I apprehend the Clock goes only 24 hours; or at least, must be drawn up every day. As much as the Barrel is lessened so much less line or cord it will spend; and the same Pendulum will still do.

As to De Moivre's Tables, I confess I do not understand them, having never studied anything of Annuities on Lives. And I only put in as much as my Bookseller desired me, with just as much explanation as he marked out in De Moivre's Book; so that I am really sorry I cannot answer your question, who am,

SIR,

Your most humble servant,

JAMES FERGUSON.

BOLT COURT, FLEET STREET,
March 27th, 1769."

COMET'S ORBIT PROJECTION.—Among drawings in the Library of George III. (now in the British Museum), is a projection, by Ferguson, of a Comet's path, on paper 14 inches by 8, entitled,

> "A Projection of that part of the Eastern half of the Heavens in which the Comet was seen at London; Wednesday, Septr. 13th, 1769, at 40 minutes after 3 in the morning; delineated by J. Ferguson."

This projection shows Cancer, Gemini, Taurus, Orion, Lepus, Canes Major, &c. As Ferguson about this period sometimes visited George III. at Kew, it is probable that he gave this projection to the King on one of those visits.

FERGUSON, and DR. JOHNSON.—The following (which may be referred to about this period) is from "Boswell's Life of Johnson:"—"1769, October 26th.—There was a pretty large circle this evening.[262] Dr. Johnson was in very good humour, lively and ready to talk upon all subjects. Mr. Ferguson, the self-taught philosopher, told him of a newly invented machine which went without horses; a man who sat in it turned a handle, which worked a spring that drove it forward. Then, Sir, said Johnson, what is gained is, the man has his choice, whether he will move himself alone, or himself and the machine too." [263]

IMAGINARY SALE OF BUSTS—Martin and Ferguson.—Glancing over Dodsley's *Annual Register* for 1769, we stumbled upon an amusing effusion in rhyme.—Subject—an imaginary sale of Busts. Several are sold; to Garrick is sold the head of *Roscius;* this disposed of, the head of Newton is put up, for which *Martin and Ferguson contended.* No doubt Ferguson would see this poetical bill of sale—would smile at the allusion to himself as one of the *bidders,* and at the idea of night compelling him to leave the sale and to *lecture home.*

These are the lines referring to "Newton's head," &c.—

"For Newton's head, whose piercing eyes
Explor'd the wonders of the skies,
Who could with rectitude declare
The size and distance of each star,
Martin and Ferguson contended;
And how the contest would have ended

262 Dr. Johnson's residence at this period was No. 7 Johnson's Court, Fleet Street. This was his residence from about 1765 to 1775, when he removed to Bolt Court (the Court in which Ferguson then resided).

263 The following paragraph is extracted from the Leeds Mercury of April 11th, 1769:—

"A correspondent writes that Mr. Moore's newly invented machine to go without horses, for which he obtained his Majesty's Patent, is not only adapted to wheel carriages, in general, such as coaches, chaises, carts, waggons, &c., but to ploughing, harrowing, and every other branch of husbandry; also, to all other machines and engines now in use throughout the kingdom; in various branches of manufacture wherein draught horses are now employed. We hear that the ingenious inventor has sold all his own horses, and by his advice many of his friends have done the same, because the price of that noble and useful animal will be so affected by their new invention, that their value will not be one-fourth of what it is at present." Vide London Mechanics' Magazine, vol. 16th, p. 135.

Also, in the "Patent List," 1769, No. 930, it is recorded—"T. Moore, Linen Draper, Several multiplying levers, or additions of power, that will expedite and extend motion and reduce vibration, by which power is gained without a proportionate loss of time,—superior to what hath been yet practised; which will be of great public utility in agriculture, carriages, navigation, mills, engines, and other things where mechanical power is or can be used," &c. It is likely that Ferguson referred to this "*new invention*" by Mr. Moore on the above occasion.

I know not, had not evening come
And called them both to lecture home;
They gone, no bidders could I see,
So light was held Philosophy!"

Vide "A Familiar Epistle to a Friend," Dodsley's *Annual Register* for 1769, p. 226.

Martin, Ferguson's rival bidder at this imaginary sale, was the celebrated Benjamin Martin, Optician and Mathematical Instrument Maker in Fleet Street, from 1750 to 1782. He was the author of many works on Natural and Experimental Philosophy, Astronomy, Biography, &c., and a renowned maker of Orreries, Telluriums and Planetariums. Brewster's Edinburgh Encyclopædia, at p. 629, refers to one of these machines of his construction. As an author, lecturer, and a maker of orreries, &c., he was Ferguson's rival. His shop was known as the *Newton's Head*, a bust of the philosopher being over his door. Martin was born at Chichester in 1704—died in London in 1782, aged 78. Ferguson, in 1769, resided in Bolt Court, Fleet Street, close by the "Newton's Head," and consequently was a near neighbour to Martin.

THE SOLAR SPOTS.—The following view of and short note on the Solar Spots, is copied from a leaf inserted between pp. 256, 257 of Ferguson's MS. "Common Place Book" (Col. Lib. Edin.).

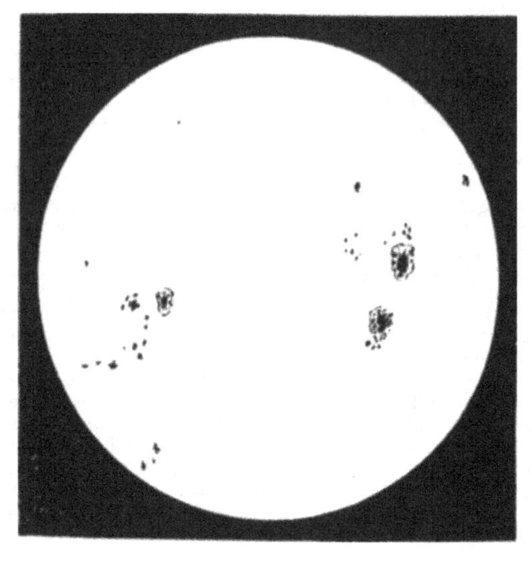

Over the view are the words, "*Spots seen on the Sun, Tuesday, Nov*r. *21st, 1769, at* 10 *o'clock.*"

Annexed to the sketch is the following note :—

"*I observed the Sun at Mr. Mudge's in Fleet Street, and made this drawing of the Spots, which are more in number than those seen at Mr. Nairne's in Cornhill on 3d December,* 1768. *But I am informed that the greatest spot now seen is not near so large as when observed some days ago.—J. F.*"

The shop of Mr. Mudge, Watchmaker, in Fleet Street, was a few doors (on the same side of the street) to the east of Ferguson's residence in Bolt Court. The above sketch is on the same interpolated leaf of the "Common Place Book" as the sketch of December 3d, 1768, is delineated on.

1770.

LECTURES at NEWCASTLE-ON-TYNE.—Early in the year 1770, Ferguson was at Newcastle-on-Tyne, &c., delivering a Course of twelve Lectures on Mechanics, Hydrostatics, Hydraulics, Pneumatics, Electricity, and Astronomy.

INTERVIEW WITH Dr. HUTTON, (NEWCASTLE).—Ferguson, immediately after his arrival in Newcastle, appears to have called on Dr. Hutton, by whom he was kindly received. The doctor gave him the use of his school for his lectures, and also interested himself among his friends by disposing of tickets for the course.

Dr. Hutton, in his Tracts, makes mention of Ferguson's professional visit to Newcastle; and particularly of a large drawing which Ferguson had—showing how to divide the area of a Circle into any number of equal parts by means of concentric circles. Dr. Hutton says, "About the year 1770, when Mr. James Ferguson, the ingenious lecturer on Astronomy and Mechanics, in his peregrinations came to Newcastle, where I was then residing, to give the usual course of his public lectures, on which occasion, with the assistance of my friends, I not only procured him a numerous and respectable audience, but also accommodated him with the free use of the new school rooms, which I had lately built, to deliver his lectures in. As Mr. F. commonly amused my family and friends at evenings, with showing his ingenious mechanical contrivances and drawings, on one of these occasions he produced a very neat and correct drawing, on a large scale, being a construction of a problem—showing how to divide the Area of a Circle into any number of equal parts, in the very prolix way as given by Hawney; but which he exhibited as a great curiosity. I ventured to remark to him that I thought a much simpler construction might be found out for this problem, which was then new to me. As Mr. F. expressed a wish to see such a thing as a simpler construction, which however he seemed to have his doubts of procuring, I was induced to consider it that evening before going to rest, and discovered the construction.

"The next morning I showed him the new and very simple construction, with its demonstration, which he seemed much pleased with, on account of the apparent simplicity, but doubted very much that it might not be correctly true. On referring him to the accompanying demonstration to satisfy himself of its geometrical truth, I was much surprised by his reply, that he could not understand that, but he would make the drawing correctly on a large scale, which was always his way to try if such things were true. In my surprise I asked where he had learned Geometry, and by what Euclid, or other book; to which he frankly replied he had never learned any Geometry, nor could ever understand the demonstration of any of Euclid's propositions.

"Accordingly the next morning, with a joyful countenance, he brought me the construction neatly drawn out on a large sheet of pasteboard, saying he esteemed it a treasure, having found it quite right, as every point and line agreed to a hair's breadth by measurement on the scale." [264] This problem and

[264] Charles Hutton, LL.D., F.R.S., &c., an eminent Mathematician, born at Newcastle-on-Tyne, 14th August, 1737. He was a teacher in his native town from about the year 1758 to 1773, when he was elected Professor of Mathematics to the Royal Military Academy at Woolwich. He contributed many papers to the Royal Society, amongst which may be mentioned his papers "*On the Force of fired Gunpowder;*" "*On the initial Velocities of Cannon Balls;*" "*On the Mean Density of the Earth.*" In 1783, "*Elements of Conic Sections.*" In 1785 he published his "*Mathematical Tracts.*" In 1786, his "*Tracts on Mathematical and Philosophical Subjects.*" In 1795, "*Mathematical and Philosophical Dictionary.*" 1796, "*A Course of Mathematics;*" and translated Montucla's work, "*Montucla's Recreations in Mathematics and Natural Philosophy.*" In 1803, he, in conjunction with Drs. Pearson and Shaw, undertook the task of abridging the Philosophical Transactions, which were completed in 1809 in 18 quarto volumes. Was for sometime foreign Secretary of Royal Society. Seceded from the R. S. in 1784. He died 27th January 1823 in the 86th year of his age.

the construction he afterwards inserted in his "Select Mechanical Exercises," p. 123, printed in 1773.

The problem alluded to by Dr. Hutton is as follows:—"To divide a given Circle into any proposed number of equal parts by means of other circles concentric with the given one." (Hutton's Tracts; Prob. 34, vol. 3, p. 379).

Ferguson proposes the question thus:—"To divide the Area of a given Circle into any required Number of equal Parts, by concentric Circles."

"In the annexed figure let A B D E be a circle, whose area is required to be divided into 5 equal parts by concentric circles, as F G H I.

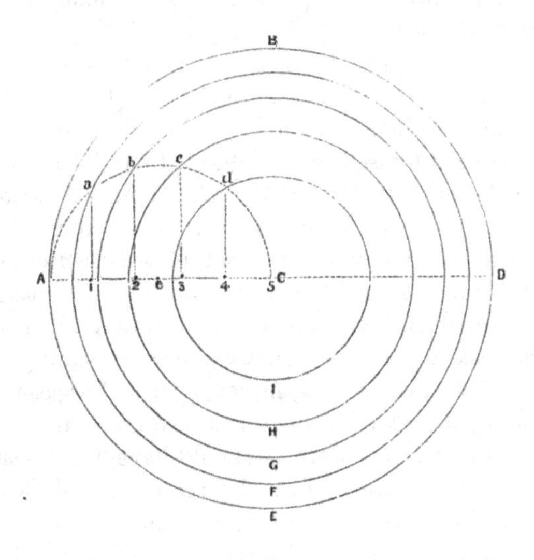

"Divide the semi-diameter A C into 5 equal parts, as A 1—1 2—2 3—3 4—4 5; and on the middle point *e* as a centre, with the radius *e* A, describe the semi-diameter A *a b c d* C. From the points of equal division at 1, 2, 3, and 4, and perpendicular to A C, raise the perpendiculars 1 *a*, 2 *b*, 3 *c*, 4 *d*, till they meet the semicircle in the points *a*, *b*, *c*, and *d*: through which points, draw the concentric circles F, G, H, I, and the thing will be done.

"Supposing that five blacksmiths should agree to buy a grinding-stone among them, each paying an equal share of the price, and that each man should therefore have the use of the stone, to wear off a fifth part of it, till it came to the last man, who was to wear it out; the first man should wear the stone from E to F, the second, from F to G, the third, from G to H, the fourth, from H to I, and the fifth, from I to the centre or axle C.

"By this easy method, which I learnt of Mr. Hutton, teacher of the Mathematics at Newcastle, the area of any circle may be divided by concentric circles into any required number of equal parts. For, into whatever number of equal parts the radius A C be divided, the area of the circle will be divided into the like number of parts, all equal among themselves." ("*Select Mechanical Exercises*," pp. 123, 124)

Dr. Hutton states that Ferguson made a drawing on a sheet of pasteboard, of the method he had given him for solving the problem in question, and on presenting it to the doctor, Ferguson expressed himself quite delighted with the new solution, as he had found that it was *quite right.* A writer in the Edinburgh Review referring to this point, remarks that "it was probably by measuring the radius of each circle, on a scale to which his figure was adapted, and thence computing the area (the rule for which he no doubt took for granted), he would find the difference of the contiguous circles constantly the same. It is a curious circumstance, however, that Ferguson, who had so strong a genius for Mechanics, and so much invention wherever machinery was concerned, should have had so singular an incapacity for comprehending the reasonings of Geometry, at the same time that he had taste sufficient to admire the beauty of its conclusions."—*Edinburgh Review*, vol. 22, p. 95, 1813–1814.

TIDAL CLOCK FOR LONDON BRIDGE.—In the year 1770, Ferguson contrived and made a Clock to show the state of the tides at London Bridge. This Clock was more simple in its construction than the one he contrived for Captain Hutchinson, in the year 1764, for showing the state of the tide at Liverpool. Ferguson describes this new Tide Clock (illustrated by drawings) in "Select Mechanical Exercises," here transcribed,—

"A Clock showing the apparent daily Motions of the Sun and Moon, the Age and Phases of the Moon, with the Time of her coming to the Meridian, and the Times of High and Low Water, by having only two Wheels and a pinion added to the common Movement.

The dial-plate of this Clock is represented by Fig. 1 in the annexed engraving. It contains all the 24 hours of the day and night.

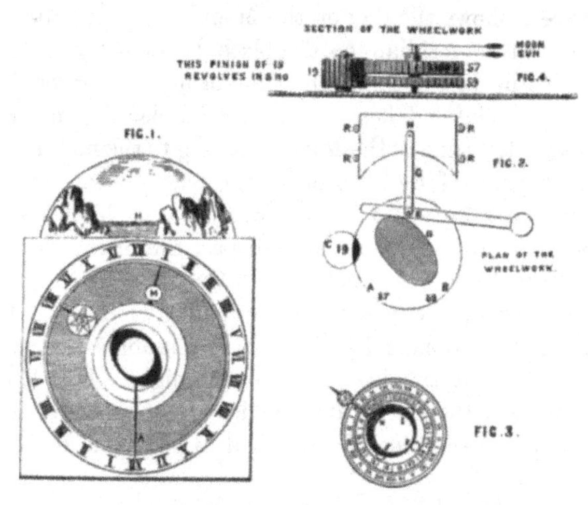

Dial-face, Wheel-work, &c., of Ferguson's Tidal Clock.

S is the Sun which serves as an hour-index, by going round the dial-plate in 24 hours; and M is the Moon which goes round in 24 hours 50½ minutes, from any point in the hour-circle to the same point again, which is equal to the time of the Moon's going round in the heavens, from the meridian of any place to the same meridian again.

The Sun is fixed to a circular plate (as in Fig. 3) and carried round by the motion of that plate, on which the 24 hours are engraven, and within them is a circle divided into 29½ equal parts for the days of the Moon's age, accounted from the time of any new Moon to the next after; and each day stands directly under the time (in the 24 hour circle) of the Moon's coming to the meridian, the XII under the Sun standing for mid-day, and the opposite XII for midnight. Thus, when the Moon is 8 days old, she comes to the meridian at half an hour past VI in the afternoon; and when she is 16 days old, she comes to the meridian at 1 o'clock in the morning. The Moon M (Fig. 1) is fixed to another circular plate, of the same diameter with that which carries the Sun; and this Moon-plate turns round in 24 hours 50½ minutes. It is cut open, so as to show some of the hours, and days of the Moon's age, on the plate below it that carries the Sun, and, across this opening, at *a* and *b*, are two short pieces of small wire in the Moon-plate. The wire *a* shows the day of the Moon's age, and time of her coming to the meridian, on the plate below it that carries the Sun; and the wire *b* shows the time of high water for that day, on the same plate. These wires must be placed as far from one another, as the time of the Moon's coming to the meridian differs from the time of high water at the place where the clock is intended to serve. At London-Bridge, it is high-water when the Moon is two hours and an half past the meridian.

Above this plate, that carries the Moon, there is a fixed plate N, supported by a wire A, the upper end of which is fixed to that plate, and the lower end is bent to a right angle, and fixed into the dial-plate at the lowermost or midnight XII. This plate may represent the Earth, and the dot at L London, or any other place at which the clock is designed to show the times of high and low water.

Around this plate is an elliptical shade upon the plate that carries the Moon M: the highest points of this shade are marked *High Water*, and the lowest points *Low Water*. As this plate turns round below the fixed plate N, the high and low water points come successively even with L, and stand just over it at the times when it is high or low water at the given place; which times are pointed out by the Sun S, among the 24 hours on the dial-plate: and, in the arch of this plate, above XII at noon, is a plate H that rises and falls as the tides does at the given place. Thus, when it is high water (suppose at London) one of the highest points of the elliptical shade stands just over L, and the tide-plate H is at its greatest height: and when it is low water at London, one of the lowest points of the elliptical shade stands over L, and the tide-plate H is quite down, so as to disappear beyond the dial-plate.

As the Sun S goes round the dial-plate in 24 hours, and the Moon in 24 hours 50½ minutes, the Moon goes so much slower than the Sun as to make $28\frac{1}{2}$ revolutions in the time the Sun makes $29\frac{1}{2}$; and therefore the Moon's distance from the Sun is continually changing: so that, at whatever time the Sun and Moon are together, or in conjunction, in $29\frac{1}{2}$ days afterwards they will be in conjunction again. Consequently, the plate that carries the Moon moves so much slower than the plate that carries the Sun, as always to make the wire *a* shift over one day of the Moon's age on the Sun's plate in 24 hours.

In the plate that carries the Moon, there is a round hole *m* through which the phase or appearance of the Moon is seen on the Sun's plate, for every day of the Moon's age from change to change. When the Sun and Moon are in conjunction, the whole space seen through the hole *m* is black; when the Moon is opposite to the Sun (or full) all that space is white: when she is in either of her quarters, the same space is half black half white: and different in all other positions, so as the white part may resemble the visible or enlightened part of the Moon for every day of her age.

To show these various appearances of the Moon there is a black shaded space (Fig. 3) as N *f* F *l*, on the plate that carries the Sun. When the Sun and Moon are in conjunction, the whole space seen through the round hole is black, as at N: when the Moon is full, opposite to the Sun, all the space seen through the round hole is white, as at F: when the Moon is in her first quarter, as at *f*, or in her last quarter, as at *l*, the hole is only half shaded; and more or less accordingly for each position of the Moon with regard to her age; as is abundantly plain by the Figure.[265]

The wheel-work and tide-work of this clock is represented by Fig. 2, in which A and B are two wheels of equal diameters. A has 57 teeth, its axis is hollow, it comes through the dial of clock, and carries the Sun-plate with the Sun (S, in Fig. 1). B has 59 teeth, its axis is a solid spindle, turning within the hollow axis of A, and carrying the Moon-plate with the Moon (M in Fig. 1). A pinion C of 19 leaves takes into the teeth of both

265 This way of showing the varied Phases of the Moon is simple and ingenious, but it will not show the varied phases correctly, only approximately. It is only by the rotation of a ball, half black half white, in $29\frac{1}{2}$ days, that the varied phases as shown in the heavens can be represented, and this method Ferguson has adopted in some of his other astronomical machinery.

the wheels, and turns them round. This pinion is turned round by the common clock-work, in 8 hours; and, as 8 is a third part of 24, so 19 is a third part of 57: and therefore the wheel A of 57 teeth, that carries the Sun, will go round in 24 hours exactly. But, as the same pinion C (that turns the wheel A of 57 teeth) turns also the wheel B of 59 teeth, this last wheel will not turn round in less than 24 hours 50½ minutes of time; for as 57 teeth are to 24 hours, so are 59 teeth to 24 hours 50½ minutes, very nearly.[266]

On the back of the Moon-wheel of 59 teeth is fixed an elliptical ring D, which, as it turns round, raises and lets down a lever E F, whose centre of motion is on a pin at F; and this, by means of an upright bar G, raises and lets down the tide-plate H, twice in the time of the Moon's revolving from the meridian to the meridian again. The upper edge of this plate is shown at H, in Fig. 1, and it moves between four rollers R, R, R, R, in Fig. 2."

In his recently discovered MS. "*Common Place Book*," we find drawings and a description of this Clock, being similar to what is here given. Along the foot of the Dial-face in the MS. are the words PEREUNT ET IMPUTANTUR in large pen-and-ink lettering, being the same motto that we find on his Card-Dial (p. 246).—MS. "*Common Place Book*," p. 95, Col. Lib. Edin.

At the conclusion of his MS. description, Ferguson adds, "*I contrived this Clock in July*, 1770."

Ferguson, after concluding his description of this Clock, adds, "I have made one of these clocks to go by the movement of an old watch, in the following manner:—

266 These are Ferguson's old and well-known numbers, already noticed when describing other pieces of clockwork—57 : 24 : : 59 = 24 h. 50 m. 31 s. 35 thds., the true mean apparent period being 24 h. 50 m. 28 s. 22 thds. 42 fourths. The relative velocities of these wheels may be viewed as a kind of "*Mechanical Paradox*," for with respect to any given points in the two wheels being brought exactly over or against each other, the point in wheel 59 will fall back continually from the point in wheel 57; and while the point in wheel 59 is thus *going back*, it is also *going forward!* (both wheels turn round in the same direction).

We may here notice that the mean apparent diurnal revolution of the Moon is said to be accomplished in 24 h. 50 m. 28 s. 22 thds. 48 fourths, this period appears to be 3 thds. 6 fourths too slow, which may be tested as follows:—Suppose the lunation to consist of 29 d. 12 h. 44 m. 2·88 s., decimally reduced = 29·5305887 d.; in that time, the Moon makes 28·5305887 apparent revolutions from the meridian to the meridian again; hence, $\frac{29 \cdot 5305887}{28 \cdot 5305887}$, and 29·5305887 ÷ 28·5305887 = 1·035050104 × 24 = 24·841202496 h. × 60 × 60 × 60 × 60 = 24 h. 50 m. 28 s. 19 thds. 42 fourths.

"The first or great wheel of a watch goes round in four hours. I put a wheel of 20 teeth on the end of the axis of that wheel to turn a wheel of 40 teeth, on the axis of a pinion C of 19 leaves, by which means *that* pinion is turned round in 8 hours." Two wheels A B of equal diameters, and having 57 and 59 teeth respectively, are turned by it,—"the wheel A in 24 hours, and the wheel B in 24 hours $50\frac{1}{2}$ minutes." [267]—"*Select Mechanical Exercises,*" 3d Edit. London, 1790, p. 19.

Plan of the Astronomical Wheel-work of the Solar and Lunar Watch.

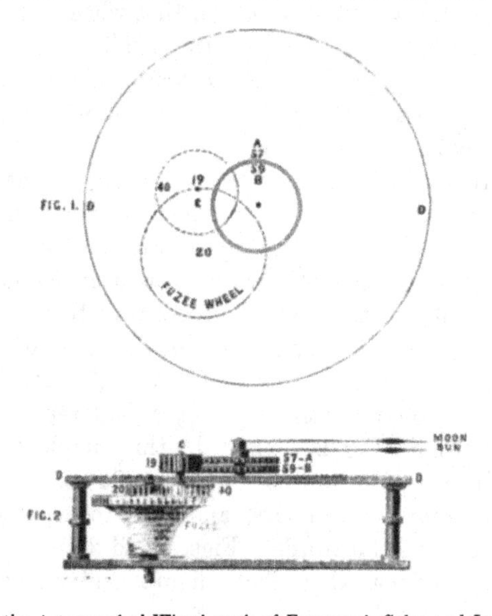

Section of the Astronomical Wheel-work of Ferguson's Solar and Lunar Watch.

As there is no engraving of the plan of the wheel-work of this solar and lunar watch, in any of Ferguson's works; we shall give a vertical and sectional plan of it, taken from our own drawings.

[267] Ferguson here mentions that he "*put a wheel of* 20 *teeth on the end of the axis of that wheel,*" viz. the fuzee wheel. There must be a mistake here; for if the small wheel 20 be put on said axis, how could the watch be wound up?—wheel 20 would turn about with the key in the act of winding; and consequently, would in its turn act upon and turn round wheels 40, 19, 57, and 59, and derange the positions of the hands. The small wheel 20 must therefore not be placed on the axis of the fuzee, *but made fast to the back of the fuzee wheel itself,* and then when the key is applied to the axis of it, and in the act of winding up, these wheels will not be in the least affected thereby.

Fig. 1 is a vertical or ground plan of the wheel-work. Fig. 2 shows it in section as applied to the watch. Description:—The fuzee wheel is assumed to turn once round in 4 hours; to the back of the fuzee wheel is made fast by pins a small wheel of 20 teeth, which drives a wheel of 40 teeth once round in 8 hours. The axis of this wheel of 40 teeth ascends through the frame D D of the watch (shown in Fig. 2 at *c*); on this axis is made fast a small wheel of 19 teeth, which takes into and drives round a wheel of 57 teeth, shown in Figs. 1 and 2 at A; now as the small wheel 19 turns once round in 8 hours, it is evident that if 19 be made to turn a wheel of 57 teeth, that wheel will turn once round in 24 hours; accordingly, the small wheel 19 drives wheel A (in Figs. 1 and 2), of 57 teeth, causing it to turn round in 24 hours precisely, which wheel is made to carry an index, having on its extremity a small ball representing the Sun; the small wheel 19, also turns a wheel *under* wheel 57, having 59 teeth; consequently, this wheel of 59 must move much slower than wheel 57, (in the ratio of $\frac{57}{59}$), so much so, that instead of turning round in 24 hours, with wheel 57, it will not turn round in less time than 24 hours 50 minutes 31 seconds 35 thirds; the hollow axis of wheel 59 ascends through the hollow axis of wheel 57, and carries an index, carrying on its extreme point a ball representing the Moon, as shown in Fig. 2. Thus, by this simple contrivance, the Sun will make a revolution round the 24 hours on the dial-plate of the watch in 24 hours, and the Moon in 24 hours 50 minutes 31 seconds 35 thirds. Figs. 1 and 2, in the above cut will be easily understood, as both figures and letters on them have reference to the same details.[268]

In concluding his description of this Clock and Watch in his MS. "Common Place Book," he says,

[268] If 57 : 24 h. : : 59 ?—59 × 24 = 1416 ÷ 57 = 24 h. 50 m. 31 s. 35 thds., which is a motion *too slow* by 3 s. 13 thds. The true mean apparent diurnal revolution of the Moon is accomplished in 24 h. 50 m. 28 s. 22 thds. 48 fourths. (See note 154).

We may here mention that the late Mr. Andrew Reid, watchmaker, London, who died at Brixton in 1835, aged 85, had often told us that he was on familiar terms with Ferguson; was often at his house in Bolt Court, Fleet Street, and had there seen several works of watches fitted up with a variety of astronomical motions, as also strong works noted for their correct time-keeping—one pair of double silver cases suited all the works. These cases, with one of the works, came into Mr. A. Reid's possession in 1801, from whom it was purchased by the writer.

"I contrived this Clock in July, 1770, and put the movement of an old Watch to it, by means of which it is now going very well (March 10th, 1775); the two wheels A and B are only of Card-paper, and yet I have reason to believe that they will last 30 years.—J. F."

No traces whatever of either this Clock or Watch are now to be found.—A writer in the "Operative Mechanic" "*thinks it extremely probable*" that Ferguson "*had the dial of the Clock at Hampton Court in his eye when he contrived this Clock.*" See the "Operative Mechanic, and British Mechanist," p. 495. We are of a different opinion; although the Clock at Hampton Court shows the apparent daily revolutions of the Sun and Moon,—it does not exhibit the rise and fall of the tides; besides, Ferguson made a Clock in 1747 which exhibited the apparent motions of the Sun and Moon. (See pp. 113—120).

TIDE ROTULA.—Sometime during this year, Ferguson contrived and made a *Tide Rotula*, which consisted of one *fixed* or *immoveable* card of about 8 inches in diameter, having round its circumference twice twelve hours in Roman characters; above this immoveable outer card is a circular moveable card,

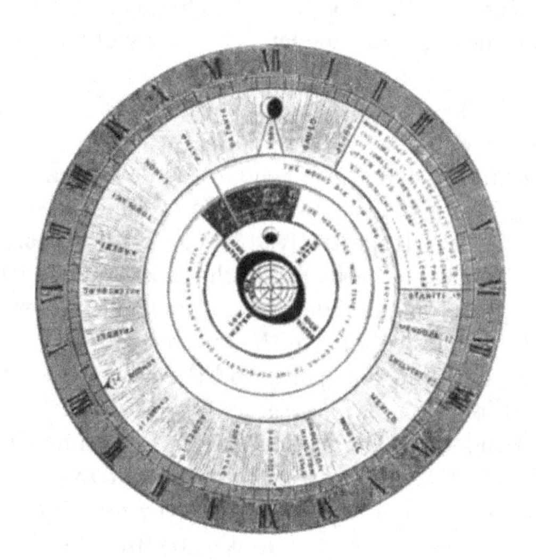

having in a broad circular space round its circumference, the names of many Seaports in their relative positions to times of High water; within this circle of names of Seaports there are two other concentric circles,—the first having round it twice twelve hours in small Roman characters,—the second, or interior circle, has round it the 29½ days and other divisions of the Moon's age, and round the centre of *this* circle there is a black elliptical shade for producing the varied phases of the Moon through a small circular opening in the card above, and done in the same way as in the Clock last described. Above this last-mentioned card there is a smaller one having a Moon fixed to it, and directly under that is the small circular opening in which are exhibited the phases of the Moon as before noticed; round the centre is the figure of the Earth and some of its several circles, surrounded by an ellipsis representing water; all which particulars are shown in the annexed woodcut of the Rotula.

Of this new Tide Rotula, the following is Ferguson's description, from which it will be seen that this instrument is only a modification or improvement of the Tidal Clock, for showing the times of High and Low water at London Bridge, just described.

"This Rotula," says Ferguson, "is only an improvement on the Clock just described, by making it show the time of the Day and Night in the most remarkable places in the world at any instant of time at London, or at any other given place, without one single additional wheel.

As it is in the scheme,—(without being made to go by means of a clock or watch, or any other method),—it is not useless; for, if you put London to the time of the day or night there, all the other places will point to the time of the day or night then opposite to them respectively. If you put the narrow black slip, under the Moon, to the day of the Moon's age (seen through the opening) on the middle plate, the same slip will cut the time of the Moon's coming to the meridian for that day, in the circle of 24 hours on that plate, and the other slip (directly opposite to the words, *High Water*) will cut the time of High water at London Bridge on that Day, in the same circle of 24 hours. And at any time, if London be put to that time, and the Moon's slip to the day of her age, you may then know the state of the Tide at London, by the position of the elliptical shade; the longest diameter of which will be even with the two XII's. in the outermost circle of 24 hours when it is High Water at London Bridge, and even with the two VI's when it is Low Water. In the intermediate times, the position of the elliptical shade shows whether the Tide is ebbing or flowing.—J. F., 1770."[269]

"INTRODUCTION TO ELECTRICITY," PUBLISHED.—In 1770, Ferguson went to press with a new work—very different from any of his previous Treatises—viz. an Elementary Treatise on Electricity, a small octavo of 140 pages, illustrated by three folding-plate engravings of the Electrical Machine and apparatus. This work has passed through several editions,—viz. in 1770,—1775,—1778,—1790,—1823,—1825, &c. The following is copy of the title-page. (The edition of 1823 is in Brewster's Ferguson's Essays).

> "An Introduction to Electricity—in Six Sections. I. Electricity in General. II. A Description of the Electrical Machine. III. A Description of the Apparatus (belonging to the Machine) for making Electrical Experiments. IV. How to know if the Machine be in good order for performing the Experiments, and how to put it in order if it be not. V. How to make the Electrical Experiments, and to preserve Buildings from Damage by Lightning. VI. Medical Electricity. Illustrated with Copper-plates. By James Ferguson, F.R.S. London: Printed for W. Strachan and T. Cadell in the Strand. MDCCLXX. Price 4s."

Plate 1 exhibits the surface of a large table containing a variety of apparatus. On one end of it is fixed a Globe Electrical Machine, with multiplying motion—in front are three large Leyden Phials,—A Florence-flask,—A Thunder-house—and the Conductor is furnished with "A ring of bells,"—Pith-balls,—Star · Wheel,—Attraction Plates,—Feathers,—Spichar, &c., probably the same machine and apparatus which he bought from Mr. Nairne, Optician, &c., Cornhill. Plate 2. View of the works and Dial-face of an Electrical Clock.—An Electrical Wheel-work Orrery.—A Mill.—Toothach instrument, &c. Plate 3. View of an Electrical Triple-pump.[270]

269 A Rotula—in the autograph of Ferguson—similar to that described, is in the possession of William Topp, Esq., Ashgrove, near Elgin, who kindly sent us a copy of it.

270 We have a beautifully-executed Indian-ink drawing, by Ferguson, entitled "An ELECTRICAL MACHINE, with a new ELECTROMETER. *Invented by Mr. John Lane in Aldersgate Street, London;*" and in the lower right-hand corner of the drawing we find, "*J. Ferguson, Delin.*"—also a wheel cut out of card—it has 45 teeth most accurately cut into it,—likewise a trundle wheel of 15 cogs, with three cranks formed on its axis, which appears to have formed part of Ferguson's Electrical Apparatus—probably part of the model of a mill, or electrical pump models.

As just observed, Section VI. of this work is devoted to "Medical Electricity." Ferguson, after describing several curious cases and experiments performed on the subject, says,

> "*For my own part, being but a young electrician, I can have very little to say with respect to the Medical part. But, as far as I have had experience, I shall here relate the facts.*" (Ferguson's Electricity, p. 120).

In this medical section, several curious cases are related. The following refer to himself and his wife :—

"I was once, at Bristol, seized with a sore throat, so that I could not swallow anything. Mr. Adlam, of that city, who is a fine electrician, came and drew many electric sparks from my throat, and in about half an hour after, he did the same again. He staid with me about an hour longer, and before he went away, I could both eat and drink without pain; and had no return of that disorder. I have relieved several persons in such cases, but never in so short a time as Mr. Adlam cured me." (*Electricity*, p. 125).

Referring to his wife's case, he says, "One time my wife happened to scald her wrist by boiling water. I set her upon the glass-footed stool directly, and took sparks from the wrist. In a short time I found the redness of the skin (occasioned by the scald) begin to disappear, and she felt immediate relief. A linen bandage was then put round her wrist, and in a few hours after, I repeated the operation, which entirely cured her, and there was not the least blister on the skin, nor any difference in its colour from what it had before the accident. If it had not been taken immediately, and before a blister had risen, perhaps electrifying would have been of little or no service." (*Electricity*, pp. 130, 131).

This concise treatise has been long out of print, and is now seldom to be seen.

LECTURE ON ELECTRICITY.—In the Catalogue of his Ap-

In looking over one of our old Catalogues we observe several articles of Ferguson's Electrical Apparatus, which belonged to a Mr. Desormeaux, probably bought by him at Mr. Ferguson's public sale, in March, 1777. Mr. Desormeaux's property was sold by public auction by Messrs. Jordan and Maxwell at their Rooms, No. 331 Strand, on 12th March, 1806. In the Catalogue, is the following Lot:—"*A Set of Ferguson's Electrical Apparatus, consisting of Sundry Models of Mills and Pump-work, with an electrical model of an Astronomical Clock, Showing the nature of an Eclipse, and pointing out the Moon's age and the Tides—this Lot sold for* £1 17s.'

paratus, appended to Tables and Tracts, given at pages 344—347, it will be observed that his VIIth Lecture is on the *Armillary Sphere, Dialling*, &c. To the "Introduction to Electricity," he appends a similar list, with this exception, that the apparatus pertaining to Lecture VII. is omitted, and a Lecture on *Electricity* substituted, thus—

VII.

"An Electrical Machine, with such an apparatus to it as is described in this Treatise." (Ferguson's Introduction to Electricity. Lond., 1770, p. 138). See list of apparatus in foregoing notice.

From this we learn that Ferguson, sometime between the publication of his "Tables and Tracts," in 1767, and that of "Introduction to Electricity," in 1770, began to lecture on Electricity. It will be recollected that he purchased an Electrical Machine and Apparatus from Mr. Nairne, Cornhill, London, in 1761, and it might have been expected he should, long before 1767—1770, have added the subject of Electricity to his Course. (See page 262).

Bridge Building.—The numerous notices already given of Ferguson's multifarious writings and inventions, show, that his mind was always ready to turn to anything and everything useful or curious.

In the present case, he leaves wheels and pinions, clocks and orreries, for Bridges and Bridge Building.

At the end of his MS. "Common Place Book," he gives a large pen-and-ink representation of a plan invented by him, about 1770, for building bridges "without Frames under the Arches." On the top of the sketch the following is written:—

"A new method of building a Stone Bridge, without any supporting Frames under the Arches; and so, that any single Arch shall support itself without Abutments at the ends; and would not fall if part of the top of it was cut out. Invented by James Ferguson."

A cut of the bridge will not be required, as the following extract from his short description of it will be sufficient to show how he proposes to construct his bridge,—

"When the Piers are finished, begin with laying the first stone of each side of the Arch a little way over its Pier, and cover the Pier with stones to the same level, then lay the next higher stone of the Arch a little way over the first one, and go on, as before, till you finish the Arch at top. I never saw nor heard of anything of this kind before." MS. "Common Place Book," p. 275 (Col. Lib. Edin.).

From this short note, the reader will see that the arches of such a bridge would have an odd appearance, and would somewhat resemble the crow steps on the gable-end of a house. Blackfriar's Bridge, London, near Ferguson's residence in Fleet Street, was opened in 1770, and it is probable that it was while the complex frames used in the construction of this bridge were being removed that led Ferguson to design one that could be built "without any supporting Frames under the Arches."

FERGUSON'S PERPETUAL MOTION SCHEME.—An old note informs us that, "between the years 1760 and 1780, London abounded with Perpetual Motion-seekers and their public Exhibitions." Ferguson himself, a mechanician and inventor, would, no doubt, visit many of them, and would see "schemes ingenious, curious, and specious." He did not believe it possible to produce perpetual motion by any contrivance which required the motive power to be *within itself* (all external agents, such as that of a never-failing stream; the ever varying pressure of the atmosphere, &c., being ignored); yet, notwithstanding that, we find him, in 1770, devising a plan for perpetual motion.

In his "Common Place Book," we find a fine pen-and-ink drawing of his scheme, which he designates as "*the most rational scheme,*" but at same time declares it to be "*downright nonsense.*" At the conclusion of his description of it he writes, "*Whoever makes it will find that it is a mere Balance;*" from this, it may be inferred that the machine was never made—never having had any other existence than that on paper.

The following is a copy of the drawing and description from the "Common Place Book."

Annexed to the drawing, in the left hand corner at top, Ferguson writes as introductory to it—

"This is the most rational Scheme for a Perpetual Motion, which, notwithstanding, is downright nonsense."

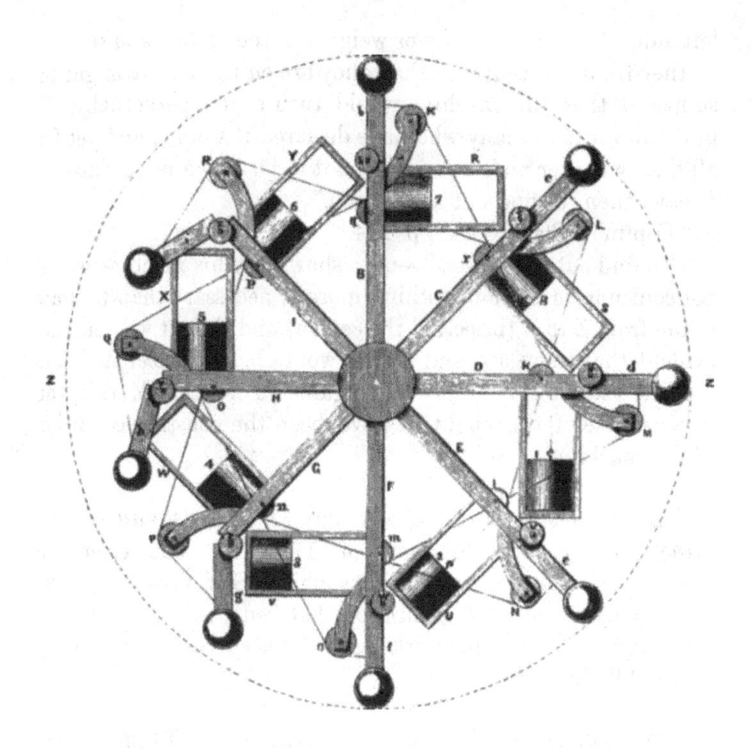

"J. Ferguson, *inv. et delin.*, Sept. 6, A.D. 1770."

DESCRIPTION.—"The axle at A is placed horizontally, and the Spokes B, C, D, &c., turn in a vertical position. They are jointed at *s*, *t*, *u*, &c., as a common Sector is, and to each of them is fixed a Frame as R, S, T, &c., in which the weights 7, 8, 9, 1, 2, &c., have liberty to move. When any spoke as D, is in a horizontal position, the weight I in it falls down, and pulls the part *b* of the then vertical spoke B straight out, by means of a cord going over the pulleys K and *k* to the weight I. The spoke C *c* was pulled straight out before, when it was vertical, by means of the weight 2, belonging to the spoke E *e*, which is in the horizontal position D *d;* and so of all the others on the right hand. But, when these spokes come about to the left hand, their weights 4, 5, 6 fall back, and cease pulling the parts *f*, *g*, *h*, *i;* so that the spokes then bend at their joints X, *y*, *z*, and the balls at their ends come nearer the centre A, all on the left side. Now, as the balls or weights at the right hand side are further from the centre A than they are on the left, it might be supposed that this machine would turn round perpetually. I have shown it to many who have declared it would; and yet for all that, whoever makes it, will find it to be only a mere Balance. I leave them to find out the reason."

"Common Place Book," p. 254.

"To find out the reason"—or to show, why his machine would not continue in motion, nothing more is necessary than to draw a line from Z to Z (bisecting the centre) and then it will become evident that there are, and always would be, *more weight below* the line than above it. To have gained Perpetual Motion, the very reverse of this ought to have been the constant condition of the machine.

ELECTED a MEMBER of the AMERICAN PHILOSOPHICAL SOCIETY—During the summer of 1770, he was elected a member of the American Philosophical Society—an honour unanimously conferred on him by that body, in consequence of their high opinion of his works, and of his success and celebrity as a public lecturer.

NEW TERMS for READING his COURSE of TWELVE LECTURES.—It will be remembered that we have already given a copy of Ferguson's terms for reading his Course of Lectures in, and at certain distances from, London (under date 1767). He this year, 1770, publishes, on the last leaf of his "INTRODUCTION TO ELECTRICITY," a new set of terms. For the same *distances* of 1767, he in 1770 solicits a greater number of subscribers—an indication that he now believed his success and reputation would secure the larger as easily as he before secured the smaller number of persons to attend his lectures. The following are the new terms alluded to.

In 1767, in London, before beginning to lecture, there must be 20 subscribers at a Guinea each.

In 1770, there must be 30 subscribers at a Guinea each.

In 1767, within 10 miles of London, 30 subscribers at a Guinea each.
In 1770, do. do. 40 do. do.

In 1767, within 100 miles of London, 60 do. do.
In 1770, do. do. 80 do. do.

The difference in the increase demanded in 1770 would pay the costs attendant on delivering his Course of Lectures, and leave him with at least as much as the nett proceeds of the lesser number. (Vide Tables and Tracts. Lond., 1767, pp. 327, 328, and compare with "Introduction to Electricity," p. 140; also under date 1767 of this work).

1771.

DEATH of JAMES GLASHAN (*Ferguson's old Master and Patron*).—In entering into this year, we have to record the death of James Glashan, Farmer, Ardneedlie, near Keith, on 9th January; the news of whose death, on reaching Ferguson, would occasion him much sorrow. The good old man is now gone, but his good deeds live after him. Ferguson remembers them, and as it were, with a golden pen writes them down in his Memoir for the admiration of future generations. (See note 14, pp. 14, 15, &c.)

The Reverend Mr. Annand, the present minister of Keith, kindly forwarded the following inscription copied from Mr. Glashan's tombstone:—

"James Glashan was born 11th December 1686, and died 9th January 1771, in the 85th year of his age."

ON THE VELOCITIES OF FALLING BODIES, AND ON APPARENT AND TRUE LEVELS.—The following are copies of two papers, by Ferguson, which we, in 1831, copied from the original belonging to Deane Walker, Esq., London; both are written on a sheet of foolscap, and dated 1771.

"1*st, Of the Velocities acquired by falling Bodies, and the Spaces they fall through in given Times.*

In successive equal parts of time, as 1, 1, 1, 1, 1, &c., the spaces through which a body falls are as 1, 3, 5, 7, 9, &c., and the acquired velocities are as 1, 2, 3, 4, 5, &c., continually; so that the velocities are as the times, and the spaces are as the squares of the times in falling.

Thus, in the first second of time (from the instant of beginning to fall) the body will fall through 16 feet; in the next second it will fall through three times 16, or 48 feet, which added to the former 16 makes 64 feet, the whole space fallen through in

2 seconds of time; in the third second of time, the body falls 5 times 16, or 80 feet, which added to 64 makes 144 feet, the whole space fallen through in 3 seconds; in the fourth second it falls 7 times 16, or 112 feet, which added to 144 feet, makes 256, the whole space fallen through in 4 seconds; in five seconds it falls 9 times 16, or 144 feet, which added to the 256 feet, makes 400 feet the whole space fallen through in 5 seconds; and so on continually, increasing as the odd numbers 1, 3, 5, 7, 9, 11, in 1, 2, 3, 4, 5, 6, seconds of time.

Whatever velocity the body acquires at the end of the first second, it will acquire twice as much at the end of the next, three times as much at the end of the third, four times as much at the end of the fourth, five times as much at the end of the fifth second, and so on continually.

In the following Table, the numbers under T denote the seconds of time, from 1 to 60, in which the body continues to fall: the numbers under S denote the spaces, in feet, through which the body falls in any second from 1 to 60: and the numbers under N denote the whole number of feet the body falls through, at the end of any number of seconds, from 1 to 60. Thus, between the end of the 59th and 60th second, the body falls 119 times 16 feet; and at the end of the 60th second it has fallen through 57600 feet, or $10\frac{10}{11}$ miles.

In a quarter of a second from the instant of beginning to fall, a body would fall one foot: at the end of half that second it will have fallen 4 feet: at the end of three quarters of that second it will have fallen through 9 feet: and at the end of that whole second, through 16. The whole spaces fallen through being as the squares of the times in which the body falls. Qu.—How many feet would it fall through in an hour?

In 60 seconds the space is 57600 feet, and the square of 60 is 3600. But 57600 multiplied by 3600 is 207360000, the number of feet the body would fall through in an hour, in a free or unresisting space; and this number being divided by 5280, the number of feet in an English mile quotes 39272·7 for the number of miles, or about ten times the diameter of the Earth.

TABLE OF FALLING BODIES.

T	S	N	T	S	N	T	S	N	T	S	N
1	1	Feet 16	16	31	4096	31	61	15376	46	91	33856
2	3	64	17	33	4624	32	63	16384	47	93	35344
3	5	144	18	35	5184	33	65	17424	48	95	36864
4	7	256	19	37	5776	34	67	18496	49	97	38416
5	9	400	20	39	6400	35	69	19600	50	99	40000
6	11	576	21	41	7056	36	71	20736	51	101	41616
7	13	784	22	43	7744	37	73	21904	52	103	43264
8	15	1024	23	45	8464	38	75	23104	53	105	44944
9	17	1296	24	47	9216	39	77	24336	54	107	46656
10	19	1600	25	49	10000	40	79	25600	55	109	48400
11	21	1936	26	51	10816	41	81	26896	56	111	50176
12	23	2304	27	53	11664	42	83	28224	57	113	51984
13	25	2704	28	55	12544	43	85	29584	58	115	53824
14	27	3136	29	57	13456	44	87	30976	59	117	55696
15	29	3600	30	59	14400	45	89	32400	60	119	57600

JAMES FERGUSON, LONDON, 1771."

See also "Common Place Book," p. 188; and "Select Mechanical Exercises," pp. 117—120.

"2d, *On the Height of the Apparent Level above the True.*

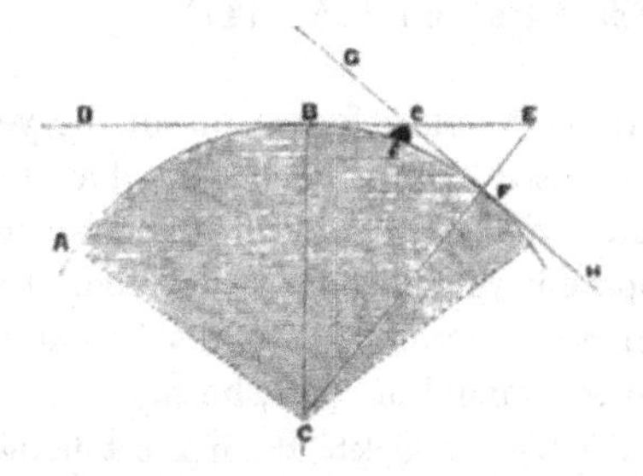

In the above figure, let A B F be part of the Earth's Spherical surface, C the Earth's centre, B C or F C its semi-diameter, D B E a tangent to the Earth's surface at B, drawn perpendicular to B C; and G F H a tangent to the Earth's surface at F, drawn perpendicular to F C. The line D B E is a true level at B. But being carried on straight toward D or E, it rises above the Earth's surface; and although it seems to be level as seen from B, it is above the true level at F, by the whole height F E; for, at the point F, the tangent G F H is the true level.

At the distance of a geographical mile or 6094 feet, from the point B, the line B D or B E will be 10·637 inches above the globular surface of the Earth; At two such miles from B, the same line will be four times as high (or 3 feet 6·548 inches) above the Earth's surface; At three miles distance, nine times as high (or 7 feet 11·733 inches) above the Earth's surface; and so on, always increasing in height according to the square of the distance; for, if F be twice the distance of *f* from B, F E must be four times as high as *f e*, if both their tops touch the right line B *e* E.

At the distance of an English mile, or 5280 feet, the apparent level is 7·90 inches higher than the true; at two miles distance, it is four times as high, or 2 feet 7·60 inches above the true; at three miles distance, nine times as high; and so on, increasing in proportion to the square of the distance.

At the distance of a degree or 60 geographical miles, which are equal to 69¼ English miles, the height of the apparent level above the true is 3191 feet and $\frac{9}{10}$ parts of an inch. And therefore, a hill, whose top was so far above the level of the sea, would be just seen at top by an eye at the surface of the sea, and 69¼ English miles from the hill.—J. F., 1771."

(See also MS. "Common Place Book," p. 249; and "Select Mechanical Exercises," pp. 115—117).

Astronomical Clock.—Sometime in the year 1771, Ferguson appears to have improved the model of Ellicott's Clock, which he made in 1757, by adding wheel-work to it for producing an apparent revolution of the Moon, thereby rendering it more useful and complete. In his "Common Place Book," in alluding to the model of 1757, he says,

> "To have this Clock complete, the apparent diurnal revolution of the Moon ought to be represented, and this I intend to do (if God will be pleased to let me live) as soon as other absolutely necessary business will allow me to carry it into execution."

He then gives a rough sectional plan of his new arrangement, of which the following is a copy:—

Section of the Dial-wheels of New Astronomical Clock.

By comparing this new arrangement with the plan of the clock on page 221, it will be seen that the only difference in the works consists in the present plan having *three additional wheels* for producing the daily revolution of the Moon, viz. wheel A 61, B 57, and C 59. Wheel A of 61 teeth turns round in 24 hours, to which is made fast wheel B of 57 teeth, turning round with it in the same time; this last wheel turns wheel C of 59 teeth (the lowermost wheel in the central group) once round in 24 hours 50 minutes 31 seconds 35 thirds, the axis of which ascends through the wheel above it, and on its top carries a hand, with a small Moon, round in the time just mentioned. From this it will be seen that wheels 57 and 59 are Ferguson's old familiar numbers, so often alluded to in this work. The rest of the wheel-work is described in pages 219—222, to which the reader is referred. The following is a vertical plan or calibre of the wheel-work :—

Plan of the Dial-wheels of New Astronomical Clock.

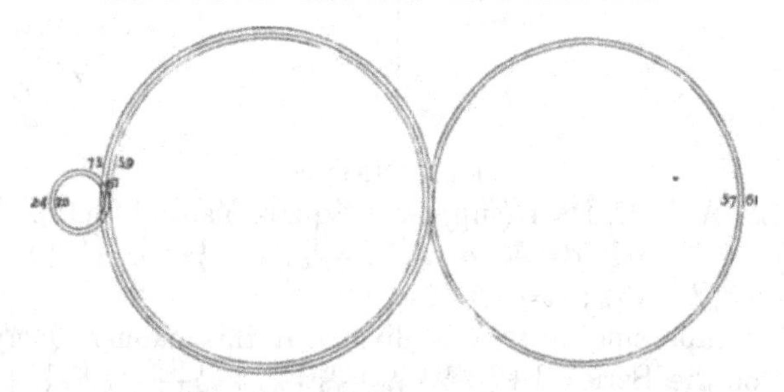

Note.—The several concentric circles in the above plan, although of different sizes, represent respectively, wheels of the *same diameter*. Thus, wheels 20 and 24 are the same in diameter; so are the wheels 73, 61, 59, and also the pair 57 and 61. See "Common Place Book," pp. 104, 105.

DIVISION OF THE SQUARE.—The following curious illustration of the Division and Sub-division of the Square, by Ferguson, is taken from his MS. "Common Place Book," and appears to have been done in the year 1771.

"A Square Divided and Subdivided into Halves, Quarters, Eighth Parts, &c."

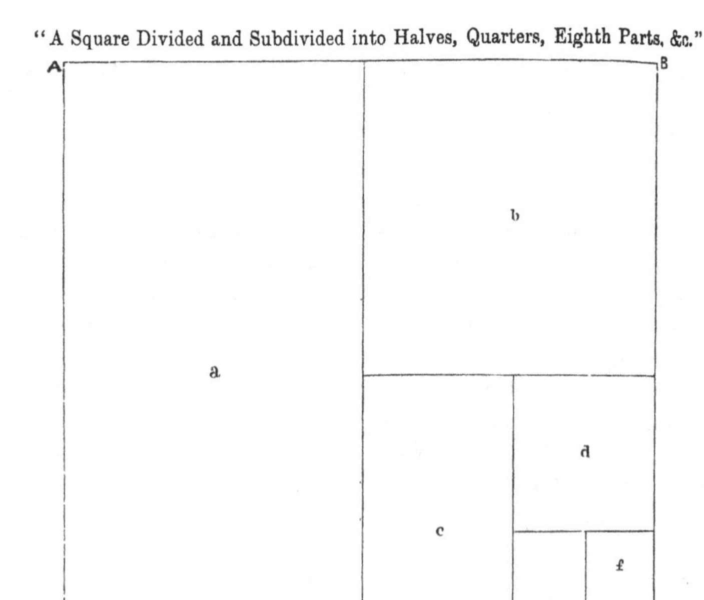

EXPLANATION.

"Let A, B, C, D=1 (suppose 1 Square Yard of Cloth). Then $a=\frac{1}{2}$; $b=\frac{1}{4}$; $c=\frac{1}{8}$; $d=\frac{1}{16}$; $e=\frac{1}{32}$; $f=\frac{1}{64}$; $g=\frac{1}{128}$: $h=\frac{1}{256}$; $i=\frac{1}{512}$; $k=\frac{1}{1024}$; $l=\frac{1}{2048}$; $m=\frac{1}{4096}$.

Now supposing unity to be divided in this manner, Query the sum of the Series $\frac{1}{2}+\frac{1}{4}+\frac{1}{8}+\frac{1}{16}+\frac{1}{32}+\frac{1}{64}+\frac{1}{128}+\frac{1}{256}+\frac{1}{512}+\frac{1}{1024}+\frac{1}{2048}+\frac{1}{4096}$? Answer, $\frac{4095}{4096}$, for the dark part (under m) which is not reckoned, but supposed to be wanting, is $\frac{1}{4096}$ part of the whole A, B, C, D."

"Common Place Book," p. 59.

WHEEL-WORK FOR A WATCH TO MAKE FOUR BEATS IN A SECOND.—Ferguson, in his "Common Place Book," has the following calculations, &c., for a Watch to go 30 hours and make four beats in a second. They appear to have been made sometime in the year 1771:—

"Calculation for a Watch that shall go just 30 Hours, and make four beats in every second of time. The second wheel of the Watch to go round in one Hour, and carry the Minute

Hand; and the fourth (or contrate) wheel to go round in one minute, and carry a hand for showing seconds.

The fusee and first wheel must go round in four hours. This wheel must have 48 teeth, and turn a pinion of 12 leaves, on whose axis is the second wheel which has 60 teeth. This wheel turns a pinion of 10 leaves, on whose axis is the third wheel of 60 teeth, which turns a pinion of 6 leaves, on whose axis is the fourth (or contrate) wheel of 48 teeth, turning a pinion of 6 leaves, on whose axis is the Crown Wheel of 15 teeth, which gives 30 beats in each revolution.

Then, as the second wheel will go round in one hour, the crown wheel will go 480 times round in an hour, and as it has 15 teeth, and makes 30 beats in each revolution, 30 times 480 will be 14400, the number of beats in an hour, which contains 3600 seconds; and 14400 divided by 3600, quotes 4, the number of beats in one second.

The Fusee must have 7½ turns in its Spiral, to let the chain go so many times round it; and as each turn is equal to 4 hours, 7½ turns are equal to 30, the number of hours that the watch will go after it has been fully wound up.

The whole calculation may be expressed as follows:—

48)12—60), 10—60), 6—48), 6—15.

This stroke) placed between the Pinion and Wheels, shows that the wheel turns the pinion; and this stroke — between any pinion and wheel, shows that the pinion and wheel are upon one and the same axis. The numeral figures show the numbers of teeth in the wheels, and the number of leaves in the pinions respectively.[271] JAMES FERGUSON."

"Common Place Book," pp. 82, 83.

GUTHRIE'S GEOGRAPHICAL GRAMMAR.—During the summer of 1771, we find Ferguson employed upon Guthrie's Geographical Grammar, he having been engaged by the publisher, Mr. Knox, of the Strand, to write the Astronomical part of that work, and otherwise to revise it. The astronomical part, by

271 The above is the old way of arranging wheel and pinion numbers; the more concise way is as follows:—

$\frac{12}{48} \times \frac{10}{60} \times \frac{6}{48} \times \frac{6}{15} = \frac{4320}{2073600}$, and $2073600 \div 4320 = 480$, that is, in one revolution of the first mover 48, the last mover of 15 teeth turns 480 times on its axis, and as wheel 15 makes 30 beats in turning round its axis, $480 \times 30 = 144000$ beats, which wheel 15 makes in an hour, as by Ferguson.

Ferguson, is an excellent epitome of the elements of Astronomy, with the planetary distances, as deduced from the then late transits of Venus. The following is an advertisement of this work, which, as revised by Ferguson, appears to have been re-published about the end of the year 1771 :—

"A Geographical, Historical, and Commercial Grammar, showing the Present state of the Kingdoms of the World. By William Guthrie. The Astronomical part by James Ferguson, F.R.S. Published by J. Knox, of the Strand, London."[272]

Curious Calculations regarding the New Assembly Room at Bath.[273]—The following curious paper is taken from the "Common Place Book," p. 185, and appears to have been written about the end of the year 1771 :—

"The Dimensions and Capacity of the New Assembly Room at Bath. Length 105 Feet 8 Inches, Width 42 Feet 8 Inches, Height 42 Feet 6 Inches. These numbers turned into Inches, and multiplied into each other, give 331100160 for the number of cubic inches of air in the Room ; which, being divided by 231 (the number of cubic inches in a wine gallon) quotes 1433334 for the number of gallons of air in the room when empty of People. And as a gallon of air weighs 64 grains, and 5760 grains make a Troy Pound, the whole weight of air in the room is 15926 Troy Pounds.

From the above number of gallons let 4334 be subtracted, on account of the space taken up by the Company and Candles; and there will remain 1429000 gallons of air for the company to breathe, and the candles to burn. Now, each person consumes a gallon of air in each minute in breathing, and so does each candle every minute while it is burning.

Supposing then, that there are 900 persons, and 100 candles,

272 William Guthrie (an old friend of Ferguson) was born in Brechin in 1708. He went to London in 1730. In 1746, a Government annuity of £200 was conferred on him for literary services rendered to the Pelham administration. In 1765 he published his "Geographical, Historical, and Commercial Grammar," the 24th edition of which was published in 1818. In 1767 he published his "History of Scotland." He died in London on 9th March, 1770, aged 62. (Black's "History of Brechin," pp. 137, 138).

273 On referring to a Bath Guide-book, we find that "the New Assembly Rooms at the east end of the Circus were opened for the reception of Company in October 1771, Richard Tyson, Esq., Master of the Ceremonies." The dimensions given in the "Guide" are precisely the same as noted by Ferguson. (Cruttwell's "New Bath Guide," for 1790, p. 21).

in the room, which in all make 1000; they will require 1000 gallons of air per minute to keep them alive.

Divide 1429000 by 1000 and the quotient will be 1429 minutes, or 23 hours 49 minutes for the time the company and candles might live in the room if all the doors and windows were so close shut as to admit no fresh air; and at the end of that time all the candles would go out, and all the company would die. Hence, each person would lose $\frac{1}{1429}$th part of his life each minute while he was there, if no fresh air was admitted."[274]

PYRAMIDAL TABLE CLOCK.—The following calculations and details of the astronomical part of a curious pyramidal or rather pagoda-formed Table Clock, is taken from the "Common Place Book." They appear to have been written in 1771. The details in the MS. are accompanied by a very fine drawing in Indian ink of the outside appearance of the Clock. Ferguson says,

> "It was given me by a Nobleman, quite out of order, much broke, and in several detached pieces; so that it cost me much labour and pains to put all the parts of it together, and I found much of the inside work so very bad, as had made it impossible that ever it could have gone, so as to come within an hour of the truth in 24 hours; and was always apt to stop. These parts I threw away, and had new work put in place of them; and now it goes very well, and it is a pretty ornamental piece of furniture in my room.—JAMES FERGUSON."

"Calculations of the Astronomical part of my Brass Clock with four Faces, which strikes the Hours and Quarters; Shows the Hours and Minutes, the Moon's Age and Phases, the motions and Places of the Sun and Moon, and nodes of the Moon's orbit, with the days of all the New and Full Moons and Eclipses,—

8 hours ⎞ 57 = The Sun's diurnal motion in 24 hours.
Pinion 19 ⎠ 59 = The Moon's daily motion in 24 h. $50\frac{1}{2}$ minutes, and from Change to Change in $29\frac{1}{2}$ days.

59)40—4)26—21) $\frac{59}{56}$ 59 = The Sun's motion round the Ecliptic in 365 d. 5 h. 43 m.
56 = The retrograde motion of the nodes round the Ecliptic in $18\frac{2}{3}$ years.

A pinion of 19 leaves is turned, by the Clock movement, round in 8 hours. This pinion takes into two wheels, one of which has 57 teeth, and the other 59. The wheel 57 carries a Sun round a Dial-plate in 24 hours, and on that plate the 24 hours are engraved.

[274] We may here remark that each person would lose *less* at first, and *more* at last, than $\frac{1}{1429}$th part.

The wheel of 59 teeth carries the Moon round from the meridian XII on the Dial-plate to the same again in 24 h. 50½ m., and from the Sun to the Sun again in 29 days 12 hours.

On the wheel of 59 teeth is fixed a small wheel of the same number, and concentrical with it, which small wheel turns a wheel of 40 teeth, on whose axis is a pinion of 4 leaves, turning a wheel of 26 teeth, on the axis of which is a pinion of 21 leaves turning two wheels, one of which has 59 teeth and the other 56.

The last wheel of 59 carries a Star-Plate with the Ecliptic and its Signs, 366 times round in 365 days 5 hours 43 minutes; so that, as the Sun goes 365 times round in a year, from the meridian to the meridian again, and the Ecliptic 366 times round in the year, the Sun's place is shifted every day in the Ecliptic so as to go quite round it in 365 days 5 hours 43 minutes; but in nature, the Sun goes round the Ecliptic in 365 d. 5 h. 49 m.; so that the Sun's motion is only 6 minutes too slow, in the Clock, in a year.

The wheel of 56 teeth (turned by the above-mentioned Pinion of 21 leaves) carries the nodes of the Moon's orbit backward round the Ecliptic in 18⅔ years, which is very near the truth.

The wheel of 40 teeth, with its pinion of 4 leaves, and wheel of 26 teeth (turned by that pinion) with its pinion of 21 teeth, are hung .upon the wheel of 57 teeth, which goes round in 24 hours." "Common Place Book," p. 90.

ANOTHER ANCIENT CLOCK.—In the "Common Place Book," we have a description and fine Indian-ink drawing of an ancient clock which he got in a present from Mr. Mudge, Watchmaker, London, in 1771. On beginning his description he says,

> "About five years ago, my good friend Mr. Thomas Mudge, Watchmaker in London, made me a present of an old German Clock, bearing date A.D. 1560. It shows Solar and Sidereal time; the motions of the Sun and Moon through the Ecliptic, with the Moon's Phases; The times of the rising and setting of the Sun and Moon, and several of the most remarkable stars; But the Moon's motion is far from being as exact in it as in mine; The Moon revolves in 29 days 15 ho. 37 m. 52 s., which is too slow by 2 ho. 53 m. 49 s. in each Lunation, and will therefore require that the Moon be set forward in the Ecliptic, or according to the order of the Signs, 19 degrees always on the first of January."

The sidereal train in this Clock is similar to that of Ferguson's own train. At the conclusion of his long description he makes the following curious observation:—

"I confess that I was very glad to find I had hit upon the same sort of motions in a Clock, as had been done 215 years ago; but of which I knew nothing till I had seen this Clock; and it is abundantly plain, that by comparing the wheel-work and numbers of teeth in this and mine, I have not (and indeed could not have) taken the hint from this ancient German.

LONDON, *Jan.* 1*st*, 1776.
JAMES FERGUSON."

Common Place Book, p. 136.

ASTRONOMICAL CLOCK.—In the end of December, 1771, Ferguson invented an Astronomical Clock for showing, by means of a *moveable horizon*, the daily rising and setting of the Sun. In his "Common Place Book," he has pen-and-ink drawings of the Dial-face and wheel-work, along with a short description of them, which is initialed and dated, thus,

"J. F.
*Dec*r. 30*th*, 1771."

Under the drawing of the Dial-face is written,

"A Clock Showing the Hours, Minutes, the Day of the Month, and the time of the Sun's Rising and Setting every Day in the Year, by means of a Moveable Horizon in the Arch at Top."

The drawing of the wheel-work of this Clock is very complex. It does not appear that he ever made it. On January 1st, 1773, he appends the following note to his description:—

"On second thoughts, I find this scheme may be carried into execution by a much less number of wheels.—*J. F.*, *Jan.* 1*st*, 1773."[275]
"Common Place Book," pp. 100, 101.

1772.

PLAIDALLI'S PERPETUAL MOTION.—As already mentioned, there were in London, about this period, several exhibitions of pretended "*perpetual motions*,"—few excited so great interest as that of Plaidalli's, which consisted of a glass globe moving on a vertical axis; and along with it, two balancing arms of brass proceeding from its upper pole,—the whole was inclosed within a

[275] The principal part of this Clock is its annual train, which results in 365 days exactly. The wheel, which goes once round in this time, has a pin near its circumference, on which the central part of the moveable horizon rests, and is by it caused to rise and fall in the arch of the dial-plate, where a secondary hand revolves in 24 hours. The annual train is derived from 24 hours,—viz. a pinion of 7 leaves turns round in 24 hours = 1 day, and turns a wheel of 35 teeth, which has a pinion of 7 leaves that turns a wheel of 42 teeth, which has a pinion of 6 leaves which turns a wheel of 73 teeth once round in 365 days,—or thus, $\frac{7}{35} \times \frac{7}{42} \times \frac{6}{73} \equiv \frac{294}{107310}$ $107310 \div 294 = 365$ days. Several unnecessary wheels are introduced for transmitting motion from one part to another. In the transmission of motion *three wheels* are used, which we find could be as effectually done by one wheel.

spacious bell-shaped glass cover to keep the instrument from being touched. Its inventor gave out that the globe and balances turned round by virtue of the power of Electricity, and it was kept in continual motion. For a considerable length of time it was "*a great puzzle.*" Ferguson went to see it, carefully inspected it, and, like the many who had preceded him, came away from it puzzled too. Referring to the matter in his "Common Place Book," where he has a drawing and description of it, he says,

> "On Monday, March 9th, 1772, I went to see Plaidalli's machine. Its glass globe and balance were turning round when I went into the room, and continued turning while I staid there, which was rather more than an hour. Plaidalli, the inventor, told me it had been going for five months. I confess I am altogether at a loss to account for this constant motion of the globe, but have no opinion that Electricity is the cause of its motion; and do at present consider it as a very ingenious deception. I see the reason why mine cannot move; but the other moves and I cannot see the reason of its motion." JAMES FERGUSON."

Shortly after Ferguson's visit, Plaidalli's "perpetual motion" was discovered to be a deception—showing that Ferguson was right in his conjecture.[276] Regarding the "discovery of the deception," he has another entry in his "Common Place Book," viz.

> "March 16th, 1772; I am now convinced that I was right in considering Plaidalli's Scheme a deception, for such indeed it is, and a very ingenious one it is too. The cause of its motion is found out, and Electricity had no hand in it. JAMES FERGUSON."
>
> "Common Place Book," pp. 255, 256.

MAZE OR LABYRINTH.—In his latter years, Ferguson appears to have exercised his ingenuity on a great variety of subjects. The following is copy of his *Maze*, distich, and description accompanying it, which appears to have been drawn sometime early in 1772. (Extracted from "Common Place Book," p. 119.

"A MAZE OF LABYRINTH.

"Into this Labyrinth one may go;
But, how to get out he never may know.—J. F."

276 The past history of Perpetual Motion Seekers is a history of *perpetual failures*, and it may be safely predicted, that all the efforts of the *seekers* in the present day, as well as in all time to come, will end in like results—in nothing. The aged *seeker* is a sad spectacle—he is the representative of misspent time—has through life been in pursuit of a *phantom*, and now grown old without having effected anything; or as Cowper has it,

"Dropping buckets into empty wells,
And grown old in drawing nothing up."

"In this Labyrinth the dark curved lines denote so many Walls or high Hedges, the white spaces between them are the Walks. A is the entrance into the Labyrinth; and B is a round Pond of water, from which twelve Walks go off. If a person goes into such a Labyrinth, he must go once round before he can come to the Pond; and then, if he turns off abruptly into either of the eleven other Walks, he may wander on and never know how to get out again. But, if he takes such a particular notice of the Walk *a*, from the Pond, so as to know it from all the rest, or has a magnet-needle, and takes its bearing from the Pond, he may by that means find his way out.—JAMES FERGUSON, *inv. et delin.*"

In Common Place Book, on same page, there is a pen-and-ink sketch of another "Maze or Labyrinth," similar to the above, but not so intricate,—annexed to which are the two following lines:—

> "Into this Labyrinth one may go
> And the way out he may easily know."

LECTURES in DERBYSHIRE.—In the autumn of 1772, we find him delivering his usual Course of Twelve Lectures on Experimental Philosophy and Astronomy, to large audiences, in Derby, and other neighbouring towns. During his sojourn, he visited *the Peak District,* and took notes of its many wonders. The Devil's Cave appears to have interested him strongly, as he soon afterwards wrote a description of it, which was read before the Royal Society on November 16th, 1772, and was shortly afterwards published in the form of a Tract, entitled,

"A Description of the Devil's Cave at Castleton in Derbyshire."
(Vide also Universal Magazine, vol. 17).

SUCCESS of FERGUSON'S WORKS—ASTRONOMY—LECTURES on SELECT SUBJECTS on MECHANICS, and INTRODUCTION to ASTRONOMY.—During the year 1772, the fifth edition of "*Astronomy, Explained upon Sir Isaac Newton's Principles,*" was published, being the fifth edition in 16 years. The "LECTURES on SELECT SUBJECTS in MECHANICS, &c., also went to press, being the fourth edition in 12 years; besides a new edition of "AN EASY INTRODUCTION to ASTRONOMY for YOUNG GENTLEMEN and LADIES," being the third in the space of 4 years—proofs of the great demand for, and popularity of, these works.

TRISECTION OF THE ANGLE.—Ferguson, in his "Common Place Book," has a pen-and-ink sketch of a method, with brief directions, showing how the famous problem, "Trisection of the Angle," may be solved, *approximately,* which appears to have been suggested to him in 1772. The following is copy of the drawing and directions:—

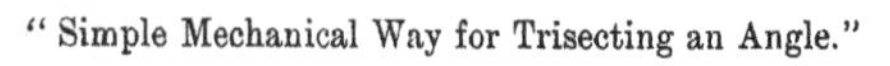
"Simple Mechanical Way for Trisecting an Angle."

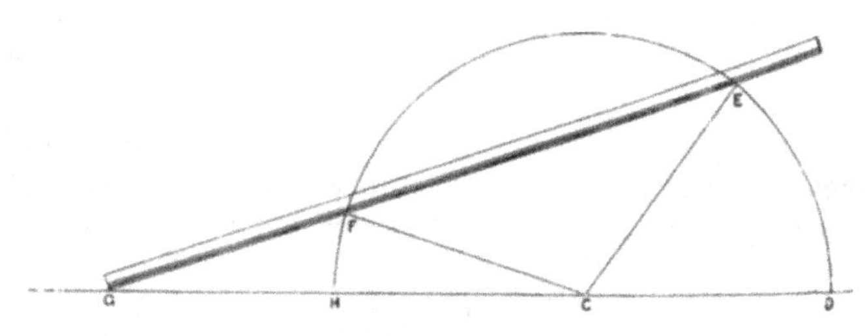

"Let E C D be the given angle which is required to be trisected. Draw the right line D C G at pleasure, and about the angular point C describe the semicircle H F E D; then stick a fine pin in the point E, and, on the edge of a Ruler G F E I, from the end G, set off G F equal to the radius C E, making a mark on the edge of the Ruler at F. This done, lay the edge of the Ruler to the pin E, and keeping the end-corner G always on the line G H C D, slide the Ruler on the pin till the mark F be in the semicircle. Then stop the Ruler in that position, and from F draw the right line F C; and the angle F C H shall be equal to a third part of the given angle E C D, as was required."

"Common Place Book," p. 192.

MERIDIAN LINE.—In several of his works, Ferguson shows how to find a meridian line. In his Common Place Book he has a pen-and-ink projection of a new method for finding one, by Mr. Aubert, done in 1774; and as it is perhaps the most correct of his methods, we give it a place, along with his short description appended to it.

"A sure and easy way of finding a true Meridian, by Alexander Aubert, Esq., of Austin Friars, London.[277]

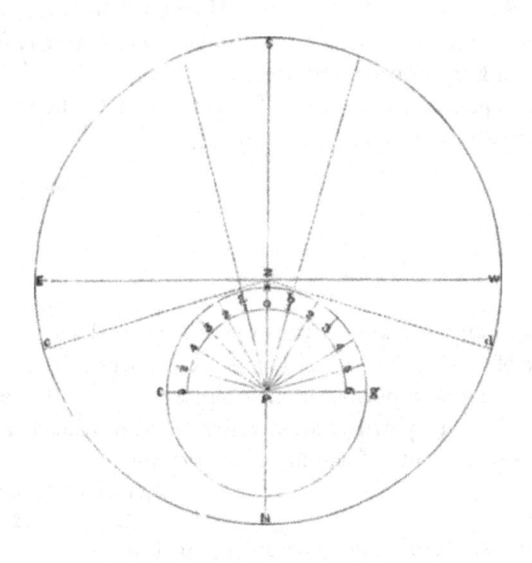

277 See note 256, p. 325.

Let * be a star going round the North Pole P, in the circle *e a b g*, according to the order of the letters, and passing over the Meridian P Z S, a little to the north of the zenith Z. When the Star is at *a* the line Z *a c* of its Azimuth is at rest, being then a Tangent at *a* to the circle in which it moves. Let the Star's Altitude be then observed, and the instant of time by a good Pendulum Clock be noted. When the Star has passed the Meridian, watch its Altitude till you find it the same at *b* as it was at *a*, and note the time exactly by the Clock; and then the Azimuth line Z *b d* of the Star will be at rest again. Divide the interval of time between the two equal Altitudes at *a* and *b* by 2, and it gives the instant of time when the Star passes the Meridian. Or the Azimuthal points *c* and *d* being ascertained, the angle *c* Z *d* is found, which being bisected in the point N, draw N Z S (which will pass through the Pole P) and you will have a true Meridian Line.

As a star that passes over the meridian near the Zenith changes its Altitude very fast, its Altitude may be taken when it is but a very little way from the Meridian, either East or West; and consequently, its apparent Altitude will not be affected sensibly." ("Common Place Book," p. 131).

Ten Squares increasing by Unity.—The following was drawn and written sometime in the year 1772, and is from the original in Ferguson's autograph in our possession. It is entitled, "Ten Squares, whose areas from the first to the last gradually increase by unity.

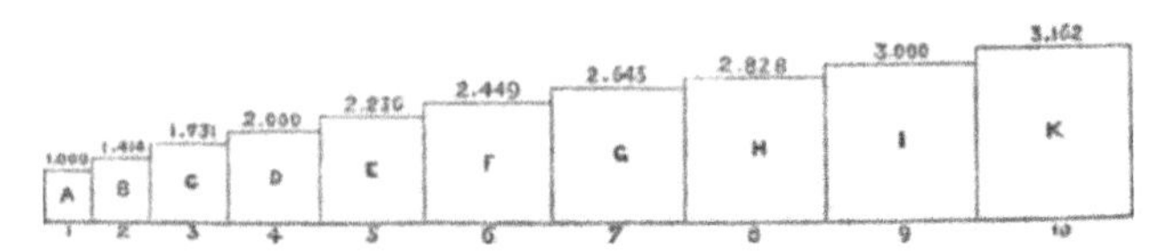

Thus, in any measure, if the area of A be 1, the area of B is 2; the area of C is 3; the area of D is 4; the area of E is 5, and so on.

The numbers underneath each square, express the length of the side of each square, against which they are placed; and they are the square roots of the areas respectively.

James Ferguson.

London, 1772."

(Vide also "Common Place Book," p. 190).

A LINE APPEARING OF THE SAME LENGTH, ALTHOUGH VIEWED FROM DIFFERENT STATIONS.—This curious problem is taken from the original diagram and explanation of date November 1774 (in the possession of Dawson Turner, Esq., Yarmouth, in 1836).

"To place a right Line so as that it shall appear of the same length if viewed from different points of distance.

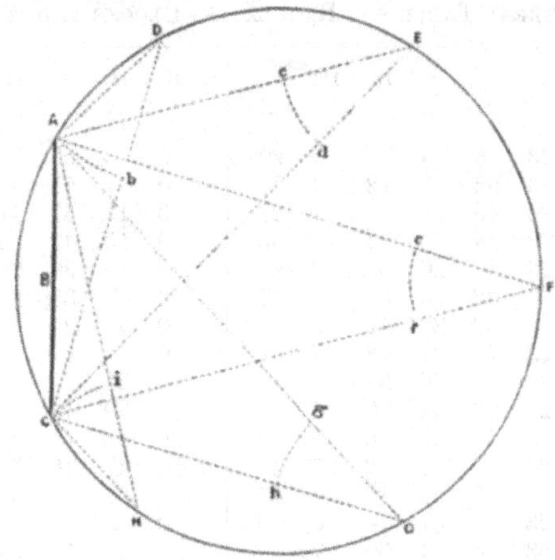

Let A B C be a given straight line, whose extremities A and C touch the periphery of a Circle. If this Line be viewed from any point of the circle between A and C, in the part A D E F G H C, it will always appear of the same length, because it will be seen under an angle of the same measure.

Thus, if the eye be at D, the line will be seen under the angle A D C, whose measure is the arc A *b*. If the eye be at E, the line will be seen under the angle A E C, whose measure is the arc *c d*. If the eye be at F (still farther off), the line will be seen under the angle A F *c*, whose measure is the arc *e f;* and so on if seen from any point between F and C.

These arcs A *b*, *c d*, *e f*, *g h*, C *i*, being all made with the same opening of the compasses, will be all found of equal lengths by measurement; which shows, that all the angles they measure are equal. If the given line A B C be equal to the radius of the Circle, it will be seen under an angle of 30 degrees by an eye placed in any point of the periphery of the Circle, in the part A D E F G H C. JAMES FERGUSON.

LONDON, *Novr.* 1774."

(Vide also "Common Place Book," p. 244).

Time Table for Regulating Clocks and Watches.—In the year 1772, Ferguson computed a Table for regulating Clocks and Watches—the following is copy of it, along with the explanatory remarks appended (taken from the original in the possession of the late Mr. Upcott, Islington, London).

" A Sidereal Table for Regulating Clocks and Watches.

Rev.	Days.	H.	M.	S.	Th.	IV.	V.	H.	M.	S.	Th.	IV.	V.
1	0	23	56	4	6	0	29	0	3	55	53	59	31
2	1	23	52	8	12	0	58	0	7	51	47	59	2
3	2	23	48	12	18	1	27	0	11	47	41	58	33
4	3	23	44	16	24	1	56	0	15	43	35	58	4
5	4	23	40	20	30	2	25	0	19	39	29	57	35
6	5	23	36	24	36	2	54	0	23	35	23	57	6
7	6	23	32	28	42	3	23	0	27	31	17	56	37
8	7	23	28	32	48	3	52	0	31	27	11	56	8
9	8	23	24	36	54	4	21	0	35	23	5	55	39
10	9	23	20	41	0	4	50	0	39	18	59	55	10
11	10	23	16	45	6	5	19	0	43	14	53	54	41
12	11	23	12	49	12	5	48	0	47	10	47	54	12
13	12	23	8	53	18	6	17	0	51	6	41	53	43
14	13	23	4	57	24	6	46	0	55	2	35	53	14
15	14	23	1	1	30	7	15	0	58	58	29	52	45
16	15	22	57	5	36	7	44	1	2	54	23	52	16
17	16	22	53	9	42	8	13	1	6	50	17	51	47
18	17	22	49	13	48	8	42	1	10	46	11	51	18
19	18	22	45	17	54	9	11	1	14	42	5	50	49
20	19	22	41	22	0	9	40	1	18	37	59	50	20
21	20	22	37	26	6	10	9	1	22	33	53	49	51
22	21	22	33	30	12	10	38	1	26	29	47	49	22
23	22	22	29	34	18	11	7	1	30	25	41	48	53
24	23	22	25	38	24	11	36	1	34	21	35	48	24
25	24	22	21	42	30	12	5	1	38	17	29	47	55
26	25	22	17	46	36	12	34	1	42	13	23	47	26
27	26	22	13	50	42	13	3	1	46	9	17	46	57
28	27	22	9	54	48	13	32	1	50	5	11	46	28
29	28	22	5	58	54	14	1	1	54	1	5	45	59
30	29	22	2	3	0	14	30	1	57	56	59	45	30
40	39	21	22	44	0	19	20	2	37	15	59	40	40
50	49	20	43	25	0	24	10	3	16	34	59	35	50
100	99	17	26	50	0	48	20	6	33	9	59	11	40
200	199	10	53	40	1	36	40	13	6	19	58	23	20
300	299	4	20	30	2	25	0	19	39	29	57	35	0
360	359	0	24	36	2	54	0	23	35	23	57	6	0
365	364	0	4	56	32	56	25	23	55	3	27	3	35
366	365	0	1	0	38	56	54	23	58	59	21	3	6

The first column shows the number of revolutions of the Stars in a year, and the next columns in the left hand division show the mean solar times in which these revolutions are performed. Those in the right hand division show the daily accelerations of the stars, or how much any given star gains upon the time shown by a well-regulated Clock true to the $\frac{1}{27993600000000}$ part of a second in the whole year.

Every star comes to the meridian 3 min. 56 s. 53 thirds 59 fourths 31 fifths sooner on any given day than it did on the day before; and consequently, if it gains so much upon the time shown by the Clock 'tis a sure sign that the Clock goes true; if otherwise, the pendulum must be lengthened, or shortened accordingly.

The Equinoctial points in the heavens recede 50″ of a degree every Julian year, which causeth the stars to have an apparent progressive motion eastward equal to 50″ in that time. And as the Sun's mean motion in the Ecliptic is only 11 Signs 29° 45′ 40″ 15‴ in 365 days, 'tis plain that at the end of that time he will be 14′ 19″ 45‴ short of that point of the Ecliptic from which he set out at the beginning; and the stars will be advanced 50″ with respect to that point. Consequently, if the Sun's centre be on the meridian with any star on any given day of the year, that star will be 14′ 19″ 45‴+50″ or 15′ 9″ 45‴ east of the Sun's centre, on the 365th day afterward, when the Sun's centre is on the meridian; and therefore, the star will not come to the meridian on that day, till the Sun's centre has passed it by 1 min. 0 sec. 38 th. 56 iv. 54 v. of mean solar time; for the meridian takes so much time to revolve through an arc of 15′ 9″ 45‴; and then, in 365 d. 0 h. 1 m. 0″ 38‴ 56 iv. 54 v. the star will have completed just 366 revolutions from the meridian to the meridian again.

In the first line of the right hand part, the 3 m. 55 s. 53‴ 59 iv. 31 v. might have been put down 3 m. 55 sec. ·8998658, or if it had been put down 3 m. 56 s. 9, it would have been near enough the truth.

This Table is better than the Table near the end of this Book (*i. e.* Common Place Book).[278]

Calculated by JAMES FERGUSON."

See also "Common Place Book," p. 57.

[278] In this Table, Ferguson starts with 23 h. 56 m. 4 s. 6 thds. 0 fourths 29 fifths as being the length of a sidereal day. In a note we sometime ago received from G. B. Airy, Esq., Astronomer Royal, he gives 23·934469 hours for the length of a sidereal day, which, reduced, is 23 h. 56 m. 4 s. 5 thds. 18 fourths 14 fifths, which is a value *minus* Ferguson's period by 42 thirds 15 fourths.

DEATH OF FERGUSON'S ELDEST SON, JAMES.—This son was of a weakly constitution. In 1763 his father apprenticed him to Mr. Nairne, Optician, Cornhill, London, to learn the practical departments of that profession, it being probable that the youth inherited his father's ingenious turn for Mechanics and Philosophy. He had been ailing for some years, but more particularly during the autumn of 1772. He, shortly afterwards, worn out by pulmonary consumption, was confined to his bed; died at 4 Bolt Court, Fleet Street on 20th November, 1772, and was interred in the churchyard of Old St. Mary-le-Bone on November 25. For other references to this son, see pages 127 and 278.

1773.

PROJECTION OF PART OF THE SPHERE.—Ferguson, in his MS. "Common Place Book," page 176, gives an elaborate projection of part of the Sphere in the latitude of London (written early in 1773); it is entitled,

> "A Projection of as much of the Sphere as is useful in the Latitude of London, for showing the Time of the Day, by the Sun's Altitude; with the times of his Rising and Setting, and his Amplitude at these times on each day of the Year for ever, by
>
> JAS. FERGUSON,
> Finished April 6th, 1773."

This projection of part of the Sphere does not appear to have been published. Ferguson made several copies of it,—we have seen two MS. copies, in his autograph, besides the one in his Common Place Book.

BLIND MAN'S CLOCK.—In his MS. "Common Place Book," is an Indian-ink drawing and short account of a very ancient German Clock, which appears to have come into his possession about this period. As this Clock has often been referred to as being one of the oldest Clocks extant, we shall give his description of it, as also a more full account extracted from Beckman's History of Inventions.

"A Clock, by which a blind Man may know the time of the Day or Night, and also the places of the Sun and Moon in the Ecliptic; made in Germany in the Year 1525, and now in the possession of James Ferguson, F.R.S. A.D. 1773.

The Hand goes round the Dial-Plate in 24 Hours, the Sun goes round through all the 12 Signs of the Ecliptic in a Year;

the Moon goes round them in 27 days 8 hours, from the Sun to the Sun again (or from Change to Change) in $29\frac{1}{2}$ Days, showing her Age on every day of the Year in a Circle divided into $29\frac{1}{2}$ equal parts.

At the Noon XII is a double knob, and a single one at every other Hour. So that, a blind Man, by feeling round from the double knob over the rest, and counting them till he comes to the Hour-hand, may by that means know the time: and as two knobs answer to a whole Sign, and as the double knob is against the beginning of Cancer, he may count the rest to the Sun or Moon, and so find their Places in the Ecliptic.

Diameter of the Dial-Plate $9\frac{1}{4}$ inches."

"Common Place Book," p. 106.

The following is Beckman's account of it:—

"An artist named Jacob Lech, a native of Prague, constructed this clock in A.D. 1525,—the wheels of which are all of iron, and retain some punched marks of divisions, which appears to support the conjecture that they were cut by manual labour. This clock is somewhat different from the one constructed by Vick in 1364, being much more improved. It has a spiral spring, and fusee of soft metal, and a screw instead of notches at the ends of the double levers of the balance, with tapped weights of lead instead of a regulator,—with the addition of some wheel-work for exhibiting the motions of the Sun and Moon in an engraven ecliptic, as also a contrivance to strike one at every hour. A cat-gut was originally the band of the fuzee, but the introduction of a modern chain unfortunately has destroyed nearly three out of the eight spiral threads at the smaller end; so that instead of going 48 hours (viz. 8 threads multiplied by 6 hours), the remaining five threads will allow it to go only 30 hours, which it at present does, though very irregularly, with vibrations of about one second each, by estimation. On examining the wheel-work, the Moon's train of wheels, it is found to be $\frac{12}{6} \times \frac{82}{6} = \frac{984}{36} = 27\frac{1}{3}$ days, or 27 days 8 hours for a periodic revolution of the Moon.[279] And the Sun's train of wheelwork $\frac{984}{36} \times \frac{81}{6} = \frac{79704}{216} = 369$ days exactly for the length of a year, being a motion in the Moon's period too slow by 16

279 This fraction $\frac{984}{36}$ produces the following result,—viz. $984 \div 36 = 27$ d. 8 h. for a periodical revolution of the Moon, the true mean period in round numbers being 27 d. 7 h. 43 m.; error in wheel-work period, 17 m. too slow in each revolution.

minutes 55 seconds 19 thirds, and in that of the solar period by 3 days 18 hours 11 minutes 10 seconds.[280] These errors, that either the art of calculating planetary mechanism at the time the clock was made, had not been understood, or that the errors shown were not esteemed of any consequence. Beckman, in his "*History of Inventions*," says, "This clock was the property of *Mr. Ferguson, Lecturer on Natural and Experimental Philosophy, No. 4 Bolt Court, London*, at the sale of whose effects (after his demise), a Mr. Peckit, of No. 50 Old Compton Street, purchased it in the year 1777, and that shortly afterwards, Mr. Peckit bequeathed it to the British Museum, where it may be seen." Another account says, "Mr. Peckit, an ingenious Apothecary of Compton Street, Soho, hath shown me an Astronomical Clock which belonged to the late *Mr. Ferguson*, and which still continues to go. The workmanship on the outside is elegant, and it appears to have been made by a German in 1525, by the subjoined in the Bohemian of the time:

Iar. da. macht. mich. Iacob Zech
Zu. Prag. Ist. bar. da. man. zalt. 1525.

The above Englished, is,

Year, when, made, me. Jacob Zech.
At Prague. is. true. when. counted. 1525.

I have transposed the words as I find them in the original; but *war* seems to have stood in the place of "*bar*," at least Barrington has translated the German word wahr—*is true*, and we must read,

Da. man. zält. 1525 jar
Da. macht. mich. Iacob. Zech. zu. Prag. Ist. wahr.

This clock is nine and one-fourth inches in diameter, and the height five inches." Beckman's "History of Inventions;" also, Henderson's "Historical Treatise on Horology," 2d Edit. London, 1836, p. 15).

We may add, that from Ferguson's drawing of this Clock it is seen that the works are enclosed in a short brass cylinder, somewhat resembling the brass case of a Ship's Chronometer. He had another similarly-shaped old German Clock, but of a much less size, without the Astronomical part, with twelve hours on the

[280] The fraction $\frac{78704}{216}$ gives a result very wide of the true period of the year; had $\frac{107}{8}$ been added to the lunar train of $\frac{984}{36}$, the result would have been $\frac{984}{36} \times \frac{107}{8} = \frac{105288}{288} \div 288 = 365 \cdot 83$ days nearly; or a still nearer approximation might have been obtained by the fraction $\frac{147}{11}$, thus,—$\frac{984}{36} \times \frac{147}{11} = \frac{144648}{396} \div 396 = 365 \cdot 27$ days, or 365 d. 6 h. 29 m.

dial, and raised knobs, as already described. This one is in the Museum of Banff.

PROJECTION of the GREAT LUNAR ECLIPSE of 30th JULY, 1776.—We have a fine pen-and-ink projection of this Eclipse, calculated by Ferguson on 10th June, 1773 (fully three years before it occurred); it is entitled,

> "A Projection of the Great total Eclipse of the Moon that will fall on the 30th of July in the Year of our Lord 1776."

Annexed are the following particulars of the Eclipse:—

"Beginning (July 30th) at	10 h.	8 m. P.M.
Total immersion at .	11	14
Middle of the Eclipse,	12	3
Beginning of emersion,	12	52
Ending, ...	13	57

Digits Eclipsed, 19."

Below the projection we find,

> "Calculated June 10th, 1773, by J. Ferguson."

This must be among the last of Ferguson's Eclipse Projections.

THE ARITHMETICAL TRIANGLE.—We have a curious Triangular Table which appears to have been written by Ferguson sometime in the year 1773,—entitled,

> "The Arithmetical Triangle for Combinations."

And annexed, on each side, is the description of it, commencing with

> "This Table shows every possible Combination in any number of things, not exceeding 20, &c."

and ending with—J. F., 1773. (See also MS. "Common Place Book," p. 87 (Col. Lib. Edin.).

THE ANGLE TRISECTOR.—In the year 1834 we copied from the original in the possession of the late Mr. Upcott, Islington, London, the following drawing and description of an ingenious Instrument for Trisecting Angles; we also observe the same drawing and description in his "Common Place Book."

"An Instrument by which an Angle may be either Trisected or equally Divided into any odd number of equal sides.

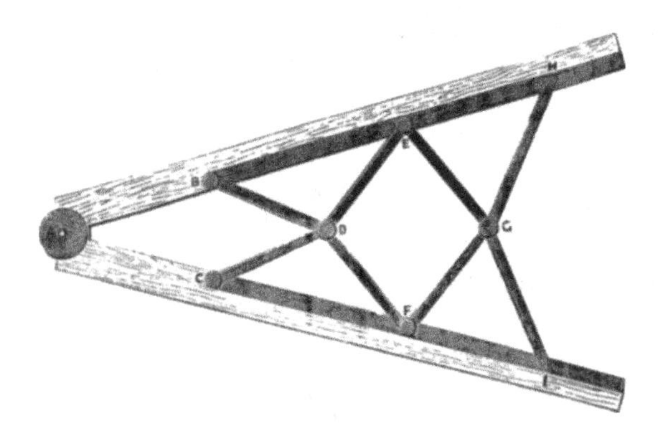

The joint A opens like that of a common Sector; but the rest D, E, F, G, are moveable, and so are the legs B D, D E, D F, F G, G H, and G I. The joints E and F, and the ends H and I slide along the edges B H and C I. The legs are all of equal length, and the distance A B or A C, is equal to the length B D or C D from centre to centre.

To trisect an angle, set off the length of one of the legs (as D E) from the given angular point on each of its containing sides; apply D to the angular point, and the innermost edges of the legs D E and D F to the sides of the angle; or, in other words, make the angle E D F equal to the angle given, and the angle B A C shall be equal to one *third* thereof.

If you want to divide the given angle into 5 equal ones, make the angle H G I equal to the given angle; and the angle B A C shall be a *fifth* part thereof.

If you want the Seventh or Ninth, &c., part of the given angle, add as many more legs, as will form so many Rhombuses as the part exceeds 5; and making the last legs form an angle equal to the given one, the angle B A C shall always be the part required. JAMES FERGUSON.

LONDON, 28*th July*, 1773."

See "Common Place Book," p. 192.

Ferguson did not confine his calculating powers to wheel-work alone; he frequently amused himself with calculations on a variety of curious subjects. The following series of Questions and Answers relative to the National Debt appear to have been written by him in or about the year 1773, when the public debt was estimated at £130,000,000.

"QUESTIONS AND ANSWERS RELATIVE TO THE NATIONAL DEBT."

"Qu. 1st.—Supposing this debt to be no more than 130 Millions of Pounds Sterling at present (although it is really much more), and that it was all to be counted in Shillings; that a man could count at the rate of 100 shillings per minute, for twelve hours each Day, till he counted the whole. How much time would elapse whilst he was doing it?

Answer.—98 years 316 days 14 hours and 40 minutes.

Qu. 2d.—The whole of this sum being 2600 millions of shillings, and the Coinage standard being 62 shillings in the Troy Pound; What is the weight of the whole?

Answer.—41 millions 935 thousand 484 Troy Pounds.

Qu. 3d.—Supposing a man could carry 100 Pound weight from London to York; How many men would it require to carry the whole?

Answer.—419 thousand 355 men.

Qu. 4.—If all these men were to walk in a straight line, keeping at two yards distance from each other; What length of Road would they all require?

Answer.—476 miles, half a mile, and 70 yards.

Qu. 5.—The breadth of a shilling being one inch, if all these shillings were laid in a straight line close to one another's edges; How long would the line be that would contain them?

Answer.—41 thousand and 35 miles; which is 16 thousand and 35 miles more than the whole circumference of the Earth.

Qu. 6.—Supposing the Interest of this Debt to be only 3½ per cent., per annum; What doth the whole annual Interest come to?

Answer.—4 millions 550 thousand Pounds sterling?

Qu. 7.—How doth the Government raise this Interest yearly?

Answer.—By taxing those who lent the Principal, and others.

Qu. 8.—When will the Government be able to pay off the Principal?

Answer.—When there is more money in England's Treasury alone than there is at present in all Europe; or perhaps in the World?

Qu. 9.—When will that be? Answer.—*Never.*

A cubic inch of Guinea Gold weighs 9 ounces 1 pennyweight 11 grains Troy, or 4355 grains; and consequently, a cubic foot (containing 1728 cubic inches) weighs 1306 pounds 6 ounces, or 7525440 grains,—now the Standard weight of a Guinea is 129 grains; divide 7525440 by 129, and the quotient will be 58336·74 for the number of Guineas that may be coined out of one Cubic Foot of Gold; which reduced into silver money is 61253·57£ (or 61253 Pounds 11sh. 4d. 3 farthings). Then, as 61253·57 Pounds is to 1 Cubic Foot, so is 130,000,000 Pounds to 2122·325 Cubic Feet, the quantity of Gold equal to the National Debt. And so much Gold would weigh 2,772,817·6 Troy Pounds.—JAMES FERGUSON, 1773." (See also "Common Place Book," page 112).

DEATH OF MRS. FERGUSON.—After a lingering illness, and confinement to her bed, Mrs. Ferguson died of consumption at 4 Bolt Court, Fleet Street, London, on September 3d, 1773, aged 52 years. She was, according to a correspondent, of "a fragile constitution—often ailing—of a timid, quiet, and retiring disposition." During the last ten years of her life she was destined to encounter much sorrow. In 1763 occurred the loss of her only daughter,—her mother, to whom she was much attached, died in 1771,—and her eldest son, a youth of much promise, died in 1772. Her remains were interred beside those of her son, James, in the churchyard of Old St. Mary-le-Bone, on September 6th.

For a photograph of Mrs. Ferguson, taken from the original Indian-ink portrait by her husband, the reader is referred to date 1743, note 73, p. 47.

SELECT MECHANICAL EXERCISES, Published.—During the summer of this year (1773) he appears to have been employed in arranging many of his loose philosophical papers for the purpose of forming them into a volume, as also in writing his Memoir prefixed to it. This work appears to have been published early in September,—the following is copy of its very full title-page:—

"Select Mechanical Exercises, Showing how to construct different Clocks, Orreries, and Sun-Dials, on Plain and Easy Principles; with several Miscellaneous Articles and New Tables.—I. For expeditiously computing the Time of New or Full Moon within the Limits of 6000 Years before and after the 18th century.—II. For graduating and examining the usual Line on the Sector, Plain Scale, and Gunter. Illustrated with Copperplates. To which is prefixed, A Short account of the Life of the Author. By James Ferguson, F.R.S. London: Printed for W. Strachan and T. Cadell, in the Strand. MDCCLXXIII." (Price 5s.)

This work is an octavo of 272 pages, illustrated with nine copperplate engravings of Astronomical Clocks, Orreries, Sun-Dials, &c., and has passed through several editions,—viz., in 1773,—1778,—1790,—1823. It is out of print, and now difficult to be procured. The edition of 1778 has a likeness of Ferguson,—that of 1823 was published by Dr. Brewster (now Sir David Brewster) in his "Ferguson's Essays."

This is a very curious work, containing 36 miscellaneous articles. It has always been a favourite with mechanics in general, particularly with clock and watchmakers. The papers composing it are somewhat similar to those in "Tables and Tracts," from which, in the present work, we have given many extracts.

Prefixed to this work is the celebrated "Short Account of the Life of the Author," comprised in 43 pages, 36 of which give the details of his Life in Scotland, from 1710 to 1743, and part only of the remaining 7 pages are made to suffice for the incidents of his life in England, from 1743 to 1773, the date when he wrote. As previously mentioned, there is every reason to conclude that Ferguson was employed on his Memoir at the time of his wife's death, and this will be sufficient to account for the Memoir coming so abruptly to a conclusion. See note 3, pp. 1, 2.

JOHN, THE YOUNGEST SON OF FERGUSON, now in his 14th year, left London about the end of October, 1773, for Aberdeen, to attend the Medical and Surgical Classes of Marischal College. He attended the classes of this University from 1773 to 1777 inclusive.[281]

[281] We are indebted to John Cruickshanks, LL.D., Professor of Mathematics, and Secretary of the College, for the following entries from the Album of Marischal College:—

"Under 'Classis Prima, 28 Februarii, 1774
Joannes Ferguson, filius Jacobi Londinensis.'

SYLLABUS OF A COURSE OF LECTURES, Republished.—About the end of the year 1773, Ferguson republished his Syllabus of a Course of Lectures on Mechanics, Hydrostatics, Hydraulics, Pneumatics, Electricity, and Astronomy, with a title-page similar to that of the same Syllabus published in 1769.

1774.

COX'S PERPETUAL MOTION CLOCK.—In the latter part of the year 1773, and for some time in 1774, a mechanical curiosity, known as "*The Perpetual Motion Clock*," was "publicly exhibited in the Museum of Mr. James Cox, of the City of London, Jeweller." Thither Ferguson was attracted, or invited, to inspect this novelty; and at the request of the inventor, Mr. Cox, he gave, in writing, his opinion of it. But let precedence be given to Mr. Cox's somewhat inflated announcement, as follows :—[282]

"THE PERPETUAL MOTION

is a mechanical and philosophical time-piece, which, after great labour, numberless trials, unwearied attention, and immense expense, is at last brought to perfection; from this piece, by an union of the mechanic and philosophic principles, a motion is obtained that will continue for ever; and although the metals of steel and brass, of which it is constructed, must in time decay, (a fate to which even *the great globe itself, yea all that it inherit*, are exposed) still the primary cause of its motion being

Under 'Tertia Classis, 14° Febrii 1776
Joannes Ferguson, filius Jacobii, Londinensis.'
Under 'Classis Quartia, Febr 1777
A.M: Joa. Ferguson.'"

Dr. Cruickshanks remarks, that "John Ferguson must have been a Student also in Session 1774-5: otherwise, he could not have become A.M. in 1777;" and adds, "by some unaccountable neglect, the class in which he would have been in, in 1774-5, had not been matriculated at all."

Although qualified as a Surgeon, he never practised. In 1777, he appears to have returned to London, and resided there until 1804, when he went to reside in Keith, Banffshire. He did not remain long there, having left Keith in 1806, where he resided until his death in 1833. (See page 237; also, Appendix notice of him).

[282] This Perpetual Motion advertisement is taken from a small quarto Tract which we have, entitled, "A DESCRIPTIVE INVENTORY of the SEVERAL EXQUISITE and MAGNIFICENT PIECES of MECHANISM and JEWELLERY, Comprised in the Schedule annexed to an Act of Parliament, made in the THIRTEENTH YEAR of the reign of His present MAJESTY, GEORGE the THIRD, for enabling Mr. James Cox, of the City of London, Jeweller, to dispose of his Museum by way of Lottery. (Growing Arts adorn Empire). London: Printed by H. Hart, Crane Court, Fleet Street, for Mr. Cox. MDCCLXXIII." pp. 72; articles described in it, 56).

constant, and the friction upon every part extremely insignificant, it will continue its action for a longer duration than any mechanical performance has ever been known to do.

"This extraordinary piece is something about the height, size, and dimensions of a common eight-day pendulum clock; the case is mahogany, in the architectural style, with columns and pilasters, cornices and mouldings of brass, finely wrought, richly gilt, and improved with the most elegantly adapted ornaments. It is glazed on every side, whereby its construction, the mode of its performance, and the masterly execution of the workmanship, may be discovered by the intelligent spectator. The time-piece is affixed to the part from whence the power is derived; it goes upon diamonds, or (to speak more technically) *is jewelled in every part*, where its friction could be lessened; nor will it require any other assistance than the common regulation, necessary for any other time-keeper, to make it perform with the utmost exactness. Besides the hour and minute, there is a seconds hand, always in motion; and to prevent the least idea of deception, as well as to keep out the dust, the whole is inclosed within frames of glass, and will be placed in the centre of the Museum, for the inspection of every curious observer.

N.B.—The very existence of motion in the time-piece is originated, continued, and perfected, from the philosophical principle by which it *alone* acts."

The foregoing is at best an unintelligible description, of which nothing can be made. Another announcement by, probably, a friend of Mr. Cox (or by Cox himself), was published, of which the following is a copy:—

"TO THE PUBLIC.

Among other great works now introduced at Mr. Cox's Museum is an immense *Barometer* of so extraordinary a construction, that by it the long-sought-for, and in all likely the only PERPETUAL MOTION that ever will be discovered is obtained. The constant revolution of wheels, moving in vertical, horizontal, and other directions, is not only physically produced, but the indication of time, from an union of the philosophic with the mechanic principles, is effected. Upon the dial, besides a minute and an hour hand, is another hand, dividing the minute into sixty parts; these hands are motionless till affixed to the primary motion, so that the motion of the time-piece (as Mr. Cox, in his descriptive inventory, judiciously expresses it) *is originated, continued, and perfected*, by the philosophic principles through which it is (solely) actuated.[283]

The encouragement Mr. Cox has for many years given to men of genius, and the perseverance with which he has pursued the great line of utility, have not only given birth to productions that have astonished all Europe, as well as the Eastern world, but have at last produced the wonderful machine above described. Several of the most eminent Philosophers and Mathematicians in this kingdom, who have examined it attentively, are of opinion, that it will lead to farther improvements both in philosophy and mechanics; and we hear that Mr. Cox intends to devote a part of every week to the gratification of such gentlemen in the scientific world as wish to be acquainted either with the construction or the mode of operation, the principles of action, or the masterly execution of so capital a performance. This article is the work of many years, during which time, numberless ineffectual and expensive trials were made, which perhaps would have damped any ardour but Mr. Cox's, and probably have prevented the world from ever being benefited by so valuable a discovery." (Cox's Descriptive Inventory Catalogue of his Museum, p. 2).

Still, nothing can be made of the foregoing description; nothing is told to explain its principle of action and internal construction. Mr. Cox appears, by letter or otherwise, to have invited the most eminent Philosophers, &c., to his Museum, to inspect this "*wonderful machine.*" Among others, we find that Ferguson examined it, and having probably been asked by Mr. Cox to give him his written opinion of its merits, gave what must be considered a very favourable opinion of the unique contrivance. Ferguson at all events sent him the following letter, which conveys information regarding the motive power employed to produce the continuous motion.

[283] A description of the motive power of this Clock, along with illustrative engravings, may be seen in "The London Mechanics' Magazine," vol. x. p. 277; vol. xi. pp. 85—88.

We have a beautifully-executed pen-and-ink drawing of this Clock, done by Ferguson in 1774, on a sheet of paper 19 inches by 12, the gift of the late J. L. Rutherfurd, Esq., Windmill Street, Edinburgh.

"SIR,

I have seen and examined the above described Clock, which is kept constantly going by the rising and falling of the quicksilver in a most extraordinary Barometer; and there is no danger of its ever failing to go, for there is always such a quantity of moving power accumulated as would keep the Clock going for a year, even if the Barometer should be quite away from it. And indeed, on examining the whole contrivance and construction, I must with truth say, that it is the most ingenious piece of Mechanism I ever saw in my life. JAMES FERGUSON."

BOLT COURT, FLEET STREET,
Jan. 28*th*, 1774.

Since Ferguson, after minutely inspecting this barometric Clock, has pronounced it to be the most ingenious piece of mechanism he ever saw in his life, there can be no doubt of its being a very extraordinary one. For a lengthened description of it, with woodcut illustrations, see Ferguson's MS. "Common Place Book," pp. 85, 88, 89, 133 (Col. Lib. Edin.); also, Dircks's "Perpetuum Mobile," pp. 123—127, 182, 183, 330—338, and Lond. Mech. Mag., as noted.

LECTURES in LONDON, at BATH and BRISTOL.—From our notes, it is seen that Ferguson, during the months of January and February of this year, was in London, and engaged in delivering several courses of Lectures to parties who had subscribed for them. At same time, his friends in Bath and Bristol were engaged in getting a subscription-list filled up for a Course in each of these cities. About the end of March, finding that the required number of subscribers had nearly been obtained, he left London for Bath; and immediately after his arrival, issued the following advertisement:—

"BATH.

MR. FERGUSON, having been favoured with a SUBSCRIPTION for a COURSE of TWELVE LECTURES on EXPERIMENTAL PHILOSOPHY and ASTRONOMY, proposes to begin them on MONDAY the 4TH APRIL next, at TWELVE O'CLOCK, at MR. OXFORD'S *great* AUCTION ROOM in BOND STREET in this City, and to continue them every WEDNESDAY, FRIDAY, and MONDAY following, at the same hour, till the Course be finished. And for the conveniency of those who cannot so well attend on the aforesaid days, he will repeat the same lectures on TUESDAYS, THURSDAYS, and SATURDAYS, to accommodate each respective company.

SUBSCRIPTIONS at ONE GUINEA will be taken at MR. FREDERICKS, BOOK-

SELLER in the GROVE; at MR. BALLS, BOOKSELLER, OPPOSITE the OLD ROOMS; MR. TENNENTS, BOOKSELLER, MILSOM STREET, and at MR. JONES', DRAPER in the Church Yard." (Bath Chronicle, 31st March, 1774).

About the middle of May, Ferguson left Bath for Bristol to deliver his Course of Twelve Lectures on Experimental Philosophy and Astronomy to 64 subscribers. This course was finished on or before May 21st, as will be seen from the following letter written by him to his publisher.[284]

"BRISTOL, *May 21st*, 1774.

DEAR SIR,

I had 118 Subscribers at Bath,—at Bristol, I have just finished a Course to 64, and it is likely that I shall begin a second Course on Thursday next to about 24 Subscribers; but my health is far from being good. Considering the great success I have so often had in this City, I cannot do less than make a present of all my Six books, in Octavo, to the public Library here; and am, with great respect,

DEAR SIR,

Your most obliged humble servant,

JAMES FERGUSON."

During his residence in Bristol, and in the interval between his first and second Courses of Lectures, he delivered a popular lecture on Electricity, to which the following advertisement refers, and is dated on same day as the foregoing letter.

"BRISTOL, *May 21st*, 1774 (Saturday).

MR. FERGUSON will begin a LECTURE on ELECTRICITY at the BELL in BROAD STREET on *Monday next*, at *Six o'clock* in *the Evening*, in which, besides the common experiments, he will set MODELS of MILLS, CLOCKS, and ORRERIES in motion by streams of ELECTRICAL MATTER, Show an artificial fire screw, a natural representation of the AURORA BOREALIS, an illuminated MOON, and a SATURN with its RING; the method of SECURING BUILDINGS from damage by Lightning; and conclude with an account of MEDICAL ELECTRICITY, so far as he has found it useful.

ADMISSION; two shillings and sixpence each person.

He will begin a second course of Lectures on Thursday next at 5 o'clock in the evening, provided twenty-five persons subscribe their names thereto, at MR. CADELL'S shop in WINE STREET, by Thursday Evening.

284 This letter was copied, in 1831, from the original in the possession of the late William Upcott, Esq., Islington, London, and appears to have been addressed to his publisher, Mr. Thomas Cadell, Strand, London. The late Mr. C. T. Partington, Lecturer on Natural Philosophy, London, published an edition of Ferguson's "Lectures on Select Subjects in Mechanics," &c., in 1825. At the end of his preface he gives a copy of this letter, and under Ferguson's likeness, in the frontispiece, is a fac-simile of the three first lines of the letter, given as a specimen of Ferguson's plain and elegant handwriting.

☞ *He has this day published*, A LECTURE on ECLIPSES of the SUN and MOON; the true year of our SAVIOUR'S CRUCIFIXION; the SUPERNATURAL DARKNESS at that time; and DANIEL'S 70 WEEKS: Price one Shilling." (Felix Farley's Bristol Journal, 21st May, 1774).

The second course, to about 25 subscribers, commenced on Thursday, 26th May, and was completed on or about the 10th June. During this course he presented his six octavo volumes to the Bristol Public Library.

FERGUSON'S WORKS—BRISTOL LIBRARY.—About twenty years ago, when in Bristol, we visited the PUBLIC LIBRARY, in King Street, to get a sight of these volumes. We were courteously received by the Librarian, who at once unshelved them. The "token of gratitude" consists of

1st, Astronomy—On Sir Isaac Newton's Principles.
2d, Lectures on Select Subjects in Mechanics, &c.
3d, Tables and Tracts, Relative to Several Arts and Sciences.
4th, Easy Introduction to Astronomy for Young Gentlemen and Ladies.
5th, An Introduction to Electricity.
6th, Select Mechanical Exercises.

In "*Lectures on Select Subjects in Mechanics*," was found, on a fly-leaf, in Ferguson's autograph, the following bequest:—

"As a Small Token of Gratitude to the Inhabitants of Bristol in general, for having repeatedly encouraged my coming among them to read Lectures on Experimental Philosophy, and for many other Instances and marks of their Favour, these Six Octavo Volumes are presented to the Library Society, with the Greatest Respect, by their most humble servant the Author, JAMES FERGUSON."

BRISTOL, *June* 10*th*, 1774.

After completing his Course to the 64 subscribers, another company less numerous subscribed for a second Course. The short interval between the Courses was filled up by preparing for publication a Tract on the Eclipses of the Sun and Moon, and in delivering a popular Lecture on Electricity.

AN ASTRONOMICAL LECTURE on ECLIPSES, PUBLISHED at BRISTOL.—Sometime about 21st May, he published, at Bristol, an Astronomical Lecture on Eclipses of the Sun and Moon, the Darkness at the Crucifixion, and on the Prophet Daniel's Seventy Weeks, now a very rare Tract, of which the following is copy:—

"An Astronomical Lecture on Eclipses of the Sun and Moon, the true Year of our Saviour's Crucifixion, the Supernatural Darkness at that Time, and the Prophet Daniel's Seventy Weeks. By James Ferguson, F.R.S. Bristol: Printed by S. Farley, in Castle-Green. Price Two Shillings."

About the middle of June, Ferguson had returned to London. Shortly afterwards was published the *eighth edition* of his "ANALYSIS of a COURSE of LECTURES on MECHANICS," &c. (the title-page of which is the same as that given under date 1763).

MURDOCH FERGUSON, the second Son, appears to have qualified for a surgeon about the end of 1774. Soon after, he left London and tried to establish himself in Bury St. Edmunds. He remained there about 8 months, when, finding himself unsuccessful, he returned to London, and for some time resided with his father at No. 4 Bolt Court, Fleet Street, and directed his attention to qualifying for an Army Surgeon.

MACHINE FOR FINDING THE WEIGHT OF A BODY IN ANY LATITUDE.—We have a pencil drawing, by Ferguson, of this curious machine, which is initialled and dated, J. F., 1774. There is no description appended to it. On referring to "Common Place Book," we find a similar drawing in pen and ink, with description, which we give as under. The drawing is designated at top as

"A machine for finding how much heavier any given weight is, when carried into great North or South Latitudes, than it is at or near the Equator."

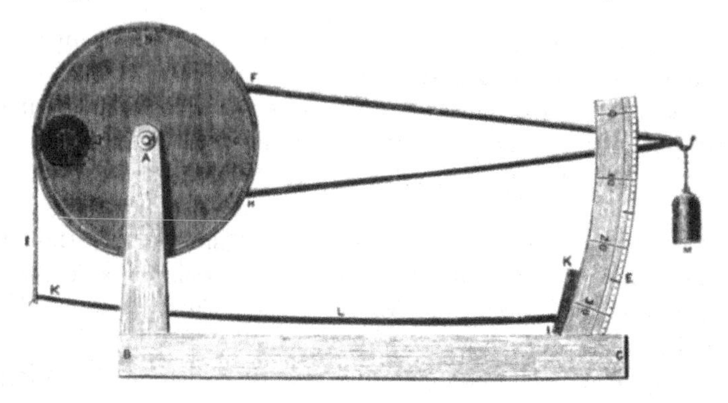

It is the power of the Earth's attraction that causes or constitutes the weight of all bodies on or near the Earth's surface; and as this attraction lays equally strong hold upon every particle of matter, those bodies which contain the greater quantities of matter, must be of so much the greater degrees of weight.

Bodies are lighter at the Equator than they would be at or near the Poles of the Earth; 1st, Because the Earth is an oblate Spheroid (its equatorial diameter being longer than its axis), a Body at the Equator is farther from the great seat or place of attraction than it would be if carried to either of the Poles; for attraction decreases in proportion as the square of the distance increases. And 2dly, Because the centrifugal force of a body at the Equator, in being carried round upwards of 1000 miles each hour, by the Earth's daily rotation on its axis; and the centrifugal force being contrary to the attractive, diminishes the weights of Bodies, most of all at the Equator. But, at the Poles there is no centrifugal force at all.

A B C is a Frame, in which the axis of the wheel D turns, a little way, on a sharp edge, in the round hole above A, like the axis of a common Balance; and E is a thin arc of brass divided into degrees, and third parts of a degree. F G and H G are two small bars of brass, fixed to the wheel D, and joined together at G, by which contrivance, they make a Lever, strong, light, and inflexible. This lever is just counter-balanced by a leaden weight in the wheel at *a*. In the edge of the wheel is a groove, in which, one end of a small silk cord I is fixed at or near *b*, and the other end of the cord is fixed to a spring K L, whose thick end is screwed fast to *k l* to the graduated arc E; and M is a weight hung on the end G of the Lever, which weight is counter-balanced, by the force of the spring acting on the wheel, and tending to raise the Lever and weight by pulling the cord I downward.

Now, supposing this machine to be in a Ship at the Equator; hang the weight upon the end of the Lever, and note the point or degree on the arc E at which the Lever rests. Then take off the weight, and when the Ship has sailed into any great northern or southern Latitude, hang on the weight again, and observe how much lower the Lever rests on the arc than it stood at the Equator, and write down the difference. This done, lay as many grains upon the top of the weight as will make the

Lever descend just as much more as is equal to the said difference; and *that* number of grains will be equal to what the weight M is heavier (by the increase of gravity) at the then place of the Ship, than it was at the Equator. As the axis of the wheel bears on its sharp edge in the hole A, there will be little or no friction on the axis; and therefore, the least quantity added to the weight M will move the Lever.

According to Theory, the weight or gravity of any body at either of the Poles, is in proportion to the weight or gravity it would have at the Equator as 289 is to 288."

"Common Place Book," p. 120.

THE ONE-WHEELED CLOCK.—On the back of one of Ferguson's papers in our collection (dated 1774), there is a pen-and-ink drawing of a One-wheeled Clock. There is no description appended to it, simply these few words,—

> "The number of Wheels in a Clock reduced to one, by means of a Double 'Scapement."

We shall give a copy of the drawing, and leave it as a problem to exercise the ingenuity of clock and watchmakers.

On glancing over the pen-and-ink sketches in "Common Place Book," we find a similar drawing, but still no description, —merely the few words we have given are annexed to it. (See "Common Place Book," p. 127).

"The Pulse Glass."

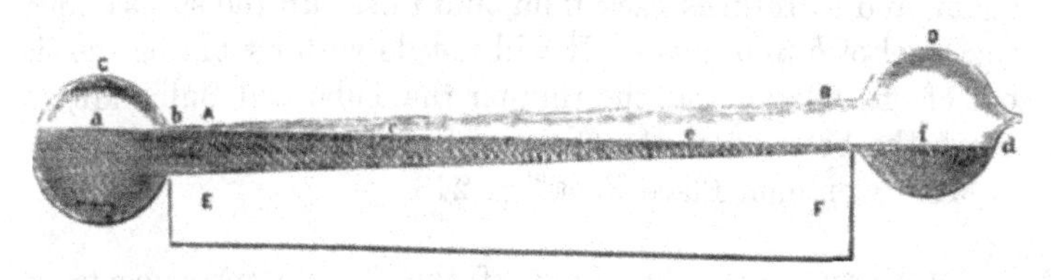

PULSE GLASS.—The above engraving of "The Pulse Glass" is taken from the pencil drawing of it on the back of an autograph letter of Ferguson's of date 1774. In "Common Place Book," we find a pen-and-ink sketch of it, with a description, of which the following is copy:—

"This glass consists of a Tube A B, with the Bulb C on one end of it, and the Bulb D on the other. D is open at *d* when the glass is blown, and the whole is filled about half full of water, or Spirits of Wine, through the small opening at *d*. Then, all the air that can be taken out of the Glass is taken out, through the part *d*, and then *that* part is hermetically sealed, under the Receiver of an Air-Pump, by the Sun's rays transmitted through a burning Glass, which collects them into the focal point at *d*, and so closes it up by melting that part.

Then, the glass is laid sloping upon the piece of wood E F, in such a manner, that when the surface of the water or Spirits of Wine in it comes to a level, *a c e f*, the part *b* of the Tube may be a little below that level, and the fluid spot at or about A. Then, if a person puts the ball of the thumb or finger gently upon the lowest Bulb at C, the fluid at *a* sinks down to *b*, and a very large Bubble goes off from *b* towards *f*; then the fluid rises to *a* again, and sinks, and throws off another Bubble toward *f*; and so on continually, as long as the finger or thumb is left upon C; and all the while, the empty space from C to *c* never grows bigger at one time than another, nor the space between D and *f* less. The same thing will happen if the Sun shines upon the Bulb C, and not upon any other part of the Glass.

The reason of this pulsation is, that the heat of the finger, or thumb, or Sun-beams, raises a vapour or steam from the surface of the fluid at *a*; that this steam presses down the fluid at *b*, and then goes off in a Bubble to the other end *f*, which being colder than at *a*, the steam is there condensed into the fluid again, and so returns back fluid, and raises up the surface from the level of *b* to *a* again. It will not do without taking the air out of the Glass; and the thinner the Tube and Bulbs are, so much the better does the Glass perform."

MS. "Common Place Book," p. 248.

ANALEMMA.—We have one of these little instruments, in Ferguson's autograph. It is made of thick card; and consists of two circular discs,—diameter of the larger one, 6 inches, of the smaller, which is partly open, 4¾ inches. They are connected by a pin in the centre upon which they turn; on both cards are a series of circles and figures, as also rules for using them. In a space near the centre is inscribed, "*J. F.*, *London*, 1774."[285]

QUADRANTS.—We have also, in Ferguson's autograph, and of same date, two Quadrants cut out in card. The sides of each respectively, are 10½ and 10⅛ inches. On the edge of the arc of the largest one are laid down the 90 degrees of the quadrant, and numbered 10, 20, 30, &c., within which are the hours in Roman characters, above which are a series of complex curves, which refer to the Ecliptic Signs, and over these are several circles, within which are the Days of the Month and Signs of the Zodiac, resembling the horizon of a small globe. In the interior are directions for using it. On the edge of the arc are "*J. F.*, 1774."[286] The second quadrant is somewhat similar in its appearance, but for a different purpose. On the arc are also laid down the 90 degrees of the quadrant, and numbered 10, 20, 30, &c., interior to which are a double set of Roman hours, above these hours are a series of curves referring to the latitude, &c., over which are written directions for use. At top are

[285] Ferguson appears to have made several of these Analemmas (orthographic projection of the Sphere upon the plane of the Meridian); there is one in the Museum of Banff, another in a loose condition in his "Common Place Book."

[286] There is one exactly the same as this in the Museum at Banff.

laid down the Days of the Month and Sun's Declination. Each quadrant has *sights* at top, formed by cutting into the card, leaving part to act as a hinge to keep them standing out, during an observation. In a small space within the arc of 90 degrees is inscribed, "J. FERGUSON, LONDON, 1774."[287]

1775.

ASTRONOMICAL ROTULA and TRACT.—About the beginning of the year 1775 were republished the Astronomical Rotula, and the small octavo Tract of 16 pages. The price of the Rotula and the descriptive Tract was 5s. 6d. The following is a copy of the title-page of this Tract:—

> "The Description and Use of the Astronomical Rotula, Showing the Change and Age of the Moon, the Motions and Places of the Sun, Moon, and Nodes in the Ecliptic; with the Times and Phases of all the Solar and Lunar Eclipses. By James Ferguson, F.R.S. London: Printed in the year MDCCLXXV."

As previously mentioned, these Rotulas and Tracts have been long out of print, and are now very scarce. We have a copy of one rotula, and also several copies of the Tracts, which we purchased some years ago at an old book and print shop in London.

REVIVAL OF THE "ASTRONOMICAL CHRONOLOGY,"—CONTROVERSY BETWEEN FERGUSON AND THE REVEREND J. KENNEDY. —In the year 1762, Kennedy published his "*Complete System of Astronomical Chronology Unfolding the Scriptures.*" In the months of May, June, and November, 1763, Ferguson attacked several portions of this work in the "Critical Review," after which latter date nothing more was heard until March in 1775, when Kennedy, after a silence of *twelve years*, came again before the public in a pamphlet, which, with its appendix, extended to 94 octavo pages. The following is a copy of its title-page:—

> "An Explanation and Proof of '*the Complete System of Astronomical Chronology, unfolding the Scriptures.*' In which the TRUTH and REALITY of the ORIGINAL LUNI-SOLAR RADIX is clearly and fully ascertained; first by Calculations a Priori; then confirmed, to the minutest Exactness, by Calculations a Posteriori, through an

287 The late Deane Walker, Esq., so often mentioned, had also a quadrant similar to this. We may here note that we purchased the Analemma and Quadrants from a party in the north who had received them from Miss Wilson of Keith, the daughter of a brother of Mrs. Ferguson.

extensive interval of 5800 years. In a Series of Letters Addressed to MR. JAMES FERGUSON, Author of A Treatise of Astronomy upon Sir *Isaac Newton's* Principles of Philosophy. (*Magna* est veritas, et prevalebit) By the Rev. JOHN KENNEDY, Rector of Bradley, in Derbyshire, and Author of the Complete System of Astronomical Chronology Unfolding the Scriptures. London: MDCCLXXV."[288]

We think any impartial reader, after a perusal of this production, will be of opinion that, like the author's "EXAMINATION of MR. FERGUSON'S REMARKS," published in the Critical Review, May 1763, it exhibits a pretty considerable amount of ignorance, arrogance, and petulance; but in charity let it be borne in mind, Mr. Kennedy was now on the verge of fourscore years.

The first part of Kennedy's present pamphlet, "EXPLANATION AND PROOF," &c., consists of four long letters on some sections of Ferguson's Astronomy; on Ferguson's Mean length of the Solar Year and Lunation Period; on the 1st year of the Mosaic Creation; on Ferguson's Astronomical Tables; on the Calculation of Ancient Eclipses by Ferguson's data, &c.; and as a matter of course, Kennedy, after *examining* these subjects in his own peculiar way, finds, to his own satisfaction, that his "Astronomical Chronology" is right, and Ferguson's Astronomy wrong. The four letters take up 45 pages of the "*Examination;*" then follows:—

An APPENDIX to the FOREGOING LETTERS, containing a Review of MR. FERGUSON'S Remarks upon the Astronomical Part of my System of Chronology. (See the Critical Review for May, June, and November, 1763).

This Appendix occupies 49 pages of the publication—is full of astronomical periods, tables, and calculations—and abounds with cavil and puerile criticism.

It may be here noticed that Kennedy does not commence his series of letters with "*Dear Sir*" or "*Sir*," but sneeringly with "*Old Friend*," and concludes it with "*Your old acquaintance and well-wisher, John Kennedy.*" The "APPENDIX" also

[288] Kennedy dates his pamphlet "October 11th, 1774." In our copy, some former possessor of the pamphlet has written under this date, "*it was dated Bradley, Septr. 27th,* 1774." The "Explanation and Proofs" do not appear to have been published until February, 1775. A copy, part of Letter 1st of "*Explanation and Proofs,*" in MS., appears to have been sent to Ferguson by a *footman* from Kennedy, for prefixed to the printed letter in our copy of the pamphlet we find written, "*This letter (nearly) was left with Mr. Ferguson by a livery servant with as much rudeness as if he had been under Kennedy's tuition all his life,*"—evidently written by one of Ferguson's friends.

begins with "*Old Friend*," but, in concluding it, he has forgotten to assure Ferguson that he was his "well-wisher," simply ending it with the initials "J. K."

Immediately after the appearance of Kennedy's pamphlet, Ferguson made a reply in print. The first of these letters is short, consisting of only four pages; the second letter occupies 28 pages octavo. Under date 1763, notice is taken of this "Astronomical Chronology" controversy, and quotations given from these letters. To meet the "*carpings and warpings*" in Kennedy's "*Explanation and Proof*," just published (March, 1775)—cavils reproduced in great part from his pamphlet of 1763, Ferguson conceived that although twelve years had elapsed since the controversy was closed, the proper answer was still the same as he had given in the Critical Review in 1763; and he accordingly reprinted the article from this paper in the form of a pamphlet of 28 pages, prefacing it with the first letter of 4 pages. Both these were issued at same time, under one cover, about the beginning of March, 1775.

As it would be tedious and uninteresting to quote from Kennedy's sophistical pamphlet, the "Explanation," it will be sufficient to give the first letter of four pages entire, with copious extracts from LETTER II. The following is copy of the title-pages to these LETTERS:—

"Two Letters to the Rev. Mr. John Kennedy, containing An ACCOUNT of many MISTAKES in the Astronomical Part of his Scriptural Chronology. And his Abusive Treatment of Astronomical Authors. By James Ferguson, F.R.S. London: Printed for T. Cadell in the Strand. MDCCLXXV."

LETTER I.

"To the Reverend John Kennedy.

REVEREND SIR,

You know, that in the Critical Review for May and June A.D. 1763, I gave some account of your *System of Chronology Unfolding the Scriptures.* But although in that Account, I adhered strictly to the truth, it appeared by a small pamphlet you published soon after that I had incurred your *high* displeasure. And in order to avoid all further disputes between us about astronomical matters, in some months afterwards (Nov. 1763) I addressed a letter to you in the Critical Review; of which, I have found that you have taken no notice of in print.

But now it seems you want to revive the old affair (of 1763) between us;

for, in the preface to your Letters to the Reverend Dr. Blair,[289] printed in the year 1773 (which I never heard of till a few months ago when they fell accidentally into my hands), you charged me with having found an eclipse of the Moon, in the year before Christ, 201, two days before the Moon was full; and with having miscalculated the days of all the eclipses on record before the Christian Æra, besides having been guilty of many other 'enormous errors' in my tables and calculations. You have likewise asserted (in the same preface) that I am 'an illiterate and incompetent judge;' and that, if there had been a *Censor Tabularum* among us, 'all my solar and lunar tables, without exception, would have been interdicted, as unfit for common use.' You, Sir, have set yourself up for this *Censor Tabularum;* for you have also said, in the same preface, that if a proficient was to calculate, even by MEYER'S Tables (although, by-the-bye, they were never found to differ one minute of time from observation), he would find that they are no solar or lunar tables at all. A very fine compliment to the government under which you live, for having given Meyer's widow such a large sum for these tables in manuscript; and to our Astronomer-royal, for calculating the Nautical Ephemeris from them, for finding the Longitude.[290]

You may abuse me as much as you please; but, however *illiterate* you take me to be, I believe I know the meaning of the few Hebrew words at the bottom of this page; which I need not explain to you, who are a good Hebrew scholar, and too well acquainted with your Bible to be at any loss where to find them. And, as many who read this may be quite ignorant of the Hebrew language, I shall neither give the English meaning of them, nor direct where to find them; because I would not willingly raise a general laugh against you.[291]

The whole reason for your finding fault with my solar and lunar tables is, that they do not agree with your calculations for ancient times or events. Indeed it would be a great wonder if they did; for they are founded upon astronomical observations, whereas the whole basis of yours is only an assumed hypothesis, which you call *Mosaic principles,* and whereby you pretend to have found out in what year of the Julian period the world was

289 The Rev. Dr. Blair, Prebendary of Westminster, published, in 1773, a Chronological work, which Mr. Kennedy at once attacked in a series of Nine Letters addressed to Dr. Blair. In his preface to them he vilified Ferguson, which Ferguson here notices. Dr. Blair paid no attention to Kennedy's letters, at which Kennedy expressed himself as neglected, and bemoans his fate.

290 Tobias Meyer, the eminent astronomical calculator, died in 1762, aged 39. His widow received £3,000 from Government, being part of the sum which they had offered to discover the longitude within certain limits.

291 "The few Hebrew words at the bottom of the page" bring out Ferguson in a new light—it shows that *the sedate astronomer* had some humour in him—that he could do a smart thing in a very quiet way, and in this case in a way which he knew to be the best for arousing public curiosity to the meaning of the cabalistic letters. *The few Hebrew words* which Ferguson has at the bottom of his page are

אל תען כסיל כאולתו פן תשוה לו גם אתה׃

That is, "*Answer not a fool according to his folly, lest thou also be like unto him.*" Proverbs 26 chap. 4 ver.

created. A thing that not only Moses and the prophets have been silent about, but even our Saviour and his apostles also; and consequently a thing that no man now can know, without an immediate revelation from Heaven, which we are not to expect. And the late Reverend Mr. Bowen of Bristol has plainly shown that by calculating upon principles similar to yours, he can prove the date of the creation to be as many years before or after your date thereof as he pleases: of which he has given Examples in the Christian's Magazine several years ago. Indeed I am glad that you have condemned my tables; for, if you had commended them, they must have agreed with your calculations; and then the merest dabbler in astronomy could have proved them to be false.

In your *Scripture Chronology* you have taken a figure from my Book on Astronomy, and *there* you insinuate that I had meant by it to amuse or deceive the unwary learner. But, in a pamphlet which you published *soon after*, you inserted the same figure: there you applauded it, and said it was taken from Dr. Long; although the doctor has no such figure in his book. This, I remember, was taken notice of by the Reviewers, who called upon you to show from which plate of the doctor's Astronomy you copied that figure, and in what part of the book he describes it: but you never complied with their desire—for this good reason, that you could not.

With respect to the number of people who can read and judge a single Critical Review can fall only into the hands of few persons; and most of those who have read my above-mentioned letter to you therein may probably have forgot it at this distance of time. I therefore now think proper to publish it, with some additional notes, subjoining it to this, as a full answer to all you have hitherto written or ever can write against me. I know, you did not like it at first, and if you dislike it still, you must thank yourself for its second appearance, which is entirely owing to what you have mentioned concerning me (about the eclipse) in the preface to your Letters to Dr. Blair.

I am,

REVEREND SIR,

Your humble servant,

JAMES FERGUSON."

The following are extracts from

LETTER II.

(First printed in the Critical Review, Nov. 1763).

Reprinted, March 1775.

"To the Reverend Mr. Kennedy.

REVEREND SIR,

As the printer of the Critical Review for the month of May, 1763, found my paper containing the remarks on the astronomical part of your System of Chronology, too long to be all inserted in the Review for that month, and promised to conclude it in his next, which he accordingly did; it was reasonable to suppose that you would have deferred your examination of these remarks till the Review for June was published, that you might have had an opportunity of examining, and, if you could, of refuting them all at once.

But, it seems, this was more than you had patience to wait for, and therefore you thought proper to attack one half of my paper before you had read the whole of it; which, in the opinion of the candid, will not, perhaps, be deemed very fair on your side.

As I expected you would likewise examine the second part, I have now waited three months for what you had further to say, that I might at once reply to the whole.

But as you have disappointed me in this, I shall reply to what you have published in relation to the first part (so far as it requires a reply), and give myself no further trouble about anything you may chuse to write against me for the future; unless I find reason to alter my present opinion." *

Letter 2d to Kennedy, pp. 5, 6.

Here Ferguson, by an asterisk, refers to a foot note, which shows, that between 1763 and 1775, Kennedy had been using endeavours to bring Ferguson's works into disrepute,—the foot note in question says,

"And now I think I have very good reason, because, in the preface to his Letters to Dr. Blair, he has endeavoured, as far as he could, to put an end to the sale of my Astronomy, Tables and Tracts, &c., at once. But this gives Mr. Cadell no concern, who has bought the copyright of them all."

Letter 2d to Kennedy, p. 6.

* * * * *

Further on in this letter

"The question between us is, whether you, or anybody else, can calculate the true times of new and full moons and eclipses by *your* tables, which you praise so highly, as well as you see astronomers do by *theirs*, which you altogether condemn? This, Sir, you know, can never be done by your tables, either with equations or without them. And therefore, however useful your book may be to the mere chronologer, and however exactly your boasted coincidences may (or rather, indeed must) agree with the principles upon which you have constructed these tables, and may thereby dazzle and deceive the superficial scholar, and conceal from him the imposture of your method, you can never deceive any one who is but tolerably qualified to judge of these matters." [292]

* * * * *

[292] This is somewhat strong language, but it must be borne in mind that Kennedy was the aggressor in this way, in not merely strong language, but in insolence, and this retort by Ferguson is as just as it was necessary. He had, it appears, an awkward predilection for calling things by their proper names, "Kennedy's Tables and Calculations were an imposture,"—the base is *supposition;* and consequently, "an imaginary starting point will terminate in imaginary results," Kennedy appears to have keenly felt Ferguson's remark as to the

"Toward the end of your *Candid Examination* of my remarks, you tell me what is very true; namely, that if I take an attentive view of your calculations, I shall find no appearance, or specification, of any radix at all, which, you say, is such a paradox as I shall never be able to solve, unless I can cordially, and with *some degree of faith* too, follow your directions, which are comprehended in these three words, *Search the Scriptures.*

But since *your* Mosaic principles are, that the moon was *full* upon the *third* day of the creation-week, just 24 hours before the Sun was created to *enlighten* her; that the *fourth* day of the original week was the sixteenth day of the original month." *

Ferguson has an asterisk here, referring to the following foot note:—

"* As we never heard that there were days or nights before the world was created, *Query*, whence does Mr. Kennedy fetch those *fifteen* days which preceded the fourth day of the original or first week?" (J. F.)

He then proceeds,

"And that, on the noon of the *fourth* day, the sun was created upon the first point of Libra, which was then vertical to a point in a meridian in the Great South Sea, 156 degrees west of Greenwich;[293] and that you have been able, from these prin-

imposture of his *method.* In the appendix to "Explanation and Proofs," he takes notice of it as follows:—"Pray Mr. Critical Examiner, what can you possibly mean by the *imposture of my method?* such vilifying, reproachful, and degrading terms call aloud for a justifying comment, and by these presents I publickly demand it, and expect to see it in your next 'Critical Review' (Kennedy's "*Explanation and Proof,*" Appendix, p. 84). No retraction of course was made by Ferguson; he simply gave utterance to what was true and well known. Kennedy, a great master in the art of wounding the feelings, winces when he is himself touched a little in that way—he never retracted any of his "vilifying" expressions regarding Ferguson, but he expects Ferguson to retract and apologize for his retort! A former possessor of our copy of Kennedy's "*Explanation and Proof*" has written on the margin of it several remarks not very complimentary to Kennedy. We shall quote one,—p. 26, on margin at foot, we find, "*If Kennedy's ignorance of astronomy had not shone forth in every line he writes we should have only branded him as an ludicrous lyar, as it is, call it by what you will.*" Pretty severe this; but as this was never before published, being only a private noting, it, along with what follows, shows that the current of opinion was against Kennedy; in fact, he had not a single supporter—everyone was against him—newspapers, magazines, and the public voice, cried him down, but he heeded not.

293 Ferguson, in a foot note, remarks, that "He places his original, or fiist meridian, in the Great South Sea, 156 degrees west of Greenwich—a mere assumption, to make it agree with his numerical measures; and one would wonder why he did not rather place his first meridian in Paradise, and adapt his numerical measures to answer thereto; which would at least have appeared more natural than to have put it almost opposite to the river Euphrates, which is still in being, and which Moses says, was one of the rivers of Eden."—(J. F.)

ciples, and the farther assistance of the Pentateuch, to deduce that the precise length of a mean lunation is 29 days 12 hours 44 minutes 1 second 45 thirds, and that the precise length of the solar tropical year is 365 days 5 hours 49 minutes.[294]

Unluckily for these principles, I happened to remember that Moses says, that both the Sun and Moon were created on the *fourth* day of the original week, without mentioning whether the moon was then new or full, or over what meridian or meridians these two *luminaries* were created (for his calling the

[294] Regarding the true mean lengths of the tropical year and of a lunation, notice has already been taken; (see notes 239, 240, pp. 275, 276)—the year consists of 365 d. 5 h. 48 m. 51·6 s.; and a lunation, 29 d. 12 h. 44 m. 2·87 s.

In Ferguson's MS. "Common Place Book" there is a humorous memorandum by him on Kennedy's absurd method for determining the length of the year. It was never published. The following is copy of it:—

"The true length of the Solar tropical year determined by Scripture data, in imitation of the Reverend John, Kennedy.

Put 3 = the Hutchinsonian Elohim, the first in order, }
,, 6 = the Hexameron, the second in order, } in respect of time.
,, 5 = the Pentateuch, the third in order, }

365 = the position altered = the years of Enoch's Life = the days in a year.

Now for the Odd Hours.

Divide 365 by 70 = Anni ætatis humanæ.

70)365(5 hours.
350
———
15

For the minutes,

Cube the number 15, then add the number of Books of Moses, both to the units and tens of the cube number, and divide by 70 as before.

15×15×15 = 3375
55
———
70)3430(49 minutes.
280
———
630
630

There being no remainder, 'tis plain that the year can neither exceed nor fall short of 365 d. 5 h. 49 m., which is Mr. Kennedy's true *Shanah* or Solar revolution." MS. "Common Place Book," on a small half sheet of paper fastened in between pages 70 and 71. As it is not on a page of the book, it is likely that Ferguson may have carried it about in his pocket-book, to show to his friends as a quiet joke.

In this *illustration*, the first figure, 3, is put down as being equal to "*the Hutchinsonian Elohim—the first in order.*" This is a hit by Ferguson on the "Hutchinsonians," of which he had a very indifferent opinion. In alluding to this body in his "Common Place Book," p. 5, he says, "*The Hutchinsonians do indeed damn Sir Isaac Newton, but everyone who attends to demonstrative truth, allow that they are the most wrong-headed set of people upon the face of the Earth, regarding neither Reason nor Demonstration.*"

moon a luminary [295] is no demonstration to me that she was created full), and that he has said as little either about the precise length of a year or of a lunation; and therefore, I had *too little faith* in Mr. Kennedy's directions to induce me to *search the Scriptures* for what I was preconvinced I should never be able to find in them; and so began to think that, instead of a *true Mosaic radix*, he had got a windmill in his head." (Pp. 12, 13).

* * * * *

"I suppose you remember very well that you told me, some years ago, you would not let me know the measure or length of your Mosaic lunation, because you was sure that if you did, I would correct all my lunar tables by it, but that you would soon publish it in a book which would surprise the world, and that you would try your Mosaic lunation by some well-vouched ancient eclipses, the times of which were distinguished by such sure characteristics, that we would be at no loss about them. But you then told me that eclipses were no part of your doctrine; and yet it seems they now are; for, in your own way, you have calculated several of them in your book." (Pp. 13, 14).

* * * * *

"I often desired you to calculate the time of full moon in September, the year before Christ 201, from your own numbers, and to let me have it under your hand; but this was a request you never thought proper to comply with, although I never denied you any calculation you asked for. But I was not cunning to catch you by guile, as you have since owned, you did me.

As to the mistake in one of my calculations, in a letter which I sent you *long ago*, and of which you have *now*, with inexpressible joy, in the 10th page of your *Candid Examination*, declared me to be the author,—this was not very fair, in print, to expose a mistake in a letter of correspondence which was never printed. You have done all in your power to expose me for it; you make the most of it you can. A mistake it was; and I am not like you, for I do not pretend to be infallible

295 The foot note in question says, "It is surprising to think how, in our English translation of the Bible (Gen. c. i., v. 16), the Sun and Moon should have been called *two great lights*, seeing the original is המארת, which signifies *luminaries*, not *lights*. אור (v. 3) signifies *light*, and המאור (v. 16) signifies a *luminary* or *instrument* for conveying the light.—J. F." (Let. 2d, p. 13).

either in constructing tables of my own, or in computing from those of others. You know full well, that my beginning a literary correspondence with you was solely owing to the request of a gentleman of distinction, who is now dead.[296] But is it possible for you to imagine that there should be no errors in the many letters and calculations which I sent you, and which I could seldom spare time to examine, on account of my business, on which my family's bread depended?[297] If you have preserved all the rest of my letters, you are welcome to print every one of them, provided you also print the copies of yours to which they are answers; and so to complete the sweet revenge which *you think* you have thus begun in your examination of my remarks on the unastronomical part of your system.

You tell me, in that *little* pamphlet, that the last lunar tables which I published are different from the first; and therefore, you say I am 'tossed about with every blast of doctrine, adopting one measure to-day and another to-morrow.'

Tossed about with every blast of doctrine! No, Sir; for, notwithstanding all the violent and most magisterial blasts of *your* doctrine, which, I daresay, you take to be *some* doctrine, I am so far from being *tossed* about as not to be in the least *shaken* thereby; nor has all your dust been able to hurt, much less to blind, my eyes.

I own that my last astronomical tables are not exactly the same with the first; for I am never ashamed to mend or improve anything I have formerly done; and am obliged to every one who assists me with proper advice and materials for that purpose. I do not pretend that these tables are yet perfect, nor that they can be brought to such a mathematical degree of exactness as you assert that all your tables and calculations are. And therefore I must be content with what improvements I can make from time to time, from the observations of astronomers: for Moses has given me no assistance at all in these matters.

But you are a perfect man in all these things. Your measures of years, lunations, sidereal and solar days, are all so *mathemati-*

[296] The gentleman here alluded to appears to have been Ferguson's distinguished friend, the Rev. Dr. Birch, who died in 1766. (See note 255, p. 325).

[297] The Reverend Mr. Kennedy was introduced to Ferguson by the Rev. Dr. Birch about the year 1755, when Ferguson had a hard struggle to maintain himself and family; and it is a matter of surprise that he could spare time for the examination of so many frivolous and absurd calculations as were sent to him by Kennedy.

cally true, that the least particle of time can neither be added to them, nor taken from them, without doing violence to nature. You will therefore, undoubtedly, abide by *your own* numerical measures; and to them I will now bind you down.

You tell us that the mean time of the new moon in April, 1764, is the 1st day, at 10 hrs. 11 min. 39 sec. 15 thirds, in the morning; and that the precise length of a mean lunation is 29 d. 12 hrs. 44 min. 1 sec. 45 thirds. The sun will be eclipsed at the time of this new moon; and you lay so great a stress upon it as to say, 'Should it be found, by a diligent observation, that we have *nearly* calculated the *middle time* of this *future eclipse* of the sun,[298] in our own meridian, it will then be demonstrably certain that the calculations were regulated,—1. By a true series of tropical years and lunations, from the autumnal equinox at the creation to the vernal equinox A.D. 1764. 2. By an exact quantity of the solar tropical year, and of a mean lunation. 3. By a true meridian distance.'

And now, from the mean time of new moon in April, 1764, as calculated by yourself from your own measure of a lunation, &c., you are here called upon to calculate backward, so as to give us,

1. The mean time of full moon at Alexandria, in September, in the year before Christ 201; which full moon, according to Ptolemy, rose eclipsed at that place.

2. The mean time of full moon at Syracuse, in September, in the year before Christ 331; which full moon also rose eclipsed at that place, according to Ptolemy.

3. The mean time of full moon at Babylon, in December, in the year before Christ 383; which full moon set eclipsed at sun-rising according to Ptolemy.

4. The mean time of new moon at Athens, in August, in the year before Christ 431, when the sun was eclipsed in the evening according to Thucydides." (Pp. 14—18).

* * * * *

"Upon calculating backward, by *your* measure of a lunation, from the mean time of new moon in April, 1764 (as given by yourself, in which you come very near the truth), through all the lunations up to the above times, I find, that in the first-

298 It will be remembered that Ferguson, afterwards, when at Liverpool, on April 1st, 1764, made observations on this eclipse. (See pp. 290—296).

mentioned of these eclipses, your numbers make the time to have been almost 22 hours after the full moon rose; so that she was then below the horizon, and within three hours of rising in the morning after the eclipse happened; and consequently, according to your measures, *that* eclipse could not be visible at Alexandria.

In the second of these eclipses, your numbers make the time to have been fifteen hours after the moon was risen; and therefore, for the time of the year, the moon was set, and the eclipse invisible at Syracuse.

In the third-mentioned eclipse, your numbers make the time to have been fifteen hours after the moon was eclipsed; and therefore, for the time of the year in which it happened, the moon, being in a high sign of the ecliptic, must not only have been risen at Babylon, but even very far advanced above the horizon, and on a wrong day for her to have been eclipsed.

In the fourth of those eclipses, your numbers make the time to have been four hours too late, and the sun to have been below the horizon of Athens; so that he must then have been invisibly eclipsed at that place.

And, upon trying to rectify your numbers, by applying the proper equations, I find all the times too late by the following quantities. The first eclipse about nine hours, the second almost ten, the third somewhat more than ten, and the fourth, ten and a quarter. So that your measures will not come near the observed times, either with or without equations; and as we find your *mean* times are too late, and the further back from the present times so much the later, this is an evident demonstration that your measure of a lunation is too short, seeing it brings down the times of all the ancient new and full moons too near to the present; and manifestly overthrows all your lunar astronomy at once. And yet, you would have us to believe that you had all your measures from the Scriptures, although every one who reads them knows that minutes, seconds, and thirds of time are never once mentioned there, and very seldom hours;[299] and hence it is too plain, that a deist,

[299] Minutes, seconds, and thirds of time are not to be found in the Scriptures, certainly, but in them there are something like equivalents to be found; for instance, "*moments*" is of frequent occurrence, which, perhaps, answers to our *second* of time. We also find "*the twinkling of an eye,*" which has been estimated at *the tenth part of a second.*" The term *hour* occurs for the first time in the Scriptures

who could persuade any unwary scholar to believe you, might draw an argument from your book specious enough to make him disbelieve the Bible." (Pp. 18—20).

* * * * *

"The truth is, that, intent upon your chronological studies, you have forgotten the very elements of logic. By changing the terms of the question, *you* bring out absurd conclusions, and then fix them on others (much like your way of treating *Meyer's* tables, as mentioned already). The definition of a solar day is as plain as any definition in Euclid; and the only question is, what is its mean length above the sidereal. This, without any proof, you assert to be four minutes precisely, and this *assertion* you call a *definition;* in consequence of which, you aver, that whatever star comes to the meridian with the sun, on any given day of the year, will come to the meridian twenty minutes before the sun on the 365th day afterward (but instead of this, it will be a minute later than the sun on coming to the meridian); and is therefore so notoriously false, as is plain to every observer, that I need not say anything farther about it. And yet, according to your own account, it is the very foundation of your astronomical tables, from which you could derive no assistance from the incongruous accounts and observations of astronomers.[300]

in the Book of Daniel, chap. 5, ver. 5th. But before the time of Daniel, about the year 720 B.C., we read of the Sun-dial of Ahaz. How was time then divided, and by what names were the divisions and subdivisions known? The Scripture "*moment*" and "*twinkling of an eye*" does not refer to time, but to something of sudden import. It is mentioned in history that Alexander the Great, on his taking Babylon, in the year B.C. 331, discovered, on slabs of brick, a list of all the eclipses of the Sun and Moon, and *times* of their taking place, for the long period of 1,903 years; thus showing that the Babylonians, 2,234 years B.C., had a method for recording divisions and subdivisions of time, which period reaches backward almost to the understood period of the Deluge.

300 This *assertion* of Kennedy's is of itself sufficient to show that he had much to learn in the science of Astronomy. It is difficult to conjecture by what process of calculation he came to the conclusion that any given star, after leaving the meridian with the Sun, would, on the 365th day afterwards, come to the same meridian 20 *minutes before the Sun!* He assumed 23 h. 56 m. as the *exact* length of the sidereal day, which, on being decimally expressed, is = 23·933334 h. × 366 = 8759·600244 h. = 8759 h. 36 m. 0 s. 52 thds. 42 fourths 14 fifths = 364 d. 23 h. 36 m. 0 s. 52 thds. 42 fourths 14 fifths, which wants 23 m. 59 s. 7 thds. 17 fourths 46 fifths of the 365 d., but Kennedy makes it 20 m. exactly! It is possible he might say to himself,—why, these 23 h. 56 m., my value of the sidereal day, is *exactly* 4 m. less than the solar day of 24 h., and this 4 m. × 365 = 1460 m., and as a day of 24 × 60 = 1440 min., this is minus the 1460 by 20 minutes exactly. There are 23 h. 56 m. 4·906 seconds in a sidereal day, which, when decimally expressed, is 23·934469 h. × 366 = 8760·015654 = 8760 h. 0 m. 56 s. 21 thds. 15 fourths 50 fifths = 365 d. 0 h. 0 m. 56 s. 21 thds. 15 fourths 50 fifths, which is

With no less absurdity you assert that all equations of time are *unastronomical,* and *ought* to be rejected. For shame! Mr. Kennedy, blush at these things; for, however well you may think yourself qualified to find out Hebrew roots, what (as you term it) '*the stream of commentators*' have not discovered before you; yet every novice in astronomy can show how grossly you are mistaken in this matter; and, without the assistance of Hebrew, Greek, or Latin, can prove that you have taken upon you to write on a science of which you know nothing at all. You might just as well condemn all theory and observation by the lump, as talk at this rate; and indeed it appears by the greatest part of your writings that you are inclined to do so." (Pp. 23—25).

* * * * *

"To satisfy the sixth query of your '*Examination,*' namely, 'whether the sun, moon, and stars militate against your system of Chronology?' I answer—They do. For, the sun militates against you, were it for nothing else than your denying the equation of time:—the moon militates against you, because your lunar numbers answer not to the times of her eclipses:—and the stars *in their courses* militate against you, because, according to your measure of a sidereal day, *they* ought to have a progressive motion of five degrees *westward,* with respect to the equinoctial points, every year; whereas, in truth, their apparent progress is *eastward* from these points, and *that* not quite the 60th part of a degree in a year." (Pp. 23—26).

* * * * *

"In the calculation you give at page 252 of your book, 'you say your conclusion is very remarkable, and very singular, because no other tables can produce such coincidence.' A most wonderful coincidence this! Strange, indeed, that if you count five inches forward from one end of a foot rule, and seven inches back from the other end, both your reckonings shall end at the same point! And of this very kind are several others of your astonishing and most accurate conclusions.

4 s. 17 thds. 41 fourths 4 fifths minus Ferguson's value in Table at page 394; the difference arises from our period—the modern value of a sidereal day—being slightly less than Ferguson's. At top of Table just mentioned, 23 h. 56 m. 4 s. 6 thds. 0 fourths 29 fifths is given as the value of a sidereal day. We give 23 h. 56 m. 4 s. 5 thds. 18 fourths 14 fifths, being the value communicated to us by G. B. Airy, Esq., Astronomer-Royal, and which is now used in all similar calculations.

And yet, amidst all this trifling, you have been cautious enough to keep within such a proper distance of the best determinations of the length of a tropical year, and synodical month, as not to affect the dates of the years in your chronological accounts, when they are not connected with eclipses: so that, unless your tables are tried by the sure test of eclipses, you are sure to be pretty safe." (p. 27).

* * * * *

"As to the twenty-three *queries* which you have raised upon my remarks, and which by a strange kind of logic you call an '*Examination*' of these remarks, they seem to me not to require any answer at all. For, only to query whether such and such a remark be true, is neither a confirmation nor a refutation of it: and this is all you have done with regard to what you call an *Examination* of them; for not one of them have you offered to refute. I still abide by the justice of these remarks, do you *query* as much as you please. But, if any judicious astronomer discovers any error in them, and will detect it (which has never been done yet since it was first published, in 1763), and put *his* name to what he writes, I will as publicly own it; but I shall take no notice of any anonymous publication.

The only reasons I had for taking no notice of this very *Candid* Examination of yours are these which follow:

First,—because you have therein told me, that if I do not take an opportunity to refute your 'confident assertions,' you shall conclude that I have given up *my own tables*, and become a convert to your scheme.

Secondly,—because you have therein charged me with purloining what you call *the most shining paragraph in my whole book* of Astronomy from you; and which you have the consummate vanity to say *does credit to my performance.*

The substance of which paragraph is, that if it could be proved from the writings of Moses, that the sun was created upon the point of the autumnal equinox, and the moon in opposition to the sun, as well as it can be proved by these writings that the sun and moon were created on the *fourth* day of the original week, there would be data enough for ascertaining the age of the world. (But you will have it that, on the *fourth* day, the moon was 24 hours past her *full* before there was a sun to enlighten her). For on account of the incommensurability of a week to a lunation, and of both a week and a lunation to a year, we might venture to say that 200,000 years would not be sufficient to bring all these three circumstances together again.

You *now* ask me 'whether I learnt it from my illustrious master, Newton? or if I happily collected it, as you did, from your divinely illustrious master, Moses? or if I did not rather deign to purloin it from you, *his humble* commentator?' as you say I certainly did, and then tell me that, be *that* as it may, it will bear testimony against its cavilling author, in support of your Chronology, as long as the sun and moon endureth. Vain man! My answer is, that I knew it long long before I knew you; and had no need to collect so plain a thing, either from the writings of Moses, or of Newton, or even to *purloin* it from Kennedy, who is a greater man than either of them; and who, by calling it *his*, and me only its purloiner, has owned *himself* to be the *cavilling* author. And indeed, I know but *one* author who is more cavilling (if possible) than himself; and *that* is an author who is despised both by him and me.[301]

It is neither to be found in the writings of Moses nor Newton; and if *you* have anything of it in your book (printed so long after mine), you have explained it as you have done your (pretended) Mosaic numbers; that is, in such a long-winded and unintelligible manner, that you scarce have occasion for a Dutch commentator to help to explain it into greater darkness—solve it if you can.

Suppose a Clock to have three hands, all going round the same way on its dial-plate, one of them in 7 days, another in 27 days 7 hours 43 minutes 5 seconds, and the third in 365 days 5 hours 48 minutes 54 seconds. If all these hands set out together from any given point of the dial-plate, and continue to go onward, *query*, How many years, months, days, hours, minutes, and seconds must revolve before all these hands can be in conjunction, or together again at the same point? If you are puzzled about how to solve this, pray collect it from the writings of Moses, where you will find it just as readily as you found the other.[302]

301 Ferguson here probably alludes to שטן, "*the Author of all Evil.*"

302 A similar question is proposed and answered in pp. 253, 254. But the result there given, viz. 1427651677822 Julian years 2 d. 10 h. 40 m. 48 s., is

Thirdly, and lastly, because you cannot *pervert* me into the belief of *your* doctrine, you have thought fit to tell me, in your *Candid Examination*, that '*historical evidence has no more weight with me, nor makes any more impression upon my mind, than the reveries of a sick man's dream; though heaven may have given a sanction to its truth.*' Disingenuous Sir! Although you and I always differed widely in our sentiments with respect to astronomy, I had a good opinion of your heart till now; and believed you to be a well-meaning man, a searcher for divine truth, and a sincere Christian. What you have here said would wound my very soul if it were true. But, as heaven knows it to be false, and you dare not lay your hand on your heart and say, before God *it is true*, seeing you never had the least ground for it, either from my actions, my writings, or my conversation, I now look upon you to be below my farther notice. Nevertheless, I sincerely wish you a better mind, and do bid you a hearty farewell.

JAMES FERGUSON (1763)."

After the appearance of Ferguson's "TWO LETTERS," the reviewers took the matter in hand, reflecting severely on Kennedy's conduct. The following is extracted from the Monthly Review for 1775, vol. 53, p. 354.

"In the year 1763, Mr. Ferguson published some remarks on the astronomical part of Mr. Kennedy's 'Complete System of Chronology;' these were submitted to the author's perusal before they were printed, and the Reviewer very candidly subjoined his name to them when they were actually published. They were so unfavourable to the credit of Mr. Kennedy, as a calculator, that he took the first opportunity of commencing a correspondence with Mr. Ferguson, which has been since carried on with a very unjustifiable degree of heat and asperity. The original question in debate is perplexed and confounded by personal reflections, and by unkind but unsuccessful attempts to depreciate the abilities and fidelity of our '*popular Astronomer*,' nor indeed have other names, however justly celebrated, escaped the lash of Mr. Kennedy's pen. 'I know you' (says he, speaking of Mr. F.) 'to be extremely deficient in the very first principles of practical astronomy; you suspect it not, because you have not been thoroughly searched into; and the reason why this is not generally known and generally regarded, is, because but few, perhaps not above *one* in 100,000, earnestly concern themselves about astronomy.' As to the famous Luna-Solar Radix, or the true position of the Sun and Moon, with respect to each other at the very instant of their creation, by which the age of the world is to be accurately determined, this, our author informs us, is a mere 'scriptural datum,' and could never have been ascertained by the

widely different from that given in a foot note at page 31 of this letter, viz. "67,403,285,211,584,724 solar or tropical years and 125 days," which, although reduced to Julian years, would be at least 47,000 greater than given at p. 254.

acutest penetration of the human mind! 'non sagacissimâ ing nii Newtoniani vi.'

But he has not told us where this unscrutable secret is disclosed, nor does he seem to possess the true key for discovering it. Mr. F. has clearly convicted both his principles and calculations of unpardonable errors in a variety of instances; we are therefore sorry to find that he has yet published only the first part of his defence, and that more is to follow, on his intentions. Should Mr. K. resume this involved and intricate inquiry to which no disciple of Hutchinson seems equal, we hope that he will discover a greater portion of the true spirit of philosophy, and make some apology for the very illiberal reflections which he has cast on his antagonist. On his integrity as well as on his judgment, we could not read the following, as Mr. F. has quoted it, without concern and indignation.—'*Historical evidence has no more weight with me, nor makes any impression upon my mind, than the reveries of a sick man's dream; though heaven may have given a sanction to its truth.*' Let our readers peruse the modest Astronomer's defence. 'Disengenuous Sir! although you and I always differed widely in our sentiments with respect to Astronomy, I had a good opinion of your heart till now, and believed you to be a well-meaning man, a searcher for divine truth, and a sincere Christian. What you have here said would wound my very soul, if it were true. But as heaven knows it to be false, and you dare not lay your hand upon your heart and say, Before God it is true; seeing you never had the least ground for it either from my actions, my writings, or my conversations, I now look upon you to be below my farther notice! Nevertheless, I sincerely wish you a better mind, and do bid you heartily farewell."

Kennedy, in the Letters appended to his pamphlet, entitled, "Explanation and Proof," replied to the foregoing reprint of 1763, it is somewhat curious that he did not do so at the time they appeared. The reply is in his usual fashion; but here quotation may be spared. Shortly after the publication of these scurrilous letters, Ferguson addressed a third letter to Kennedy on account of his abusive treatment of Astronomical Authors. This third and last letter to Kennedy was written and published in July, 1775, in the form of an octavo pamphlet of 16 pages. The following is copy of its title-page,—we give the letter in full.

"A Third Letter to the Rev. Mr. John Kennedy, on Account of his abusive treatment of Astronomical Authors. By James Ferguson, F.R.S. London: Printed in the year M.DCCLXXV."
(Dated July 19th, 1775).

"REVEREND SIR,

In your two last pamphlets of letters, the one to the Rev. Dr. Blair, and the other to me, you have been pleased to call my Lunar Tables and Calculations a mere ridiculous farce, and shameful imposition upon the intelligent Reader: you have complained of your great unhappiness in having such a weak, illiterate, and incompetent Antagonist as me to deal with; whom, you say, must have been either delirious or intoxicated when I calculated from your lunar measures, and found fault with them; you have told me that I have *a scrap of reputation to lose:*' that you have much to reprove me for, and much to teach me: that I am fallen very low in your esteem; and that I have found an Eclipse of the Moon in September, the year before Christ 201, *two* days before the Moon was full; besides having fallen into many other enormous errors and blunders, with which whole pages might be filled.

But it would be well for you if the *errors* and *blunders* you charge me with could not, with the strictest justice, be retorted upon yourself, by every person who understands Astronomy; although, perhaps, *some* of those who do not may be staggered by your bold assertions. However, this I am pretty sure of, that no person, who has deigned to read your two absurd pamphlets, will reckon me so weak as you have termed me, for having taken notice of these abusive and illiberal publications; by which you have quite debased your own character as a gentleman, a disciple of Moses, and a Minister of the Gospel of Christ.

Whatever was the matter with you when you charged me with the above-mentioned error of *two days* in calculating the Eclipse of the Moon in September, the year before Christ 201, I shall not pretend to determine; but can with truth assert, that, if you trust to *your own* measure of a mean Lunation, your charge against me was altogether false and groundless. For, the whole difference between your measures and mine amounts not quite to 26 hours in 5770 years; How then is it possible that there should be *two days* between us in 1964 years? If you have wilfully falsified your own measures, in order to make people believe mine are wrong, as you formerly did Meyer's and Ptolemy's, by mixing some of your own numbers with them; or, if you have calculated the time of that full Moon and Eclipse from a radix different from that of your original Mosaic full Moon, which, according to your account, *was 24 hours before there was a Sun to enlighten her*, I cannot help it. Be that as it may, I think proper to inform you that the time of the said full Moon and Eclipse (which *you* say is two days wrong by my calculation) has very lately been calculated by an able hand, from Meyer's Lunar Tables, published by the Board of Longitude, and which, from the most accurate observations, have been very seldom found to err above one minute of a degree in the Moon's place: and they make the time of *that very* Eclipse to have been but two minutes and a half sooner than I have made it from my own Tables and Calculations. 'Tis true, that because Meyer's Tables differ from your measures, you have said *they* are no Lunar Tables at all.[300] But are you weak enough to believe that *your* bare word will be taken for this, by any man who has the least knowledge of Astronomy; or will it not rather tend to sink all your publications into a general disrepute?

I hope you are not quite so ignorant as not to know that the length or measure of a Lunation depends entirely upon the times in which the Sun and Moon go round the Ecliptic. The former of these you have *asserted* to be 365 days 5 hours 49 minutes; and numberless observations have *proved* the latter to be 27 days 7 hours 43 minutes 5 seconds (misprinted 4 seconds in my book of Astronomy).

Now, if you understand enough of common Arithmetic (which I have great reason to doubt of) to enable you to calculate from these numbers, you will find that the length of the mean Lunation must be 29 days 12 hours 44 minutes 3 seconds 7 thirds and 26 fourths, which is 1 second 22 thirds and 26 fourths more than your assumed and unscientific measure makes it.[304]

303 Meyer's Tables.—We have, in Ferguson's autograph, on a large sheet of paper, memoranda relative to these Tables, which is entitled, "PRECEPTS FOR FINDING THE MOON'S PLACE BY MEYER'S TABLES;" it is divided into four *sections.* Section 1st is entitled, "*Preparatory Arguments,*" of which there are 14. Section 2d, "*Arguments for the Moon's Latitude,*" of which there are 2. Section 3d, "*Arguments for the Moon's Equatorial Parallax,*" of which there are 3; and Section 4th, "*On the Equatorial Parallax, horizontal diameter,*" &c., being an explanation.

304 To solve this question,—Rule—Multiply the times together and divide the product by their difference. Kennedy assumed the length of a solar or tropical year to be 365 days 5 hours 49 min. *exactly*, which, when decimally expressed = 365·24236111 days. The periodical revolution of the Moon is done in 27 d. 7 h. 43 m. 5 s.; decimally expressed, is = 27·32158565,—then,

$$\overset{\text{d.}}{365 \cdot 24236111} \times \overset{\text{d.}}{27 \cdot 32158565} = \frac{9979 \cdot 0004520750940715}{337 \cdot 9207754600000000}$$
$$\text{and } 27 \cdot 32158565 - 365 \cdot 24236111 =$$

9979·0004520750940715÷337·9207754600000000 = 29·530591714 days = 29 d. 12 h. 44 m. 3 s. 7 thds. 26·7225 fourths, which agrees with Ferguson's period above, with the exception of ·7225 dec. of a fourth; perhaps the difference arises from Ferguson having a lower ratio of decimals than we have used. It is not very likely that Kennedy was equal to the solving of such a question, at least he never offered to do so in print.

Suppose the length of the year to = 365 d. 5 h. 48 m. 55 s. = 365·24230324 d., and of the Moon, 27 d. 7 h. 43 m. 5 s. = 27 d. ·32158565,—365 d. ·24230324 = 337·92071759d. for a divisor. 365·24230324 × 27·32158565 = 9989988709749325060

And herein you differ so far from yourself, as to prove, that, if you have given the true length of the solar tropical year, your measure of the Lunation is absolutely false, and can have no foundation in nature.

It appears that the late eminent Astronomer, Dr. Pound, did not find the length of the Solar tropical Year to be 365 days 5 hours and quite 49 minutes; and therefore, according to him, the length or measure of the mean Lunation is 29 days 12 hours 44 minutes 3 seconds 2 thirds 58 fourths. This measure I have kept by, in making my Lunar Tables, and it is 1 second 17 thirds 58 fourths more than you have assumed for the length of it; which difference between your lunation measure and mine, multiplied by the number of Lunations in 5770 years, amounts to 1 day 1 hour 45 minutes in that time.[305] But as, in your pamphlet just published in this present year, you assert that there are *two days'* difference between us in 1964 years (viz. from the year before Christ 201 to A.D. 1764), which, of course.

÷337·92071759 = 29d. ·5305929, or, 29 d. 12 h. 44 m. 3 s. 13·6 thds.; again, take the length of the year at 365 d. 5 h. 48 min. 57 sec. = 365·24232639, and that of the sidereal period of the Moon, 27 d. ·32158565; then, 27·32158565—365·24232639 = 337·92074074 for a divisor, and 365·24232639 × 27·32158565 = 9989995035496403035 ÷ 337·92074074 = 29 d. ·53059193 = 29 d. 12 h. 44 m. 3 s. 8·6 fourths. In this letter, Ferguson mentions that in his calculations he adopted Dr. Pound's measure of a lunation, viz. 29 d. 12 h. 44 m. 3 s. 2 thds. 58 fourths. It is evident that such a measure was not obtained by Dr. Pound from any of these values of a year, but from a period greater than 365 d. 5 h. 48 m. 57 s.

The mean tropical revolution of the Earth is now found to consist of 365 d. 5 h. 48 m. 51·6 s. = 365·242264 d., and a sidereal revolution of the Moon, 27 d. 7 h. 43 m. 4·6675 s. = 27·321582 days. Therefore, 365·242264 × 27·321582 = 9978996465741648, and 27·321582—365·242264 = 337·920682 for a divisor; thus, 9978996465741648 ÷ 337·920682 = 29·530588 days = 29 d. 12 h. 44 m. 2·88 s., the mean standard length of the lunation, as used in modern calculations.

[305] The more accurate observations of recent times show, that a mean lunation consists of 29 d. 12 h. 44 m. 2 s. 52 thirds 49 fourths. Therefore, Ferguson's period of 29 d. 12 h. 44 m. 3 s. 7 thirds 26 fourths is a period *too slow* of the mean average period by 14 thirds 38 fourths = 877 fourths *slow* in a lunation. Kennedy's assumed period consists of 29 d. 12 h. 44 m. 1 s. 45 thirds nett, and is therefore a period *too fast* in a mean lunation by 1 s. 7 thirds 49 fourths = 4069 fourths *fast* in a mean lunation. From this it is obvious that neither Ferguson nor Kennedy had the true mean average period of a lunation.

The true mean period of the lunation lies nearly in the middle of the Ferguson-

Ferguson's period. Kennedy's period.
29 d. 12 h. 44 m. 3 s. 7 thds. 26 fourths + 29 d. 12 h. 44 m. 1 s. 45 thds. 0 fourths = 59 d. 12 h. 28 m. 4 s. 52 thirds 26 fourths ÷ 2 = 29 d. 12 h. 44 m. 2 s. 26 thds. 13 fourths = mean of these two periods, which is only 26 thirds 36 fourths minus the true mean average length.

It is singular that Kennedy should have had such a very minute period as 29 d. 12 h. 44 m. 1 s. 45 thirds for his lunation, and an abrupt period of 365 d. 5 h. 49 m. for his tropical year—a period ending in minutes.

would amount to 5 days and about 21 hours in 5770 years, it is plain that you have *wilfully* falsified your own Lunation-measure, in order to make the few readers of *that* abusive pamphlet believe that I have made this egregious error. But you have acknowledged, in your Scripture Chronology, that you was obliged, in one case, to throw off '*a redundant day*,' without which, it seems you found it impossible to make your measures agree with the times. This, in fact, is *redundant nonsense*: for, can you possibly imagine that the celestial Luminaries had no motion throughout the whole of *that* day? and I believe you will not *say* that it was the Day on which the Sun and Moon stood still at the word of Joshua; for I can *swear* that I heard you say your Scheme could not be any way affected by that event.

And now 'tis plain, that as your Lunation-measure agrees neither with your own length of the solar-tropical year, nor with the Astronomers' length thereof, it has no foundation in nature; and consequently (to use your own words concerning the Lunar Tables of Astronomers), 'can claim no higher title than that of—*Speciosæ Nugæ*.'

But I might have saved myself the time and trouble of proving your lunation-measure to be false, seeing that you yourself have demonstrated it to be so in your *System of Astronomical Chronology Unfolding the Scriptures*. For, in the Calculation you have there given of the great total Eclipse of the Sun, in the year before Christ 603, which put an end to the long war between the Medes and Lydians (as related by Herodotus), by over-spreading both the Armies, and frightening them with a sudden darkness, you make the Sun to have been some hours set below the horizon of the field of battle when that Eclipse became total. But my Tables (which you altogether despise) make the time of that Eclipse, at the same place, to have been at 15 minutes past 11 o'clock in the Forenoon, on the 18th of May, the very day on which the said battle was to be fought, according to your Chronology. So that, where you keep by your own numbers, we differ not quite 10 hours in 2366 years. If you had done so in the former Case, the difference would have been still less; although you have called it *two days*, which is your own wilful blunder.

Upwards of fifty years ago, which was long before you wrote

for the press, SIR JONAS MORE and DR. KEILL adopted 365 days 5 hours 49 minutes for the length of the solar-tropical year, and framed Tables of mean motion agreeable thereto And because (as you say) I have mentioned the same length in my Book of Astronomy, you have had the assurance to tell me that I *pirated* this length from you. I confess I cannot find the place: and if I have done so, it must have been done to avoid [always] tiring the reader's patience with seconds, thirds, and fourths; having at Section 251 and 371 mentioned the length to be 365 days 5 hours 48 minutes 57 seconds.[306] But this is much the same with a former printed charge you brought against me, about having *purloined* something from you into my Book of Astronomy; although that Book was printed several years before your Scriptural Chronology was put to the press.[307]

Fie upon it! miserable, and desperately bad must your Cause be, when, instead of endeavouring to defend it by truth (which indeed is out of your power) you thus descend to falsehood and abuse. I apprehend that you yourself are sensible it is so; although the meanness of your soul, and greatness of your pride, will not suffer you to own it: and now, for these reasons, I am not,

REVEREND SIR,

Your humble Servant,

JAMES FERGUSON."

BOLT COURT, FLEET STREET, LONDON,
July 19, 1775."

"P.S.—I do not find you yet have answered the *Scrupulous Calculator* touching your blunders in computing two Lunations in the Egyptian and Julian Styles, in Lloyd's Evening Post, No. 2663. Indeed, Sir, there is a vulgar proverb against you. And although you was afterwards solemnly called upon, for

[306] It is surprising that Ferguson should here note that he could not find out the place in his Astronomy where he had given 365 d. 5 h. 49 m. as the length of the tropical year. Such a length is given by him in *Section* 47 of his Astronomy. Our copy, the quarto edition of 1764, gives the tropical period of the Earth round the Sun of various lengths,—viz.

At page 13 and Section 47th of our quarto edition, the length of the year is stated to be 365 d. 5 h. 49 m.
,, 197 do. 353, the length given is 365 d. 5 h. 48 m. 55 s.
,, 108 do. 246 } do. 365 d. 5 h. 48 m. 57 s.
,, 255 do. 371 }

[307] Ferguson's Astronomy was published in 1756. Kennedy's "Complete System of Chronology Unfolding the Scriptures," in 1762; hence, Ferguson dates 6 years before Kennedy.

good reasons, by the Editor of Lloyd's Evening Post, to let the world know from whence you took your Quotation of a Letter (as you say) from RICCIOLUS to KEPLER, cited in your Epistle to Dr. Blair, you thought proper to be silent; and why? because the Scrupulous Calculator has absolutely foiled you.[308]

But, I hope you will not refuse to answer the *Lover* of *Astronomy's* Letter to you in Lloyd's Evening Post, No. 2802, for Wednesday, the 14th of June this year; because, no doubt, you *think* of yourself what you have *said* of me; namely, that *you have a scrap of reputation to lose.* Perhaps you will say you have not seen this Evening Post; and therefore, that you may not pretend ignorance, I shall subjoin a copy of that Letter."

'To the Reverend Mr. John Kennedy.

REVEREND SIR,

As you have hit upon such an exact measure of a Lunation, that you can thereby ascertain the true time of any New or Full Moon, past or to come, without any of those numerous Equations used by Astronomers, which you have affirmed are unnecessary, unastronomical, and intended by

308 In Lloyd's Evening Post for May 13—16th, 1774, there is a long letter addressed to the editor by "*A Scrupulous Calculator,*" reflecting on Kennedy's treatment of him,—he begins,

"SIR,

In my Letter published in this Paper, No. 2615, p. 313, addressed to the Rev. Mr. Kennedy, I pointed out some mistakes made by him in a pamphlet lately published, in his Calculations, especially in his Computation of the mean Full Moon in the month Thoth, anno Nabonass. 27, as being connected with the Julian Style, by using a different Radix from whence he intended to compute. Since which, I find a letter inserted in your Paper, No. 2625, p. 395, wherein the Author does not show the justness or unjustness of my Remarks, but instead of defending his computation, proposes for me to calculate that mean lunation in the Egyptian and Julian styles *as he has set me an example.** This I shall prudently decline; first, because if I follow his example, I must likewise blunder; and secondly, because I am truly sensible, unless I *plough with his Heifer,* he will not permit me to be right,† and I shall undergo the hard fate of Ptolemy, Meyer, and others, whose mean lunations were in a most astonishing and absurd manner tortured, tried, condemned, and executed, in his Astron. Chron., p. 252 to 273. I therefore choose to take his advice, and make my appeal to the Public, and take my chance with other Scribblers for a favourable decision."

Then follows three closely-printed columns of abstruse calculations to the complete overthrow of Kennedy, and is signed, "*A Scrupulous Calculator.*" We have the leaf of Lloyd's Paper on which this learned letter appears. At foot we find written, "Mr. Fisher, Attorney-at-Law," with an asterisk referring to one above, showing that he was the "Scrupulous Calculator."

* This appears to have been Kennedy's way to get out of a dilemma when pressed.
† Another instance of Kennedy's "*way*"—every one was wrong who differed from him.

Meyer, and the other Table-makers, only to correct the deficiency of their mean measures; you would oblige the purchasers of your Chronology, and others, if you would send to the Printer of this Paper, as soon as you can, the true time of Full Moon in July, 1776, deduced from your own Epocha, Lunation, and Tables, together with the *Calculus*, in order to show that you have not mixed the numbers of any other person with your own. If your predicted time answers to the Observation as near as the time deduced from other Tables, it will be the means of proving the superior value of your Radix and Measures above all others; but, if you decline this Proof, your former assertions will be treated as mere Chimeras.

A LOVER OF ASTRONOMY.'

To this Kennedy made no reply. (See his treatment of "A Scrupulous Calculator," note 308).

And from this date, July 19th, 1775, we hear no more of Kennedy, until his death in 1782.[309]

THE ART of DRAWING in PERSPECTIVE, &c.—Although Ferguson was now in indifferent health, he could not remain idle. During the summer of this year, he wrote and published "*The Art of Drawing Made Easy.*" This work is an octavo of 124 pages, embellished with nine folding copperplate engravings, and, like all of Ferguson's other works, this one on Perspective is written in plain and simple language, and is perhaps still the best introductory Treatise on Perspective that can be put into the hands of the young student. The editions of this work are of the years 1775,—1778,—1803,—1807,—1810,—1823, &c.; that of 1823 was published by Sir David Brewster in his "Ferguson's Essays." The annexed is a copy of its title-page:—

"The Art of Drawing in Perspective Made Easy to those who have no previous knowledge of the Mathematics. By James Ferguson.

309 Kennedy, from every source of information we have received, appears to have been "a perfect specimen of arrogance, pride, vanity, and self-conceit":—another account assures us that he knew little or nothing of Practical Astronomy, although he assumed to be a great master therein; another informs us, that "he was a man most fluent and flippant in speech, never at a loss for a word," and that "his vocabulary required to be corrected and reprinted."

On applying to the present rector of Bradley for notes on Mr. Kennedy, he very kindly suplied us with the following, which is copied from the stone covering Mr. Kennedy's grave at Bradley, Derbyshire,—viz.

"*The Rev. Mr. John Kennedy Died Feb. 4th,* 1782, *aged* 84. *He was Rector of the Parish of Bradley* 48 *years. If thou wouldst know more of this good and learned man consult his Book.*"

Illustrated with Plates. London: Printed for W. Strachan and T. Cadell in the Strand. MDCCLXXV."

Ferguson's constitution, "always feeble" (according to the late Mr. Andrew Reid, so often quoted), "now began to show symptoms of breaking up—he was now often unwell and confined to his room;" and when he was able to walk abroad, "it was with an unsteady and feeble step." He was now also "greatly afflicted with gravel, so much so, that many a night he was kept from sleeping by the severity of his pains."

Notwithstanding his impaired health, he was always engaged in some favourite study or pursuit. At commencement of his preface to this work on PERSPECTIVE, he thus feelingly alludes to the state of his health, and its consequent influence on the mind:—

"In my infirm state of health, a situation that is very apt to affect the mental faculties, I thought my last book on Mechanical Exercises would have been the last book I should ever publish. But, as I have been constantly accustomed to an active life, and to consider idleness as an unsupportable burden, I have of late amused myself at intervals, as my usual business would permit, with studying *Perspective*, &c."

Thus, at intervals of his usual business, and it may be supposed when free from pain, he amused himself in writing this Treatise.[310]

EQUATION ROTULA.—During the year 1775, Ferguson made several Rotulas for showing the Equation of Time; he also computed an Equation Table for the second year after leap years. We have one of the Rotulas, a small one, in an unfinished state (on a card 4¾ by 3⅛ inches); so far as it goes, it consists of a series of concentric circles—round and within the exterior one in unequally divided spaces—are the names of the Days of the Month; in next circle, the days of the Month irregularly numbered, against which, in another circle, are the minutes of time to be added or subtracted as the Clock or Watch is *Faster* or *Slower;* and in the interior circle are, in unequally divided spaces, "*Sun Faster than the Clock,*"—"*Sun Slower than the Clock,*"—"*Sun Faster,*" and "*Sun Slower,*" showing how the Clock or Watch is to be set. The interior of the circle is blank,

310 This will remind the reader of many recorded instances of men of letters and science, who, like Ferguson, worked up to the last;—of Archimedes and his Problem; of Copernicus and his celebrated work on Astronomy; of Sir Isaac Newton, Dr. Dalton, &c.

as also a space on the card below it, evidently intended for Explanation and Directions. "J. F., 1775" is in the lower right hand corner of the card.

MURDOCH FERGUSON.—To the kindness of E. Balfour, Esq., Secretary of the Royal College of Surgeons, we are indebted for the following extract from the College Record:—

> "Mr. Murdoch Ferguson obtained, on the 6th July, 1775, from the Court of Examiners, the Certificate of Qualification of Surgeon to a Regiment; also, Mr. Murdoch Ferguson, on the 7th November, 1776, passed as Surgeon to a Ship of the first rate."

Shortly after Murdoch received his Certificate of Qualification as a Surgeon to a Regiment, we find, by a notice in the London Gazetteer, that he was married to a Miss Vincent of Royal Hill, Greenwich, on 16th August. The following letter from his father to the Rev. Mr. Cooper of Glass, Banffshire, refers, among other matters, to this union.

LETTER FROM FERGUSON TO THE REV. MR. COOPER.—The original of the following interesting letter from Ferguson to his friend the Rev. Mr. Cooper of Glass, Banffshire, is in our possession:[311]—

"REVEREND SIR,—I now take the pleasure of sending you, by my youngest son,[312] an Analysis of my Course of Lectures, my Pamphlet on the Doctrine of Eclipses, and my three printed letters to Mr. Kennedy, whom 'tis as impossible to silence as to still the waves of the raging sea.[313] On sending him my printed Letter, he wrote me a very abusive and threatening one, the purport of which you will easily guess at when you read the inclosed copy of the answer I wrote thereto.

My eldest son (a Surgeon) was married about three weeks ago, and I have been obliged to give him his Portion, which has left me so little money, that I am afraid I shall not be able to send you any Books this year.

I thank you for your kind inquiry about my health, which is rather better than when I wrote last to you; but I am still much afflicted with the Gravel, which, sometimes for a fortnight together, does not allow me to sleep a quarter of an hour at any time through the whole night.

311 "The Reverend Mr. Cooper was inducted minister of the Parish of Glass in the year 1755, and died in 1795, aged 78, and in the 40th year of his ministry. He was an old and much-valued friend of Ferguson's." (Extracted from a letter from the Rev. Mr. Duguid, present minister of Glass).

312 This refers to his son John, who would be about taking his departure at the time for his Classes at Marischal College, Aberdeen.

313 About 7 months after the date of this letter, Ferguson again wrote Mr. Cooper, and presented him with his three printed letters to Kennedy. If Ferguson had forgotten he had already sent copies of these letters to his Reverend friend, it shows that his memory was giving way in the spring of 1776. (See page 145 for second letter).

My Sister is very well here, and desires her best respects to you.[314] Be so good as to send for my sister Elspeth, and slip the three Guineas quietly into her hand which you will find here inclosed; and tell her that two of them are for herself, and the third for her daughter Jensey.—I am, with great respect,

REVEREND SIR,

Your most obliged humble servant,

JAMES FERGUSON."[315]

No. 4 in BOLT COURT, FLEET STREET,
LONDON, *Sept.* 12*th*, 1775.

The above shows that Ferguson sent to Kennedy a copy of his last printed letter (letter third); that in return he had received an "abusive and threatening one;" that he answered Kennedy's rough letter, and had inclosed in this letter a copy of the answer for his Reverend friend. We have that copy, perhaps now the only one in existence—it is very smart, and would, no doubt, prove a bitter dose to one so singularly self-conceited and arrogant. It is as follows:—

"REVEREND SIR,

I received your letter dated the 2d instant, and do confess, that before I could be qualified to answer it properly, I ought to go a week to Billingsgate in order to learn proper language. I am now called *Poor James Ferguson, an ignorant, illiterate, stupid Blockhead and Dunce;* with many other such vilifying Epithets as any one but *poor proud John Kennedy* would have been ashamed of.

You now tell me that you think yourself *somebody*—that you have put a *gag* in my mouth—that you shall be a *Gad-fly* to pursue, gall, and sting me —that you will exalt your *puissant arm* and *sharp-edged Sword,* bravely to fight your way!!!

A *Gad-fly* to pretend to be *somebody!* I have indeed heard that *that little pestiferous Insect* has a *sting* in its *tail;* but never knew before that it had a *puissant arm,* or that it could wield a *sharp-edged Sword.* A *Flapper* is preparing for this Gad-fly.

The many errors and blunders I have hitherto charged you with you have thought proper to pass over in silence—not one of them have you offered to deny the justness of, or to refute; but have, as you say of me, *cowardly declined,* and *flown off to some other thing.* You now want me to calculate Ptolemy's Eclipses from Ptolemy's *data* according to the Egyptian Style. When you have cleared yourself of what I have formerly charged you with, and have convinced the world that you have truly reduced the days of the Egyp-

[314] This was Ferguson's sister Janet. She appears to have left Banffshire for London in July, 1773, to nurse Mrs. Ferguson, who was then unwell, and who shortly afterwards died. She remained after her death, and had charge of Ferguson's household affairs till his death, when she returned to the north.

[315] For this letter, and for the copy of the one Ferguson sent to the Rev. Mr. Kennedy, we are indebted to the kindness of the Rev. Mr. Thomson of Fetteresso, in Kincardineshire, who, in 1861, kindly presented them to us.

tian months to the days of the Julian months; and that, without cooking and mutilating even *your own* numerical measures you can make your Calculations agree with the recorded times of these Eclipses, I shall be willing to comply with your desire.

All that I have hitherto charged you with is but little in comparison with what is yet to come out. The way you have treated the Lunations of *Ptolemy* and *Meyer* shall be exposed to the public more fully than has yet been done; and (what is new) the way that you have cooked, mutilated, and tortured *your own* Lunation-measure shall be detected and exposed. You have said that 29 days 12 hours 44 min. 1 sec. 45 thirds is the exact mathematical measure of a mean Lunation,[316] and that the least particle of time can neither be added thereto, nor subtracted therefrom, without doing violence to nature, as the Lunations follow one another in an uninterrupted succession, like the equal links of a chain. If so, what in the name of that *Stranger* called *common sense* could induce you to throw off a *redundant day* at the end of 49680 Lunations, reckoned downward from your original Radix, and *there* to make a short link in your Lunation-Chain? according to this Doctrine of yours, the whole is less than all its parts put together.

It shall be proved that you have thrown off this redundant day in your *Practice* almost 757 years before the time appointed by your *Theory;* and that, by no less than 21 well-vouched Eclipses, your Theory differs from your coaxed Practice. These Eclipses you will acknowledge to be disinterested evidence, if you have even the least regard for truth, although they undoubtedly prove that all your errors and blunders have arisen from your having fixed your original Full Moon too late, and made your Lunation-measure too short, and then it will be manifest that the errors and blunders you charge me with are all your own. The Calculations in your Scripture Chronology shall be analyzed, and Kennedy shall be set against Kennedy, like a House divided against itself. I thought proper to inform you of all this before-hand, in order to convince you that you have neither *gagged my mouth* (as you say you have)—that I have not like *a dastardly Coward* (as you say in your true Billingsgate style) *quitted the Field with precipitate Flight;* and that I am neither afraid of your *Gad-fly sting*, nor of your *puissant arm* and *sharp-edged Sword.*[317]

Remember that I shall not answer any written Letter of yours after this date —it is neither worth my while, nor have I time to throw away to so little purpose, who am,

REVEREND SIR, (I should have said
most Reverend Gad-fly)

LONDON, *Aug.* 16*th*, 1775.

your well wisher, tho' not
your humble Servant,
JAMES FERGUSON."

The above is the Copy of a Letter to the Rev. Mr. John Kennedy, Rector of Bradley, in Derbyshire."

316 Regarding the length of the mean lunation, see former notes on Kennedy, &c.

317 Ferguson's state of health prevented him from publishing his observations, and "*the Reverend Gad-fly*" therefore escaped the infliction of the "*flapper.*"

FERGUSON'S HOUSEHOLD.—At the close of this year, his household was reduced to three persons,—viz. Ferguson himself, his sister Janet, and a domestic servant. His son Murdoch, then lately married, had an establishment of his own. John, the youngest son, was attending the Classes at Marischal College, Aberdeen. And, as previously mentioned, his sister Janet came to London, in the summer of 1773, to attend on Mrs. Ferguson, who was then on her death-bed. This sister remained with Ferguson until his death in Nov. 1776.

1776.

MONEY BEQUEATHED TO FERGUSON.—Our first note for 1776 refers to a considerable sum of money which had been bequeathed to Ferguson by a relative who died about the beginning of this year, but by whom, and the sum left, we have been unable to learn. The Artisan or Mechanics' Instructor, vol. 1st, p. 235, notes, "that he left behind him a sum to the amount of about Six thousand pounds." The Practical Mechanics and Engineers' Magazine, vol. 1., p. 176, says, "he left upwards of £6000 at his death,"—and one of the obituary notices in the Appendix, that "he bequeathed to different Legatees, at his death, above £5000." And "A Plain honest man" who defends him (in the Appendix), and who knew Ferguson, speaks of "the few thousands he left behind him, which came to him a very short time before he died, by means of the death of a near relation." From these accounts it appears that Ferguson, about ten months before his death, had bequeathed to him by a near relative, "a few thousand pounds," but which he did not live to enjoy.[318]

TABLE OF THE EQUATION OF TIME.—In the month of March,

[318] If Ferguson received these "few thousand pounds" only about ten months before his death, it is obvious that he must have been possessed of a considerable sum of his own before that; for, in his WILL, he mentions that he had expended upwards of £1100 on Murdoch. He received his "*portion*" at the time of his marriage, in August, 1775 (see letter of date Sept. 12th, 1775, pp. 441, 442), but it is possible that the "*portion*" may have been a small one, and supplemented on receipt of the legacy. In the WILL we find no sum stated as to the gross amount he had, but £200 was to be set aside for his daughter Agnes, should she cast up in distressed circumstances. Three of his sisters had £290 among them. John, the son, on reaching 21, was to get £1100, and then Murdoch and he was to get equal halves of the balance, and the three executors had £30 among them as a mark of esteem; thus, all these sums (omitting the £1100) amount to £1620,—"*the rest and residue*" sum noted was to be equally divided between Murdoch and John, on the latter attaining his 21st year, in February, 1780.

1776, appeared Ferguson's last publication, a sheet Table of the Equation of Time, probably printed from his equation computations in the previous year. Along the top of the sheet in a long line is the following :—

"Table of the Equation of Time, Showing the Equation of Time for Leap Year, and the First, Second, and Third Year thereafter. By James Ferguson, F.R.S., 1776."

And at foot of the sheet we find,

"Sold by the Author at No. 4 Bolt Court, Fleet Street. Price One Shilling."

This is a very scarce publication, so much so, that when it is to be had, from five to ten shillings is asked for it. The copy which we had long in our possession is now in the Museum of Banff.

THE NORTHCOTE LIKENESS OF FERGUSON Published.—This likeness of Ferguson, the first ever published, was painted by Northcote, and published on March 20th, 1776, just about eight months before his death, when in the 66th year of his age. It has always appeared to us that this likeness could not possibly be correct—much too youthful and gay for a careworn man of 66. When his publisher, Mr. Cadell, brought out the second edition of Ferguson's "Select Mechanical Exercises," on 1st January, 1778, he *rejected* this likeness for the frontispiece to this work, and inserted one taken from the Townsend print, published in December, 1776. Mr. Cadell having done this, shows that the Townsend likeness was the correct one. We have a copy of this likeness by Northcote. Within a broad rimmed oval, on an engraved oblong ground 12⅞ by 10 inches, is the likeness of Ferguson facing the right, draped in a fur-necked cloak or mantle, with book in left hand, and another close by lettered "*Newton Phil.*" and underneath as follows :—

"James Northcote pinxt.—F. Haward fecit." [319]
Published March 20th, 1776, for W. Shropshire, No. 158 New Bond Street.

[319] James Northcote, the celebrated Painter, was born at Plymouth in 1746. He came to London in 1771, being then in his 25th year, and was for five years after that time the pupil of Sir Joshua Reynolds. In 1776, he set up for himself as a Portrait painter. He died in his house, Argyle Street, London, on 13th July, 1831, aged 85. Ferguson's likeness, by Northcote, was done in March, 1776. It must have been his earliest, or one of his earliest efforts on his own account.

A half size copy from this likeness is given in the Library of Entertaining Knowledge, vol. i., p. 208. The publishers of which volume do not seem to have been aware of its rejection by Ferguson's publisher. (See note on the Townsend likeness of Ferguson in the Appendix).

FERGUSON'S (supposed) LAST LETTER extant.—On April 10th, 1776, Ferguson wrote a long and interesting letter to the Rev. Mr. Cooper of Glass, Banffshire, regarding the origin of his Mechanical Paradox, &c., which, so far as known, is the last letter extant written by him (being dated seven months before his death). Instead of placing it here, under the year of its date, it was made to follow the description of the Mechanical Paradox, under date 1750 (pages 145—148); that was considered the more appropriate place, and to which the reader is referred; but as there are some remarks at the beginning of this letter unconnected with the Paradox, we shall here reproduce the first ten lines of it, and give notes on them,—

"REVEREND AND DEAR SIR,—I am glad that my last letter came safe into your hands, and do return you my sincere thanks for delivering the one inclosed in it to my Sister, begging that you will now repeat the same favour, as it is exactly on a similar occasion, and she may still be in need of a small supply.[320] I thank God that I am now much recovered of my gravel, and in hope of getting quite well again.[321]

I herewith send you an account of my Mechanical Paradox,[322] and my three Letters to parson Kennedy,[323] who is now very quiet. He has been sadly trimmed by the reviewers, &c." (See letter of date April 10th, 1776, pp. 145—148).

COMMON PLACE BOOK.—This MS. folio, so frequently alluded

[320] The "*last letter*" here alluded to refers to the one in our possession, dated Sept. 12th, 1775, and inserted at pages 441, 442. It is probable that other *three guineas* would also be inclosed in this last letter for his sister Elspeth, showing Ferguson's tender regard and attention to her wants without solicitation.

[321] By referring to letter of date 12th Sept. 1775, pp. 441, 442, it will be observed that about the time he wrote it he was in great distress with gravel. In this last letter, written seven months afterwards, he expresses himself as having got much better of it, and "in hope of getting quite well again,"—this hope was illusory. In June he was again laid prostrate in bed, to which he was mostly confined till his death on 16th November.

[322] This refers to his pamphlet on the Mechanical Paradox, published in 1764, and noticed at page 302.

[323] It is singular that Ferguson should again present Mr. Cooper with copies of his *Three Letters* to "parson Kennedy," seeing that he sent him copies by his son John on Sept. 12th, 1775. Probably Ferguson's memory was beginning to fail, and had forgotten the circumstance.

to in these pages, and from which we have given so many curious extracts, was, with other two MS. volumes of Ferguson's (noticed at pages 224 and 235, 236) discovered in Edinburgh towards the end of 1865, and with them was bequeathed to, and deposited in the College Library, Edinburgh, shortly afterwards, by the late James Lauder Rutherfurd, Esq., agent for the trustees of the late John Ferguson, Esq., the youngest and then only surviving son of Ferguson.

It would appear that this MS. folio, with the other two quarto MS. volumes, and also a box containing a small orrery, drawings and writings which had belonged to Ferguson, were, shortly after the son's death, in 1833, deposited in the Writing Chambers of his solicitor, Mr. Balderstone, in Edinburgh, and it was in these premises, subsequently occupied by his successors, Messrs. Scott, Moncrieff, & Dalgetty, W.S., that these were found.

As formerly noticed (note 194), we were early, in 1864, presented with the box and its contents; the existence of these MS. was not then known, otherwise they would probably have been presented to us by Mr. Rutherfurd along with the box.

The "Common Place Book" is 14½ by 9½ inches, and 1½-inch thick, is half bound, and appears to have been originally intended for either a Merchant's Day Book, or Journal, as it is ruled throughout with £ s. d. columns, in red ink, similar to such books,—has 276 pages, with several interleaved ones of a quarto size, 184 Miscellaneous Articles, and 108 pen-and-ink drawings, many of which are very beautiful. These writings and drawings are not entered on consecutive pages in the order of their dates, but appear to have been inserted at random, without any regard to order or arrangement, and many of them are to be found in his printed works. Near the top of the first fly-leaf we find, in Ferguson's autograph,

"James Ferguson's
Common Place Book,"

but without date. It does not, however, appear to have been commenced before 1756. Ferguson died in 1776, so that it would be his "Common Place Book" for about 20 years.

To Sir David Brewster, Principal of the University of Edinburgh, to whom we are indebted for the loan of the MSS., and for the liberty of taking extracts therefrom, we take the opportunity here afforded to acknowledge our sense of his kindness, with many thanks.

MURDOCH FERGUSON SHIPWRECKED.—According to a memorandum, Murdoch Ferguson was shipwrecked in September, 1776, narrowly escaping with his life.[324]

FERGUSON ON HIS DEATH-BED.—Of the closing scene, we learnt from the youngest son (John), and also from Mr. Andrew Reid, so often quoted, that Ferguson was confined to the house from the end of June till November, 1776, and that during the greater part of these five months he was confined to bed, occasionally suffering great pain from gravel and other ailments.

DEATH OF FERGUSON.—Died at his residence, 4 Bolt Court, Fleet-street, London, on Saturday morning, 16th November, 1776, at a quarter before Six o'clock, the great Self-taught Astronomer and Philosopher, JAMES FERGUSON, F.R.S., aged 66 years and 7 months. He died as he lived, "an upright, worthy man, and a sincere believer in the Christian faith." His death was deeply regretted in this and other countries. His son John, in 1831, informed us that his father's "intellect, during his long confinement, was clear and unclouded to the last;" that his "last words were faltering, and in part inaudible; but so far as they could be understood, they were in prayer, resigning his soul into the hands of his Maker." (For Obituary Notices, see Appendix).[325]

FERGUSON'S FUNERAL.—"Ferguson was interred in the Churchyard of Old St. Mary-le-Bone, London, on Saturday

324 Several writers allude to this disaster, but do not mention *where* it occurred. In the Appendix to Nichol's Life of Bowyer, p. 596, is the following reference:—"His son was a surgeon, and attempted to settle at Bury—stayed but a little while there—went to sea—was cast away, and lost his all a little before his father's death; but finds himself in no bad plight since." This is a singular expression, "*no bad plight since.*" Can this have reference to the property to which he succeeded on the death of his father? Partington, in the Addenda to Ferguson's Memoir, in his Edition to Lectures on Select Subjects, notes, "*his eldest son suffered shipwreck, narrowly escaping with his life.*" Our note of it is taken from an allusion made in a letter which belonged to Mr. Adam Walker, a friend of Ferguson, which was shown to us by his son, Mr. Deane Walker, so frequently quoted. The disaster is noted in the letter as having occurred in September, about 2 months before his father's death, but does not allude to where it occurred.

325 In our conversation with John Ferguson, Esq., in 1831, relative to his father's death, he mentioned that there were only four persons present at his death,—viz. himself, his brother Murdoch, his aunt Janet Ferguson, and the domestic servant.

morning, 23d November, 1776." [326] The following "Order of the Funeral" is copied from an old manuscript, long in the possession of the late Miss Wilson of Keith, a near relative of Ferguson, who, before her death, presented it to Mr. Robert Sim (lately deceased).

"THE ORDER OF MR. FERGUSON'S FUNERAL.

A Hearse and Six Horses with Black Feathers.

Three Mourning Coaches and Four Horses each.

PALL BEARERS.

Messrs. Strachan.	Messrs. Mackenzie.
—— Jeffery's.	—— Nairne.
—— Allen.	—— Wyndlow Hodgson.

CHIEF MOURNER,—MURDOCH FERGUSON.

MOURNERS.

Messrs. J. Wilson.	Messrs. P. Wilson.
—— T. Cowan.	—— J. Ferguson.
—— T. Cadell.	—— T. Mackie." [327]

Shortly after the funeral, an altar-tomb, as here represented by the woodcut, was placed over the grave, on the south side of which lie the remains of his wife and eldest son James.

Ferguson's Tomb, Old St. Mary-le-Bone Churchyard, London.

In the summer of 1835 we visited the tomb and made this sketch of it,—the inscription was then much weather-worn, so

326 MS. Records of the parish of Old St. Mary-le-Bone, London.

327 There appears to have been 13 at the funeral,—viz. Messrs. Strachan and Cadell, Booksellers; Mr. Nairne, Optician; J. Wilson, Westminster (a relative); P. Wilson (supposed to be a brother-in-law); T. Mackie, Grocer; Mr. Mackenzie, Grocer; Murdoch Ferguson, eldest son; and J. Ferguson, youngest son. We have no note of Messrs. Jeffery, Allen, and Hodgson.

much so, that it was read with difficulty. A friend who visited the tomb in 1864 informs us that he found the lettering on the cover-stone so much wasted as to be scarcely traceable. That it may be preserved, we here give it from our copy taken on the occasion of our visit.[328]

Here

is interred the body of

JAMES FERGUSON,

F. R. S.

Who, blessed with a fine natural Genius,

by unwearied application (without a Master),

attained the Sciences.

Astronomy and Mechanics he taught

with singular success and reputation.

He was modest, sober, humble, and religious,

and

His works will immortalize his Memory,

When this small Monument is no more.

He died 16th Nov. 1776, aged 66.

On the south side of this Tomb lies interred the body of

ISABELLA, his wife, who died 3rd Sept. 1773, aged 53.

And by her side lies the body of

JAMES FERGUSON, their eldest son, who died

20th Nov. 1772, aged 24 years.

[328] In the year 1864, Mr. George Page, of 30 Wardour Street, Soho, London, at our request, made for us a very neat model, in beech, of the tomb, on a scale of an inch to the foot. It rests on a broad base of wood which is covered with green cloth. The inscription is printed on paper, and pasted upon the cover.

Ferguson's Residence, 4 Bolt Court.—The following is a south view of the house No. 4 Bolt Court, Fleet Street, London, the residence of Ferguson from 1765 till his death, November 16th, 1776.

This view is from a photograph which was taken at our request, in 1862, from near the centre of the Court, about 30 yards to the south of the house, which, in the view, is shaded, to distinguish it from houses adjacent.[329]

329 The house, No. 4 Bolt Court, in which Ferguson so long resided, and died, is the property of Lord Calthorpe, and is now tenanted by Messrs. Henry Wilkinson & Co., Sheffield, and used by them as their "Sheffield Warehouse." Mr. J. Brashier, their agent, sent us the following particulars regarding the internal arrangement, for which we return thanks:—"Below the level of the pavement there is a Kitchen and Cellar—two rooms on basement floor—two rooms on second floor—two on third floor—three bed-rooms on fourth floor, and two attics, in all 13 apartments. The two windows on right hand above the door, and large window on same level in gable wall, looking into the Court, is generally supposed to have been Ferguson's principal parlour.

FERGUSON'S WILL.—The following is copy of Ferguson's Will, taken from the original in the Wills' Office, Doctors' Commons, London.

IN THE NAME OF GOD Whom I humbly adore, and into whose merciful hands I resign my Soul hoping to be saved by the attonement of Christ my Redeemer; being at present weak in Body but perfectly sound in Memory and Judgement, I make this my last Will and Testament, as follows; first,—I desire that my funeral charges and just debts may be duly paid,—secondly,—I give and bequeath to my Sister, Elspeth Ferguson, Twenty pounds sterling, and the like sum to my Sister, Elizabeth Ferguson, to be paid within a month after my decease; or to their heirs, if they themselves or either of them die before that time;—Thirdly,—I give and bequeath to my Sister, Janet Ferguson, now living with me, two hundred and fifty pounds sterling, with all my household furniture, including any one of the Clocks that she likes best, and as many of my Books as she chuses to take; but no part of the machinery[330] on which I have read Lectures on Experimental Philosophy and Astronomy;—The rest of my Books, Clocks with my two Watches I leave to be equally divided according to their value, between my two Sons, Murdoch and John Ferguson,—absolutely prohibiting my said sons from selling or giving away or printing any of my manuscripts, because they are not sufficiently correct to bear printing;—fourthly,—I desire my Executors after named, to lay out Two hundred pounds Sterling of my money, in the three per cent. annuities or any other way where they think it will be more safe, which I give and bequeath to my Daughter, Agnes Ferguson, with the Interest that may accrue thereon, in case she shall demand the same within seven years after my decease; but if she doth not demand it within that time, my Will is, that my Executors do equally divide the said two hundred pounds with the Interest thereof, equally between my said two Sons, Murdoch and John;—Hoping that

330 All Ferguson's sisters, so far as known, are here named, excepting Margaret—probably she was then dead. It is remarkable that his brother John is not in the WILL;—he was a small farmer in 1776; and, as previously noticed, died at Relashes (Rothiemay) in 1796, aged 88.

they will have humanity enough to assist and support her afterwards, if she applies to them and they be assured that she is in necessitous circumstances;—fifthly,—I desire my Executors to sell all the machinery on which I had read my Lectures on Experimental Philosophy and Astronomy as soon after my decease as may be; and to put all the moneys arising from the sale thereof to what money of mine remains in their hands, after all the above deductions are paid;—I reckon all this Machinery to be well worth Two hundred and ninety pounds exclusive of the Globes and Universal Dialling Machine; but my Executors are not hereby restricted from selling them for a less sum,——I desire that they may be first tendered to Doctor Buchan[331] at Edinburgh for One hundred and ninety pounds, and if he will give that sum for them, let him have the Machine for showing that a Mixture of all the Colours makes a White, and the Universal Dialling Machine and Brass Quadrants into the Bargain; although I have not used them in my Lectures;—Sixthly,—I request that my Executors will accept of Ten Guineas each, not as a Reward or compensation for their trouble,—but as a small token of my respect and esteem;—seventhly,—as my said Son, Murdoch has had full one thousand one hundred pounds sterling of my money since his education was finished, and as my Son John has as yet had nothing from me, but what I have paid for his maintenance and education which is not yet over,—my Will is, that after all the above mentioned deductions are made by my Executors, the remaining money arising from the sale of my machinery, together with the rest and residue of all that I die possessed of be put out to Interest by my Executors, which Interest is to be applied by them towards my said Son John's maintenance and education till he arrives at the age of Twenty one years; and then, my Executors are to pay him One thousand one hundred pounds of the principal to be entirely at his own disposal, and that then, whatever money of mine still remains in their hands they are to divide it equally among my said two Sons Murdoch and John Ferguson, Share and Share

331 Dr. William Buchan, author of the celebrated work, "Domestic Medicine," was born at Ancrum, Roxburghshire, in 1719,—died in London, 25th February, 1805, aged 76. In 1775 he, it seems, contemplated delivering lectures on Natural Philosophy in Edinburgh, but it is not known if he did so,—most likely not. Ferguson's apparatus in whole, or in greater part, was disposed of by public sale in March, 1777.

alike,—my papers Vouchers in the same parcel with this my last Will and Testament, will show what money I die possessed of;—Eighthly; I nominate and appoint Mr. Thomas Cadell, Bookseller in the Strand,—Mr. Thomas Mackie, Grocer, in Princes Street, Soho, and Mr. Thomas Cowan, lately a Grocer, in Oxford Street,—all now living in London, to be the Executors of this my last Will and Testament, which is written, signed and sealed by my own hand,—on the fifteenth day of August, in the year of our Lord, One thousand seven hundred and seventy six——James Ferguson."

25th November 1776.

Appeared personally George Adams, of the Parish of Saint Dunstan in the West, London, — Mathematical Instrument Maker,—and James Wilson of the Parish of Saint James, Westminster, in the County of Middlesex, Gentleman,—and respectively made oath that they, these Deponents, knew and were acquainted with James Ferguson, late of the Parish of Saint Dunstan in the West, London, deceased, for several years preceding, and to the time of his death, which happened in the present month of November, as they, these Deponents have been informed and verily believe,—and also with the deceased's hand-writing, and way and manner of subscription,—having had frequent occasion of seeing the deceased write and subscribe his name,—and they, these Deponents, having carefully viewed and perused the paper writing hereunto annexed, purporting to be the last Will and Testament of the said deceased—beginning thus, "In the name of God whom I humbly adore, and into whose merciful hands I resign my Soul, hoping to be saved by the attonement of Christ my Redeemer, being at present weak in Body but perfectly sound in mind memory and Judgment, I make this my last Will and Testament, as follows,"—and ending thus "Eighthly. I nominate and appoint Mr. Thomas Cadell, Bookseller in the Strand,—Mr. Thomas Mackie, Grocer in Princes Street, Soho, and Mr. Thomas Cowan, lately a Grocer in Oxford Street, all now living in London, to be my Executors, of this my last Will and Testament, which is written, signed, and sealed by my own hand on the fifteenth day of August, in the year of our Lord, One thousand, seven hundred and seventy six," and thus subscribed "James Ferguson"—do depone and verily, and in their conscience believe, as well whole series, Body, and contents of the said Will, as the name "James Ferguson" appearing thereto, set, and subscribed, to be in the handwriting and usual manner of subscription of him, the said James Ferguson, deceased.[233]

George Adams,—James Wilson;—Same day the said George Adams and James Wilson, were duly sworn to the truth of this affidavit before me Geo. Harris,—Surrogate:——Present Mark Holman N. P. PROVED at London 27th Nov. 1776, before the Worshipful Andrew Coltee Ducarel, Doctor of Laws, Surrogate,—lawfully constituted by the Oaths of Thomas Cadell,—Thomas Mackie and Thomas Cowan the Executors named in the said Will to whom admon was granted, having been first sworn duly to administer."

Stamp'd OFFICE COPY

COURT OF PROBATE.

458 Bellat
fol. 16 T. M.

332 As mentioned in note 318, the sums named in the WILL amount to £1620; but *the rest* or *residue* is not named, although it must have been considerable, besides what would be received from sale of apparatus, household effects, &c.

APPENDIX.

APPENDIX.

NEWSPAPER OBITUARY NOTICES OF FERGUSON'S DEATH.

"On Saturday Morning, died at his house in Bolt Court, Fleet Street, of a lingering illness, the justly celebrated Mr. James Ferguson, Lecturer in Natural Philosophy and Astronomy, one of the greatest geniuses in Mechanics that ever appeared in this or any other country, as his several publications abundantly testify,—in addition to the above, with a very sagacious mind, and the most engaging and primitive simplicity of manners, he possessed an uncommon share of good nature and humility. His whole deportment was indeed highly becoming of what he professed himself to be—a sincere believer in the Christian faith—so that it may be justly said of him '*that he was an Israelite indeed in whom there was no guile.*'"—The New Morning Post and General Advertiser, London, Monday, November 18th, 1776.

The Caledonian Mercury, of date Edinburgh, November 20th, 1776, has precisely the same announcement; as also many other newspapers during this month.

"On Saturday morning, about 6 o'clock, died at his house in Bolt Court, Fleet Street, Mr. James Ferguson the Astronomer. Mr. Ferguson was more than 66 years of age; and though generally thought to be in distressed circumstances, has bequeathed to different legatees above £5000."—The Gazetteer and New Daily Advertiser, London, Monday. 18th November. 1776.

1776.—"Died on November 16th, Mr. James Ferguson, Lecturer in Natural Philosophy and Astronomy, an excellent mechanic, and no bad miniature painter, at his house in Bolt Court, Fleet Street. He was a man who, by mere force of genius, made considerable progress in mathematical arts and sciences, wrote several useful works, and both projected and executed a great number of ingenious instruments and machines."—Dodsley's Annual Register, vol. 19, p. 194; Gentleman's Magazine, vol. 46, p. 397.

POSTHUMOUS DETRACTOR AND DEFENDERS.

"MR. URBAN,

The posthumous fame of Mr. Ferguson has suffered much from his affectation of poverty and distress, while he was secretly possessed of thousands. His seeming humility was as much put on. Many mistakes may be found in his Lectures on Physical Subjects to which he was in no wise adequate. Some of these were pointed out to him privately, that the sale of his works might not be hindered, and the errors amended in a future edition; but he always received these private intimations with ill humour, and rejected such friendly corrections with disdain, &c.—W. L." [333] — Gentleman's Magazine, Dec., 1776.

"Mr. Urban,

The strictures of W. L. on Ferguson's Select Lectures, so far as they relate to science, gave me pleasure; because, it is from the joint efforts of a number of individuals only, that we can ever hope for any tolerable degree of precision. But really the sarcastic, invidious manner they are introduced in, gave me pain; because the censure seems wanton and unprovoked. What have the public to do with the moral rectitude of a man, whose only pretensions were to read lectures on Natural Philosophy?—or, How are we to judge of the truth or consistency of the accusation? He affected the appearance of poverty, yet he died possessed of what some would call a competency. Did any one give a supernumerary guinea? Verily they had their reward, in having relieved supposed distress. But are there not family circumstances that might apologize for such conduct? Such may be suggested, and I am told actually did exist. He wore the mask of humility, but rejected well-meant counsel with disdain. Is it at all strange that a man, who merely by the strength of his own genius raised himself to deserved estimation in the walks of science, should be sensible he possessed superior talents? Or, is it not possible that amendment may be proposed in such a manner as to give just cause of offence to the most abject? Indeed I think it may; and that W. L.'s *Essay* is one proof of it; 'many mistakes are to be found in his Lectures on Physical Science,' and so there are in most, if not all authors; this, therefore, is no proof that he in particular 'was in no wise adequate.'

I would not wish to write a panegyric on Mr. Ferguson, yet thus much I thought due to the merit of a man, that, so far as I know, seems to have filled his station in life with a degree of clearness and utility that has seldom been exceeded.—Simplex."

"Mr. Urban,

It was with the greatest concern that I read in your Magazine for January a most malignant abuse of a good man, an ingenious and sound philosopher, Mr. James Ferguson.

[333] We have looked into a great many Newspapers and Magazines for November, 1776, for obituary notices of Ferguson,—all pay the highest tribute of respect to his worth and genius, with the exception of this vituperous writer under the cover of the initials W. L. Can it be supposed that these initials came into existence through the agency of the pen of the pugnacious Rector of Bradley—the vilifier of Ferguson—the Reverend John Kennedy?

'His posthumous fame' (says this writer) 'has suffered much from his affectation of poverty and distress, while he was secretly possessed of thousands. His seeming humility was as much put on. Many mistakes may be found in his Lectures on Physical Subjects, to which he was in no wise adequate. Some of them were pointed out to him privately that the sale of his works might not be hindered, but he always received these private intimations with ill humour, and rejected such friendly corrections with disdain.' And this worthy gentleman, however, it seems, wishes to point them out publicly, after his death, when he is no longer able to defend himself; but his efforts are as futile and contemptible as they are ill-timed. His insinuations are false, and only show how much malice and venom this animadverter adds to his ignorance.

I was well acquainted with Mr. Ferguson for many years, and altho' I knew that he was oftentimes really in great distress, notwithstanding his honest labours and industry, owing to an unfortunate connection, yet he always took pains to conceal his misery and was backward and scrupulous to receive the benevolence of his friends who wished to contribute to his necessities. And when His Majesty was graciously pleased to bestow upon him a pension of fifty pounds a-year, he expressed himself as under the deepest obligations of gratitude, and always spoke of it, as making him easy and comfortable. And, as to the few thousands he left behind him, I can very truly affirm, they came to him only a very short time before he died, by means of the death of a relation, from whom he had never before received anything. As to his humility (which this writer would give us to understand was merely affected and hypocritical), all who were acquainted with Mr. Ferguson know it was on the contrary most unfeigned; and so excessive, as to have been, in some instances, very prejudicial to him, when he met with minds ungenerous enough to take advantage of it to his detriment.

Justice to the memory of an old departed friend demands this vindication of his injured character; and I hope, Mr. Urban, you will with your usual impartiality insert it in your magazine, especially after having given place to such an unprovoked and virulent abuse of his good name and character.

Base indeed must that mind be which can find no better employment than to load with reproaches the memory of a man deceased, who was in his manners most simple and irreproachable, and in the pursuit of his studies, one of the most useful, candid, and ingenious men of the age in which he lived, and to whom the world is indebted for great improvements in the science of Astronomy, and for great and elegant elucidation of it, and for many most useful mechanical inventions.—A Plain honest Man."

Thus, these two writers, as well as Dr. Houlston, in his notice at page 462, allude to some "unfortunate family connection." The "*Plain Honest Man*" in particular, who was for many years well acquainted with Ferguson, notices the "unfortunate connection," but does not refer to its source. We are sorry to note that it appears to point to Mrs. Ferguson. The following notice from the *Mirror* has been frequently alluded to and quoted, and it therefore requires to be reproduced here, although we have some scruples as to its correctness.

"James Ferguson, F.R.S.—Somewhere about the year 1770, whilst Ferguson was delivering a lecture on Astronomy to a London audience, his wife entered and maliciously overturned several pieces of his apparatus. Mr. F. observing the catastrophe, only remarked the event by saying, 'Ladies and Gentlemen, I have the misfortune to be married to this woman.' It would seem that Mr. F. and his wife lived unhappily together. She is never alluded to in the excellent Memoir written by himself.[334] Mr. Ferguson seems to have been somewhat penurious in his habits; for it is on record that he used very frequently to borrow small sums of money from his friends for the purpose of its being understood that he was in poverty, and so elude the chance of his friends borrowing from him.[335] He died in 1776, with property and money to the value of about £6000. Shortly after the death of Mr. Ferguson a Scotch publican had the philosopher's likeness put over his door, and the house went by the name of 'The Ferguson's Head,' where was long to be seen one of Mr. Ferguson's large orreries. —W. S."—Mirror, London, Saturday, 25th February, 1837, vol. 29, No. 822, p. 128.

Dr. Houlston of Liverpool on Ferguson.—* * * * "Mr. Ferguson was universally considered as at the head of Astronomy and Mechanics in this nation of philosophers, and he might justly be styled self-taught, or rather heaven-taught; for in his whole life he had not above half-a-year's instruction at school. He was a man of the clearest judgment, and of the most unwearied application to study; benevolent, meek, and innocent in his manners as a child; humble, courteous, and communicative; instead of pedantry, philosophy seemed to produce in him only diffidence and urbanity, a love for mankind and for his Maker. His whole life was an example of resignation and Christian piety. He might be said to be an enthusiast in his love of God, if religion, founded on such substantial and enlightened grounds as his was, could be called enthusiasm. After a long and useful life, unhappy in his family connections, in a feeble and precarious state of health, worn out with study, age, and infirmities, he was at length permitted to attain that heaven on which his thoughts and views had long been fixed, and which is the ultimate reward of learning, virtue, patience, and piety."—Annual Register, vol. 19, 1776.[336]

Capel Lofft, Esq., on Ferguson.—"Mr. Ferguson died on the

[334] It is needless to inform the reader that Ferguson *does* allude to his wife in his "*excellent Memoir;*" even although he had not, it would have signified nothing.

[335] We rather think that this may refer to the doings of Murdoch the son. See note on Murdoch; also, letters of "Simplex" and "A Plain Honest Man," in the Appendix.

[336] Dr. Houlston, an eminent surgeon and author, one of Ferguson's most intimate friends, resided in Chapel Street, Liverpool. Died about the year 1784.

sixteenth of November, 1776, having struggled with a constitution naturally infirm longer than could have been reasonably expected. * * * Some manuscript tables, diagrams, and a philosophical correspondence of this heaven-taught philosopher are in my hands, which were given by him to my EUDOSIA before our marriage; nor can I reprove myself for the pride which I often feel in reading over his letters to Miss *Emblin*, written to her before our marriage." [337]

SIR DAVID BREWSTER ON FERGUSON.—" Mr. Ferguson may in some degree be regarded as the first elementary writer on Natural Philosophy, and to his labours we must attribute that general diffusion of scientific knowledge among the practical mechanics of this country, which has in a great measure banished those antiquated prejudices and erroneous manners of construction that perpetually misled the unlettered artist. But it is not merely to the praise of a popular writer that Mr. Ferguson is entitled; while he is illustrating the discoveries of others, and accommodating them to the capacities of his readers, we are frequently introduced to inventions and improvements of his own; many of these are well known to the public; and while some of them have been of great service to experimental philosophy, they all evince a considerable share of mechanical genius. To a still higher commendation, however, our author may justly lay claim; it has long been fashionable with a certain class of philosophers to keep the Creator totally out of view when describing the noblest of his works. But Mr. Ferguson has not imbibed those gloomy principles which steel the heart against its earliest and strongest impressions, and prompt to suppress those feelings of devotion and gratitude which the structure and harmony of the universe are so fitted to inspire. When benevolence and design are particularly exhibited in the works or in the phenomena of nature, he dwells with delight upon the goodness and wisdom of their Author; and never fails to impress upon the reader, what is apt to escape his notice, that the wonders of creation, and the various changes which the material world displays, are the result of that unerring wisdom and boundless goodness which are unceasingly exerted for the comfort and happiness of man."—Brewster's Preface to his Edition of Ferguson's " Lectures on Select Subjects." [338]

" FERGUSON'S PERSONAL APPEARANCE."—The following remarks on Ferguson's personal appearance were obligingly sent in a letter of date June 7th, 1833, by Mr. Andrew Reid—so often referred to. He was well acquainted with Ferguson; and as previously noted, died in 1834, aged 84 years.

" Personal appearance and Dress of Mr. Ferguson about the year 1774."——" Mr. Ferguson had a very sedate appearance, face and

337 The author of Eudosia, or a *Poem on the Universe*, so frequently alluded to.

338 The name of this philosopher is universally known. He, in 1805, edited an edition of Ferguson's Mechanics; in 1811, an edition of Ferguson's Astronomy; and in 1823 collected and published Ferguson's other works under the title of " Ferguson's Essays."

brow a little wrinkled; he wore a large full stuff wig, which gave him a venerable look, and made him to appear older than he really was. He usually wore a white neckerchief, especially when delivering his lectures. His coat had no neck, was of large dimensions, reaching down below the knee, and coming full round in front; was decorated with large buttons, and of course had the usual huge pockets and double folded-up sleeves fenced with shirt wrist ruffles. His waistcoat was also large; had likewise no neck; large pockets; and reached down to near his thighs. He wore knee breeches, generally of black velvet, or plush, and fastened at the knee with silver buckles; generally wore black stockings, full shoes with buckles. When walking about he wore the cocked hat of that day slightly trimmed with lace; and, in these, the latter days of his life, he walked about with the aid of a staff."

LIKENESSES OF FERGUSON.—There are at least *four* original and large likenesses of Ferguson extant; from two of which miniatures have been taken:—

THE BEATSON LIKENESS, No. 1.—The earliest likeness of Ferguson extant, is now in the possession of Mrs. Thomson of Nether Cluny, Dufftown, by Craigellachie, Banffshire:—It is on canvas 23 inches by 19, and on the back is the painter's name and date, viz., "JOHN BEATSON, 1756." Mrs. Thomson has very kindly sent us a photograph from the original, in an oval $6\frac{1}{4}$ by $5\frac{1}{4}$ inches. This is a very plain likeness—the plainest of any we have seen—almost rustic. The face, with its fine high and massive brow, looks to the left; the part of the coat and vest shown represent garments of a very common description. This bears not the slightest resemblance to any of the other likenesses. As it is dated in 1756, it is likely that it was painted shortly after Ferguson published his great work, "*Astronomy Explained upon Sir Isaac Newton's Principles*," which was first issued about midsummer 1756. The publication of this work greatly increased Ferguson's fame and popularity, and might have induced Beatson, and perhaps other painters, to take portraits of "*the rising man*." In 1756 Ferguson was about 46 years old. (See also date 1756.)

THE CASBORNE LIKENESS, No. 2.—Is a likeness of Ferguson done in chalk, within an oval of 22 inches by 17. This likeness was by him given to a lady of title at Windsor, who, in her turn, presented it to Mr. Emblin of Windsor, the father of Miss Emblin, the Eudosia of Ferguson's "Young Gentlemen and Ladies' Astronomy," and is now in the possession of Mrs. Casborne of New House, Pakenham, Bury St. Edmunds, the daughter of Ferguson's "Eudosia." As this one belonged to Ferguson, it is extremely probable that he esteemed it a good likeness,—and worthy of being presented to a titled friend. Mrs. Casborne remarks that it is so faded that a photograph cannot be taken from it. There is neither name, initials, nor date attached to this likeness.

THE NORTHCOTE LIKENESS, No. 3.—is a large engraved likeness of Ferguson, from Northcote, engraved in March 1776, about 8 months before his death, as mentioned in note 319, p.445. This is a three-quarter face likeness; is too youthful, both in the face and in the dress, for one of 66 years. It is singular that the publishers of the "Library of Entertaining Knowledge" should have selected it for one of their embellishments, when Ferguson's own publisher rejected it. For particulars of this likeness, see p. 445.

THE TOWNSEND LIKENESS, No. 4.—This is esteemed the best of all the likenesses of Ferguson; the appearance in every way is quite in keeping with his age at the time it was done. From this one his publisher, Mr. Cadell, had a reduced engraving for the frontispiece of the second edition of "Select Mechanical Exercises," published on January 1st, 1778, the only difference being, that instead of looking to the right as in the original, the face looks to the left. It is a front likeness (very nearly); the head is covered with a large and full bag-wig, the face careworn and venerable; part of the body is visible, enveloped in the huge dress-coat and vest of last century. The right hand rests on a celestial globe. Underneath is in scribed,

John Townsend, pinx^{t}. Published Dec. 7th 1776.

JAMES FERGUSON, F.R.S.

Printed for Robert Stewart, Engraver and Modeller of Portraits in Wax. No. 15 Millman Street, Bedford-row, Holborn.

This likeness is very scarce; we have a copy framed and glazed. Size of the engraving,—the oblong part of the ground, $14\frac{3}{8}$ by 10 inches; the oval upon the ground, and within which the likeness is engraved, $13\frac{1}{4}$ by 10 inches. Partington has a frontispiece from this likeness in his edition of Ferguson's 'Lectures on Select Subjects,' published in 1825.

We may notice that a likeness, apparently from either the Townsend or Cadell one, was published by "W. Bent, London, Dec. 21st, 1785." We have another, within an oval, quite a caricature, beneath which is the surface of a table on which are representations of an armillary sphere, a meridian instrument, diagram of a Solar Eclipse, and a book entitled "*Lectures on Astronomy*," and over the top are "James Ferguson, F.R.S." This engraving is from some magazine. It has no indication to show what book it had embellished, neither has it a date.

FAMILY OF FERGUSON'S PARENTS.—As far as can now be ascertained, the family consisted of six children,—*two sons* and *four daughters.* John appears to have been the eldest son, and Margaret the eldest daughter. We shall write down the sons' names first, and then give the names of the daughters, following each other, as Ferguson has them in his WILL, which is most likely in the order of time of their ages. Before tabling the children's names, it may be mentioned that it is not known when Ferguson's parents were born, consequently their ages at their marriage and deaths are a blank. It is, however, very probable that they were both born sometime about the years 1666 or 1667. They were both alive in 1736, as Ferguson in his own Memoir tells us that he about this period, when in Edinburgh, had saved a little money: out of which he was enabled to spare, he says, "what was sufficient to help to supply my *father* and *mother* in their *old age*." At this period his parents would be about 70 years old; and it appears they were both dead before 1742. They died respected by all who knew them. It is not known where the parents were buried; probably at Rothiemay.

Ferguson's father's name was JOHN FERGUSON, his mother's name was ELSPETH LOBBAN—and, as far as is now known, had the following children:—

John Ferguson, the eldest son, born 12th March, 1708.[339]
James, (the subject of this memoir,) born 25th April, 1710.[340]
Margaret,[341]
Elspeth,[342]
Elizabeth,[343]
Janet,[344]
} Not known when born nor where interred; probably at Rothiemay.

The following is extracted from the fly leaf of a pocket Bible which belonged to Mr. Ferguson about 1758—the entries are in his autograph. This Bible, so often referred to, is now in the possession of Doctor George at Keith. We may here note, that Ferguson gave this Bible to his son James about the year 1758; that James gave it to his sister; that she presented it to her aunt Janet, after whose death in 1773 it came into the possession of the late Miss Wilson of Keith —who gave it in a present to her physician, Dr George of Keith. (See note 121, p. 72).

339 A daughter of John Ferguson, named Isabella, was married to Alexander Humphrey, in Pitlurg, a few miles south of Keith. John Ferguson's farm, or croft, in his latter days, was at Relashes, about 1½ mile north of Rothiemay. He was long an elder of the parish of Rothiemay. After a life of toil, he died at Relashes, 1796, aged 88, and was buried in Rothiemay churchyard.

340 James died in Bolt Court, Fleet Street, London, 16 Nov., 1776, aged 66 years.

341 Margaret Ferguson:—The Rev. Mr. James Mackie, minister of Alves, near Elgin, informs us, that Margaret Ferguson was his great-grandmother, and that she was the eldest daughter of the family. When she died is not known; probably before 1776, as Ferguson, in his WILL, names his other sisters Elspeth, Elizabeth, and Janet, but no notice is there taken of Margaret. Margaret died at Rothiemay, and was buried there.

342 Nothing is known of Elspeth Ferguson. It will be remembered that Ferguson, in his letter to the Rev. Mr. Cooper of Glass, sent a present to her of *two guineas*, and also *a guinea* to her daughter "Jensy."

343 Of Elizabeth Ferguson nothing is known.

344 Dr. Cruickshanks, University of Aberdeen, informs us that Janet Ferguson was married to James Brown, a small crofter at Barnhills, near Rothiemay, that he knew both Janet and her husband, and that she died suddenly in 1793, leaving three daughters, named Margaret, Isabella, and Janet. This sister appears to have been a favourite of Ferguson's. He left her more money than any of his other

What follows, refers to the family of Ferguson's father-in-law, George Wilson of Cantley, near Keith. [345]

"George Wilson, Cantley, married to Elspeth Grant, daughter of Arch^d. Grant of Edin Valley, the 28th of August 1712; [346] of which marriage there were—

Alexander, born	Sunday ye 5th of June	1714.	
Margaret, „	Monday 27th July	1717.	
Isobel, „	Monday 21st Decemr.	1720.	Died 3d Sept. 1773. [347]
William, „	Tuesday 23d January	1722.	
Peter, „	Monday 15th February	1725.	Died 12th July 1745.
George, „	Monday 17th June	1728.	Died 6 Oct. 1763.
James, „	Friday 2d April	1731.	
John, „	Wednesday 26th June	1734.	Died 8th March 1764.
Robert, „	Friday 6th August	1736.	Died 20 March 1739.

George, the father, died 22d March, 1742, at ½ past 9 P.M., aged 63 years 6 months. Elspeth Grant, the mother, died 29th January, 1771."

On the fly leaf of the Bible, are the following entries in Ferguson's autograph:—

"The present of James Ferguson to his sister, Agnes Ferguson, Friday, April 7th, 1758;" and underneath this is written, (also in Ferguson's hand):—

"This Bible is the present of Agnes Ferguson, to her uncle James Wilson, 19th April, 1760."

On the back of the title page of the New Testament of the Bible, are the following entries, (also in Ferguson's autograph):—

"FERGUSON'S FAMILY.—James Ferguson and Isabella Wilson were married on Thursday, May the 31st, 1739, and had the following children:"— [348]

sisters, besides household property. She resided with Ferguson for the last three years of his life, and was with him when he died on 16 Nov., 1776, after which she appears to have returned to Rothiemay, Banffshire.

[345] As mentioned in a former note, the Wilson family were interred in the churchyard of GRANGE. On the monumental stone of the family there, we find—"Sacrum Memoria, George Wilson, nuper in Cantly, qui mortem obit 22d die Martis, A.D. 1742, ætatis suæ 64to. Hoc amoris et doloris monumentum Uxor superstes et moerens posuit." For this extract, we are indebted to the kindness of the Rev. James Allan, present minister of Grange.

[346] The Archibald Grant here mentioned was an ancestor of the present Grants of Arndilly, about 10 miles from Keith.

[347] As recorded in a former note, "*Isobel,*" the second daughter and third child of this marriage, became in 1739 the wife of Ferguson.

[348] As noted elsewhere, James Ferguson the father, was born at the Core of Mayen, near Rothiemay in Banffshire, on 25th April 1710; died in No. 4 Bolt Court, Fleet Street, London, on 16th Nov. 1776, and was interred in Old St. Mary-le-Bone churchyard, London:—His wife, Isabella Wilson, was born at Grange, near Keith, Banffshire, on 21st Dec. 1720; died in 4 Bolt Court, Fleet Street, London, on 3d Sept. 1773, aged 53, and was interred also in the churchyard of Old St. Mary-le-Bone, London; beside their oldest son James.

Agnes,	born	Thursday 29th Augt.,	1745. [349]
James,	„	Tuesday 11th October,	1748. [350]
Murdoch,	„	Friday 3d Nov.,	1752. N.S. [351]
John,	„	Tuesday 27 Feby.,	1759. N.S. [352]

349 AGNES FERGUSON:—As recorded in the foregoing list of births, Agnes Ferguson was born in London, on Aug. 29th, 1745. She was remarkable for her beauty and intelligence; "she suddenly disappeared" about the end of July or early in August 1763, and was never more seen by her parents. Our late researches regarding her show that she was decoyed by a young nobleman and taken to Italy. He abandoned her, and she, being probably ashamed to return to her parents whom she had disgraced, to maintain herself, wrote articles for the Magazines. She afterwards became an actress, for a brief period. She ultimately led an irregular life, and died in poverty in a miserable garret, in Old Round Court, Strand, (now removed) in January 1792, aged 47 years. (See pp. 72, 279—285.)

350 JAMES FERGUSON, the eldest son, as shown in the foregoing list, was born in London on 11th October 1748. Little is known of him; it is highly probable he inherited a share of his father's mechanical genius, as he, in 1763, was apprenticed to Mr. E. Nairne, Optician and Philosophical Instrument Maker, Cornhill, London. He died of consumption (4 Bolt Court, Fleet Street), on 20th November 1772, aged 24, and was buried in the churchvard of Old Mary-le-Bone, London. (See pp. 127, 278, 396.)

351 MURDOCH FERGUSON—(the second son)—according to the foregoing list, was born in London on 3d November 1752, new style. Nothing is now known of his early years;—he studied surgery and medicine for one session only at Edinburgh University in 1767; where he completed his studies is now not known. In July 1775, he obtained from the Court of Examiners of the late corporation of surgeons of London, his Certificate of Qualification of Surgeon to a Regiment. In August 1775, he married a Miss Vincent of Greenwich,—attempted to settle in Bury-St. Edmunds as a Surgeon, but did not succeed. He entered the Service of the Royal Navy as Surgeon, in H. M. S. *Thunderer*, on 30th May 1778. Discharged 30th January 1779, being superseded. He next entered H. M. S. *Emerald*, on 25th November 1780. Discharged 16th July 1781, to Sixth Quarter at Hasels—superseded; and lastly, on 19th January 1782, he entered H.M.S *Africa* as Surgeon. Discharged and superseded on 2d July 1783, after which date, the War Office authorities inform us, his name is not in their lists. In a "List of Officers in His Majesty's Fleet, with the date of their Commissions—British Museum, p. 2486," we, however, find Murdoch Ferguson in the Lists as surgeon in 1796, 1797, 1798, and his Seniority is dated 21st March 1781.—In looking into that rare publication—Kent's London Directory—we find "Murdoch Ferguson, Surgeon, Paternoster Row, Spitalfields, for 1789."—In "1792, 1793, 1794, 1795, 1796, 1797, 1798 and 1799, Murdoch resided in Burr Street, St. Catherines beyond Tower Hill."—In the Books of the Parish of St. Catherine Tower Hill, it is recorded that Murdoch Ferguson was summoned for poor-rates in arrear for the house in Burr Street, in the latter year (1799), when he declared his inability to pay, and gave it up.—Murdoch died about the year 1803, aged 51.—In 1836, we accidentally met a gentleman in London who knew Murdoch—he informed us that he and his wife were rather given to intemperate habits, and that he sometimes delivered lectures in a school-room in the East end of London. It would seem that Murdoch *feigned* poverty when he had plenty and to spare—and that several of his own friends, and friends of his father, on the supposition that he was in straitened circumstances, supported him for some time before his death;—and at their expense he was buried. After his death, on an examination of his papers, it was discovered that he was comparatively wealthy.—In vol. viii. p. 411 of Nichols' Literary Anecdotes, we find the following—"*Murdoch Ferguson was the person in part maintained, and at last buried at the expence of his friends, and since found to have died rich.—This would lower any character*"—and in the Index to this volume, we find "*Murdoch Ferguson a Miser.*"—It is not known where Murdoch was interred—neither is it known when

SALE OF FERGUSON'S EFFECTS.—Nichols, in his "Literary Anecdotes," states that "Ferguson's valuable Library of Mathematical

or where his wife died; nor if she had any family. (See pp. 177, 353, 410, 441, 448, and 473.)

352 JOHN FERGUSON—as recorded in the foregoing list—was born in London on 27th February 1759, new style. Nothing is known of his boyhood. He attended the medical classes, &c., of Marischal College, Aberdeen, during the Sessions of 1773–4, 1775–6, 1776–7. It is not known what course John pursued after completing his College studies. It is certain that he never followed out the medical profession. His father, at his death, left him a large sum of money, which, along with other sums bequeathed to him by his maternal uncle, Lieutenant James Wilson, after his father's death, put him in easy circumstances, and enabled him to indulge in that 'indolent ease' in which he so much delighted.—There is a blank of 26 years in the life of John;—nothing whatever is known of him from the time he left college in 1777 until the year 1803. In 1803 we find him residing in Keith, Banffshire, and conveying by a title-deed certain houses and grounds, on the west side of the New Street in the Kirkton of Keith, to his maternal cousin, "Miss Margaret Wilson" (being the houses and grounds which her relation had left by will to him). How long John continued to reside in Keith is not known. According to several of our memoranda, John, early in the present century removed from Keith and took up his residence in Edinburgh. In his latter years he resided with the Misses Moir, 10 Windmill Street, Edinburgh, and there he died on Sunday, 13th October 1833, aged 74 years and 8 months. For a considerable length of time before his death, John was very frail and confined to his room and bed. On Saturday, October 12th, his death appearing evidently at hand, he desired his *Will* to be drawn up and attested, which was done, the following is a copy of it:—

"I, John Ferguson, residing in Windmill Street, Edinburgh, being resolved to settle my affairs as to prevent all disputes after my death, in regard to the succession to my moveable means and estate, do hereby nominate and appoint Christian Moir, Charlotte Moir and Sarah Moir, all residing in Windmill street aforesaid, to be my Executors and Universal Legatones, hereby leaving and bequeathing to them, the whole goods, gear, debts, sums of money, household furniture and other moveables whatsoever, that may pertain or be resting owing to me, at the time of my decease, together with all bonds, bills, promissory notes, and other vouchers and instructions, of the said debts, and all that has followed or may be competent to follow thereon. But always with and under the burden of the payment of all my just and lawful debts, sick bed and funeral charges, and also of the Legacy herein after appointed to be paid:—with full power to the said Christian, Charlotte and Sarah Moir, to intromit with the whole of my foresaid moveable Estate, to give up Inventories thereof, to confirm the same, and generally to do everything in the premises competent to Executors. And ordain my said Executors to pay and deliver the sum of Three hundred pounds sterling to Miss Margaret Wilson my Cousin, residing in Keith, in the County of Banff, which Legacy shall be payable six months after my decease.—And I consent to the Registration hereof in the Books of Council and Session, or others competent, therein to remain for preservation, and to that effect constitute
my Procurators. In witness hereof, I have subscribed these presents written by Alexander Brand, Clerk to Messrs. Scott and Balderstone, Writers to the Signet, also, the marginal addition hereto, likewise written by the said Alexander Brand at Edinburgh the twelfth day of October, Eighteen hundred and thirty-three, before these witnesses, William Brand, also Clerk to the said Messrs. Scott and Balderstone, and the said Alexander Brand, writer hereof."

"By desire of the said John Ferguson, who declares he cannot write, by reason of bodily infirmity, and he having touched my pen, and authorized me, I William Scott Notary Public do subscribe for him the above written Testament with the Marginal addition thereto, the

Books, Manuscripts, and Instruments, was sold by auction by Leigh and Sotheby, Nov. 15–23, 1802;" and as this has frequently been

same having been previously read over to him, by me in presence of the said Witnesses

Signed { Wm Scott, N. P. Wm Brand, witness.
Alexr. Brand, witness."

John Ferguson was interred in Old Grayfriars burying-ground, Edinburgh. The following is extracted from the Grayfriars parish record.

"*John Ferguson, Esquire, died at Windmill Street,* 13*th October* 1833, *aged* 75 *years;—Recorded,* 15*th October* 1833.—*Place of Interment close to the north-east corner of Stark's Ground.*" There is no stone to mark the last resting-place of John.

James J. Rutherfurd, Esq., of 10 Windmill Street, Edinburgh—(the house in which John Ferguson lived and died) was well acquainted with John. In a letter received, of date 1861, he says, "John Ferguson was of sedate temperment, courteous in his manners, and steady in his friendship; as is shown by his domestication with the old Ladies, the Misses Moir, with whom he resided from about the beginning of the century to his death in 1833. He paid them a very liberal board, and ultimately kept a carriage for their express use; of his personal property they received a bequest of several thousand pounds—but by a delay in making his *Will*, his heritable property, which was said to be considerable—went to a second or third cousin in Aberdeenshire. Whatever property the Astronomer his father may have left John, was chiefly derived through his mother and relations of hers.—With a considerable knowledge of History, and a slight tinge of Latinity, and also an appreciation of several of the most elegant of the French writers, John probably never even read his father's popular Work '*the Young Gentlemen and Ladies' Astronomy*,' for he had no taste for Mathematical Science — and was indifferent to all matters of Arithmetical computation.—There never was seen in his house a single work of his father's." One of our notes mentions, that "John Ferguson was a very amiable gentleman, and highly respectable, but eccentric and of no great calibre of mind." A friend who knew John Ferguson intimately—mentions that he, (John) had often told him, that his father used to attempt to teach him the Problems in Euclid by models cut by his own hand in wood and in card;—but in vain!—I think," (our friend adds) "that I see John just now taking a sad pinch of snuff as a wind-up to his confession."—Another friend informs us, that "John Ferguson had a great delight in playing on the flute—and that during his latter days it was his sole amusement."

In the autumn of the year 1831, when in Edinburgh, on hearing that a son of the great self-taught astronomer was living at 10 Windmill Street, we lost no time in going to see him. We took leave to introduce ourself, and was cordially received; he told us he had just been playing a favourite air on his flute. Besides having the pleasure of seeing the only surviving son of so celebrated a man as Ferguson, we were then desirous of obtaining information, to put a few questions to him regarding his father and family, as also regarding several pieces of apparatus, &c., which had belonged to his father. Regarding his lost sister, he appeared to know nothing. As to when and where his paternal grandfather and grandmother died, their ages and where buried, he could give us no certain information; he however thought that they would be interred in the churchyard of Rothiemay. He produced several articles which had belonged to his late father, such as a small orrery, an astronomical dial on a large thick oblong card, astronomical and mechanical drawings, writings, &c. He informed us that he had sent several articles as presents to his cousin, Miss Wilson of Keith (now dead), viz., a telescope which his father had made during his younger years, also scales and compasses, astronomical quadrants, on cards, &c. Several articles of his father's which we had seen, were mentioned, of which he had a clear remembrance. During our short stay in Edinburgh at the time mentioned, we had two interviews with him, when he communicated many little incidents, which might have been otherwise lost, and which are now interwoven with other details in the pre-

noticed by writers when referring to Ferguson, we shall here show that Ferguson's effects were sold by auction shortly after his death, in 1777, and not in 1802, a date 26 years after his death.

When we first read Nichols' note, we were of opinion that it was wrong, and subsequent research confirmed it—

1st. John T. Graves, Esq., Cheltenham, some years ago, kindly sent to us the catalogue of this sale of 1802—it is entitled

> "A Catalogue of the valuable Library of Mathematical Books, Manuscripts, and Instruments of the late James Fergusson, Esq., F.R.S., Astronomical Examiner, which will be sold by auction by Leigh and Sotheby, at their House, York Street, Covent Garden, on Monday November 15th, 1802, and seven following days at 12 o'clock."

The name Ferguson in the above is printed Fergusson—a spelling never adopted by the subject of our memoir. "As Astronomical Examiner," Ferguson never used such a designation; but there was a James Fergusson, teacher of Navigation and the Sciences in Hermitage Row, Tower Hill, who spelt his name Fergusson, and adopted the title of "Astronomical Examiner." Date 15th Nov., 1802, is a date 26 years after Ferguson's death.

2d. In looking over the catalogue we find many books which Ferguson was not likely to possess—as also many that were published after his death, and therefore could not belong to Ferguson.

3d. We have a pamphlet entitled

> "Observations on the present state of the Art of Navigation, by James Fergusson: London, 1787."

This is a date 11 years after Ferguson's death, and is the production of James Fergusson of Hermitage Row, just noticed. From all which it is evident that the sale of 1802 was the sale of the effects of James Fergusson, mistaken by Nichols for the Ferguson of our Memoir.

Again—Ferguson in his will orders his effects to be sold without delay—the executors would never *wait* 26 *years to do so;* we find in Beckman's "History of Inventions," that the sale of the effects of Ferguson took place shortly after his death, in 1777. See p. 398 of the Memoir. Therefore, Ferguson's effects were sold by public auction in 1777, and that of James Fergusson in 1802.

It is somewhat singular that two persons nearly of the same name, and of nearly the same profession, should be living in London at the same time; we have in our researches frequently come into contact with this *duplicate*, and occasionally have had considerable trouble in unraveling references which may have pointed to either party.

sent extended Memoir. (For notes on John Ferguson, see pp. 237, 403, 404, 448.)

The articles which he sent to the late Miss Wilson of Keith are now in the Banff museum.

Ferguson's Limnings.—The oldest Limning by Ferguson extant, is a miniature likeness of his early friend and patron, William Baird, Esq. of Auchmedden, now in the possession of F. G. Fraser, Esq., Findrack, Aberdeenshire, (his great-grandson). As formerly mentioned, this likeness is still in excellent preservation. Mr. Fraser, in his note, mentions that the Indian ink portrait of Mr. W. Baird is 6 inches long, by 4½ inches broad, and that "J. F. pinxt," is in the lower right hand corner, in front, and under it "William Baird of Auchmedden, Summer of 1733." The portrait is placed in a wooden frame with a gilt border round it. (See date 1733 and the photograph of this portrait).

The next oldest set of Limnings by Ferguson extant, we take to be those now in the possession of Cosmo Innes, Esq., Professor of Universal History in the University of Edinburgh. The set consists of nine portraits; seven of these, of former Professors of Edinburgh University. None of them are dated—but it is certain that they were done by Ferguson between the years 1736–1743. Mr. Innes, in his note regarding these portraits, intimates, that they are those of

John Stewart, Professor of Natural History, University of Edinburgh.
Mathew Stewart, „ of Mathematics, „ „
Robert Hunter, „ of Greek, „ „
John Stevenson, „ of "*Philosophice*," „ „
George Stewart, „ of Humanity (Latin), „ „
William Wallace, „ of History, „ „
James Balfour, „ of Moral Philosophy, „ „

James Boswell and Temple, reading Cicero; with the mottoes—"Idem velle atque idem nolle ea demum fermia amicitice est;" and, "Unius amor erat pariterque musarum amice."

R. Dundas of Arniston, President of the Court of Session—(an unfinished head).

These portraits are arranged in one frame, and may be seen at Mr. Innes's house, Inverleith Row, Edinburgh.

The next Limning by Ferguson known, is a miniature portrait of Simon Lord Lovat, done in the year 1740, now in the possession of A. T. F. Fraser, Esq. of Abertarf, near Inverness.

This unfortunate nobleman was *out* in 1745, and in consequence, his estates were confiscated. He suffered on Tower Hill, London, early in 1746.

Mr. Watson, Preston Place, Edinburgh, informs us that he has several miniature portraits in Indian ink by Ferguson. Drawn, probably, between the years 1736 and 1743.

The miniature portrait of Mrs. Ferguson appears to have been done by Ferguson, in London, somewhere about the year 1750, when she was in the 30th year of her age. This portrait is noticed under date 1743, along with a miniature photograph from the original. It now belongs to Mr. Gordon, Leith. (See date 1743, note 73, pp. 46, 47).

The miniature portrait of George Wilson of Grange, near Keith, father-in-law of Ferguson, is in the possession of the Editor, (gifted by the Rev. Dr. Bowie, Kinghorn, in May 1865). This is a very fine portrait, Indian ink on vellum, pleasant countenance, broad forehead, with large powdered wig, and in his right hand an open Bible; mounted in a black oval frame. George Wilson died in 1742; this portrait must, therefore, have been taken *before* 1742, between the years 1733–1742.

The miniature portrait of MURDOCH FERGUSON, second son of Ferguson, is in the possession of the Rev. Dr. Bowie of Kinghorn. It represents Murdoch, when a child about three years old. This is an exquisite miniature; nothing can exceed in beauty the face of this child. As Murdoch was born at the end of the year 1752, this portrait would be taken sometime in the year 1756. It is mounted in a black oval frame.

Dr. Bowie has also two fine pen-and-ink drawings of "*A New Crane with four different powers, Invt. by James Ferguson*;"—(See p. 286)—as also of "*The Quadruple Pump Mill, made by James Ferguson.*" These drawings are about 12 inches by 8, in black frames and glazed.

The miniature portrait of the Rev. Anthony Dow of Fettercairn, in the Mearns, and of his daughter Jane Dow, are two beautiful portraits, and are now in the possession of the Rev. Dr. Robert Trail, Free church minister at Boyndie, Presbytery of Fordyce, Banffshire; (the great-grandson of the Rev. A. Dow, and grandson of his daughter Jane); the daughter, Jane, died in 1805, aged 78, showing that she was born in 1727; the portrait has a youthful appearance, and must have been done when the lady was in her teens—say about the year 1741, when she was in her 15th year.

FERGUSON'S ORRERIES:—After Ferguson's death in 1776, the London Philosophical Instrument Makers made many very superb orreries, all in brass, on the plan as laid down by Ferguson in plates VI. and VII. of his "*Select Mechanical Exercises.*" Such orreries in brass-work are now generally thought to have been made by *Ferguson;* but they were not so; they are "*according to Ferguson's* method," but not *made by him.* There are now few Orreries made by him to be met with.—The following is a list of those we have seen and heard of.

1st. The small Orrery made in Edinburgh, early in the year 1743, and which he, shortly after his arrival in London, sold to Sir Dudley Rider. The Earl of Harrowby informs us, that this Orrery is still in existence at his seat, Sandon Hall, near Stone in Staffordshire, but that it was much damaged by the fire which destroyed Sandon Hall in 1846.

2d. At the public sale of the Mathematical and Philosophical Instruments of Thomas Hawys, Esq., late of Charterhouse Square, London, there was sold on Tuesday, October 13th, 1807, by Messrs. King and Lochee, at their great Rooms, 38 King Street, Covent Garden, Lot 68, A large Orrery, with glass shade, made by Ferguson.—(See pp. 320—322, and note 257, p. 341.)

3d. Mr. Bartlett, watchmaker, Maidstone, had in his possession, long before 1841, a large wooden orrery made by Ferguson—the same sort of orrery as is commonly represented in the frontispiece of Ferguson's Astronomy—the horizon was also of wood, divided by hand; and the names of the signs, months of the year, &c., were done in a very neat manner by Ferguson's pen. On a visit to Maidstone, in 1841, we were shown this orrery by Mr. Bartlett. He then informed us that he had bought it in Old Compton Street, Soho, London, many years before, from a dealer in "Philosophical Instruments" and Curiosities, and that the price he paid for it was £4 14s. 6d. We wished to buy it from Mr. Bartlett, but at that time he was not disposed to part with it. Many years afterwards, we wrote Mr. Bartlett, inquiring if he was yet disposed to part with the orrery. In his reply, he informed us, that he had some years before sent it as a present to the Committee of the Anti-Corn-Law League Bazaar, Manchester, as his contribution to it. On our applying to the secretary for the name of the party who purchased it at the Bazaar, we were informed that the *sums* received for articles were alone entered in their books, not the names of the purchasers. So, all trace of this orrery is for the present lost. (See p. 45.) [353]

4th. Professor Millington, for sometime a Lecturer on Astronomy and Natural Philosophy in London, had a large wooden orrery which was made by Ferguson, and which he frequently exhibited in his lecture-room. It showed the rotation of the Sun, the motions of Mercury, Venus, the Earth and Moon, as also the, retrograde motion of the Moon's orbit, and consequently, all the eclipses of the Sun and Moon. About forty years ago, the Professor emigrated to America and took this orrery with him.—(See page 45.)

5th. In the year 1851, Mr. George Walker of Port Louis, bequeathed a brass orrery, made by Ferguson, to University College, London, particulars of which are given in pages 321, 322.

6th, MECHANICAL PARADOX and ORRERY UNITED:—This is a small orrery, made by Ferguson in 1760, and which was in the possession of Ferguson's son, John, until his death in 1833. In consequence of some dispute about the son's Will, this orrery was sent by the Executors, along with other items, to the office of Messrs.

[353] In the Anti-Corn-Law Bazaar Catalogue, entitled,—"Catalogue of Paintings, Models, Articles of Virtu, Curiosities and Specimens of Natural History, on sale at the Anti-Corn-Law Bazaar, Manchester, October 1845," No. 221, pages 12-13, we find—

> 221 "ORRERY, made by Ferguson, the celebrated Astronomer.
> This machine shows the motions of the Sun, Mercury, Venus, Earth, and Moon; and occasionally the superior planets, Mars, Jupiter, and Saturn, may be put on. Jupiter's Satellites are moved round him in their proper times by a small winch, and Saturn has his five Satellites and his Ring, which keeps its parallelism round the Sun, and by a lamp put in the Sun's place, the Ring shows all the phases described in the 204th article of the volume accompanying the Orrery."

This is the same Orrery, which was, after Ferguson's death, exhibited by the Scotch publican in the London Inn, called "*The Ferguson's Head.*" (See pp. 45 and 462).

Scott and Balderstone, W.S. (now represented by Messrs. Scott, Moncrieff, and Dalgetty, W.S., Edinburgh.) See description of this one, under date 1760.—This orrery has just been presented to us, by James Lauder Rutherfurd, Esq., 10 Windmill Street, Edinburgh,—along with several pen-and-ink drawings and writings by Ferguson.—(See pp. 193—200.)

We have made several inquiries after the large wooden orrery which Ferguson, with assistance, made in 1744–1745, but without any result. It would appear that he presented this orrery to a daughter of the Right Honourable Sir Stephen Poyntz, about the year 1756. (See Tables and Tracts, p. 167; also note 111, pp. 162, 163.)

We may here remark, that Ferguson had several eminent rivals in the art of constructing orreries; of whom may be mentioned—

1st. EDWARD WRIGHT, Optician and Mathematical Instrument Maker: London.

2d. JOHN NEAL. 1745: In which year he published a description of his Patent Planetary Machine, 40 pp. and 4 plates.

3d. J. FIDDLER, Mathematical Instrument Maker.

4th. RANDAL JACKSON, Clockmaker, 8 Chapel Row, Spa-Fields, London, was also a maker of Orreries, and made "*Complete Orreries, Planetariums, &c., as invented by Messrs. Ferguson and Martin.*"[354] Jackson was the author of a small pamphlet of 14 pp., on a "*New invented Orrery.*"

5th. GEORGE ADAMS, Optician, &c., Fleet Street, London, made a many fine Orreries and Cometary Machines.

6th. BENJAMIN MARTIN, Optician, Orrery Maker, &c., Fleet Street, London, was the maker of several complicated Orreries, &c., and in 1771, published an Octavo pamphlet "*On the Description and Use of an Orrery,*" 28 pp. and 1 plate.

The learned REV. DR. WILLIAM PEARSON, the writer of the article "*Planetary Machines,*" in the Edinburgh Encyclopædia, in reviewing the wheel-work results of Orreries made during the last century, gives the preference to Ferguson's: he says, "*the small Orrery of Ferguson is more correct than any of its predecessors.*"

Since Ferguson's time, the Rev. Dr. William Pearson in England, and M. Antide Janvier in France, stand in the front rank of makers with accurate, and also equated motions. The late DAVID FORRESTER, Teacher, Kirkcaldy; and JOHN FULTON, shoemaker at Fenwick, near Kilmarnock, may also be mentioned as the makers of complicated Orreries and Planetaria. According to his catalogue,—"Mr. Benjamin Gorrill, *Orrery Machinist,* 15 Edward Street, Parade, Birmingham, makes at the present time, very accurate Orreries at prices from £25 to £100 and upwards."

The following is a list of Relics of Ferguson, collected by purchase, between the years 1830 and 1861, by the Editor of this Memoir,

354 This advertisement is taken from the title-page of Jackson's pamphlet, from which it is seen that this was one of several manufactories where *orreries according to Ferguson's* plan were made.

and disposed of by him to the Committee of the Banff Museum in 1864.

1. An Astronomical Clock.—This clock is in the form of a ship's chronometer, on which is mounted a small terrestrial globe, on an axis inclined at an angle of 23½ degrees, on which it daily turns from west to east. The clock has a dial-plate on its cover having 12 hours, and an index connected with the works points out the time of the day. Also, under the globe there is an oblique-lying dial-plate, and an index on the inclined axis points out the hours on it (this dial-plate has 24 hours engraven on it). This time-piece is about 5 inches in diameter, 3 inches thick, and 7 inches in height from the base to the North Pole of the globe. On the inside of the winding-plate are engraven the following names, showing to whom it had successively belonged,—viz., John T. Desagulier, LL.D., 1729, Lect. on Nat. et Exp. Phil., London; Benjamin Franklin, LL.D., 1757 (the American statesman and philosopher); James Ferguson, F.R.S., 1766; Kenneth M'Culloch, 1774; and the initial letters G. W.

2. An Old Watch, double silver cases, with silver turn-piece over the key-hole of inner case. The works are very strong, and still in good going order. On the inside of the rim of the outer case are engraven the following names:—James Ferguson, 1746, and Andrew Reid, 1801.

3. A Wheel-cutting Engine, with which Ferguson cut the teeth of the wheels of his orreries, and other astronomical machinery; full of dotted circles for almost any number, odd or even; strong, and in good condition.

4. A Telescope, made by Ferguson about the year 1732. It has three pulls or draws, is about 12 inches long when closed, and 33 inches when in focus. It appears to be made of parchment or sheep-skin, is ornamented with horn rings, and covered with a black sort of skin.

5. An Enamelled Case, gold mounted.—This is a small tubular case. In the inside, it is inlaid with tortoise-shell; on the outside, it is ornamented with four gold hoops or rings; use unknown.

6. An Old Black Case and Brass Scale.—The case contains a thin brass scale graduated, and has a brass slider; supposed to have been Ferguson's dialling scale.

7. A Large Engraving of Ferguson's First Astronomical Rotula, published at Edinburgh in 1741; part wanting;—an exceedingly scarce print.

8. An Autograph Rotula by Ferguson, most beautifully written and figured, consists of three large moveable circles in thick Bristol board, quite filled with a mass of figures and small perforated circles, and, according to its title, 'shows the times of all the new and full moons and eclipses for 12,000 years.' This is the original Rotula from which Ferguson published his subsequent ones.

9. An Autograph Tract by Ferguson, finely written, descriptive of the Rotula, being original manuscript from which he published his Rotula Tracts.

10. 'The Universal Analemma,' in the autograph of Ferguson; made of strong Bristol board; moveable horizon, &c.

11. An Autograph Astronomical Quadrant, made of thick Bristol board, full of curve lines and figures in Ferguson's autograph.

12. A Printed Astronomical Rotula; three moveable circles.

13. Ferguson's Last Publication, entitled 'A Table of the Equation of Time, London, 1776.'

14. Ferguson's Likeness.—A fine large print, by Townsend, with Ferguson's left hand resting on a celestial globe. Proof before letters.

The following articles are in a glazed gilded case:—(1) a Boxwood Wheel of 33 teeth, cut by Ferguson, being the wheel which carried round the second satellite in his Satellite Machine (see Ferguson's 'Tables and Tracts,' p. 158). (2) A Hollow Paper Axis, being the axis of one of the wheels of the Satellite Machine. (3) The Boxwood Handle or Winch which set in motion the wheel-work of ditto. (4) A Semicircular Graduated Card, in the autograph of Ferguson. (5) A Wheel of 45 teeth, made of thick Bristol board, beautifully cut by Ferguson. (6) A Trundle Wheel of 15 cogs, mounted on a triple crank axis, which was driven by the above wheel of 45 teeth; both belonged to Ferguson's Electrical Apparatus. (7) The Apogee Pointer of Ferguson's 'Mechanical Paradox' (see Ferguson's 'Select Mechanical Exercises,' p. 45.)

Relics of Ferguson in the possession of the Editor:—We have the numerous papers, pen-and-ink drawings, rare prints, tracts, rotulas, &c., noticed in the Memoir;—Also, the Mechanical Paradox Orrery, the MS. Card Sun-dial and miniature portrait of George Wilson of Grange, Ferguson's father-in-law. See pp. 193—200; 244—246, and 474.

Sundry other relics, mentioned in the Memoir.

Snuff Box, made by Ferguson.—In a letter lately received from the Rev. Mr. Moir of Rothiemay, he says—"About three months ago I received a snuff-box made by the astronomer—from a James Forsyth, whose father's mother was Anne Ferguson, eldest daughter of John Ferguson, the astronomer's brother. It is a very curious piece of workmanship. It is mounted in silver. The box is composed of small staves of black horn and bones alternately. The snuff-box contains a very small quantity of cotton with the seeds in it,—and also, a small quantity of wool of a brown colour which Ferguson is said to have procured at Forres, where there was a manufactory for making cloth of the colour similar to that of the wool in the box. James Forsyth stated that the box had been in the Ferguson family from the days of the astronomer." This is certainly a great curiosity, and a most interesting relic.

Gold Signet Ring.—We have been informed that Mr. Gordon, Innkeeper, Keith, has a gold signet-ring which belonged to John

Ferguson, Esq.—the youngest son of Ferguson. It has the initials "J. F." cut on a Cornelian stone, set in a jointed gold frame, which serves as a lid to cover a recess. On the lid being opened, the recess is shown, glazed over, and under the glass is "a little hair of a light colour, slightly tinged with grey, and plaited three-fold"—very probably the hair of Ferguson—if so, then this is a most interesting and valuable relic.

Astronomical Rotula, in Culloden House, near Inverness (p. 40.)

Tide Rotula, in the autograph of Ferguson, in the possession of William Topp, Esq., Ashgrove, near Elgin (p. 371).

Lunarium Rotula, in the autograph of Ferguson, in the possession of Mr. Robert Sim, Keith.

Letter:—The Letter written by Ferguson to the Rev. Mr. Cooper of Glass, relative to the origin of the Mechanical Paradox, in the possession of Mr. Robert Sim, Keith (p. 145, 146).

Letter:—The Letter written by Ferguson to Rev. Alexander Irvine of Elgin; now in the Museum of Elgin (p. 225).

Letters:—The Letters written by Ferguson to Rev. Dr. Birch, in the British Museum; also, the letters in the possession of the Editor, &c.

Watch Movement:—In the catalogue of the Antiquarian Society of Edinburgh—published in 1863—at page 76 and No. 89, we find "Astronomical Watch Movement, said to have been made for James Ferguson the Astronomer."

The Beatson Likeness:—in oil—in the possession of Mrs. Thompson, Nethercluny, Mortlach, Banffshire (p. 215).

The Casborne Likeness:—The Likeness done in chalk, in the possession of Mrs. Casborne, New House, Pakenham, near Bury St. Edmunds (p. 13).

Bible:—The Bible, so often referred to, now in the possession of Dr. George, at Keith (p. 72).

Manuscript Volumes:—The three MS. Volumes, viz.: "Astronomical Tables and Precepts," p. 224,—"Astronomy," pp. 235, 236, and the "Common Place Book, pp. 446, 447; in the College Library, Edinburgh.

The Ferguson Prize; Grammar School, Keith:—In a letter received from the Rev. Dr. F. S. Gordon of St. Andrew's Episcopal Church, Glasgow—a native of Keith,—he says, regarding the prize:—

"I instituted the prize in 1862, to be competed for at the yearly examination by the Presbytery of the Parochial School of Keith, my native town, and where I was early taught. I give £1 annually, for the encouragement of natural talent in Mechanical Art of any kind," &c.

With Dr. Gordon's letter is one received by him from Mr. Smith, teacher of the school; the following is copy of part of it referring to the prize:—

"At last we have been able to comply with your wish in the matter of the Ferguson prize, as regards mechanical models. Some six months before the examination, I intimated to the scholars that you wished it to be understood that in competing for the Ferguson prize, any model or mechanical contrivance would, in future, be preferred to trials in Geographical or Astronomical problems; informing them, at the same time, that the prize would accordingly be given for a model, provided three candidates appeared. In a day or two, no less then twelve gave in their names, and on the day of examination they all appeared with their models; consisting of harrows, ploughs, boats, ships, one wheelbarrow, one locomotive and tender, (model of the outside appearance only), one steam mill, (not in working order), and a working model of a thrashing machine, with a fan inclosed;—The Judges unanimously awarded the prize to the maker of the last-named article, made by James M'Connachie from Nether Ballendy, Glenrinnes. This model is a very perfect one—well got up—and done principally with the boy's knife. Several of the visitors expressed a wish to purchase it, but the maker seemed unwilling to part with it. The successful candidates from the commencement of the first trial for the prize in 1863 to the present year (1866) are as under:—

1863.—Robert Mantach from Rothes, aged 14. Subject, Drawing Maps.

1864.—William Gordon, Fife-Keith, aged 15. Subject, Geographical Problems.

1865.—Charles Wilson, and Robert Munro, Keith, *equal*, aged respectively, 16 and 15. Subject, General Arithmetic (prize divided).

1866.—James M'Connachie from Glenrinnes, aged 14. Subject, Mechanics,—for a working model of a Thrashing Machine and Fan.

JAMES SMITH,
GRAMMAR SCHOOL, KEITH, *9th April* 1866."[355]

We cordially wish that success may attend this praiseworthy scheme; it may be the means of bringing out a modern Ferguson. Might not some of the future models be made to show the motions of the Earth, the Moon, or those of other celestial bodies?—matters in which Ferguson so much delighted, and was so great a master.

EUCLID, PROP. XLVII. B. I.—Under date 1743, p. 60, will be found a mechanical or rather ocular demonstration of Euclid, Book 1st., Prop. 47th, being a copy of a card which Ferguson kept, cut out into

[355] In this communication the school of Keith is designated the Parochial School, and the Grammar School, but they both refer to one and the same school. It will be remembered that Ferguson, in or shortly after 1717, had "*about three months*'" instruction at the Grammar School of Keith. (P. 7.)

sections for showing to his friends. Subjoined are two other demonstrations, equally simple, taken from an old M.S. Euclid in our possession.

In Fig. 1. let the sides of the right-angled triangle on which the small squares are to be raised be to each other as the numbers 3, 4, and 5; then, upon such sides, raise the divided squares as shown in the figure—and it will be obvious, that the square raised on the side or hypothenuse 5, will contain 25 equal squares, and the two squares raised on the sides 3 and 4 will contain respectively 9 and 16 similar squares. Hence 9 + 16 = 25 squares, the sum of the squares 3 and

Fig. 1.

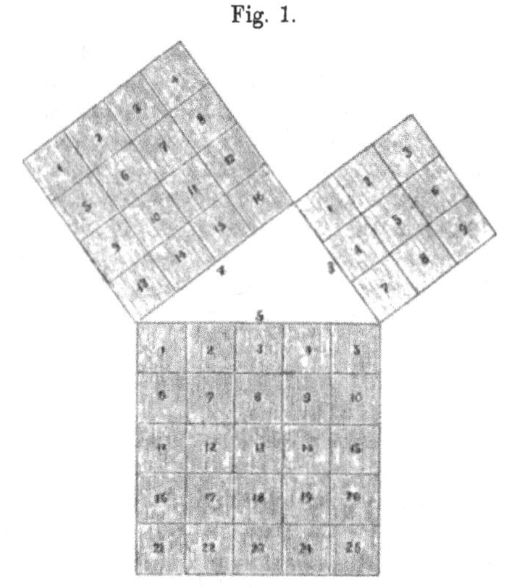

Fig. 2.

4, which are equal to the sum of the squares contained in the square raised on side 5,—the hypothenuse. (See also "Common Place Book," p. 189).

Fig. 2. shows another illustration—Let the hypothenuse, or base A, or base of the triangle be of any given length—on the length fixed on describe a square. Let the sides of the triangle B and C be of equal length, as in the figure—then upon the sides raise the squares B and C:—To demonstrate ocularly, that the area of these two small squares are precisely equal to the area of the larger square below, on the hypothenuse—draw a line horizontally across them, dividing them into two, through their diagonals, as shown. Then divide the large square through the diagonals also, and it will become evident that the large square has by the process been divided into the *four* equal triangles, 1, 2, 3, 4, and that they are of exactly the same size as are the four triangles, 1, 2, 3, 4, contained in the two squares above, described on the sides B. C.

REMARKABLE FISH; (see p. 278.)—The following letter from Ferguson to Dr. Birch, relative to a singular fish caught in King Road near Bristol, in 1763, is extracted from the Philosophical Transactions, but was received too late for insertion under its proper date.

BRISTOL, *May* 5*th*, 1763.

REVEREND SIR,

I herewith take the liberty of sending you a drawing of a very uncommon kind of fish which was lately caught in King Road,[356] a few miles from this city; and is now shown at Hot-Wells.[357] It fought violently against the fishermen's boat after they got it into their net, and was killed with great difficulty. Nobody here can tell what fish it is, only some say it is a Sea Lion; but to the best of my remembrance it answers not to the description or figure of the Sea Lion, that is given in Lord Anson's voyage. I took the drawing on the spot, and do wish I had had my Indian ink and pencils, by which it might have been better shaded; but I hope you will excuse the roughness of the draught, as it is the best I ever made with a pen. The length of the fish is four feet nine inches, and the thickness in proportion, as in the figure. The mouth is a foot in width and of a squarish form; it has three rows of sharp small teeth, very irregularly set, and at some distance from each other; it has no tongue, nor narrow gullet, but is all the way down, as far as one can see, like a great hollow tube; in the back of the mouth within, there are two openings like nostrils, and about nine inches below the jaw, and under these openings are two large knobs, from which proceed several short teeth; a little below which, on the breast side, is another knob with such teeth. On each side within, and about a foot below the jaws, there are three cross-ribs, somewhat resembling the straight bars of a chimney-grate, about an inch distant from each other, through which we see into a great cavity within the skin towards the breast; and under the skin, these cavities are kept distended by longitudinal ribs

356 Hot-Wells—a well-known locality about a mile below Bristol, on the banks of the Avon, and in the immediate vicinity of St. Vincent's Rocks, and new Suspension Bridge.

357 King Road is nearly 10 miles below Bristol, in the river Severn, near to where it receives the waters of the Avon.

3 P

plain to the touch on the outside. I put my arm down through the mouth quite to my shoulder, but could feel nothing in the way, so that its heart, stomach, and bowels must lie in a very little compass near its tail, the body thereabout being very small. From the neck proceed two small horns, hard and very elastic, not jointed by rings as in lobsters; and on each side of the back there are two considerable sharp-edged risings of a black and long substance; between each eye and the breast there is a cavity somewhat like the inside of the human ear, but it doth not penetrate to the inside. From each shoulder proceeds a strong muscular fin, close by which, towards the breast, is an opening through which one may thrust his hand and arm quite up through the mouth, and between these fins proceed from the breast two short

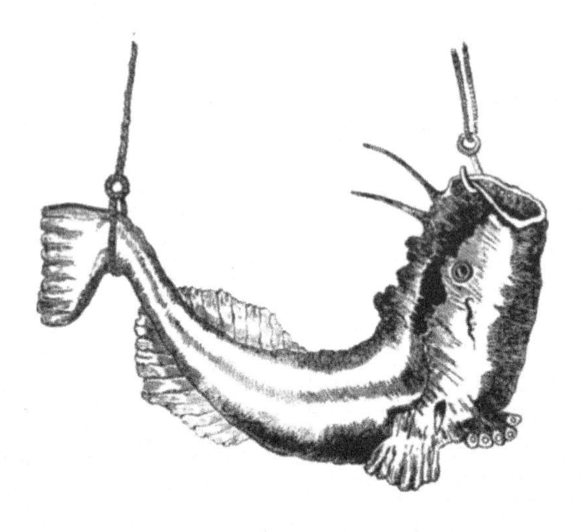

Sketch of the Remarkable Fish from Abrid. Phil. Trans.

paws, somewhat like the fore-half of a human foot, with five toes joined together, having the appearance of nails; near the tail are two large fins, one on the back, the other under the belly; the skin is of a dark brown colour, but darker spotted in several places, and entirely without scales.[358] If you think this in any way deserves notice of the Royal Society, I shall be very glad of your communicating it, and am, with the greatest esteem, Reverend Sir, your most obliged and humble servant,

James Ferguson.

Abrid. Phil. Trans., Vol. LIII.,
p. 170; year 1763.

Ferguson Memorial Pillar:—It has of late been frequently suggested that a Memorial Pillar or Obelisk ought to be erected at Core-of-Mayen on the site of the cottage where Ferguson was born. Let the men of Banffshire—Ferguson's native county—take the start and

358 As noted at page 278, this fish is the *Long Angler* of Pennant, or *Sophicus Connbicus* of Shaw.

the lead in this meritorious object: let them form a working committee and issue circulars soliciting subscriptions, and the call would no doubt be responded to widely and liberally. We shall be happy to contribute our mite:—should any surplus remain, it might be expended on renovating the tomb in Old St. Mary-le-Bone churchyard, London.—(See pages 449, 450).

Biographical Notices:—Since his death in 1776, Biographical notices of Ferguson have appeared in the Encyclopædias and other works in this and other countries—particularly in the Edinburgh Encyclopædia,—Eudosia, a poem on the Universe,—Nichols' Life of Bowyer,—Mitchel's Scotsman's Library,—Chambers' Lives of Eminent Scotsmen, &c. But all the notices are simply abridgments of his excellent memoir, prefixed to his "Select Mechanical Exercises."—In 1857 Mr. Henry Mayhew published an interesting volume—founded on the early life of Ferguson—entitled "The Story of the Peasant-Boy Philosopher."

Theological Papers.—At pages 206 and 241, are inserted Papers on Theological subjects; in the former page other papers are mentioned, and although the following one is not named, it evidently belongs to the series.—It is taken from "Common Place Book," p. 5.

"The Mosaic account of the Creation (Gen. i.) shown to agree with the Newtonian Philosophy.

This is denied by many, because Moses tells us the first thing that was made was light, and that there were three days and three nights before the Sun was created, which was not till the fourth day of the original week. They cannot reconcile this with true philosophy, because all our light is from the Sun, our day being turned toward him, and our night from being turned away from him; and therefore, they say that Moses must have wrote his account from tradition only, and not from Revelation, so that from this part of Scripture the Deists ridicule the whole.

It is recorded in the last verse of the 24th chapter of Exodus, that Moses was forty days and nights in the Mount with God (probably our Saviour). And although we have an account of a great many instructions which were given him during his long stay there, (see from ch. 24th to ch. 32) yet, undoubtedly, there was time for many more. May we not therefore suppose that he then received an account of the Creation, which was upwards of two thousand years before that time? My opinion is, that Moses's account of the Creation is philosophically true, which I explain as follows:—

In the beginning, God brought all the particles of matter into being in those different parts of open space where the Sun and planets were to be formed, in sufficient quantities for each body, and endued each particle with an attractive power, by which these neighbouring (and at first detached) particles would in time come together, in their respective parts of space; and so would form the different bodies of the solar system.

The particles which now compose the body of the Sun would shine when separate from each other, and give as much light all round them, as when they were afterward collected into one body by their mutual attractions, which might require four days. So that, if these were the first particles of matter in our system which were brought into being by the Almighty Power, light must have been the effect of that powerful word, before the Earth or any other planet existed; and three days before these shining particles were collected into a globular body called the Sun.

On the first day of the creation week, before the particles which now compose the Earth had come together, so as to form a regular solid globe (Gen. i. 2), the Deity gave the whole unformed mass of them a rotatory motion:—and as they could not all at once come together into a solid mass, they would, by their rotation, acquire a spheroidal figure or shape, such as the Earth has had from the beginning of time. The planet Jupiter, by its quick rotation, is also a spheroid.

The earthly mass having received its rotatory motion on the first day of the creation week, and being enlightened by the detached solar particles of matter, day and night would begin immediately thereon, and continue to be successive on all sides of the Earth.

On the fourth day of the week, the luminous particles were collected together by their mutual attraction, into one globular body; and then they became the Sun.

The attraction of all the solar particles taken together, would be as great before they were collected into one body as afterwards; and consequently, would reach through all the bounds of the solar system. It is therefore highly probable, that, to keep the earthly mass from being drawn toward these particles, the Deity impressed the annual motion upon the Earth on the first day of the original week, when he gave the Earth its rotatory motion; and so, the day, the week, and the year began together.

As these particles of the earthly mass gravitated toward its centre, the light aerial particles would swim above on its surface, and form an atmosphere all around it; and the Sun would exhale moist vapours from the earth into its atmosphere, which would then divide the water on the Earth from those which had risen above it. These vapours would form a mist in rising (Gen. ii. 6), and afterwards descend in dew, to water the surface of the ground.

By the firmament, called heaven (ch. i. 6), which Moses tells us divided the waters from the waters, I imagine he meant only that atmosphere which bears up the watery clouds till they fall down in rain; for we usually say, It rains from the heavens.

On the whole, then, it appears that Moses wrote from Revelation, and not from tradition, seeing that his account of the creation is so strictly conformable to the Newtonian Philosophy, (and who can say that Newton was not God Almighty's scholar?)[359] although it was written upwards of 2,000 years before Sir Isaac Newton was born.[360]

James Ferguson."

The following is copy of a "Geometrical Card" received too late for insertion under its date.—It is entitled

A plain proof, that the three angles of any plain right-lined Triangle contain 180 degrees.

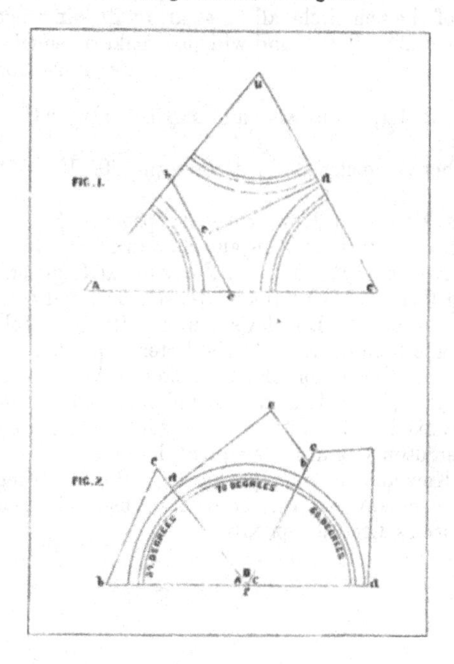

The explanation—in Ferguson's autograph—on the back of the Card, is as follows :—

"EXPLANATION.

Cut a common card into any right-lined triangular shape, as A B C, (Fig. 1.) at random, and mark the angular points by three letters, as in the figure.

Then, cut off the angle at A, any how, by a pair of scissors, as in the line *b e c*, and separate the angles B and C from one another, by cutting them asunder, as in the line *e d*.

This done, lay all the three parts (so separated from each other) upon a semicircle, divided into 180 degrees, as in Fig. 2, so as the angular points A, B, C, may meet at the centre F; and you will find that the three angular pieces will just fill the semicircle. For

359 Ferguson, by an asterisk here, refers to a foot note which says—"The Hutchinsonians do indeed damn Sir Isaac; but every one who attends to demonstrative truth, allows, that they are the most wrongheaded sett of people upon the face of the earth; regarding neither reason nor demonstration." J. F.—"Common Place Book," p. 5. (See also foot note, p. 422).

360 This is a singular, and original exposition of the Creation.—(See Moderr Theories of the Earth, &c.)

the piece *A b c* of Fig. 1. will make *A b c* of Fig. 2. The piece *B d e b* of Fig. 1. will make *B d e b* of Fig. 2, and the piece *C c e d* of Fig. 1. will make *C c e d* of Fig. 2. And moreover, if the angular points of Fig. 1. be made the centres of arcs, the radius of each equal to the radius of the semicircle, all these arcs will join when the pieces are put together as in Fig. 2, and will just make a semicircle, Q. E. D.

JAMES FERGUSON, 1756."

See Euclid, B. 1st, Prop. xiv. and xxxii.;—also, "Common Place Book," p. 186.

For the other Geometrical Cards—see pp. 60, 170–175, and 361.

FERGUSON'S VISITS TO KEW, &c.—At pages 353, 354, notice is taken of Ferguson's visits to Kew and St. James's:—Dr. Houlston, in his Obituary Notice of Ferguson in the Annual Register, informs us that the King (George III.) "frequently sent for, and conversed with him on curious topics."—Dodsley's Annual Register, vol. 19. 1776. This should have been copied into the letter at p. 462.

Mr. Partington, the editor of an edition of Ferguson's "Lectures on Select Subjects," mentions in his addenda to the autobiographical memoir prefixed to the edition, that—"for some years prior to his death, Mr. Ferguson was in the frequent habit of visiting the royal residences at Kew and St. James's, his late Majesty being much attached to the company and conversation of men of science."—Partington's Ferguson's Lectures, p. xliv.

INDEX.

C

D

F

G

M

N

Q

R

T

U

V

W

EDINBURGH:
FULLARTON AND MACNAB, PRINTERS, LEITH WALK.

Printed in the United States
By Bookmasters